Higher Excited States
of Polyatomic Molecules

VOLUME II

Higher Excited States of Polyatomic Molecules

VOLUME II

Melvin B. Robin

Bell Laboratories
Murray Hill, New Jersey

ACADEMIC PRESS   New York San Francisco   London   1975
A Subsidiary of Harcourt Brace Jovanovich, Publishers

COPYRIGHT © 1975, BY BELL TELEPHONE LABORATORIES, INC.
ALL RIGHTS RESERVED.
NO PART OF THIS PUBLICATION MAY BE REPRODUCED OR
TRANSMITTED IN ANY FORM OR BY ANY MEANS, ELECTRONIC
OR MECHANICAL, INCLUDING PHOTOCOPY, RECORDING, OR ANY
INFORMATION STORAGE AND RETRIEVAL SYSTEM, WITHOUT
PERMISSION IN WRITING FROM THE PUBLISHER.

ACADEMIC PRESS, INC.
111 Fifth Avenue, New York, New York 10003

United Kingdom Edition published by
ACADEMIC PRESS, INC. (LONDON) LTD.
24/28 Oval Road, London NW1

Library of Congress Cataloging in Publication Data

Robin, Melvin B
    Higher excited states of polyatomic molecules.

    Includes bibliographies.
    1. Molecular spectra.   2. Sprectrum, Ultra-violet.
I. Title.
QC454.M5R62   539'.6   73-9446
ISBN 0–12–589902–5 (v.2)

PRINTED IN THE UNITED STATES OF AMERICA

Contents

Preface to Volume II	vii
Acknowledgments	ix
Contents of Volume I	xi

Chapter IV **Two-Center Unsaturates**

IV.A.	Olefins		2
	IV.A-1.	Ethylene	2
	IV.A-2.	Alkyl Olefins	22
	IV.A-3.	Haloethylenes	50
IV.B.	Azo and Imine Compounds		68
IV.C.	Aldehydes and Ketones		75
IV.D.	Acetylenes		106
IV.E.	Nitriles		117

Chapter V **Nonaromatic Unsaturates**

V.A.	Amides, Acids, Esters, and Acyl Halides		122
	V.A-1.	Amides	122
	V.A-2.	Polymeric Amides	140
	V.A-3.	Acids, Esters, and Acyl Halides	146
V.B.	Oxides of Nitrogen		160
V.C.	Dienes and Higher Polyenes		166
	V.C-1.	Dienes	166
	V.C-2.	Heterocyclic Dienes	180
	V.C-3.	Higher Polyenes	189
V.D.	The Cumulenes		194

Chapter VI **Aromatic Compounds**

VI.A. Phenyl Compounds   209

    VI.A-1. Benzene   210
    VI.A-2. Alkyl Benzenes   223
    VI.A-3. Halobenzenes   230
    VI.A-4. Azabenzenes   237
    VI.A-5. Substituent Effects in Benzene   249

VI.B. Higher Aromatics   257

Chapter VII **Inorganic Systems**

VII.A. Nonmetals   269
VII.B. Metals   275

Chapter VIII **Biochemical Systems**   286

Addendum   290

References   347

Index   391

Preface to Volume II

The Addendum included in this volume functions for both Volumes I and II, and covers work appearing in print up to January, 1974, and beyond in the case of preprints sent to me. The size of the addendum reflects further material, new and old, that has been brought to my attention. That over 100 relevant papers were published in 1973 is gratifying, for it measures directly the large amount of activity in the field, and would seem to justify a need for books such as these. May I invite the interested reader to send me whatever reprints and preprints he feels are relevant, for I am maintaining a library on the subject, in the event revised editions or successive volumes of these books are warranted.

A new, higher level of activity in vacuum ultraviolet spectroscopy is about to make its debut. On the one hand, both scanable vacuum ultraviolet lasers and very intense synchrotron radiation from storage rings are available for refined and extended experimental studies of higher excitations, and on the other hand, the *ab initio* calculations of these higher states are becoming very sophisticated, and confidently carry us far beyond our intuitive feel for the subject. Though we cannot yet offer vacuum ultraviolet spectroscopy as a general tool of great value to people presently outside the field, it nonetheless generates unique data, the practical value of which must surface eventually. Paradoxically, this will probably happen first in an area completely neglected in this work, i.e., in the area of vacuum ultraviolet photochemistry.

Acknowledgments

In addition to the acknowledgments expressed in credit lines accompanying tables and illustrations in the text, permission to reproduce tables and illustrations from the sources listed is gratefully acknowledged:

Reproduced by permission of the National Research Council of Canada from the *Canadian Journal of Physics*:

Table I.A-1 R. B. Caton and A. E. Douglas. *Can. J. Phys.* **48,** 432–452 (1970);
Figures VI.A-4 and VI.A-5 P. G. Wilkinson, *Can. J. Phys.* **34,** 596–615 (1956).

Reprinted with permission from *Journal of the American Chemical Society, Journal of Physical Chemistry, Chemical Reviews,* and *Macromolecules;* copyright by the American Chemical Society:

Figure I.A-8 W. H. Adams, *J. Amer. Chem. Soc.* **92,** 2198 (1970);
Figure IV.A-2 A. J. Merer and R. S. Mulliken, *Chem. Rev.* **69,** 639 (1969);
Figure IV.A-23 J. D. Scott and B. R. Russell, *J. Amer. Chem. Soc.* **94,** 2634 (1972);
Figures IV.E-1, IV.D-3, and V.D-4 J. W. Rabalais *et al., Chem. Rev.* **71,** 73 (1971);
Figure V.A-4 D. L. Peterson and W. T. Simpson, *J. Amer. Chem. Soc.* **79,** 2375 (1957);
Figure V.A-7 J. A. Schellman and E. B. Nielsen, *J. Phys. Chem.* **71,** 3914 (1967);
Figure V.A-10 J. L. Bensing and E. S. Pysh, *Macromolecules* **4,** 659 (1971);
Figure VII.B-3 S. Foster *et al., J. Amer. Chem. Soc.* **95,** 6578 (1973);
Figure AD-4 B. R. Russell *et al., J. Amer. Chem. Soc.* **95,** 2129 (1973).

Reprinted by permission from *Journal of Physics B* and *C*; copyright by The Institute of Physics:

Figure III.A-12 R. A. George *et al.*, *J. Phys. C* **5,** 871 (1972);
Figure III.C-4 W. Hayes and F. C. Brown, *J. Phys. B* **4,** 185 (1971).

Contents of Volume I

        I Theoretical Aspects
      II Experimental Techniques
     III Saturated Absorbers
Appendix Rydberg Term Table
            References

CHAPTER IV

Two-Center Unsaturates

The compounds in Chapter III were grouped together in a natural way, for almost all of them gave spectra which consisted of prominent Rydberg excitations, usually originating with a lone-pair orbital, and a set of valence shell excitations which were generally difficult to identify and of rather low intensity. The compounds discussed in this chapter differ from those in Chapter III in having two adjacent atoms each bearing one or more pi orbitals involved in pi bonds. In the first row, this definition encompasses olefins, azo and imine compounds, ketones, acetylenes, and nitriles. Such pi bonds do not form between pairs of atoms in the second or higher rows of the periodic table, but do form readily between these heavier atoms and the first-row atoms, e.g., $(CH_3)_2S{=}O$.

As in the saturated molecules bearing lone-pair electrons, the two-center unsaturates usually display sharp Rydberg transitions originating with the pi electrons (if not too heavily alkylated), and their term values follow the trends depicted in Section I.C-1 for saturated molecules. Additionally, when halogen atoms are present, the valence shell A bands appear at very close to the same frequencies observed for the saturated halides. The unsaturated double bond, however, adds another dimension to the spectrum by virtue of transitions such as $n \rightarrow \pi^*$, $\pi \rightarrow \pi^*$, $\pi \rightarrow \sigma^*$, and $\sigma \rightarrow \pi^*$. The $n \rightarrow \pi^*$ transitions occur in systems in which one of the atoms participating in a pi bond also carries a lone pair, e.g., as in ketones. This transition may be found anywhere between 15 000 and 65 000 cm^{-1} and is always rather weak, even when formally allowed electronically. The $\pi \to \sigma^*$ and $\sigma \to \pi^*$ excitations are also rather weak, but are restricted to the vacuum-ultraviolet region, where they are very difficult to identify. The $\pi \to \pi^*$ transitions are always beyond 50 000 cm^{-1} and can be very broad and intense. One might think that the allowed $\pi \to \pi^*$ band would dominate the spectrum, and in olefins that is the case. However, in the other classes of unsaturates, it is very difficult to know just where the $\pi \to \pi^*$ transition lies. Thus, $\pi \to \pi^*$ was correctly recognized in ethylene over 30 years ago, but $\pi \to \pi^*$ in formaldehyde is still to be found.

The *ab initio* calculation of the $\pi \to \pi^*$ frequencies in the two-center unsaturates has been very difficult for several reasons. First, there is the unsettled question as to the extent of the mixing of the (π, π^*) valence shell singlet configuration and its Rydberg conjugate $(\pi, 3d\pi)$. Second, variable amounts of the Rydberg component appear depending upon the extent to which (σ, σ^*) configurations are mixed with (π, π^*). Finally, very large changes of geometry are usually involved in $\pi \to \pi^*$ excitations, so that the experimental vertical excitation energies are uncertain, as are the upper-state geometries. Thus there is considerable work remaining to be done on these, the simplest of the pi-electron chromophores.

IV.A. Olefins

Of the various two-center unsaturates, the olefins are perhaps the best understood since their Rydberg spectra are usually easy to identify, and the $\pi \to \pi^*$ band is generally found between 50 000 and 70 000 cm^{-1} with an oscillator strength of at least a few tenths. This broad, intense feature serves to bury the worrisome $\pi \to \sigma^*$ and $\sigma \to \pi^*$ transitions, but under special circumstances these weak bands can precede the $\pi \to \pi^*$ excitation and be mistaken for $\pi \to$ 3s Rydberg excitations instead. The proper identification can be made either from the term value or from a condensed-phase experiment (Sections II.B and II.C).

IV.A-1. *Ethylene*

Interpretation of the electronic spectrum of ethylene has proved to be one of the greatest challenges to molecular spectroscopists and quantum theoreticians. Though the molecule is absolutely fundamental in both molecular spectroscopy and electronic structure calculations, and has been studied intensively from both points of view for 40 years, there are still several unsolved problems associated with this molecule which are of great interest. Difficulties with the ethylene problem are a great embarrassment, for if ethylene presents fundamental unsolved problems, how can we claim to understand pi-electron molecules of even greater complexity? Though many problems remain, progress is being made in this area through a combination of new data, theoretical calculations, and intuitive reasoning.

Until the photoelectron spectrum of ethylene was determined [B2, B41, B55, B58], the electronic structure of the ground state of the planar ethylene molecule was calculated repeatedly with very little relationship to any physical measurements. However, the photoelectron spectrum gives a direct presentation of the orbital energy ladder, and, as can be seen in Fig. III.C-2 and Table II.A-I, strongly supports the predictions of the all-electron calculations. Taking the x axis as along the C—C bond, with the molecule in the xy plane, the electronic configuration of ethylene in its planar ground state (D_{2h}) is

$$(\sigma 1a_g)^2(\sigma 1b_{1u})^2(\sigma 2a_g)^2(\sigma 2b_{1u})^2(\sigma 1b_{3u})^2(\sigma 3a_g)^2(\sigma 1b_{2g})^2(\pi 1b_{2u})^2(\pi^* 1b_{3g})^0.$$

In the notation of Mulliken [M60], this state of ethylene is called the N state. Promotion of an electron from the π-bonding MO $1b_{2u}$ to the π^*-antibonding MO $1b_{3g}$ yields both a triplet and a singlet state, called T and V, respectively, whereas excitation of a $1b_{2u}$ electron into Rydberg orbitals gives R states; state I is the $^2B_{2u}$ ionic ground state and state Z has both pi electrons in the π^* MO.

The overall electronic spectrum of ethylene determined both by electron impact and optical absorption is shown in Fig. IV.A-1, and is seen to consist of several distinct regions. Absorption in ethylene begins near 50 000 cm^{-1} with a poorly structured step which quickly rises to meet a series of sharp bands beginning at 57 000 cm^{-1}, which, in turn, rest upon a continuum centered at about 61 000 cm^{-1}. Sharp bands commence again at ~70 000 cm^{-1} and another continuum peaks at ~77 000 cm^{-1}. It is generally agreed that the weak step and the first continuum form a single transition to the V state and that the transition to the lowest R state corresponds to the sharp bands at 57 000 cm^{-1}.

One of the more interesting aspects of the excited electronic states of ethylene is their equilibrium geometries, for these will determine the vibronic structures of the electronic transitions between them. More specifically, it is the C—C bond distance r_{C-C} and the C—C twisting coordinate θ that are of greatest concern, the most popular mapping of the torsion coordinate being that of Merer and Mulliken (Fig. IV.A-2) [M26]. (See also the theoretical curves of Kaldor and Shavitt [K2] and Buenker et al. [B64, B69].) The curves in Fig. IV.A-2 are thought to

IV. TWO-CENTER UNSATURATES

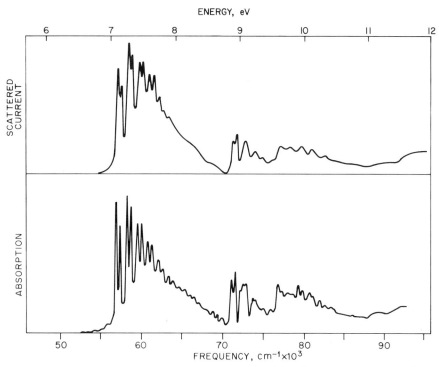

Fig. IV.A-1. The ethylene spectrum as determined by electron-impact energy loss spectroscopy (upper, 33 keV, $\theta = \simeq 0°$), and optically (lower) [G4].

be quantitatively correct in general, with the proviso that the C—C distance is different for each state at a given θ, and does not remain fixed in a given state as θ changes; these complications are not accommodated in the drawing. More detailed calculations by Buenker et al. [B69] suggest that the V-state twisting curve is grossly altered from that of Fig. IV.A-2 by interaction with the Rydberg configuration $(\pi, 3p_y)$, which is calculated to be the lower of the two configurations in the strongly twisted geometry.

The N → V band of ethylene is the prototype $\pi \rightarrow \pi^*$ excitation for pi-electron molecules, yet nothing but the most general features of this band are agreed upon by all. It was earlier thought that this excitation was really two bands, a structured one and a continuum, but there is now agreement that both features are part of the same transition.† Still, some contend on the basis of symmetry and from calculations of the π^*

---

† However, there are experimental and theoretical arguments that a second, much weaker symmetry-forbidden band is hidden within the N → V profile.

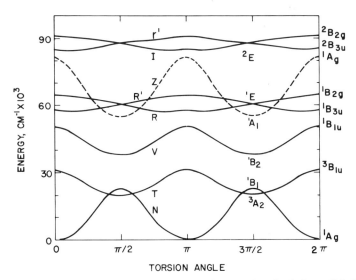

Fig. IV.A-2. Twisting diagram for the various states of ethylene [M26].

wave function that the (π, π^*) singlet state is Rydberg, while others claim it is valence shell. Estimates of its $(0, 0)$ frequency differ by 10 000 cm^{-1}, and a half-dozen different vibrational assignments have been presented for its poorly resolved vibrational structure. On the theoretical side, some claim that the Rydberg nature of the V state can be proved with a Hartree–Fock calculation, whereas others feel that Hartree–Fock is in error on this point. The only two calculations on twisted ethylene give opposite signs to the rotatory strength of the $\pi \to \pi^*$ excitation. Some suggest the V state is twisted by 90° about the C—C axis while maintaining trigonal hybridization, whereas others claim tetrahedral hybridization, with and without the 90° twist. The upper-state twisting potential is either parabolic or sinusoidal. Upon this we have built the pi-electron theory of organic molecular spectra!

Experimentally, vibronic structure is observed to commence at 48 330 cm^{-1} in the N → V transition of ethylene and to consist of 11 quanta of about 800 cm^{-1} which appear to merge into a continuum extending to about 71 000 cm^{-1} (Fig. IV.A-3). The intensity of each of the broad vibronic components increases by about a factor of three on going up the series [M10]. In the gas-phase spectrum, this vibrational structure is no longer evident beyond the beginning of the N → R absorption, but a few more quanta can be seen in matrix spectra in which the N → R absorption is strongly shifted to higher frequencies [R19]. The oscillator strength measured for the N → V transition using photoelectric detection

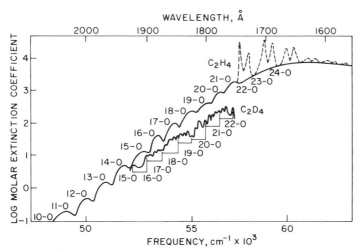

Fig. IV.A-3. Absorption spectra of ethylene-h_4 and ethylene-d_4 in the $N \to V$ region. The $N \to R$ absorption band in ethylene-h_4 is indicated by the dashed line [W25].

is 0.34, and the frequency of maximum absorption is approximately 61 700 cm^{-1} [Z2]. The $\pi \to \pi^*$ vibrational structure is more clearly resolved in the spectrum of ethylene-d_4 (Fig. IV.A-3), but is complicated by the presence of hot bands. McDiarmid and Charney place the average frequency at 550 ± 20 cm^{-1} in the ethylene-d_4 progression, with intervals of about 150 cm^{-1} also evident between the larger intervals in the deuterated compound [M10]. Wilkinson and Mulliken [W25] first analyzed the approximately 800-cm^{-1} intervals in ethylene-h_4 in terms of a single, slightly irregular progression in v_2', the C—C stretching vibration, but with hidden complications due to torsional motions involving v_4'. Such an assignment is strongly suggested by the analogous Schumann–Runge bands of oxygen, which show a large increase in the O—O distance in the upper state and a long progression of stretching with a very weak $(0, 0)$ band. Using arguments based upon isotope shifts in the ethylene-h_4–ethylene-d_4 spectra, Wilkinson and Mulliken tentatively assign the first observed vibronic feature to $v_2' = 10$ and estimate the origin to be near 40 000 cm^{-1}. If the ratio of adjacent intensities (about three) can be extended to the proposed origin, then the origin will have an intensity only 10^{-12} of that at the maximum ($v_2' = 24$) and will be very difficult to find. Merer and Mulliken [M26] take as evidence for the low frequency of the $N \to V$ origin the absorption reported to begin at $\sim 38\,000$ cm^{-1} in the spectrum of a column of liquified ethylene [R5]. However, McDiarmid has shown that this is due to a trace of oxygen and that in "pure" liquid ethylene, the absorption begins at about 48 000 cm^{-1}, close to the frequency of the first vibronic feature in the gas-phase spectrum [M15]. In agreement with the suggestion that the N → V origin lies in the region of 48 000 cm^{-1}, a 1-mm-thick pure crystal of ethylene at 20°K shows only two bands, at 48 100 and 48 920 cm^{-1} which correspond to the gas-phase bands at 48 330 and 49 140 cm^{-1} [L37].

Both theoretical calculations and reasoning by analogy lead one to expect an increased C—C distance in the V state of ethylene. Similar reasoning suggests that the minimum-energy configuration in this state has the two halves of the molecule twisted 90° with respect to one another (Fig. IV.A-2). In fact, McDiarmid and Charney [M10, M11] interpreted the N → V vibronic envelope in terms of a single progression of v_4', the torsional mode, leading to a skewed upper state. In these experimental and theoretical studies, a second, hot progression at 280 ± 50 cm^{-1} lower frequency was uncovered by studying the spectra at 373 and 200°K. Since the interval does not correspond to any ground-state frequency of ethylene-h_4, the origin of this second progression was taken to be the (1, 1) sequence band of the hot torsional vibration. The corresponding hot frequency interval in ethylene-d_4 is 200 ± 20 cm^{-1}. Consideration of the isotope effects on the zero-point energy differences led to an assignment of the vibrational quantum numbers; $v_4' = 0$ comes at 46 700 cm^{-1} and 46 950 cm^{-1} in ethylene-h_4 and -d_4, respectively, and the strongest vibrational feature which is observed has $v_4' = 12$ in ethylene-h_4 (56 300 cm^{-1}) and $v_4' = 17$ in ethylene-d_4 (56 250 cm^{-1}). On the basis of the supposed regularity of the observed intervals and by relative intensity arguments as well, it was concluded that the H$_2$C—CH$_2$ twisting potential in the upper state is parabolic

$$V = V_0 \theta^2, \tag{IV.1}$$

with $V_0 = 8360 \pm 600$ cm^{-1} and 7700 ± 560 cm^{-1} in ethylene-h_4 and -d_4, respectively. This scheme offers no explanation for the ubiquitous 150-cm^{-1} interval in the deuterated compound.

We do not present more of McDiarmid and Charney's interesting analysis because Merer and Mulliken [M26] point out that the vibrational intervals are really too irregular in ethylene-h_4 and -d_4 for such a simple potential, and because one *does* expect the C—C stretching vibration to be excited. Taking a middle course, Merer et al. [M25, M28] have computed the N → V vibronic envelope under the assumption that both v_2' and v_4' are excited in the N → V transition, with a V-state equilibrium geometry in which $\theta = 90°$ and $r'_{C-C} = 1.44$ Å. It is said that such an approach does a satisfactory job of explaining the observed envelopes [M25, M26]. For example, the striking difference in the ethylene-

h_4 and $-d_4$ patterns (Fig. IV.A-3) is said to be due to the two vibrations having very nearly the same frequency in ethylene-h_4 and thereby being badly overlapped, whereas the frequency of ν_4' is much lower than that of ν_2' in the deuterated molecule, so that they are easily separated. This analysis leads to a (0, 0) band at 38 500 cm^{-1} and a barrier height of 16 000 ± 3000 cm^{-1} in the V state. The (0, 0) band was said by Merer *et al.* to correspond to the rapid increase of absorption intensity observed by Reid in long paths of liquid ethylene at about 38 000 cm^{-1} [R5]; however, McDiarmid has convincingly shown that this is an artifact due to the presence of oxygen in the sample [M15].

Warshel and Karplus have attacked the problem of the N → V vibronic structure using a direct calculation of the Franck–Condon factors and fitting this to the observed spectrum. They find that the torsional progression is the dominant excitation, with C—C stretching being of secondary importance. The electronic origins are estimated to be 46 772 cm^{-1} (C_2H_4) and 46 832 cm^{-1} (C_2D_4), in grave disagreement with the conclusions of Merer *et al.* Again differing with the conclusions of Merer *et al.*, Warshel and Karplus feel that in C_2D_4, two quanta of C—C stretch are appended to each quantum of torsion, making the spectrum very complex, whereas in C_2H_4, these secondary stretching progressions have much lower Franck–Condon factors and so the spectrum does not appear to be so complicated [W13].

Theoretically, one expects that the (π, π^*) electronic configuration in planar ethylene will have a greatly lengthened C—C distance as compared with the ground state in which both electrons are C—C bonding. However, twisting about the C—C axis turns the antibonding π^* MO (and the bonding π MO) into effectively nonbonding MOs. Moreover, there is a bonding hyperconjugation between the CH_2 orbitals of one half of the molecule and the pi AO of the other half which will also act to shorten the C—C bond in the (π, π^*) twisted state. Thus the C—C distance and the interplanar angle are intimately coupled, and the value of ν_4' will depend strongly upon how many quanta of ν_2' are simultaneously excited, and vice versa. For example, the values of ν_4' derived from the near-vertical part of the spectrum are those appropriate for strong excitation of ν_2', and may be very different from that found near $\nu_2' = 0$, where the C—C distance is shorter. Ogilvie's [O1] suggestion that the active vibration is neither the stretch nor the twist, but CH_2 out-of-plane wagging, seems not to accord with any of the theoretical expectations.

Recent quantitative calculations of the geometry of the ethylene V state substantiate most but not all of the empirical arguments of Merer and Mulliken. Thus in calculations of varying sophistication, Kaldor and

Shavitt [K2], Kirby and Miller [K19], and Basch and McKoy [B13] all conclude that the V state has minimum energy at the 90° twisted configuration. The latter two studies also find the C—C distance expanded to 1.38 Å in the 90° configuration, as anticipated by Merer and Mulliken. Kirby and Miller also investigated the possibility of CH_2 wagging distortions in the 90°-twisted V state and found that this does not occur, contrary to the suggestions of Walsh [W7] and Ogilvie [O1], and that in the *planar* V state of ethylene, the equilibrium C—C bond distance is only 1.43 Å, rather than 1.8 Å as deduced by Merer and Mulliken.

Mulliken long ago pointed out that the π^* MO of ethylene has the same symmetry properties as the $3d_{xz}$ Rydberg AO and that π^* should "show some tendency to resemble a Rydberg orbital" [M60]. In the present work, such pairs of orbitals are called "Rydberg/valence shell conjugates" (Section I.A-1). Dunning et al. were led much further in this direction by their Hartree–Fock calculations on the excited states of ethylene [D28]. Using a valence shell basis set augmented with a number of diffuse $p\pi$-type AOs, they computed the Hartree–Fock wave functions for the N, T, and V states of planar ethylene, and the expectation values of x^2, y^2, and z^2 for each of the occupied orbitals in each of the states (Table IV.A-I). It was found that the diffuse orbitals were completely rejected by the occupied MOs in the N and T states, but that the planar V state incorporated large amounts of these diffuse orbitals into π^*. Therefore, they state, "the planar V state is just *not* a valence state," but the corresponding triplet state is. In support of their claim that the V state of ethylene is Rydberg, Dunning, et al. also point out that their computed N → V excitation energy of 59 900 cm^{-1} is in very good agreement with the experimental value of 61 700 cm^{-1} at the absorption maxi-

TABLE IV.A-I
EXTENT OF THE PI ORBITALS IN ETHYLENE[a]

Component[b]	N state	T state	V state
$\langle \pi \| x^2 \| \pi \rangle$	2.154	2.063	2.029
$\langle \pi \| y^2 \| \pi \rangle$	0.881	0.756	0.725
$\langle \pi \| z^2 \| \pi \rangle$	2.643	2.268	2.174
$\langle \pi^* \| x^2 \| \pi^* \rangle$	—	3.847	43.019
$\langle \pi^* \| y^2 \| \pi^* \rangle$	—	0.915	14.082
$\langle \pi^* \| z^2 \| \pi^* \rangle$	—	2.745	42.082

[a] From Reference [D28].
[b] Molecule is in the xy plane, with x aligned along the C—C axis. Matrix elements in units of square Bohrs.

mum, whereas in the previous valence shell calculations, which did not include the inflated pπ-type basis functions (see, for example, [R12]), N → V was predicted to come at about 75 000 cm^{-1}. In these near-Hartree-Fock calculations in restricted basis sets one had to presume that the correlation energy error was larger in the V state than in the N state in order to explain the calculated value being higher than the observed.

Such a revolutionary claim for the Rydberg nature of the V state of ethylene did not go unchallenged for long; Basch and McKoy [B13] discovered first that the Dunning calculation leads to a calculated ionization potential of only 72 600 cm^{-1}, whereas 84 750 cm^{-1} (advert.) is observed, and stress that it is very suspicious that the $\pi \to 3d$ Rydberg transition frequency could agree with the experimental value, while the limit of that series is underestimated by 12 150 cm^{-1}. In fact, knowing what we do about 3d term values (Section I.C-1), the $(\pi, 3d)$ Rydberg state should come 13 000 cm^{-1} below the lowest ionization potential, i.e., at about 72 000 cm^{-1}, rather than at 61 700 cm^{-1}. Going deeper into the problem, Basch and McKoy conclude that at the Hartree–Fock level, the Dunning calculation for the ethylene V state converges upon the $(\pi, 3d_{xz})$ Rydberg configuration, but that this is *not* the spectroscopic state observed at 61 700 cm^{-1}. Graphically, they present the situation as shown in Fig. IV.A-4, where the state calculated by Dunning *et al.* correlates with an observed Rydberg at 73 000 cm^{-1}, rather than with the V state at 61 700 cm^{-1}. Morokuma and Konishi [M52] present calculations which demonstrate that a similar set of circumstances obtain for calculations on the lowest (π, π^*) triplet state of oxygen when expanded 2pπ AOs are used in the basis set. Experimental evidence to be cited is in complete agreement with Basch and McKoy's description of the V state of ethylene as purely valence shell. The apparent failure of Hartree–Fock theory to converge upon a proper valence shell V state for ethylene will not be a problem in larger pi-electron molecules, where the V states are far below the (π, nR) Rydberg states.

The conflict over the Rydberg/valence shell character of the singlet (π, π^*) state of ethylene has triggered a flood of papers on the subject. Rose *et al.* [R24], using the equations-of-motion method, calculate that the vertical N → V excitation energy is 63 700 cm^{-1} with an oscillator strength of 0.40. In the V state, the π^* MO is somewhat more diffuse than that in the T state, but is still very much valence shell. They have also performed a calculation for the Rydberg conjugate $(\pi, 3d_{xz})$ state, and in contrast to the N → V excitation, N → R$(3d_{xz})$ comes at 71 800 cm^{-1} (13 100 cm^{-1} term value) with an oscillator strength of only 0.02. Ryan and Whitten [R29] found an expanded π^* orbital in an ethylene calculation, but also found that the mixing of (σ, σ^*) configurations into

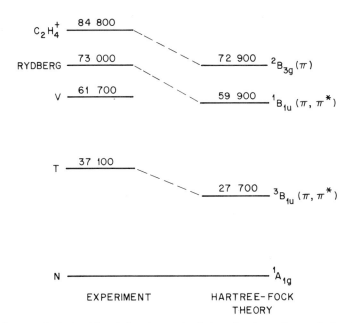

Fig. IV.A-4. Relationship of the observed ethylene levels to those calculated from Hartree–Fock theory [B13].

the singlet (π, π^*) configuration served to contract the π^* MO, so that an essentially valence shell configuration resulted 64 700 cm^{-1} above the ground state. Since their original suggestion, Dunning and co-workers [B23, D28] have reworked the problem with multiconfigurational wave functions with results which support their original contention, i.e., $\langle \pi^* | x^2 | \pi^* \rangle$ in the T state is only 12.0 square Bohrs, but in the V state, it is 35 square Bohrs. This latter value is somewhat reduced from the Hartree–Fock value (Table IV.A-1), but still leads them to call the V state "Rydberg-like"; a full-blown 3p Rydberg state has $\langle 3p\pi | x^2 | 3p\pi \rangle$ equal to approximately 65 square Bohrs. It seems undeniable that the V state of ethylene is more diffuse than the T state, but the extent of the π^* MO in the V state remains to be settled to everyone's satisfaction.

As regards the N → V transition of ethylene, two interesting complications have been pointed out by Buenker et al. [B68, B69, P15]. Ordinarily, the electric transition moment is assumed to be a constant multiplier in the calculation of Franck–Condon factors, so that the band maximum corresponds to the vertical frequency. However, Buenker et al. point out that as ethylene is twisted (ν_4'), the π^* MO changes from diffuse (Rydberglike) to valence shell-contracted, and so the moment $\langle \pi | \mu_e | \pi^* \rangle$ must be dependent upon the twist angle. As a consequence of this, the geometrically vertical transition may not correspond to the frequency of the apparently largest Franck–Condon factor, and they suggest that the true vertical value in ethylene is closer to 65 000 cm^{-1} rather than 61 700 cm^{-1}. Second, Buenker *et al.* point out that in the antiparallel 90°-twisted form of ethylene, the (π, π^*) valence shell configuration of the planar molecule becomes the $(\pi, 3p_y)$ Rydberg configuration of the twisted molecule, and that these states are close together over a long range of θ. However, their suggestion that the vibronic structure observed in the vicinity of the origin is due to transitions to the $(\pi, 3p_y)$ Rydberg state is not supported by the spectra in condensed phases (as will be discussed later).

The lowest-frequency Rydberg band of ethylene $N \rightarrow R(3s)$ has its origin at 57 340 cm^{-1} (Fig. IV.A-5), and can be fit as the first member of a five-term Rydberg series with a limit of 84 750 cm^{-1} and $\delta = 1.09$ [P42]. The $(\pi 1b_{2u}, 3sa_g)$ state of ethylene has a term value of 27 400 cm^{-1} (advert.), which is close to those of the Rydberg excitations terminating at 3s in ethane (29 500 cm^{-1}, vert.) and acetylene (26 000 cm^{-1}, advert.). Vibronic analysis of the $\pi \rightarrow 3sa_g$ band of ethylene has pro-

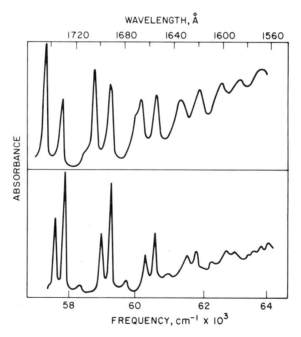

Fig. IV.A-5. Vibronic band profiles of the $N \rightarrow R(3s)$ Rydberg transitions in ethylene-h_4 (upper) and ethylene-d_4 (lower) [W25].

ceeded in an orderly way. Price and Tutte [P42] first determined the vibrational frequencies within the transition for various isotopically substituted molecules and report excitation of a progression in the C—C stretch ν_2', each member of which is an origin for what appears to be a few quanta of ν_4'. A summary of the vibrational frequencies found by them is given in Table IV.A.-II. Wilkinson and Mulliken [W25] later studied these bands under higher resolution, using an extensively purified sample. They identified five quanta of ν_2' and pointed out that the appended torsional vibrations are required by the symmetry selection rules to appear in units of double quanta, i.e., $2\nu_4'$, $4\nu_4'$, etc. Using the empirical frequency–distance relationship

$$\omega(r'_{\text{C-C}})^{2.88} = \text{const} , \qquad (\text{IV.2})$$

it was found that $r'_{\text{C-C}}$ in the R state is 1.45 Å. However, a satisfactory quantitative analysis could not be carried out, and it was concluded that vibronic interations between the R(3s) state and the underlying V state were responsible for the extreme anharmonicity of the ν_4' vibration (Table IV.A-II), as well as for peculiar isotopic–frequency ratios and Franck–Condon factors. Answers to many of the questions posed by Wilkinson and Mulliken's work are contained in the later papers of Merer and Schoonveld [M24, M27], who studied the $\pi \to 3s$ bands of ethylene-h_4 and -d_4 under high resolution and at temperatures up to 450°C. Their vibrational analysis confirmed that of the earlier workers, and added

TABLE IV.A-II

VIBRATIONAL FREQUENCIES IN THE VARIOUS ELECTRONIC STATES OF ETHYLENE

	State	$1\nu_2$	$1\nu_4$	$2\nu_4$	$4\nu_4$
C_2H_4	^1A_g	1623	1023	—	—
	V	800	—	—	—
	R(3s)	1370	96	468—	1084
	R(4s)	1450	—	—	—
	I	1290 ± 30	—	405 ± 30	—
C_2H_3D	^1A_g	—	—	—	—
	V	—	—	—	—
	R(3s)	1350	—	415	—
	I	—	—	—	—
C_2D_4	^1A_g	1515	726	—	—
	V	550	—	—	—
	R(3s)	1290	41	286	715
	R(4s)	1360	—	—	—
	R(5s)	1330	—	—	—
	I	1370 ± 30	—	280 ± 30	770 ± 30

much to this, as, for example, the forbidden $1\nu_4'$ frequency, deduced from the $\nu_4(1, 1)$ hot band. They also found that the 50-cm^{-1} splittings in certain of the vibronic bands reported by Wilkinson and Mulliken are due to rotational effects rather than some sort of vibronic splitting. Details of the rotational envelope point directly to a C-type transition and consequent out-of-plane polarization, as predicted group-theoretically for a $\pi(1b_{2u}) \to R(3sa_g)$ excitation.

On several occasions [M62, W25], Mulliken has suggested that in the configuration of minimum energy, the R state of ethylene was slightly twisted, perhaps by about 30°. In fact, it is just this feature of the R-state twisting potential that is the cause for abnormalities in the spectrum. Following Lorquet and Lorquet [L35], Merer and Schoonveld point out that in the 90°-twisted geometric configuration, the $(\pi, 3s)$ and $(\pi^*, 3s)$ electronic configurations are the two components of a doubly degenerate state which is Jahn–Teller unstable with respect to the ν_4 torsional motion. A large static Jahn–Teller distortion can then lead to a double-minimum potential of the sort shown in Fig. IV.A-2. A quantitative fitting of the ν_4' frequencies (Table IV.A-II) to such a potential led them to minima at $\theta = 25 \pm 1°$ and a central barrier at $\theta = 0°$ of 289 ± 20 cm^{-1}. Upon excitation of one or two quanta of ν_4', the molecule is essentially planar, though not so in the lowest vibrational level of the R state. In addition to explaining the ν_4' frequency intervals, the double-minimum potential also explains the relative vibronic intensities in both ethylene-h_4 and -d_4. The decrease of the rotational constant A in the R(3s) state suggests a possible slight lengthening of the C—H bonds (0.01–0.02 Å) and a slight increase of the

angles ($\sim 3°$), but not enough of a change in these dimensions to lead to the excitation of the appropriate vibrations.

Prior to the experimental determination of the R(3s) state geometry, there were several theoretical studies which led to $\theta = \sim 30°$, with a barrier amounting to several hundred cm^{-1} in the R state [B64, L35, M62, R12]. These values were derived quite independently of any arguments involving the Jahn–Teller effect, and appear to be appropriate for the corresponding ionic state ($^2B_{2u}$) of ethylene as well.

The photoelectron spectra of ethylene-h_4 and ethylene-d_4 have been recorded at high resolution by several investigators [B2, B41, B55, B58]. In the first band of ethylene-h_4 (Fig. IV.A-6), the twisting doublets are once again evident (405 ± 30 cm^{-1} spacing), attached to successive quanta of ν_2', the C=C stretch (1290 ± 30 cm^{-1}) [B58]. The broadness

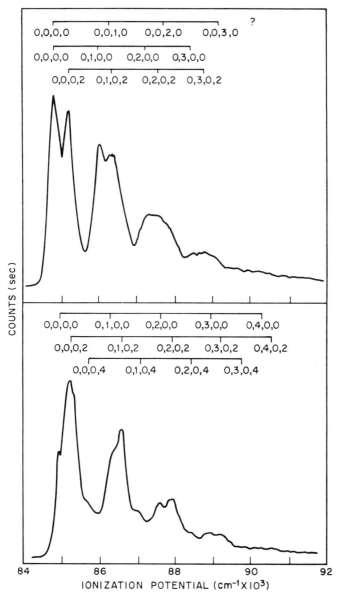

Fig. IV.A-6. Vibrational structure in the $1b_{2u}$ photoelectron bands of ethylene-h_4 (upper) and ethylene-d_4 (lower) [B58].

of certain of the peaks suggests that v_3', the

scissors, may also be excited here, though there is no evidence for it in the high-resolution optical study of the N → R(3s) and higher Rydberg bands [M24, M27, W27]. If v_3' is present in the photoelectron spectrum, then the ground-state ion has a different

angle than the $(\pi, 3s)$ state. In the photoelectron spectrum of ethylene-d_4, the relative intensities of the two components of the twisting doublets invert as compared with the -h_4 spectrum, and a fourth quantum of v_4' is also clearly visible. Just these differences distinguish the $\pi \to 3s$ optical bands of ethylene-h_4 and ethylene-d_4, and in fact the v_2' and nv_4' frequencies are very similar in the optical and photoelectron spectra of both species (Table IV.A-II and Fig. II.A-2). However, as with ethylene-h_4, there is the possibility of v_3' excitation in the -d_4 photoelectron spectrum, but not in its optical spectrum. It seems undeniable that the geometries of the R(ns) and I states of ethylene are very nearly identical, with a double-minimum potential in the twisting coordinate in each of them.

A most peculiar feature of the $\pi \to 3s$ twisting doublets is the relative intensities of the component bands. In ethylene-h_4, the $(v_2' = 0, v_4' = 0)/(v_2' = 0, v_4' = 2)$ intensity ratio is 1.41 [M24, M27], but as one progresses up the v_2' ladder, the intensities of the two components seem to approach equality. In the same band of ethylene-d_4, the intensity ratio is 0.80 [M27] with a slower approach to equality. Looking at the higher $\pi \to ns$ transitions (Fig. IV.A-1), the same ratio of about 1.4 can be seen for $n = 5$, 6, and 7 in ethylene-h_4; however, the $n = 4$ band has a ratio significantly smaller than 1.0. In the appropriate photoelectron bands [B58], the ratio of the two components is larger than 1.0 in ethylene-h_4, but smaller in ethylene-d_4. Only detailed calculations of the sort carried out by Merer and Schoonveld are able to explain these seemingly anomalous intensity ratios.

Because of the small overlap between the 3s Rydberg orbital and π MO, the singlet–triplet splitting of the $(\pi, 3s)$ configuration should be small, and almost certainly the triplet will lie upon the low-frequency tail of the N → V excitation. Nicolai [N13] has observed the energy-loss spectrum of ethylene using 30-keV Li$^+$ projectiles rather than electrons, a technique which heightens singlet → triplet probabilities. He observed a prominent band at 53 200 cm^{-1} (vert.) where only a very weak N → V tail is observed with 25-keV electrons, and assigned this to the π → 3s Rydberg triplet. Evidence of the reality of a transition at 53 200 cm^{-1} in ethylene is provided by the SF$_6$-scavenger spectrum of ethylene (Section II.D) [H33], which also shows a weak band at that frequency. The proposed singlet–triplet split of 4100 cm^{-1} is reasonable, being at the upper end of the range of values expected for Rydberg configurations (Section I.A-1). It is amusing to note that Kuppermann and Raff originally claimed to have found a band at this frequency in the electron-impact spectrum, but that subsequently a barrage of evidence was presented to show it was spurious. Now evidence supporting the presence of a band at this frequency can be cited.

In a detailed farther-ultraviolet study, Wilkinson [W27] found several Rydberg origins in the 65 000–85 000 cm^{-1} region. Following the π → 3s transition discussed previously, leading members of two other series converging upon the lowest ionization potential were found at 66 607 and 69 516 cm^{-1} [P42, W27]. These bands are especially prominent in the electron impact spectrum of Rendina and Grojean [R6]. The 18 143- and 15 234-cm^{-1} terms for these transitions may be taken as evidence for π → 3p assignments, the splitting being the result of the aspherical symmetry of the ionic core. On the basis of a theoretical perturbation calculation, Liehr [L26] has made such as assignment. However, as he points out, such transitions are $u \rightarrow u$ parity forbidden and can only appear in the optical spectrum when assisted by asymmetric vibrations. The transitions are appropriately weak (Fig. IV.A-1), but according to Wilkinson's analysis, have relatively strong origins. This is especially true of the 69 516-cm^{-1} band, which appears to have a strong origin accompanied by the excitation of ν_2' and double quanta of ν_4'. It is at first tempting to assign this band instead as π → 3d, $u \rightarrow g$ allowed, but the fact that its intensity is only about 1% that of the π → 3s excitation argues against this. Tentatively, it must be assumed that these two bands were measured from false origins, with the true origins being parity forbidden. In both ethylene-h_4 and -d_4, the lower-frequency component of the twisting doublets is stronger in the π → 3p bands. This is in line with the relative intensities found for the first band in the photoelectron spectrum of ethylene-h_4, but contrary to that for ethylene-d_4.

Yet another origin is reported by Wilkinson in ethylene-h_4 at 73 011 cm^{-1}. Like the others, this, too, is followed by only a few quanta of ν_2' and double quanta of ν_4', the origin being rather intense. It seems more certain that this is an allowed excitation to a 3d Rydberg orbital, though the term, 11 740 cm^{-1}, is smaller than that of the hydrogen atom, 12 193 cm^{-1}. Such a situation is imaginable if the C$_2$H$_4^+$ core splits the hydrogenic levels symmetrically about the hydrogen atom term value, or if the 3d level has been mixed by the core with deeper 3s and/or 3p levels as in NO [J20].

In all cases, Wilkinson presents a vibronic analysis very much like that of the $\pi \to 3s$ band; thus it seems that the geometries of all the Rydberg states are very nearly the same, regardless of the symmetry of the upper Rydberg state. Liehr argues that the $\pi \to 3p$ bands are allowed because the upper states are 90°-twisted as in the V state, but the vibronics in these states are much more like that of $N \to R(3s)$ than that of $N \to V$. In some instances, Rydberg transitions converging upon ionization potentials beyond the first can be identified by their autoionizing characteristics in photoionization experiments. In the case of ethylene, there seems to be very little autoionization, however, and so higher Rydberg series cannot be found [C14].

In a relatively low-resolution study of ethylene using the trapped-electron method [B38, B49], a distinct peak was found at 74 100 cm^{-1}, which is a region of minimum absorption in the optical spectrum (Fig. IV.A-1), followed by a minimum at 80 000 cm^{-1}, which is a maximum in the optical spectrum. It may be that an underlying valence shell transition has been uncovered.

Up to this point in the discussion, the theoretical calculations have been of relatively little help in explaining the electronic spectrum of ethylene. In fact, the spectrum has been of great help in explaining the calculations! But there are still other bands in the ethylene spectrum, and for these, the calculations are of use. According to *ab initio* GTO calculations on ethylene [R12, S20], valence shell excitations of the sort $\pi \to \sigma^*$ and $\sigma \to \pi^*$ are to be expected about 15 000 cm^{-1} beyond the $N \to V$ absorption. The two lowest $\sigma \to \pi^*$ bands ($3a_g, 1b_{2g} \to 1b_{3g}$) are symmetry forbidden, and the lowest $\pi \to \sigma^*$ excitation ($1b_{2u} \to 4a_g$) has a calculated oscillator strength of only 0.009. The only other strong band to be expected in the ethylene spectrum is $\sigma \to \sigma^*$ ($4a_g \to 3b_{3u}$), which should be about as strong as the $\pi \to \pi^*$ transition.† Zelikoff and Watanabe [Z2] report a continuum absorption in ethylene centered at 80 000 cm^{-1} with an appreciable oscillator strength; this could be the allowed $\sigma \to \sigma^*$ excitation. The absorption cross sections and photoioniza-

---

† The expected similarity of the $\pi \to \pi^*$ and $\sigma(C-C) \to \sigma^*(C-C)$ oscillator strengths could be badly upset by extensive configuration interaction between these configurations, which is, after all, a prominent aspect of the calculations on the V state of ethylene. Hopefully, more attention will be focused on the (σ, σ^*) state by the theoreticians, for if its frequency and oscillator strength can be calculated as accurately as was the transition to (π, π^*), such calculations would be extremely useful in assigning the higher excitations.

tion yields of ethylene and ethylene-d_4 from their ionization potentials to 95 100 cm^{-1} have been measured [P9].

Adapting a phenomenon of solid-state physics to molecular spectroscopy, Herzenberg et al. [H21] have considered the possibility of a multielectron plasma resonance (plasmon, Section I.A-3) in the far-ultraviolet spectrum of ethylene. Using highly questionable single-particle excitation energies, these authors conclude that such a collective excitation will be found at $\sim$50 eV with an oscillator strength over ten. The excitation is said to be strongly allowed in the electron energy-loss spectrum, but has not been observed yet. Indeed, the existence of plasmons in molecules has not been demonstrated, though it has been suggested for several molecular solids and liquids. In a more refined calculation, Crocker and Herzenberg find a transition to a B_{3u} state of ethylene at 282 000 cm^{-1} (35 eV) which again has a small optical oscillator strength but a very large generalized oscillator strength [Eq. (IV.3)] for electron-impact excitation [C29]. It is this large generalized oscillator strength which characterizes collective excitations.

The absorption spectrum of ethylene has been studied under high pressure and in condensed phases with interesting results. The spectral consequences of pressurizing ethylene with nitrogen gas [E8, R15] are shown in Fig. IV.A-7. The first figure in the upper left shows the details of both

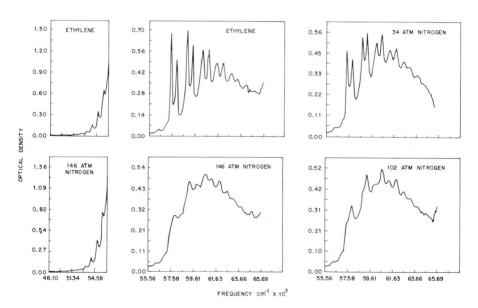

Fig. IV.A-7. Effects of nitrogen pressurization on the N $\rightarrow$ V and N $\rightarrow$ R(3s) bands of ethylene [R15].

the N → V low-frequency wing and the N → R(3s) Rydberg excitation resting upon it; the two spectra were determined with different pressures of ethylene. Upon pressurizing with nitrogen, there is a noticeable broadening of the Rydberg components to the high-frequency side, such that at 102 atm of perturbing gas (lower right), the relative intensities of the twisting doublets appear to be reversed [R15]. As explained in Section II.B, this is not really so, for the *integrated* intensities retain their original relationship as the pressure is increased, though the intensities at the maxima are altered due to the overlapping of the absorptions. At 146 atm nitrogen pressure (lower left), the previously sharp N → R(3s) Rydberg excitations are reduced to broad, badly overlapped peaks, whereas the vibronic features of the N → V transition are in no way affected by the same perturbation. This offers an excellent comparison of the relative behaviors of valence shell and Rydberg excitations under high-pressure perturbation, and also argues strongly against any appreciable Rydberg character ($3d_{xz}$) in the π^* MO. The valence shell nature of the N → V band of ethylene was also demonstrated by Miron *et al.* [M43], who found that there was a very small frequency shift of the N → V spectrum on going from a liquid solution in krypton to a solid solution, whereas the π → 3s Rydberg band under the same conditions showed a large shift to higher frequencies. A slight bump in the spectrum of ethylene in liquid argon at 67 600 cm^{-1} (vert.) is interpreted by Miron *et al.* [M44] as an excitation to an $n = 2$ Wannier exciton, a transition without an analog in the free-molecule spectrum.

In pure polycrystalline ethylene, vestiges of the N → V vibronic structure are apparent, but there is no trace here of the Rydberg transition N → R(3s) [R19]. However, such bands were found by Katz and Jortner in a 1% solid solution of ethylene-d_4 in krypton at 20°K [K6, R19]. The bands in question are shifted 4040 ± 30 cm^{-1} to higher frequencies compared with the frequencies in the gas phase, and the half-widths increase to approximately 350 cm^{-1} in the matrix, though the vibrational intervals (1300 ± 30 cm^{-1}) are very close to the gas-phase value, 1307 cm^{-1}. The shift of 4000 cm^{-1} to higher frequencies is in line with those observed for the lowest Rydberg bands in several rare gas/organic molecule studies (Section II.C). However, the bandwidth appears to be strongly dependent upon the host/guest ratio, for the N → R(3s) bands of ethylene in krypton appear optimally at 0.1% ethylene-h_4 [R19] (Fig. IV.A-8) but could not be found in the spectrum of 2% ethylene-h_4 in the same matrix [R12].

Since the inelastic electron scattering process obeys selection rules which in general are less restrictive than those for optical absorption, such spectra can often uncover transitions which are otherwise unobservable

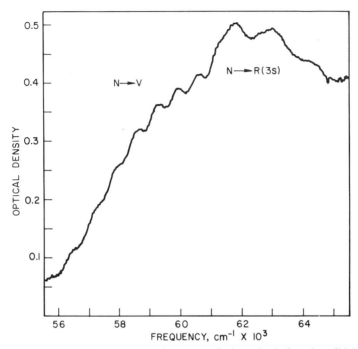

Fig. IV.A-8. Absorption spectrum of a 0.1% solution of ethylene in solid krypton at 24.5°K [R19].

optically. In ethylene, such a band was reported at 52 500 cm^{-1} in the energy loss spectrum [K41, K42], but later work by refined methods convincingly disproved its reality [D19].† One of the features of an electric-quadrupole-allowed excitation in the electron-impact spectrum is that its cross section is independent of the initial electron energy. Since the relative Franck–Condon factors within an electric-dipole-allowed transition are independent of the initial electron energy while the absolute intensity is not, the presence of an electric-quadrupole-allowed transition beneath the vibronic structure of an electric-dipole-allowed one can be revealed by an apparent change of Franck–Condon factors with changing incident energy. Thus Ross and Lassettre [R27] compared the relative intensities of the vibronic features in the 57 300–62 000 cm^{-1} region of ethylene determined by them using 200-eV electrons with those reported by Geiger and Wittmaack [G4] determined using 33 000-eV electrons (Fig. IV.A-1) and found discrepancies of up to 15%. On this basis, they

† However, such a band was later found in the scattering spectrum of Li$^+$ ions incident at high energies upon ethylene [N13]. It would appear to be the triplet component of the $\pi \rightarrow 3s$ excitation.

postulated an electric-quadrupole-allowed transition in the region of the dipole-allowed N → V and N → R(3s) transitions. Such an argument also requires that the relative intensities of the N → V and N → R(3s) transitions remain constant over the range 200–33 000 eV of incident electron energy. Siezing upon this doubtful band, Yaris et al. [Y6] have attempted to explain the CD spectrum of *trans*-cyclooctene; however, there is as yet no real direct evidence for such a quadrupole-allowed band in the ethylene spectrum, though it is predicted to come in the N → V region [R12].

In a novel investigation, Miller has calculated the angular dependence of the generalized oscillator strength for inelastic electron scattering

$$f_n(K) = (2 \Delta E/K^2) \left| \left\langle \Psi_n \left| \sum_i [\exp(iKr_i)] \right| \Psi_0 \right\rangle \right|^2 \quad (IV.3)$$

for several of the transitions in ethylene [M41]. In this expression, K is the momentum transferred in the transition from Ψ_0 to Ψ_n, their energy separation being ΔE. Comparing the $f_n(K)$ versus K^2 curves computed for the $\pi \to \pi^*$ and $\sigma \to \pi^*$ ($1b_{2g} \to 1b_{3g}$) valence shell excitations with those from the π $1b_{2u}$ MO up to five different Rydberg upper orbitals, he found that all Rydberg curves are characterized by minima, which are lacking in the valence shell excitations. An experimental check of these predictions was offered by Krauss and Mielczarek [K39], who determined the $f_n(K)$ versus K^2 curve at four frequencies, 57 750 ($\pi \to 3s$), 64 500 ($\pi \to \pi^*$), 66 700 ($\pi \to 3p$), and 73 560 cm^{-1} ($\pi \to 4s$), with an energy resolution of 800 cm^{-1}. As predicted, all of the Rydberg excitations do show minima at approximately the calculated momenta. However, a deep minimum was also found for the N → V transition. Since this is contrary to the behavior predicted for the $\pi \to \pi^*$ valence shell transition, it was concluded that the π^* orbital was largely Rydberg. Such a conclusion is at variance with other lines of evidence, and one is prompted to point out that the Rydberg N → R(3s) twisting doublets are still very much evident at 64 500 cm^{-1} in the optical spectrum, though it was presumed that all of the energy loss at this frequency was $\pi \to \pi^*$.

IV.A-2. *Alkyl Olefins*

The addition of one or more alkyl groups to ethylene affects the vacuum-ultraviolet spectrum only in a quantitative way, with many of the features of the ethylene spectrum more or less recognizable. However, new features appear in certain cyclic and exocyclic systems that are strained, and it seems likely that the same extraneous features are also present in the simpler, acyclic olefins, too, but at higher frequencies where they are not as obvious. Much of the pertinent data on alkyl olefins are summarized in Tables IV.A-III–IV.A-V.

IV.A. OLEFINS

As in ethylene itself, the $\pi \rightarrow \pi^*$ transition of an alkyl olefin is most easily recognized as the first strong band in the spectrum ($\epsilon = 3000\text{--}5000$), and, as in ethylene, the (0, 0) frequency is uncertain, and the vibrational structure is badly blurred. Thus the $\pi \rightarrow \pi^*$ excitations correspond to the intense bands near 55 000 cm^{-1} (vert.) in the alkylated ethylenes (Fig. IV.A-9), and to the bands of approximately

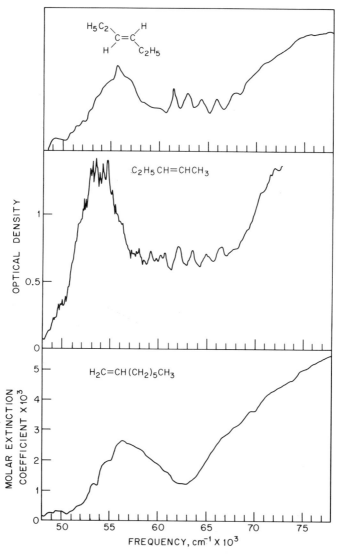

Fig. IV.A-9. Optical absorption spectra of several alkylated olefins in the gas phase [D2].

TABLE IV.A-III
Optical Properties of the Alkyl Olefins

Compound	N→V Frequency[a]	N→V Intensity[b]	Ionization potential[a]	N→R Frequency[a]	N→R Intensity[b]	N→R Term value	Ref.
$H_2C=CH_2$ [c]	61 700	0.34, 0.29, 0.34 ± 0.15	84 750	57 340 66 607 73 011	0.04, 0.03	27 410 (3s) 18 140 (3p) 11 740 (3d)	[H7, P28, P42, R6, W25, W27, Z2]
$CH_3CH=CH_2$	58 000	0.32	80 000	53 100 61 200	0.02	26 900 (3s) 18 800 (3p)	[P42, S6, W17]
$H_2C=C(CH_3)_2$ [d]	54 350	0.39	75 900	49 800 56 300 57 200	0.02	26 100 (3s) 19 600 (3p) 18 700 (3p)	[C3, G1, L32, M13, W17, W18]
cis-$CH_3CH=CHCH_3$ [e]	57 470	0.59	74 900	48 300 51 140 59 800	0.02	26 600 (3s) 23 760 (3p) 15 100 (3d)	[C3, C4, G1, G2, L15, M12, R18, W17]
trans-$CH_3CH=CHCH_3$ [e]	57 140	0.32	75 100	49 420 52 100 59 830	0.03	25 680 (3s) 23 000 (3p) 15 270 (3d)	[C3, C4, G1, G2, M12, M51, R18, W17, W18]
$CH_3CH=C(CH_3)_2$	57 840 54 000[e]	0.44 0.34 ± 0.09	71 200	48 660 52 000 59 200	0.02	22 540 (3s) 19 200 (3p) 12 000 (3d)	[C3, L32, P32, P50, R19, W18]
$(CH_3)_2C=C(CH_3)_2$	53 500 53 750[e]	0.45 0.33	67 900	45 850 48 640 52 200 57 000 62 000	0.02	22 050 (3s) 19 260 (3p) — — 10 900 (3d)	[C2, C3, C20, G2, P32, R19, W17, W18]
$C_2H_5CH=CH_2$	57 800	0.36, 0.39	77 200[f]	53 300 60 000	—	23 900 (3s) 17 200 (3p)	[C3, G1, P50, S6]

Compound						
$C_3H_7CH=CH_2$	57 470	0.38	76 600^f	53 000	—	23 600 (3s)	[C3, P50, S30]
$C_4H_9CH=CH_2$	56 100	0.34	76 300	60 000	—	16 600 (3p)	[G2, J12, P32, P50]
$C_6H_{13}HC=CH_2$	56 500^e	0.29 ± 0.08	—	—	—	—	[D2, L15, P28]
$(CH_3)_2CHCH=CH_2$ ^{g}	56 400	0.29 ± 0.08	76 000^f	55 160	—	21 440 (3s)	[C3, J12, P50, S30]
	57 800	0.38	76 600^f	60 000		16 600 (3p)	
				64 500		12 100 (3d)	
cis-$CH_3CH=CHC_2H_5$	56 500	0.45	—	51 360	—	—	[C3, C4, J12, S30]
				53 800			
$trans$-$CH_3CH=CHC_2H_5$	55 500	0.32	—	49 800	—	—	[C4, G2, J12, S30]
				59 500			
				62 200			
cis-$C_2H_5CH=CHC_2H_5$	55 800	0.48	—	51 100	—	—	[C3, G2, J12]
				54 000			
$trans$-$C_2H_5CH=CHC_2H_5$	55 300	—	72 100^f	51 300	—	20 800 (3s)	[D3, J12, L15]
				58 800		13 300 (3d)	
$(C_2H_5)CH_3C=CH_2$	53 200	0.26	73 500^f	49 600	—	23 900 (3s)	[C3, D3, J12, L32, P50, S30]
				57 700		15 800 (3p)	
				62 500		11 000 (3d)	
cis-$C_5H_{11}CH=CHCH_3$	54 600	0.29 ± 0.08	—	—	—	—	[P28]
$trans$-$C_5H_{11}CH=CHCH_3$	55 900	0.34 ± 0.08	—	—	—	—	[P28]
$[(CH_3)_2CH]CH_3C=CH_2$	53 200	0.40	—	57 500	—	—	[G2]
$(\triangle-)_2C=C(-\triangle)_2$ ^{h}	45 600	12 000	—	—	—	—	[N15]
$C_2H_5CH=CHC_3H_7$	56 400	—	—	50 120	—	—	[C2]
				51 200			
$(CH_3)_3CCH=CH_2$	57 400	—	—	53 000	—	—	[J12]
cis-$(CH_3)_3CCH=CHC(CH_3)_3$ ^{i}	54 600	—	71 600	50 500	—	21 100 (3s)	[A2, L15, R18]
$trans$-$(CH_3)_3CCH=CHC(CH_3)_3$	54 400	—	71 600	50 500	—	21 100 (3s)	[A2, L15, R18]
$\triangle$	58 000	~3000	79 500	55 500	—	24 000 (3s)	[R13, R17]
				58 000		21 500 (3p)	
				68 000		11 500 (3d)	

TABLE IV.A-III (Continued)

Compound	N → V Frequency[a]	N → V Intensity[b]	Ionization potential[a]	N → R Frequency[a]	N → R Intensity[b]	N → R Term value	Ref.
(bicyclobutane)	58 000(?)	~3000	—	—	—	—	[R13]
(cyclobutane)	56 700	0.28	76 000	52 000 61 800	—	23 500 (3s) 14 500 (3d)	[L32, R19]
(methylenecyclopropane)	55 000	0.29	—	52 500 61 500	—	—	[L32]
(cyclopentene)	56 500	0.32	74 200	47 700 57 000	—	—	[L32]
(cyclohexene)	54 900 54 000[j] 55 300[e]	0.38 ± 0.09, 0.23 ± 0.05, 0.40, 0.19, 0.20	73 500	48 750 53 140	0.004	25 450 (3s) 21 060 (3p)	[C5, P20, R19, W18]
trans-cyclo-C_8H_{14}	50 700	7200	71 000	49 200 52 000 55 435	0.01	24 300 (3s) 21 500 (3p) 18 100 (3p)	[C4, C5, P20, P28, P32, S47, T19, W18]
(methylenecyclobutane, =CH$_2$)	51 900	0.29	73 700	48 300 56 800 62 000	—	22 700 (3s) 14 200 (3d)	[B1, M6, R18, S19]
(methylenecyclopentane, =CH$_2$)	50 700	—	73 600	54 500 60 500	—	19 200 (3p) 13 200 (3d)	[B61, C20, L15, L32]
(ethylidenecyclopentane, =CHCH$_3$)	50 000	4500	68 400[f]	46 000	—	27 600 (3s)	[C20, L15]
				46 000	—	22 400 (3s)	[C20, D3, L15]

Structure					
=CH₂ (methylenecyclohexane)	52 500 52 100[e]	10 500	73 500	49 000 55 900	—	24 500 (3s) 17 600 (3p)	[C20, R19, L15]
=CHCH₃ (ethylidenecyclohexane)	54 000	—	69 700	46 000 55 800	—	23 700 (3s) 13 900 (3d)	[C20, R19]
cyclohexylidenecyclohexane	47 500? 51 500[e]	0.19	65 600	43 500 52 600	—	22 100 (3s) 13 000 (3d)	[C20, R19, S42]
norbornene	51 100	0.15	72 200	48 150 57 240	—	24 050 (3s) 14 960 (3p)	[L15, R10, R19, S49]
[k]	63 000	—	—	44 780 52 720 58 840	—	—	[R12]
α-pinene	50 000 47 600[e]	5000	65 100[f]	42 800 53 700	—	22 300 (3s) 11 400 (3d)	[A4, M6, R9]
β-pinene	50 000	—	—	45 500 55 000	—	—	[M6, R9, T19]

[a] Gas-phase vertical frequencies, cm⁻¹, listed unless otherwise noted.
[b] Oscillator strength listed if known, otherwise, the figure is the molar extinction coefficient at the absorption band maximum.
[c] Corresponding data on the deuterated ethylenes are given in references [M28, P42, W25, W27].
[d] Spectra of the deuterated derivatives of isobutene are given in reference [M13].
[e] Frequency in paraffin solution.
[f] Adiabatic value.
[g] The spectra of many other branched acyclic olefins are presented in reference [J12].
[h] The spectra of several other cyclopropyl derivatives of ethylene are given in references [N15] and [H13].
[i] The spectra of other crowded olefins are presented in references [A2] and [L15].
[j] Pure liquid.
[k] The spectra of many other polycyclic olefins are given in reference [L15].

TABLE IV.A-IV
VIBRATIONAL FREQUENCIES IN VARIOUS STATES OF THE OLEFINS[a]

Molecule	Ground state	N → R(3s)	Ion	Vibration
$CH_3CH=CH_2$	1647	3q 1360	—	C=C stretch
	578	—	1q 527	C=C twist
trans-$CH_3CH=CHCH_3$[b]	764	1q 653	—	=CH wag
	—	1q 820	—	C—C stretch
	—	1q 956	—	—CH_3 wag
	1043	1q 1255	—	=C—H bend
	1068	1q 1585	—	C=C stretch
cis-$CH_3CH=CHCH_3$	304	1q 235	—	Skeletal def.
	876	1q 700	—	C—C stretch
	1018	1q 965	—	—CH_3 wag
	1267	1q 1250	—	C—C wag
	1672	1q 1505	—	C=C stretch
	~3000	1q 2800	—	C—H stretch
$C_2H_5HC=CH_2$	—	1q 1640	—	C=C stretch
$C_3H_7HC=CH_2$	—	1q 1560	—	C=C stretch
$C_5H_{11}HC=CH_2$	—	1q 1590	—	C=C stretch
$(CH_3)_2C=CHCH_3$	—	—	4q 1370	C=C stretch
$(CH_3)_2C=C(CH_3)_2$	1670	2q 1350	3q 1330	C=C stretch
$(C_2H_5)_2C=CHCH_3$	—	2q 1340	—	C=C stretch
cyclohexyl=CH_2	—	—	3q 1340	C=C stretch
cyclohexyl=$CHCH_3$	—	—	4q 1520	C=C stretch
cyclopropyl=CH_2	—	—	2q 1450	C=C stretch
cyclopropene	1656	—	4q 1300	C=C stretch
cyclobutane	—	1q 900	—	—
cyclobutene	—	—	4q 1370	C=C stretch
methylenecyclobutane—CH_3	—	2q 1200	—	C=C stretch
cyclohexene	1646	2q 1480	—	C=C stretch
	2912	1q 2670	—	C—H stretch
bicyclohexylidene	—	1q 1500	5q 1340	C=C stretch
norbornene	—	1q 295	—	Skeletal def.
	—	1q 1140	—	—
	—	1q 1465	—	C=C stretch
	—	1q 2630	—	C—H stretch

[a] The notation "3q 980" represents the excitation of three quanta of a 980-cm^{-1} vibration.
[b] Parallel assignments have been made for trans-butene-2-d_8, with most frequencies decreased 15–25%, except for the C=C stretch.

IV.A. OLEFINS

TABLE IV.A-V
$\pi \to \pi^*$ TERM VALUES[a] IN THE METHYL ETHYLENES

Compound	Ionization potential	(π, π^*) Term value
$H_2C{=}CH_2$	84 750	23 050
$CH_3HC{=}CH_2$	80 000	22 000
trans-$CH_3HC{=}CHCH_3$	75 130	18 860
cis-$CH_3HC{=}CHCH_3$	74 920	17 920
$(CH_3)_2C{=}CH_2$	75 870	22 770
$(CH_3)_2C{=}CHCH_3$	71 240	14 660
$(CH_3)_2C{=}C(CH_3)_2$	67 940	14 440

[a] Term values in cm^{-1} (vert.).

the same intensity at $\sim$52 000 cm^{-1} (vert.) in the methylene cycloalkanes (Fig. IV.A-10) and at 50 000–55 000 cm^{-1} in the cyclic and polycyclic olefins (Fig. IV.A-11). Note, however, that in certain strained olefins such as norbornene (Fig. IV.A-11), the $\pi \to \pi^*$ molar extinction coefficient ($\epsilon = \sim 500$) can be considerably lower than in ethylene itself ($\epsilon = 6000$). It is also to be noted that as alkyl groups are appended to the olefin chromophore, the $\pi \to \pi^*$ absorption maximum shows a steady shift to lower frequencies (about 2000 cm^{-1} per alkyl group; Table IV.A-III), and it is generally agreed that all or most of the shift is due to hyperconjugation in the pi system; however, a change in the effective nuclear charge of carbon due to charge transfer within the sigma system cannot be discounted [M60]. Though semiempirical calculations on the methyl ethylenes reproduce the trend of $\pi \to \pi^*$ frequencies, including the correct ordering within the dimethyl compounds [C21, W17], the calculations have not been analyzed to reveal the common factor responsible for the "alkyl red-shift." Such an analysis could perhaps be better carried out with calculations of the sort Zeeck performed for propylene [Z1].

Using the observed $\pi \to \pi^*$ vertical excitation frequencies of Table IV.A-III and the vertical π ionization potentials determined from photoelectron spectra [R19], the (π, π^*) valence shell term values can be computed (Table IV.A-V). Of course, such term values can be taken as (π, π^*) upper-state ionization potentials, but they are not to be equated with the first members of Rydberg series (Section I.A-1). First, one notes that the ionization potential of the ground-state decreases faster with added methyl groups than does the $\pi \to \pi^*$ absorption frequency, so that the (π, π^*) term values decrease in the series. The relative rates of decrease are such that the upper-state ionization potentials decrease by only half as much as those of the ground state on going from ethylene to tetramethyl ethylene. Rephrased, with respect to the ground states, the methyl

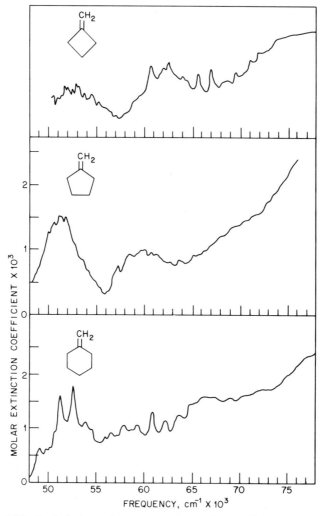

Fig. IV.A-10. Optical absorption spectra of several methylene cycloalkanes in the gas phase [D2].

groups stabilize the ion by about twice as much as they stabilize the (π, π^*) states. Also, the (π, π^*) term value for isobutene, $(CH_3)_2C=CH_2$, appears anomalously high. Almost without exception, the molecular $(\pi, 3d)$ term values are found to be $13\,000 \pm 1000$ cm^{-1}. In view of this, the (π, π^*) term values listed in Table IV.A-V ($14\,500$–$23\,000$ cm^{-1}) can be taken as substantiating the position that the π^* MO in olefins is largely

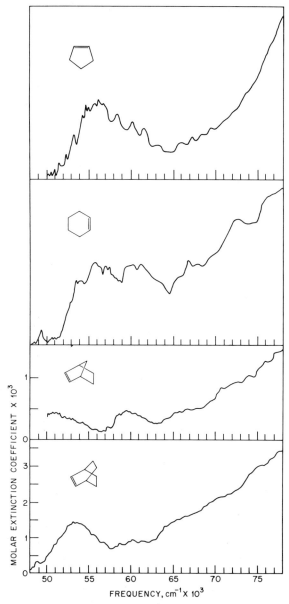

Fig. IV.A-11. Optical absorption spectra of several cycloalkenes in the gas phase [D2].

valence shell, and has very little 3d Rydberg character (see Section I.C) in the (π, π^*) singlet state.

Though it has not been measured yet, there can be no doubt that the $\pi \to \pi^*$ transition in ethylene is polarized along the C—C line. However, in the monoalkyl, *trans*-dialkyl, and trialkyl olefins, the polarization of the $\pi \to \pi^*$ transition is not determined totally by symmetry, and could be tipped considerably from the C=C line. This interesting feature of alkyl olefin spectra has not been investigated, but would be a factor intimately related to the extent of pi-electron delocalization onto the alkyl groups. In an experimental study of the polarization of the $N \to V_1$ band of a long-chain *trans*-polyene, Anex et al. [P4] found that the independent oscillator model (Section III.A-3) could be applied successfully only if the moment within the individual oscillator (*trans*-butene-2) were tipped away from the C=C line toward the line connecting the alkyl groups.

Delocalization will also have an effect on the $N \to V$ oscillator strength, acting to either increase it or decrease it depending upon the relative extent of delocalization in the two states. Unfortunately, the $N \to V$ transitions in most olefins are overlapped on the high-frequency side by other bands, or are observed only somewhat beyond their maxima, so that the experimental oscillator strengths presented in Table IV.A-III must be taken with reservations. It certainly does seem that the $N \to V$ oscillator strengths in the alkyl olefins cluster rather tightly about that of ethylene; however, one exception is that of cyclopropene [R13]. In this molecule, the π MO is approximately equally distributed among the three carbon atoms of the ring, whereas π^* is restricted by symmetry to the two olefin carbon atoms. Such a disadvantageous transition density results in a calculated oscillator strength just one-half that of ethylene, and the experimental spectrum confirms this. Watson et al. [W17] have calculated the $N \to V$ oscillator strengths in the methyl-substituted ethylenes and predict an 18% increase in this quantity in tetramethyl ethylene. This is in semiquantitative agreement with experiment (Table IV.A-III), but again it must be mentioned that the experimental values are approximate.

The full width at half-height of the $N \to V$ bands of ethylene and the acyclic alkyl ethylenes are uniformly 6000–7000 cm^{-1}. Because the minimum in the V-state potential has the two ends of the double bond twisted by 90°, with the C=C bond length increased by about 0.2 Å, the $N \to V$ transitions in these molecules will be very nonvertical and will involve long progressions of C=C twisting and C=C stretching motions. In cyclic olefins, the C=C twisting is prohibited by the inflexibility of the ring, and a long progression in the twisting will not appear, though the same cannot be said for the stretching motion. Thus the full-width at half-height for the N → V band in cyclic olefins may be noticeably smaller than in acyclic olefins. Contrary to expectations, the half-width of the N → V excitation in ethylene (9000 cm^{-1}, Fig. IV.A-1) is only about 10% or so larger than those in the cyclic olefins (Fig. IV.A-11), except for norbornene [R10], in which the width appears to be 6000 cm^{-1}.

As was discussed in Section IV.A-1, the analysis of the N → V vibronic structure of ethylene is still unsettled, and one can hope for no better in the alkyl ethylenes, where the structure is even less well-defined. Examination of the figures reveals a ubiquitous 900–1100-cm^{-1} vibration which is probably the C=C stretch in the V state. However, as in acetone (Section IV.C), this could be a CH$_3$ deformation instead, and should be tested by deuteration.

One can be quite sure of having identified $\pi \to 3s$ in ethylene itself. The same transition in the alkyl ethylenes is most readily characterized by the fact that its frequency will follow the molecular ionization potentials in a predictable way (Section I.C-1). As ethylene is increasingly alkylated, the $(\pi, 3s)$ term value will decrease from 27 410 cm^{-1} (the value in ethylene), and will approach 21 000 cm^{-1} (the alkyl limit). Since the (π, π^*) term values in heavily alkylated olefins drop as low as 15 000 cm^{-1} (Table IV.A-V), this means that in those compounds with low (π, π^*) term values, the $\pi \to R(3s)$ transition will precede the $\pi \to \pi^*$ transition by some 6000 cm^{-1} and so will appear as a weak band somewhat below 50 000 cm^{-1}. These predictions are amply verified by the spectra presented in Figs. IV.A-9–IV.A-11. In cases of intermediate alkylation, the $\pi \to \pi^*$ and $\pi \to 3s$ bands overlap more or less badly, but $\pi \to \pi^*$ remains rather broad and structureless, whereas $\pi \to 3s$ is sharper and can be located rather easily even though it is a weak band resting upon a much stronger one. That the weak, sharp bands of the alkyl olefins have Rydberg upper states is readily demonstrated using external perturbation (Sections II.B and II.C). For example, in Fig. IV.A-12, the gas-phase spectrum of tetramethyl ethylene is compared with that of a thin polycrystalline film at 23°K [R9]. Clearly, the band at about 45 000 cm^{-1} has been severely perturbed by going into the condensed phase, thus revealing the Rydberg nature of its upper state. High-pressure experiments by Evans [E8] on tetramethyl ethylene and cyclohexene also led him to assign these weak, low-lying bands as having Rydberg upper states. The condensed-phase effect on the $\pi \to 3s$ transitions of alkyl olefins is evident as well in the work of Sowers *et al.* [S46, S47], who compared the spectrum of cyclohexene vapor with that of the liquid, and in the work of Potts [P32], who studied the alkyl olefin spectra in glassy matrices at 77°K.

Since the term value of the $\pi \to 3s$ transition of ethylene itself is only

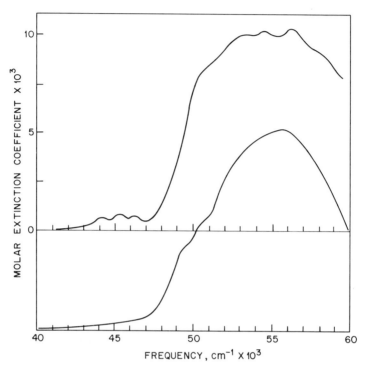

Fig. IV.A-12. The absorption spectrum of tetramethyl ethylene in the gas phase at room temperature (upper), and as a polycrystalline film at 23°K (lower) [R9].

27 410 cm^{-1}, and since this quantity rarely goes below 21 000 cm^{-1} in any compound, one reasonably expects only a small shift of the $(\pi, 3s)$ term value with the addition of alkyl groups to ethylene. The data assembled in Table IV.A-III confirm this expectation, the $(\pi, 3s)$ term of tetramethyl ethylene, for example, being only 4200 cm^{-1} less than that of ethylene. A term decrease of about 1000 cm^{-1} per methyl group is apparent in the other methylated ethylenes. As was discussed in Section I.A-1, the deviation of the 3s term value upward from the purely hydrogenic value of 12 193 cm^{-1} is dependent upon the presence in the molecule of atoms of high nuclear charge. Since the alkylation of ethylene simply adds carbon atoms to a carbon-atom chromophore, the 3s terms will change only slightly with the introduction of alkyl groups. Moreover, since delocalization of the 3s orbital over neutral alkyl groups results in less penetration than when the 3s electron is on the positively charged core, the shift of the $(\pi, 3s)$ term will be to lower values with increasing alkylation, as observed (Fig. IV.A-13).

The transition from π to 3p is $u \leftrightarrow u$ forbidden in the centrosymmetric

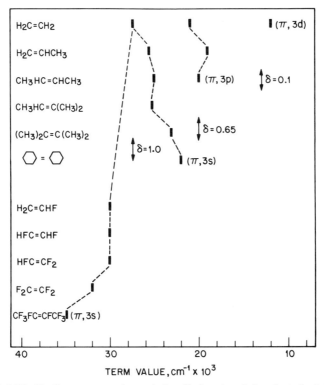

Fig. IV.A-13. Rydberg term values of the alkylated and fluorinated olefins, listed vertically in order of increasing substitution.

olefins, and is not expected to be too much stronger in the noncentrosymmetric ones. Further, our experience with saturated chromophores shows that the 3p term value is much less sensitive to substitution than is 3s, and can be expected to be 18 000–21 000 cm^{-1}. Transitions to 3d are allowed by parity, and should have term values of about 13 000 cm^{-1}, but rarely lower than the hydrogen-atom value of 12 193 cm^{-1}. The expected term values are frequently met in Table IV.A-III, but anomalies (16 000 cm^{-1} term) do appear. Many term values are approximate, however, since band origins in the optical spectra are difficult to deduce in many cases. It is also easy in these broad spectra to mistake $\pi \to 4s$ for $\pi \to 3d$, etc, leading to errors of several thousand cm^{-1} in the term values. Ideally, we should determine the positions of the $\pi \to 3p$ and $\pi \to 3d$ bands independently and thereby show that their term values are relatively constant. At present, it is more practical to argue that the $(\pi, 3s)$ term is regular, and therefore the $(\pi, 3p)$ and $(\pi, 3d)$ terms must be also.

From this, and the appropriate empirical correction for substituent effects, one can estimate where $\pi \to 3p$ and $\pi \to 3d$ transitions will fall, thus facilitating the task of spectral assignment. In line with this, we feel confident that the Rydberg transitions in olefins with terms of 27 000–22 000, 22 000–18 000, and 14 000–12 000 cm^{-1}, have 3s, 3p, and 3d upper states, respectively.

Moore's study of the ion-impact energy-loss spectra of alkyl olefins is interesting [M51]; using 3.0-keV He$^+$ ions as projectiles, the spectra of ethylene, butene-1, and *cis*- and *trans*-butene-2 all show a very intense N → T band at 33 800 cm^{-1} (vert.), followed by much weaker N → V excitations at approximately the frequencies observed optically (Table IV.A-III). However, when recorded using 3.0-keV protons instead, excitation to the triplet is missing, and the N → V band is intense and is followed by one or more bands that are even more intense. These latter bands correlate roughly with Rydberg absorptions in the optical spectra, but are definitely absent in the He$^+$-impact spectra.

It has been noted in several other chromophores (ketones, Section IV.C; oxides, Section III.E; amines, Section III.D-1) that, symmetry permitting, the $\phi_i \to 3s$ transition is relatively strong in the unsubstituted chromophore, but as more and more or larger and larger alkyl groups are appended, the $\phi_i \to 3s$ intensity falls, whereas $\phi_i \to 3p$ and $\phi_i \to 3d$ become relatively more intense. Overlap of these transitions by the $\pi \to \pi^*$ band in the olefins makes Rydberg intensity measurements precarious, but Watson *et al.* [W17] list experimental $\pi \to 3s$ oscillator strengths determined by them for the methylated ethylenes. There appears to be a general decrease of the $\pi \to 3s$ oscillator strength with increasing methylation, as prophesied, but the differences are quite small, and may not be real. In general, it is to be noted that the $\pi \to 3s$ oscillator strength in the delocalized ethylenic chromophore (~ 0.03) is very nearly the same as that found for n → 3s in the saturated chromophores. Inasmuch as the $\pi \to 3s$ transition is an electronically allowed one in ethylene itself, it will be an electronically allowed one in any olefin of lower symmetry, though there may be a vibronically induced component as well [W17].

The polarization of the $\pi \to 3s$ band in planar olefins is dictated by the symmetry selection rules as being out of plane, z. This aspect of the spectrum of perdeutero-*trans*-butene-2 has been studied by McDiarmid [M12], who finds that the $\pi \to 3s$ origin (49 677 cm^{-1}) has a C-type rotational envelope, implicating a z-axis polarization for the purely electronic transition. Merer and Schoonveld [M27] reach the same conclusion after studying the rotational envelope of the $\pi \to 3s$ band origin in ethylene.

Like its counterpart in ethylene, the $\pi \to 3s$ transition in the alkyl olefins is strongly vertical, with an upper-state geometry close to that of the ground state. In fact, it is the generally vertical nature of the $\pi \to 3s$ bands in the substituted olefins which makes them visible among the stronger $\pi \to \pi^*$ transitions. The vibronic structure of the $\pi \to 3s$ band has been analyzed and assigned in only a few alkyl olefins. These are reported in Table IV.A-IV, together with fragments of vibrational progressions observed both in the $\pi \to 3s$ optical transition and the first band in the photoelectron spectrum of several other olefins. Examples of the $\pi \to 3s$ Rydberg excitations in typical olefins are shown in Figs. IV.A-14 and IV.A-15. As mentioned by Merer and Mulliken [M26], the appearance of 1–3 quanta of a 1350–1600 cm^{-1} vibration, the C=C stretch, reduced from ground-state values of 1600–1700 cm^{-1} is a ubiquitous feature of the $\pi \to 3s$ transition in olefins. That this vibration is excited in the $\pi \to 3s$ band is reasonable, for the transition almost completely removes an electron which otherwise is C=C bonding. In fact, just this vibrational frequency is excited in the photoelectron spectra of olefins

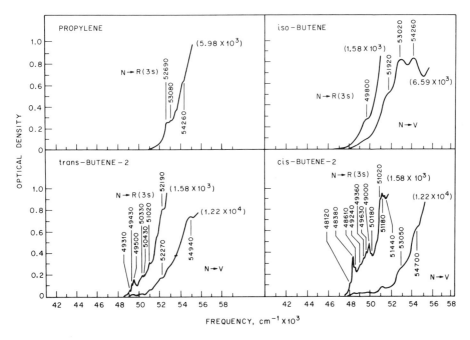

Fig. IV.A-14. Details of the $N \to R(3s)$ and $N \to V$ absorptions of several alkyl olefins in the gas phase. The molar extinction coefficients for the various curves can be obtained by multiplying the optical density by the factors given in parentheses [W18].

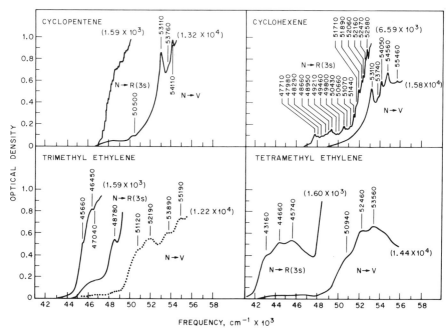

Fig. IV.A-15. Details of the N → R(3s) and N → V absorptions of several heavily alkylated olefins in the gas phase. The molar extinction coefficients can be obtained by multiplying the optical density by the factors given in parentheses [W18].

when a pi electron is ionized (Table IV.A-IV). However, in ethylene [B58] and in cis-butene-2 [M12] and trans-butene-2 [M12], the vibronic features are sharp, and analysis shows that other vibrations besides the C=C stretch are also excited in these molecules. Most likely, these other vibrations are excited in the photoelectron and π → 3s bands of all olefins, but cannot be resolved in most. One striking exception is the π → 3s band of cyclopentene (Fig. IV.A-16), in which 20 members appear with an average spacing of 130 cm^{-1} [C5, W18]. The strong possibility that both sequences and progressions appear in the ring puckering and/or C=C twisting modes, both of which may have double-minima potentials, together with some C=C stretching is mentioned by Merer and Mulliken [M26]. In a tentative analysis of this fascinating band of cyclopentene, Watson and McGlynn [W18] have placed the origin at 47 571 cm^{-1}, identified four hot bands on its low-frequency side, and found a second false origin 1457 cm^{-1} removed. This latter frequency may correspond to the C=C stretch in the upper state. Since 130 cm^{-1} is far beyond the resolution of present-day photoelectron spectroscopy, one might ex-

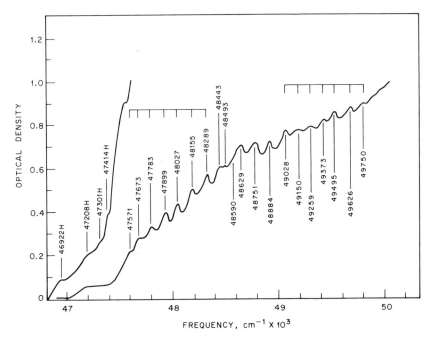

Fig. IV.A-16. Details of the N → R(3s) Rydberg excitation in cyclopentene vapor [W18].

pect the first photoelectron band of cyclopentene to show the C=C stretching vibration unencumbered by the finer vibration. It shows three quanta of the C=C stretch, 1260 cm^{-1} [P34].

In addition to photoelectron spectroscopy, another technique which holds promise for use as a tool in understanding the spectra of olefins is the rotation spectrum of optically active systems. This is an old subject in the visible region, but it only recently has been extended into the vacuum ultraviolet, the incentive coming largely from the study of optically active biological systems. In a conventional optical absorption experiment, the integral of the absorption curve is related to the oscillator strength

$$f_{0m} = (0.6667/\Delta E_{0m})|\langle \Psi_0|\boldsymbol{\nabla}|\Psi_m\rangle|^2 \qquad (IV.4)$$

and is a measure of the electric-dipole matrix element. From this, one can obtain information about the symmetries of Ψ_0 and Ψ_m, and also something about their extent in space and relative overlap. Similar useful information can be derived from the rotatory strength

$$R_{0m} = (471.38/\Delta E_{0m})\langle \Psi_0|\boldsymbol{\nabla}|\Psi_m\rangle \cdot \langle \Psi_0|\mathbf{r} \times \boldsymbol{\nabla}|\Psi_m\rangle \qquad (IV.5)$$

IV. TWO-CENTER UNSATURATES

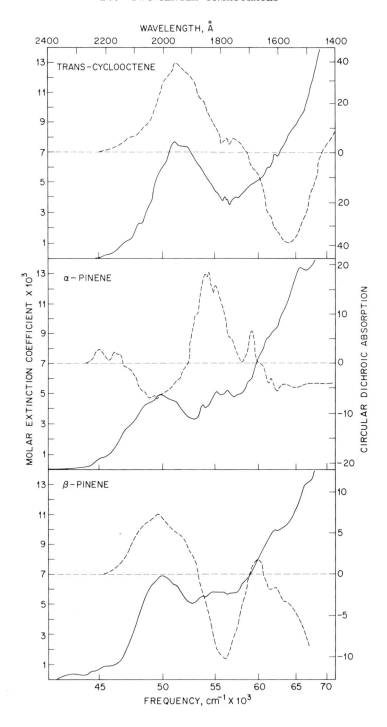

which involves a vector product of the electric and magnetic transition moments. In both Eqs. (IV.4) and (IV.5), the transition energy ΔE_{0m} is taken in atomic units (1 a.u. = 27.2 eV). Experimentally, R_{0m} is obtained from the integral of either the circular dichroism (CD) or optical rotatory dispersion (ORD) spectrum, and has both a magnitude and a sign. A third quantity of some value is the dissymmetry parameter g, defined as $4R_{0m}/D_{0m}$. The quantity D_{0m} is the dipole strength, and according to Mason [M7], may be taken as $9.8 \times 10^{-39} \epsilon \, \Delta\nu/\nu$, where ϵ is the molar extinction coefficient at the absorption maximum of frequency ν, and $\Delta\nu$ is the half-width of the band. Mason points out that if the $\Psi_0 \to \Psi_m$ transition has a large magnetic moment, as, for example, in $2p\sigma \to 2p\pi^*$, then g can be as large as 0.3, whereas if the magnetic moment is inherently small, as in $\pi \to \pi^*$ and $2s\sigma \to 2p\pi^*$ transitions, then g is orders of magnitude smaller. Though the determination of g from the absorption and CD spectra obviously can be of considerable value in determining orbital compositions, as yet there are very little data of this sort for optically active olefins which do not contain interfering chromophores such as C=O, C_6H_5, etc.

Though highly strained, the olefin *trans*-cyclooctene appears to have a regular absorption spectrum (Fig. IV.A-17), with a $\pi \to 3s$ band peaked at 47 200 cm^{-1} ($\epsilon \simeq 2000$, but strongly overlapped), followed by the $\pi \to \pi^*$ band at 50 700 cm^{-1} (vert.; $\epsilon = 7200$) [S19]. There are suggestions of irregular vibrational structure throughout the spectrum. The ring strain resulting from the trans configuration about the double bond produces a torque about the double bond which twists it away from planarity [R18]. Such a dissymmetric chromophore will be strongly optically active, and the circular dichroism spectrum of *trans*-cyclooctene has been recorded in the vacuum-ultraviolet region (Fig. IV.A-17) [B1, M6, S19]. The CD spectrum clearly shows the presence of the $\pi \to 3s$ band within the low-frequency wing of the $\pi \to \pi^*$ rotation, even though the $\pi \to \pi^*$ rotation is about ten times stronger than that of $\pi \to 3s$, and is of the same sign. In the lower-frequency region in n-heptane solution, only the $\pi \to \pi^*$ band is evident in the CD spectrum (50 500 cm^{-1}), the $\pi \to 3s$ band having been shifted upward and broadened considerably due to its Rydberg nature [Y6]. Whereas the absorption spectrum might easily be taken to show only one transition in the 55 000–60 000-cm^{-1} region, the CD spectrum clearly reveals a positively rotating transition centered at 56 200 cm^{-1}, and a stronger, negatively rotating transition with a maximum at 64 100 cm^{-1} (vert.).

Fig. IV.A-17. Optical absorption spectra (solid line) and circular dichroism spectra (dashed line) of several optically active olefins in the gas phase [M6].

The *trans*-cyclooctene molecule is too large for high-level electronic structure calculations, and its geometry can only be guessed from molecular models. However, since alkylation does not radically alter the electronic spectrum of ethylene, the chromophoric group of *trans*-cyclooctene might well be mimicked by the ethylene molecule twisted an unknown amount about the C=C double bond. This model has been used by Yaris *et al.* [Y6], who performed Pariser–Parr–Pople calculations on twisted ethylene using a Slater orbital basis, and by Robin *et al.* [R12], who used a double-zeta basis of Gaussian orbitals. In both calculations, the $\pi \to \pi^*$ promotion has a large electric moment along the C—C bond direction, and, through twisting, acquires a small magnetic moment as well in the same direction. As a result, the $\pi \to \pi^*$ transition has an appreciable rotatory strength, even at small angles of twist. However, Yaris *et al.* find the *sign* of the predicted $\pi \to \pi^*$ rotation to be opposite to that observed for the strong 50 700-cm^{-1} band. In order to get around this difficulty, they propose that $\pi \to \pi^*$ is strong in absorption, but not as strongly rotating as the $^1A_g \to {}^1B_{1g}$, $\pi \to \sigma^*$ transition, which could come in the $\pi \to \pi^*$ region and which has a rotatory strength of the opposite sign. This band is electric-dipole forbidden in planar ethylene and presumably is too weak to appear in the absorption spectrum of *trans*-cyclooctene. As pointed out by Schnepp *et al.* [M6, S19], an explanation in which there are two overlapping transitions, one dominating the absorption spectrum and the other dominating the CD spectrum, will have a difficult time rationalizing the fact that the band shapes in absorption and rotation are very much alike. In this regard, also see [S22, S23].

On the other hand, in the Gaussian orbital calculation [R12], the $\pi \to \pi^*$ transition is predicted to be strongly rotating, with the sign observed experimentally. Additionally, at the arbitrary dihedral angle of 10° twist, the $\pi \to 3s$ transition has a rotatory strength $\frac{1}{15}$ that of the $\pi \to \pi^*$ transition, and of the same sign, as observed. The magnitudes of the predicted rotatory strengths are too large, but could be corrected by decreasing the twist angle. Consideration of the integrals of the absorption and CD curves of *trans*-cyclooctene allows one to calculate a dissymmetry factor g equal to 7.4×10^{-3} for the band at 50 700 cm^{-1} (Mason and Schnepp [M6] report 4.8×10^{-3}); this is within an order of magnitude of that found for the $\pi \to \pi^*$ transition in α-pinene (discussed later) and definitely shows that the 50 700-cm^{-1} band has an inherently low magnetic transition moment. Scott and Wrixon [S23] and Levin and Hoffman [L20] concur in the view that the CD band at 50 700 cm^{-1} in *trans*-cyclooctene is due to the $\pi \to \pi^*$ transition. Using our knowledge of the behavior of Rydberg term values, we can make some educated guesses as to where these transitions should occur in *trans*-cyclooctene. The pi ionization potential in this molecule is 70 400 cm^{-1} (vert.) [R18], which leads to a reasonable $(\pi, 3s)$ term value of 23 200 cm^{-1}, just slightly above the alkyl limit. Since the limiting term values for the $(\pi, 3p)$ and $(\pi, 3d)$ configurations are 19 000 and 13 000 cm^{-1}, these transitions are expected at 51 400 and 57 400 cm^{-1}, respectively. The $\pi \to 3p$ transition will be weak in near-centrosymmetric molecules, and falling upon the stronger $\pi \to \pi^*$, it is no suprise that it cannot be seen. The $\pi \to 3d$ transition corresponds to the weak bands observed at 56 200 cm^{-1} (vert.). This analysis leaves the band at 63 300 cm^{-1} unassigned, but most certainly it is not a Rydberg excitation originating at π. Of course, $\pi \leftrightarrow \sigma$ type valence shell transitions immediately come to mind, but the possibility of a transition originating with the π orbitals of the α-methylene groups and going to π^* must not be overlooked.

There are several other reports in the literature on the properties of optically active olefins, though few go much beyond 50 000 cm^{-1}. Mason and Vane [M8] report 3-methyl isopropylene cyclopentane to show an absorption spectrum having a $\pi \to 3s$ transition at 45 000 cm^{-1}, resting upon the low-frequency edge of the $\pi \to \pi^*$ absorption. The CD spectrum reproduces the rising $\pi \to \pi^*$ absorption, but shows nothing of the $\pi \to 3s$ band. Thus, the $\pi \to \pi^*$ rotatory strength in this compound must be orders of magnitude larger than that for $\pi \to 3s$ excitation. Yogev et al. [Y9] report Cotton effects in several steroidal olefins at about 48 000 cm^{-1} in cyclohexane solution, and attribute them to $\pi \to \sigma^*$ (3s) transitions. The corresponding CD bands are reported by Legrand and Viennet [L16]. These transitions, which are not apparent in Turner's absorption spectra in cyclohexane solution [T19], should be explored further, since Rydberg transitions should be at higher frequencies and broadened considerably in solution spectra. Later, Fetizon et al. [F4, F5] concluded that in the methylene steroids and related compounds in cyclohexane solution, there are two transitions at $\sim$50 000 and 52 500 cm^{-1} of opposite rotatory power, the first of which is $\pi \to \pi^*$ and obeys an octant rule. However, Yogev et al. [Y10] present evidence for both bands having the same rotatory sense and violating the octant rule in certain compounds.

Thanks to our clearer understanding of the regularities of olefin spectra, we can now correct an otherwise awkward explanation of the ORD and absorption spectra of α-D-pinene [R9],

The recent work of Mason and Schnepp [M6] has expanded the data on this optically active olefin (Fig. IV.A-17). On the basis of intensity, the absorption centered at 49 500 cm^{-1} (vert.; $\epsilon = 5000$) must be assigned to the $\pi \to \pi^*$ excitation of the C=C double bond. In the absorption spectrum, the transition from π to 3s is barely visible as the stepout beginning at 42 500 cm^{-1}, but is very obvious in the CD spectrum. Another excitation begins at 52 600 cm^{-1} (adiab.) in the gas phase, and, as appropriate for a Rydberg excitation, this band is missing from Turner's absorption spectrum in cyclohexane solution [T19]. The adiabatic ionization potential of α-pinene is 65 100 cm^{-1} [A4], so the $(\pi, 3s)$ state has the reasonable term value 22 600 cm^{-1}, much like those of trimethyl ethylene, *trans*-cyclooctene, and norbornene (Table IV.A-III). Similarly, the Rydberg absorption at 52 600 cm^{-1} has an adiabatic term value of 12 500 cm^{-1}, identifying it as $\pi \to$ 3d.

The small experimental g value of 8.6×10^{-4} measured for the 49 500-cm^{-1} Cotton effect of α-D-pinene is quite to be expected, since the $\pi \to \pi^*$ transition has an inherently large electric moment, but an inherently small magnetic moment. In the older interpretation [R9], the 49 500-cm^{-1} band was assigned to $\pi \to$ 3s, in which case one had to argue that the dispersion of the $\pi \to \pi^*$ rotation did not carry any strength into the region below 54 500 cm^{-1}, which would be rather unusual. An extinction coefficient of 5000 is also far above that ordinarily observed for $\pi \to$ 3s in olefins, especially in the heavily alkylated olefins, where $\pi \to$ 3s is noticeably weaker than usual.

The absorption and CD spectra of β-pinene

(see Fig. IV.A-17) closely resemble those of the other highly alkylated, optically active olefins. Though absent in the CD spectrum, the $\pi \to$ 3s Rydberg excitation appears as 45 500 cm^{-1} (vert.) in the absorption spectrum, which implies a first ionization potential of 68 500 cm^{-1} (vert.). The corresponding Rydberg transition to 3d is centered at 55 800 cm^{-1} (vert.) with a term value of 12 700 cm^{-1} (vert.). The more intense $\pi \to \pi^*$ transition is located at 50 000 cm^{-1} in β-pinene. As in the other compounds, several bands are observed beyond the $\pi \to$ 3d excitation, but we cannot yet assign them.

Though the evidence is not extensive, it does appear that the $\pi \to \pi^*$ transition in an optically active olefin will be ten to several hundred times more strongly rotating than $\pi \to$ 3s, whether the chromophore is inherently dissymmetric or not. This is not unexpected, since the $\pi \to \pi^*$ promotion has an inherently large electric moment associated with it, and needs only an admixture of magnetic moment to give a nonzero rotatory strength, whereas $\pi \to$ 3s has a small electric moment and no magnetic moment. With the recent construction of CD instruments operating in the vacuum-ultraviolet region, the olefins are certain to be probed again with interesting results.

After one has combed through the alkyl olefin spectra and accounted for the $N \rightarrow V$ and $\pi \rightarrow 3s$, $3p$, $3d$ excitations, there are several bands at lower frequencies in several compounds still lacking an explanation. Such low-frequency bands are undoubtedly valence shell and are most often encountered in strained systems. The explanation here is that the strain raises the occupied sigma manifold with respect to the pi levels, with the result that valence shell $\sigma \rightarrow \pi^*$ excitations come at relatively low frequencies. The first of these to be uncovered was in

(see Fig. IV.A-18) [R12], in which a transition with its maximum at 45 000 cm^{-1} ($\epsilon \simeq 1000$) was shown to be a valence shell excitation by its spectrum in a thin film at 23°K. As can be seen from Fig. IV.A-18, the other bands in the 45 000–60 000 cm^{-1} region are missing from the solid-film spectrum, showing they have Rydberg upper states, and the $\pi \rightarrow \pi^*$ excitation comes at 63 000 cm^{-1} (vert.). Since the explanation of the low-frequency valence shell band as due to strain could not be checked theoretically due to the size of this tricyclic olefin, attention next turned to cyclopropene, a highly strained olefin of a manageable size [R13]. Once again, a valence shell excitation having its origin at 45 000–46 000 cm^{-1} and having $\epsilon = 1300$ was readily observed. Theoretically, the Gaussian orbital calculations on cyclopropene predict the lowest allowed excitation to be $3b_1\sigma \rightarrow 1a_2\pi^*$ at 61 600 cm^{-1} [P13, R13]. In an identical calculation on ethylene, the lowest $\sigma \rightarrow \pi^*$ excitation is computed to come at 77 400 cm^{-1}. Such calculations do suggest that sigma MOs are closer to pi in cyclopropene than in ethylene. Additional, more direct confirmation comes from the photoelectron spectra [B59], which show the highest sigma orbital in ethylene to be 15 700 cm^{-1} below the occupied pi MO, whereas in cyclopropene, this difference is reduced to 9700 cm^{-1}. The $3b_1\sigma \rightarrow 1a_2\pi^*$ excitation in cyclopropene is the olefin analog of the $n_+ \rightarrow \pi^*$ excitation of the isoelectronic molecule diazirine (Section IV.B), and corresponds to a type of transition first discussed by Berry [B26]. In fluorinated olefins (Section IV.A-3), an effect opposite to that of ring strain occurs, where the inductive effect of the fluorine substituents *lowers* the sigma manifold with respect to the pi MOs, so that $\pi \rightarrow \sigma^*$ valence shell excitations can be observed at low frequencies,

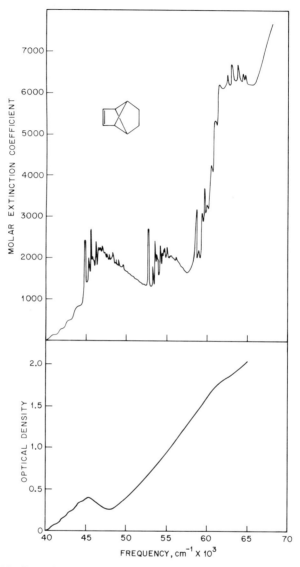

Fig. IV.A-18. The electronic spectrum of tricyclo [3.3.0.0^{2,6}]oct-3-ene in the gas phase at room temperature (upper) and as a polycrystalline film at 23°K (lower) [R12].

whereas in olefins multiply-substituted by Si(CH$_3$)$_3$ groups, the pi manifold is lowered so that a low-frequency $\sigma \to \pi^*$ valence shell transition results, Section III.G.

It appears that the $\sigma \rightarrow \pi^*$ excitations of olefins are observed at low frequencies if the strain is sufficient, but their oscillator strengths do not exceed 0.02. "Extraneous" transitions of ten times this oscillator strength appear on the high-frequency side of the $\pi \rightarrow \pi^*$ transition in several other olefins. As a prime example, Snyder and Clark [S42] report the spectrum of bicyclohexylidene

as showing two strong valence shell transitions at 48 000 and 55 000 cm^{-1} ($\epsilon \simeq 7000$), both with long-axis polarization (Fig. IV.A-19). Needless to say, neither of the bands under consideration here falls in the region of

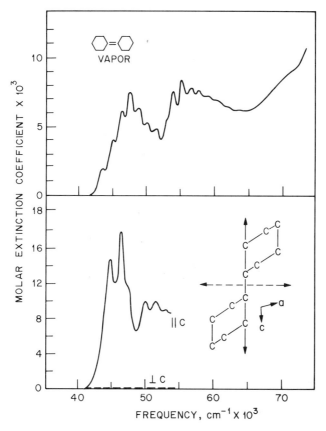

Fig. IV.A-19. Absorption spectrum of bicyclohexylidene in the gas phase (upper), and through the (010) crystalline face (lower) using light polarized parallel to the c axis (solid curve), and perpendicular to the c axis (dashed curve) [S42].

alkane absorption. Equally clear examples of such "extraneous" strong bands are visible in the spectra of 2-methyl pentene-2 [J12], 2-methyl butene-2 [C3, J12], tetramethyl ethylene† [R9], cyclohexene [P20] and cyclopentene (Fig. IV.A-11) [P20], diisobutylene [C3], 1-methyl cyclobutene [L32], and methylene cyclobutane (Fig. IV.A-10) [L32]. Snyder and Clark [S42] feel that since the only possible long-axis allowed transition besides $\pi \to \pi^*$ in ethylene is $\sigma \to \sigma^*$, this latter assignment must hold for one of the two bands of bicyclohexylidene. However, Watson *et al.* [W17], in semiempirical calculations, and Zeeck, in an *ab initio* Gaussian orbital calculation [Z1], show that pi AOs on CH_3 and/or CH_2 groups mix readily with the pi MO of the C=C double bond, in certain cases making tightly grouped sets of π and π^* MOs. Semiempirical calculations on cyclobutene, for example, predict *three* $\pi \to \pi^*$ transitions in the 49 000–62 000 cm^{-1} region. Of these, one is mostly the regular $\pi \to \pi^*$ excitation of the double bond, while the other two are mostly $CH_2 \to \pi^*$ (C=C) charge transfer excitations. From this work, it seems most likely that Synder and Clark have observed the two long-axis allowed $\pi \to \pi^*$ excitations which are possible in bicyclohexylidene, though their original $\sigma \to \sigma^*$ assignment cannot yet be discounted. Along the same lines, Yogev *et al.* [Y11] studied the polarized spectra of olefinic steroids and claim to have found two bands in the 47 000–56 000-cm^{-1} region, the first polarized parallel to the C=C line, and the second somewhat canted from that line and somewhat canted out of plane.

Combination of the vinyl group with several of the chromophoric groups discussed in the other chapters leads to very interesting, but complicated spectra. That of nitroethylene is discussed in Section V.B, that of acrylonitrile in Section IV.E, that of acrolein in Section IV.C, that of styrene in Section VI.A-5, that of the haloethylenes in Section IV.A-3, and those of the aminoethylenes are described below.

Tetrakisdimethylamino ethylene (TDAE) has the C—C lines of its

$$(CH_3)_2N \diagdown \diagup N(CH_3)_2$$
$$C=C$$
$$(CH_3)_2N \diagup \diagdown N(CH_3)_2$$

TDAE

TMBI

—$N(CH_3)_2$ groups turned almost perpendicular to the $N_2C=CN_2$ plane, thereby making the nitrogen lone pairs almost sigma orbitals, rather than pi. Presumably, the steric crowding is no less in 1,1′,3,3′-tetramethyl-

---

† There appears to be no firm evidence to support the suggestion that the carbon skeleton of tetramethyl ethylene is nonplanar [K14].

$\Delta^{2,2'}$-bisimidazolidine (TMBI). A second unique feature of these substances is the very strong splitting of the nitrogen lone pairs, resulting in an ionization potential of only 48 000 cm^{-1} (vert.) for TDAE and 48 900 cm^{-1} (vert.) for TMBI [C9, N7]. With such exceptionally low ionization potentials, correspondingly low Rydberg excitations are expected, while among the valence shell excitations, the —N(CH$_3$)$_2$ → π^* charge transfer spectrum will also be low lying, but the olefinic $\pi \to \pi^*$ band would be expected near that for, say, tetramethyl ethylene (53 500 cm^{-1} vert.).

Vapor-phase spectra of TDAE and TMBI are shown in Fig. IV.A-20

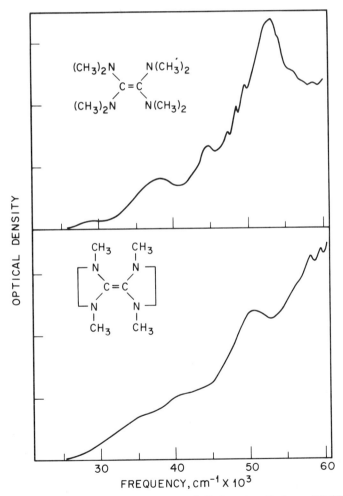

Fig. IV.A-20. Absorption spectra of the dialkyl aminoethylenes TDAE (upper) and TMBI (lower) in the gas phase [N8].

[H28, N8, N9]. Spectra of these molecules in paraffin solutions at lower frequencies are also available and show that the weak feature at 29 000 cm^{-1} (vert.) in TDAE is wiped out, as is characteristic of a Rydberg excitation, but that at 38 400 cm^{-1} (vert.) is a valence shell excitation. In TMBI, the situation is not so clear as to which bands are Rydberg. No intensity data in the gas phase are reported, but the molar extinction of the band at 38 400 cm^{-1} in heptane solution is 2400. If we apply this figure to the gas-phase spectrum, then the band at 53 400 cm^{-1} (vert.) has an extinction coefficient of 10 000 at its maximum.

The Rydberg excitation centered at 29 000 cm^{-1} in TDAE has a term value of 19 000 cm^{-1} (vert.), which suggests that it is either a transition to 3s at the alkyl limit (a low term value is expected in view of the ten carbon atoms in the molecule) with that to 3p forbidden by symmetry, or it is the Rydberg excitation to 3p with that to 3s forbidden. According to Cetinkaya et al. [C9], the uppermost orbital in TDAE has u symmetry, making the Rydberg excitation allowed to 3s and forbidden to 3p, but the orbital assignment is not too secure. The valence shell excitation at 38 400 cm^{-1} is too low and too weak to be the $\pi \rightarrow \pi^*$ excitation, and so we assign it instead as a change transfer from the dimethylamino groups to π^*. From this, it is only natural then to assign the 53 400-cm^{-1} band as $\pi \rightarrow \pi^*$ on the basis of its frequency and absorption strength. Nakato et al. [N8] point out that the frequencies and intensities of the bands in this type of system will depend upon the angle of twist of the dimethylamino groups, and so it is reasonable that the correlation between the spectra of TDAE and TMBI is obscure.

IV.A-3. *Haloethylenes*

The fluoroethylene and chloroethylene series have been studied at some length by both optical and photoelectron spectroscopy, but corresponding detailed studies on the bromoethylenes and iodoethylenes remain to be done. Still, what we have in these series is most interesting, perhaps even more so than are the alkyl olefins (Section IV.A-2). The optical spectra of the various fluoroethylenes have been determined by Bélanger and Sandorfy (Fig. IV.A-21) [B19, B20], while Lake and Thompson report the photoelectron spectra of fluoroethylene, 1,1-difluoroethylene, and tetrafluoroethylene [L2], and Brundle et al. [B59] report those for the remaining members of the fluoroethylene series. For all molecules except tetrafluoroethylene, only the first ionization potential is of concern to us at present. As regards these molecules, the theoretical and experimental work of Brundle et al. on tetrafluoroethylene is also of interest. Spectro-

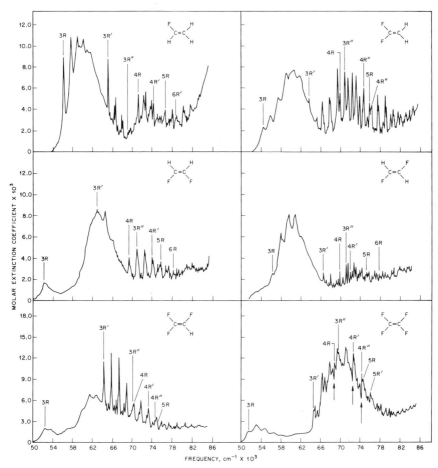

Fig. IV.A-21. Optical absorption spectra of the fluoroethylenes in the gas phase [B20].

scopic data on the various transitions identifiable in the fluoroethylene spectra are summarized in Table IV.A-VI.

It is rather surprising, but quite clear, that the vertical ionization potential of a pi electron in the C=C double bond remains very nearly constant in the series from ethylene to tetrafluoroethylene. Further studies show that the sigma MOs in the same series of molecules are strongly stabilized on fluorination. This selective stabilization of the sigma MOs upon fluorination, called the "perfluoro effect" [B59], has been uncovered in several systems, and has been discussed by Bralsford et al. [B40] and Brundle et al. [B59]. The latter workers investigated this effect in a num-

TABLE IV.A-VI
Spectroscopic Quantities for the Haloethylenes[a]

Molecule	Ionization potential	$\pi \to 3s$ Frequency	$(\pi, 3s)$ Term	$\pi \to 3p$ Frequency	$(\pi, 3p)$ Term	$N \to V$ Frequency	$N \to V$ ϵ
$H_2C{=}CH_2$	84 750	57 340	27 410	—	—	61 700	—
$FHC{=}CH_2$	85 340	56 290	29 050	65 250	20 090	60 000	10 000
$trans\text{-}FHC{=}CHF$	83 720	~54 800	~28 900	66 580	17 140	61 000	—
$cis\text{-}FHC{=}CHF$	84 120	52 360	31 760	64 000	20 120	63 000	7 500
$F_2C{=}CH_2$	86 460	~55 000	~31 500	66 300	20 160	60 500	7 900
$F_2C{=}CHF$	84 980	52 420	32 560	64 290	20 690	61 400	6 800
$F_2C{=}CF_2$	84 850	~52 800	~32 000	64 700	20 150	71 600	11 400
$cis\text{-}CF_3CF{=}CFCF_3$	92 000	57 000	35 000	—	—	61 000	4 800
$ClHC{=}CH_2$	80 660*	57 330*	23 330*	61 520*	19 140*	54 000	—
				63 048*	17 610*		
$trans\text{-}ClHC{=}CHCl$	77 750*	—	—	—	—	51 300	—
$cis\text{-}ClHC{=}CHCl$	77 830*	—	—	59 130*	18 700*	52 600	—
				59 734*	18 100*		
				60 958*	16 870*		
$Cl_2C{=}CH_2$	79 290*	—	—	60 500*	18 790*	51 800	—
				63 349*	15 940*		
$Cl_2C{=}CHCl$	76 460*	—	—	59 310*	17 150*	51 000	—
				59 590*	16 870*		
				60 610*	15 850*		
$Cl_2C{=}CCl_2$	75 300*	53 100*	22 200*	—	—	50 800	—
$FClC{=}CH_2$	—	52 600	—	61 500	—	56 560	—
$F_2C{=}CHCl$	80 980	53 200	27 800	62 290	18 690	57 880	—
$F_2C{=}CCl_2$	79 400	51 200	28 200	59 240	20 160	57 400	—
$F_2C{=}CFCl$	82 590	52 600	30 000	—	—	62 830	—
$IHC{=}CH_2$	93 000	—	—	—	—	65 000	10 000

[a] All frequencies are in cm^{-1}, and are vertical unless denoted by an asterisk, in which case they are adiabatic.

ber of perhydro molecules and their perfluoro analogs, using Gaussian orbital calculations, and concluded that the insensitivity of the pi-electron ionization potentials was a consequence of the fact that the fluorination tends to stabilize the pi MO due to the high electronegativity of that atom, but that this stabilization is countered by a strong C—F pi antibonding contribution which is of the same magnitude as the electronegativity effect. On the other hand, the two effects operate in the same direction in general for sigma MOs. Near-constancy of the highest pi MO ionization potential in the fluoroethylenes leads to a near-constancy of the $\pi \rightarrow n$s and $\pi \rightarrow n$p absorption frequencies as well, and because of this, it is rather easy to pick out the transitions terminating at 3s and 3p. The former have term values of about 30 000 cm^{-1}, and appear in the 50 000–54 000-cm^{-1} region in Fig. IV.A-21, and the latter have terms of about 18 000 cm^{-1}, appearing near 65 000 cm^{-1}. In fact, the electron-impact spectrum of tetrafluoroethylene far beyond the first ionization potential can be interpreted using only these term values and the experimental higher ionization potentials (Section I.C-1). The N → V band in the fluoroethylenes is once again made prominent by its relatively high intensity, with an absorption maximum in the 60 000–70 000-cm^{-1} region. Significantly, the N → V frequency also is seen to remain rather constant from ethylene to trifluoroethylene, but the shift between trifluoroethylene and tetrafluoroethylene is 9000 cm^{-1}. This is a most interesting effect which is absent in the fluorobenzenes (Section VI.A-3).

Vibronic structure in the $\pi \rightarrow 3$s transition of the fluoroethylenes is sharp in fluoroethylene itself, but is broad and ill-defined in the other compounds. In those compounds in which vibrational structure can be recognized in the $\pi \rightarrow 3$s band, the intervals are uniformly of 1600–1700 cm^{-1} spacing, corresponding to the C═C stretch in the upper state, reduced by 100–200 cm^{-1} from the ground-state values. For example, a progression of 1670 cm^{-1} (C═C stretch) is excited in the (π, 3s) upper state of tetrafluoroethylene (Fig. IV.A-21), which is reduced from the ground-state value of 1872 cm^{-1}; the corresponding pi ionization is accompanied by 1690-cm^{-1} (C═C stretch) and 740-cm^{-1} vibrations (totally symmetric C—F stretch, 778 cm^{-1} in the neutral-molecule ground state) [L2]. Judging from the molar absorption coefficients at the (π, 3s) maxima, the oscillator strengths for these transitions in the fluoroethylenes are between 0.05 and 0.1, favoring the lower limit.

The term values of the (π, 3s) states in the fluoroethylenes are also of interest. Since the fluorine atom is isoelectronic with the methyl group, one might at first expect the (π, 3s) terms of the fluoroethylene series to decrease with increasing substitution, as they do in the methyl ethylenes (Section IV.A-2). Table IV.A-VI indicates, however, that the term values actually increase on fluorination, going from 27 410 cm^{-1} in ethylene to 32 000 cm^{-1} in tetrafluoroethylene. As explained in Section I.C-1, this is understandable in terms of the delocalization of the 3s upper orbital over the substituent groups. In the fluoroethylenes, the optical electron resides partially upon the fluorine atoms and is bound much more tightly to the core than is a 3s electron residing upon a methyl group, due to the larger effective nuclear charge of the former. Of course, the increased penetration and tighter binding in the $(\pi, 3s)$ state of the fluoroethylenes is apparent as an enhanced term value. A similar effect is observed in the Rydberg spectra of virtually all fluorinated chromophores.

In contrast to excitations to $(\pi, 3s)$, those to $(\pi, 3p)$ in the fluoroethylenes are uniformly sharp and richly structured. This aspect of the fluoroethylene spectra parallels that in the chloroethylenes (to be discussed later) in which the transitions to $(\pi, 3s)$ are broad and poorly defined when they can be located at all, whereas excitations to $(\pi, 3p)$ are quite sharp. Though the vibronic structures of the $\pi \to 3p$ bands in the fluoroethylenes have not been analyzed yet, those in the chloroethylenes have, and provide a basis for hypothesis. Arguing by analogy with the chloroethylenes, the $\pi \to 3p$ transitions in the fluoroethylenes will first of all show two or three electronic origins corresponding to the lifting of the 3p degeneracy by the asymmetry of the core. Further, the transitions will be rather vertical, with the excitation of a C=C stretching progression together with many fewer quanta of C—F stretching and

bending. Appropriately, just these vibrations are excited in the first photoelectron bands of the fluoroethylenes [L2]. According to the symmetry selection rules, the $\pi \to 3p$ excitation will be $u \leftrightarrow u$ forbidden in all centrosymmetric haloethylenes, and this rule is clearly obeyed in the chloroethylenes (discussed later). Looking at the spectra of the centrosymmetric *trans*-difluoroethylene and tetrafluoroethylene (Fig. IV.A-21), we see that the $\pi \to 3p$ intensity is comparatively low in the former, and conclude that the transition is electronically forbidden but probably vibronically allowed due to the excitation of a nontotally symmetric vibration. On the other hand, in tetrafluoroethylene, the excitation we assign as $\pi \to 3p$ beginning at 64 700 cm^{-1} is seen to be rather strong, due consideration given to its resting upon the N $\to$ V band. It is possible that the near-degeneracy of the $\pi \to 3p$ and $\pi \to \pi^*$ excitations in tetrafluoroethylene gives the former a large intensity through vibronic mixing, or again, perhaps the strong features are really part of the N $\to$ V band rather than the $\pi \to 3p$. However, the value of the C=C stretching frequency observed here ($\sim$1600 cm^{-1}) strongly suggests that the vibronic structure belongs to $\pi \rightarrow$ 3p rather than $\pi \rightarrow \pi^*$. A vibrational analysis is sorely needed here, and a high-pressure experiment (Section II.B) would also be welcome. Jortner has pointed out the possibility that the sharp excitations between 70 000 and 74 000 cm^{-1} mix with the broad continuum of the $\pi \rightarrow \pi^*$ transition, producing the antiresonances indicated by the arrows in Fig. IV.A-21 [J16]. However, the antiresonance phenomenon will not occur for the (π, π^*) and $(\pi, 3p)$ purely electronic configurations since they do not have the same electronic symmetry, but if $\pi \rightarrow$ 3p gains its intensity vibronically from $\pi \rightarrow \pi^*$, then the vibronic mixing may result in antiresonances.

Taking a reasonable value of 13 500 cm^{-1} for the $(\pi, 3d)$ term leads to the expectation that these bands will appear in the 70 000–66 000-cm^{-1} region in the fluoroethylene–tetrafluoroethylene series. Thus the bands beginning at about 70 000 cm^{-1} in fluoroethylene (Fig. IV.A-21) are probably $\pi \rightarrow$ 3d excitations, intertwined with an overlapping $\pi \rightarrow$ 4s band.

The N $\rightarrow$ V transition of the fluoroethylenes would make an interesting theoretical study, for, as mentioned, the pi MO is constant in energy, as shown by the photoelectron spectra, and the $\pi \rightarrow \pi^*$ transition frequency is, too, except in tetrafluoroethylene, where it is 9000 cm^{-1} higher. The intensities Table (IV.A-VI) show a peculiar variation in the series, and Bélanger and Sandorfy suggest that because the N $\rightarrow$ V transition of fluoroethylene is much more vertical than that of ethylene, the (π, π^*) state in fluoroethylene may not be twisted as it is in ethylene.

Having accounted for the $\pi \rightarrow$ 3s, $\pi \rightarrow$ 3p, and $\pi \rightarrow \pi^*$ excitations, it is evident that there are still unexplained bands in the lower-frequency parts of the spectra of trifluoroethylene and tetrafluoroethylene, centered at 57 000 and 62 000 cm^{-1} (vert.), respectively. A related band may be present at 57 000 cm^{-1} in *trans*-difluoroethylene (Fig. IV.A-21). Since no other Rydberg excitation originating at the pi MO could fall between the $(\pi, 3s)$ and the lowest $(\pi, 3p)$ states, and since the second ionization potentials in these compounds are $\sim$20 000 cm^{-1} above the first, the transitions in question cannot be Rydberg, and hence must be part of the valence shell spectra. Remembering that the effect of fluorination in ethylene is to depress the occupied sigma MO manifold with respect to the pi MO manifold (see [B59] for experimental proof of this), it is reasonable to presume that $\pi \rightarrow \sigma^*$ valence shell excitations may be low lying in trifluoroethylene and tetrafluoroethylene, even lower than $\pi \rightarrow \pi^*$. This is our tentative explanation for the low-lying valence shell excitations in the highly fluorinated ethylenes. According to the Gaussian orbital calculations on tetrafluoroethylene, the lowest-frequency $\pi \rightarrow \sigma^*$ excitation $(b_{2u} \rightarrow b_{1u})$ is parity forbidden, thus accounting for the low intensity observed for the 62 000-cm^{-1} band. Note that the relative shift of the π and σ manifolds in the fluoroethylenes is similar to the explanation given for low-lying $\sigma \rightarrow \pi^*$ transitions in highly strained olefins such as cyclopropene (Section IV.A-2), with the difference that in the cycloolefin, the strain raises the sigma MO manifold with respect to the pi manifold, rather than lowering it. This lowering of the σ^* levels upon fluorination is evident as well in the ketones, Section IV.C.

The SF$_6$-scavenger spectrum (Section II.D) of fluoroethylene has been published by O'Malley and Jennings [O6], who found the triplet and singlet $\pi \rightarrow \pi^*$ transitions centered at 35 400 and 58 000 cm^{-1} (vert.), respectively. Another intense feature appeared at 77 400 cm^{-1} (vert.), which is rather unexpected, since there is no prominent feature at that frequency in the optical spectrum. Perhaps it is a transition to an upper triplet state. Moore has studied the energy-loss spectrum of 1,1-difluoroethylene, using H$^+$ and He$^+$ ions as projectiles rather than electrons as normally done [M51]. With He$^+$ ions incident at 3.0 keV, an intense $\pi \rightarrow \pi^*$ triplet was found at 37 000 cm^{-1} (vert.) and a weaker $\pi \rightarrow \pi^*$ singlet appeared at 61 200 cm^{-1} (vert.), whereas with H$^+$ ions at 3.0 keV, only the excitation to the singlet is observed.

Since the progressive fluorination of a molecule acts to move the $(\phi_i, 3s)$ term value toward the fluorine limit of 36 000 cm^{-1}, the spectra of cis- and trans-perfluorobutene-2 should show $\pi \rightarrow 3s$ transitions with unusually large terms. This appears to be the case, for in the cis compound (Table IV.A-VI), a weak band ($\epsilon = 2400$) is centered at 57 000 cm^{-1}, with a term value of 35 000 cm^{-1}. The $\pi \rightarrow \pi^*$ transition in this material has an extinction coefficient of 4800 to 61 000 cm^{-1} (vert.) [R18].

Assignments of the bands in the chloroethylene spectra seem straightforward if one compares the spectra as in Fig. IV.A-22. The spectral data are largely the work of Walsh and his collaborators [H36, T8, W1, W8, W9, W10, W11], but with contributions as well from Mahncke and Noyes [M1] and Goto [G17, G19]. The optical data are nicely complemented by the photoelectron spectroscopy of these compounds by Lake and Thompson [L2] and Jonathan et al. [J11]. Since the term values change so slowly on chlorination of ethylene, the correlation of the Rydberg levels is fairly obvious (Fig. IV.A-22), and one can confidently predict the frequencies of certain bands which otherwise have not been reported as yet. These are shown as horizontal dashed lines in the figure.

The $\pi \rightarrow \pi^*$ (N $\rightarrow$ V) transitions in the chloroethylenes have their intensity maxima in the 52 000–54 000-cm^{-1} region, the frequency being highest for chloroethylene and lowest for tetrachloroethylene. The trend of decreasing N $\rightarrow$ V frequency with increasing halogenation in the chloro-

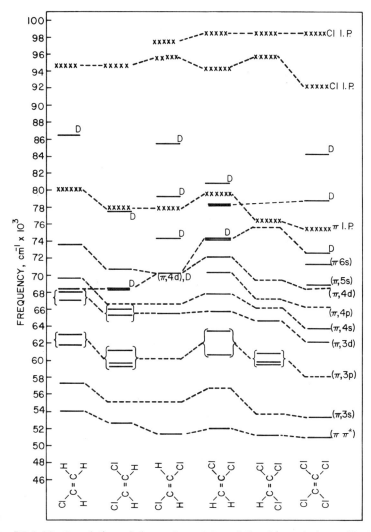

Fig. IV.A-22. Correlation of the various states of the chloroethylenes. Horizontal crosshatched lines represent ionization potentials, and horizontal dashed lines represent levels predicted but not observed.

ethylenes is counter to that found for the fluoroethylenes and the frequency shift is largest for the substitution of the first and second chlorines and smallest for the third and fourth chlorines, whereas just the opposite is observed in the fluoroethylenes. The N → V bands of the chloroethylenes are broad and structureless in every case, but occasionally have other structured bands ($\pi \to 3s$, $\pi \to 3p$) superposed upon them. Goto has measured the absorption intensities in *trans*-dichloroethylene [G17] and in tetrachloroethylene [G19] and finds the N → V transition in the latter to be somewhat stronger than in the former. Considerations of his absorption coefficients at the absorption maxima lead to oscillator strengths of about 0.4 for the N → V bands in these compounds. The increase of this quantity over the value of 0.34 in ethylene itself can be explained as a consequence of the delocalization of the π and π^* MOs over the chlorine $3p\pi$ AOs.† Sood and Watanabe [S45] report intensity data for chloroethylene, the molar extinction coefficient being 8000 at the maximum of the $\pi \to \pi^*$ band. The very broad nature of the N → V transition in the chloroethylenes (see Goto's spectrum of tetrachloroethylene, for example [G19]) once again suggests a large change in geometry in the V state; there is probably a large extension of the C=C distance, and possibly some twisting about the C=C bond.

The N → V transition is the only valence shell band which can be identified in the chloroethylene spectra; all others are assigned to Rydberg transitions converging upon either the first or second ionization potentials. Spectra of the alkyl chlorides (Section III.B-2) reveal a broad, weak band at about 57 000 cm^{-1}, called the A band, which can be assigned as a valence shell chlorine $3p\pi \to \sigma^*$ (C—Cl) excitation. The analogous excitation in the chloroethylenes is undoubtedly present in the vicinity of 60 000 cm^{-1}, but has not been identified as yet. Inasmuch as the ionization potentials for the chlorine 3p lone pairs in the chloromethanes are in the 90 000–100 000-cm^{-1} region and the pi ionization potential of ethylene is at 84 750 cm^{-1}, these assignments are evidently those to be preferred for the bands observed at 90 000–98 000 and 72 000–80 000 cm^{-1}, respectively, in the photoelectron spectra of the chloroethylenes [B4, L2]. These adiabatic ionization potentials are represented in Fig. IV.A-22 by the crosshatched lines; in this series of compounds, Rydberg transitions originating at the pi MO should follow the lower set of ionization potentials, whereas those originating with the chlorine lone-pair orbitals should follow the upper set of ionization potentials.

In chloroethylene, Walsh [W1] reports a faint, diffuse series of bands beginning at 57 330 cm^{-1} with 1300–1400-cm^{-1} spacings. This is the $\pi \to 3s$ transition, and the vibration is the C=C stretch, reduced from 1608 cm^{-1} in the ground state. The corresponding frequency in the $(\pi, 3s)$ state of ethylene is 1370 cm^{-1}. Walsh finds this band to be the first member of an ns Rydberg series characterized by $\delta = 0.85$. The $\pi \to 3s$ transition is not seen again in the chloroethylenes until tetrachlo-

† Note however, that the pi orbitals are similarly delocalized in the fluoroethylenes, yet the $\pi \to \pi^*$ oscillator strengths are somewhat lower than in ethylene.

roethylene. Unfortunately, the spectra of Walsh [W1] and of Goto [G19] do not agree too closely for this compound. Goto's spectrum of tetrachloroethylene shows the $\pi \to 3s$ band beginning at 53 100 cm^{-1}, with a single vibrational progression of 1350 ± 50 cm^{-1} attached. Walsh lists these bands, but several others to the low-frequency side as well, claiming the electronic origin is below 50 000 cm^{-1}. Tentatively, it seems that these might be impurity bands in Walsh's spectrum, and we shall accept Goto's origin, pending confirmation of the spectrum. The wisdom of this choice is supported by the term values, for the choice of origin at 53 100 cm^{-1} results in a $(\pi, 3s)$ term value of 22 200 cm^{-1}, which is the perchloro limiting term value, whereas placing the origin below 50 000 cm^{-1} in tetrachloroethylene gives a term value larger than 25 300 cm^{-1}. Though the $\pi \to 3s$ transition is ill-defined in tetrachloroethylene, the corresponding transitions to 4s and 5s are quite sharp, and have been analyzed by Humphries et al. [H36]. Having the $(\pi, 3s)$ term values (23 330 and 22 200 cm^{-1}) for the end members of the chloroethylene series allows one to interpolate values for the $\pi \to 3s$ frequencies for the remaining molecules (dashed lines in Fig. IV.A-22). Since $\pi \to 3s$ is allowed with out-of-plane polarization in both ethylene and tetrachloroethylene, it will also be allowed out-of-plane in each of the intermediate members of lower symmetry.

The large difference between the $(\pi, 3s)$ terms in the chloroethylenes (~22 000 cm^{-1}) and in the corresponding fluoroethylenes (~30 000 cm^{-1}) is a sure indication that penetration at the fluorine atoms results in much tighter binding than does penetration at the chlorine atoms. Comparison of the corresponding term values in the isolated chlorine and fluorine atoms, 32 000 cm^{-1} versus 42 000 cm^{-1}, confirms this view. In fact, we see again that the effect of chlorine atoms as substituents is just like that of added methyl groups, the $(\pi, 3s)$ limiting term value being ~22 000 cm^{-1} in both cases. The similarity is also seen in the $N \to V$ excitations (compare Table IV.A-III with Table IV.A-VI), where the substitution of the first group results in the largest incremental shift of the maximum to lower frequency, and with the fully substituted molecules absorbing at ~52 000 cm^{-1} (vert.) in both cases.

In the chloroethylenes, the $\pi \to 3p$ transitions will be $u \leftrightarrow u$ forbidden in trans-dichloroethylene and tetrachloroethylene, but more or less allowed in the other members of the series. Appropriately, this band is missing in the centrosymmetric molecules (Fig. IV.A-22), but appears in the others; it is especially intense in trichloroethylene [W10], indicating that in this molecule, the chlorines are especially effective in destroying the local symmetry of the C=C group. Among the fluoroethylenes, the $\pi \to 3p$ transition is also strongest for trifluoroethylene. One sees from

Fig. IV.A-22 that the observed $(\pi, 3p)$ frequencies follow the π ionization potentials rather nicely, with near-constant term values of 17 000 cm^{-1}. The $(\pi, 3p)$ terms in the chloroethylenes are very nearly equal to those in the fluoroethylenes (18 000 cm^{-1}), since the penetration is much reduced in the $(\pi, 3p)$ states as compared to those in the $(\pi, 3s)$ states. Careful vibrational analysis by Walsh, Warsop, et al. in the cases of cis-dichloroethylene [W9], 1,1-dichloroethylene [W11], and trichloroethylene [W10] revealed that the $\pi \to 3p$ absorption bands contain two or three electronic origins, corresponding, no doubt, to the core splitting of the otherwise degenerate 3p AOs, with a split of 2000–3000 cm^{-1} from the lowest to the highest components.

In chloroethylene, two components of the $\pi \to 3p$ transition were found at 63 048 and 61 520 cm^{-1}, each sporting several quanta of the C═C stretch, ~ 1375 cm^{-1}. The series have quantum effects of 0.595 and 0.65, respectively. The band with origin at 63 048 cm^{-1} also shows extensive excitation of a 420-cm^{-1} motion, which is described as a bending mode [S45, W1]. An analogous band in cis-dichloroethylene ($\delta \simeq 0.5$) is resolved by Walsh and Warsop [W9] into three overlapping bands with origins at 59 130, 59 734, and 60 958 cm^{-1}. The last two of these display vibrational progressions of ~ 1400 cm^{-1} (in which the frequency decreases with increasing vibrational quantum number) and a quantum or two of 810 cm^{-1}. These $(\pi, 3p)$ upper-state vibrational frequencies correspond to the totally symmetric ν_2' C═C stretch (1587 cm^{-1} in the ground state) and the totally symmetric ν_4' C—Cl stretch (711 cm^{-1} in the ground state), respectively. That ν_2' should decrease but ν_4' increases in the transition to the $(\pi, 3p)$ state can be understood in the following way [C27]. First, the photoelectron spectrum of cis-dichloroethylene [L2] shows that in the lowest ionic state, the ν_2' and ν_4' frequencies are 1400 and 840 cm^{-1}, respectively. Since these frequencies are close to those in the $(\pi, 3p)$ state, the 3p orbital is seen to be relatively nonbonding, and one has only to consider the originating π MO to explain the vibrational frequency shifts. The photoelectron spectra also show that before mixing, the chlorine 3pπ AOs are deeper in energy than the π MO of the C═C double bond. Consequently, on mixing of the two orbitals, the C═C π MO will become destabilized (this is reflected in the decrease of the π ionization potential with increasing chlorination, Fig. IV.A-22) and the chlorine 3pπ AOs will mix with it in an antibonding combination, though the MO is still C—C bonding. Consequently, it is seen that the removal of an electron from the π MO of cis-dichloroethylene will lower the pi-electron bond order in the C—C bond, but increase it in the C—Cl bond, and the corresponding stretching frequencies will change in a parallel manner, as observed. In those compounds having two chlorine atoms on the same carbon, the small amount of Cl—Cl interaction will result in a very small change in the

angle upon excitation, and the tell-tale excitation of a quantum or two of the appropriate angle bending. Geometrically, one expects the C—C distance to increase slightly in the (π, np) upper states, whereas the C—Cl distances would decrease slightly. The changes will be slight since the transitions do not display long progressions in any vibrational mode, and the excitation of only totally symmetric motions implies that the geometric symmetries of the ground and (π, np) excited states are the same. The argument quoted earlier with regard to the decrease of the C=C stretching frequency and increase of the C—Cl stretching frequency in the (π, np) upper states is in no way unique to *cis*-dichlorethylene, and should apply equally well to the (π, np) states of all the chloroethylenes. One might also try applying this argument as well to the fluoroethylenes. Though the optical transitions of these have not been analyzed yet, the photoelectron spectra have, and in these systems, Lake and Thompson do find a decrease in the C=C stretching frequency on ionization, but the C—F stretch, which is also excited, has a frequency in the ion which is slightly lower than that in the molecule, rather than higher as in the chloroethylenes. Since MO calculations on tetrafluoroethylene [B59] also show that the uppermost occupied pi MO is strongly C—F antibonding, it would seem that either the simple picture of ionization from antibonding MOs leading to increased vibrational frequencies is too naive, or that the vibrations in the fluoroethylenes have been misassigned.

As mentioned before, the $\pi \rightarrow$ 3p transition has not been located in tetrachloroethylene, but those to 4s and 5s are sharp and have been analyzed [H36]. As expected in these Rydberg states, the C=C stretch, C—Cl stretch, and

bend are all excited in the transition from the ground state. Interestingly, in the corresponding transition in ethylene, the C=C stretch is again excited, but there is no evidence for C—H stretching or

bending. The difference rests in the fact that the originating π MO is delocalized over the chlorine atoms in tetrachloroethylene, but not over the hydrogen atoms in ethylene itself.

In 1,1-dichloroethylene [W11], two components of the $(\pi, 3p)$ upper states are found, with origins at 60 500 and 63 349 cm^{-1} (adiab.), accompanied by short progressions of 1300 cm^{-1} (C=C stretch, decreased from 1620 cm^{-1} in the ground state) and 650 cm^{-1} (C—Cl stretch, increased from 605 cm^{-1} in the ground state). The corresponding frequencies in the lowest state of the ion are 1320 and 650 cm^{-1} [L2]. The two components of the $(\pi, 3p)$ manifold are the $n = 3$ members of Rydberg series having $\delta = 0.56$ and $\delta = 0.52$. The $\pi \rightarrow 3p$ band also appears in trichloroethylene [W10]; Walsh and Warsop's analysis yields three electronic origins and vibrational intervals of 1406, 670, and 400 cm^{-1}. The last is thought to be due to

in-plane bending. The vibrations which are quoted as prominent in the ion are 1320 and 330 cm^{-1} [L2]. The constant term value of 18 000 cm^{-1} for the $(\pi, 3p)$ configuration leads to expected $\pi \rightarrow 3p$ origins at 60 000 cm^{-1} in *trans*-dichloroethylene and at 57 500 cm^{-1} in tetrachloroethylene.

The excitation to the $(\pi, 3d)$ upper state is formally allowed in all of the chloroethylenes and has been unambiguously identified in all but one. Walsh [W1] and Humphries *et al.* [H36] both report a continuous absorption centered at 61 900 cm^{-1} in tetrachloroethylene, close to the frequency expected for the $(\pi, 3d)$ upper state, judging from the ionization potential. However, Goto's more recent spectrum of this substance does not suggest any transition in this spectral region. The $\pi \rightarrow 3d$ transitions in the other chloroethylenes have term values of 13 000 ± 1000 cm^{-1} and all display, simultaneously, quanta of C=C stretching (1300–1420 cm^{-1}), C—Cl stretching (640–930 cm^{-1}), and

bending (200–420 cm^{-1}). In both *cis*-dichloroethylene and 1,1-dichloroethylene, complete (π, nd) Rydberg series were uncovered, with $\delta = 0.09$ and $\delta = 0.18$, respectively, while in chloroethylene, two nd series with quantum defects of 0.05 and 0.13 were delineated [S45, W1].

Rydberg series originating with the lone-pair electrons on the chlorine atom are prominent in the chloroalkane spectra (Section III.B-2), and are to be expected in the chloroethylenes as well. In methyl chloride, the 3p $\rightarrow$ ns transitions show only a weak $n = 4$ member, but the 3p $\rightarrow$ np series is a strong one which can be followed out to the ionization limit. The 3p $\rightarrow$ 4p band, called the D band, is observed not only in the chloro-

IV.A. OLEFINS 63

alkanes [P39, P40], but in chloroprene [P43] and chlorobenzene [P48] as well, as a strong, very vertical doublet ($\sim$600 cm^{-1} splitting) having a term value of about 22 000 cm^{-1}. Just such a band is described by Walsh [W1] to appear at 68 300 cm^{-1} in the chloroethylene spectrum; measured from the second ionization potential (94 600 cm^{-1} [L2]), its term value is 26 300 cm^{-1}. A similar band appears with split components at 68 300 and 68 100 cm^{-1} in cis-dichloroethylene, the term value being 26 200 cm^{-1}. A strong feature lying just below the first ionization potential has a term of 23 100 cm^{-1} with respect to the fifth ionization potential at 100 900 cm^{-1} in cis-dichloroethylene. This pattern persists in the remaining chloroethylenes; the strong bands at 70 100, 74 200, 79 100, and possibly at 85 700 cm^{-1} in trans-dichloroethylene appear to be D bands corresponding to ionizations at 95 660, 97 270, 101 700, and possibly 111 700 cm^{-1}, respectively, as determined by photoelectron spectroscopy.† In 1,1-dichloroethylene, the two definite D bands at 74 120 and 78 130 cm^{-1} would seem to correlate with the 94 130- and 98 160-cm^{-1} ionizations, respectively, and what is probably a D band at 80 700 cm^{-1} has a proper term value for convergence to the ionization potential at about 100 900 cm^{-1}. No D bands appear in trichloroethylene, but the strong bands at 72 400, 78 700, and 84 000 cm^{-1} in tetrachloroethylene correlate well with ionization potentials at 91 790, 100 300, and 108 700 cm^{-1}, respectively. The D-band term values in the various chloro compounds are compared in Table IV.A-VII.

In Fig. IV.A-22, an attempt has been made to correlate the various D states serially without regard for the symmetries of the originating orbitals. It is found that the lowest D states in chloroethylene, cis-dichloroethylene, and trans-dichloroethylene follow the lowest chlorine 3p ionization potential nicely, but with a term value of about 26 000 cm^{-1}, which is 5000 cm^{-1} higher than one expects for this transition. However, in 1,1-dichloroethylene and tetrachloroethylene, the term values of the first two D bands in each compound have assumed the more normal value of $\sim$20 000 cm^{-1}. While one does expect the D-band term value to decrease as the olefin is chlorinated, its extraordinarily high value (more like that of a B—C band) in chloroethylene and cis-dichloroethylene is suspicious, and the assignment should be studied more closely in these compounds.

Actually, each chlorine atom in a chloroethylene molecule contributes two 3p lone pairs to the electronic structure, one in plane and one out of plane. The lone pairs discussed earlier as giving rise to the D bands

† More than likely, the fifth ionization potential (111 700 cm^{-1}) does not originate on chlorine, but is from a σ MO within the ethylene framework. In that case, the band at 85 700 cm^{-1} is assigned as $\sigma \rightarrow$ 3s.

TABLE IV.A-VII
D Bands in the Chloroethylenes

Molecule	Ionization potential	Absorption frequency	Term value
$H_2C{=}CHCl$	94 600	68 300	26 300
	109 400	86 500	22 900
cis-ClHC=CHCl	94 450	68 200	26 250
	100 900	77 800	23 100
trans-ClHC=CHCl	95 660	70 100	25 500
	97 270	74 200	23 100
	101 700	79 100	22 600
	111 700	85 700	26 000
$H_2C{=}CCl_2$	94 130	74 120	20 000
	98 200	78 130	20 000
	100 900	80 700	20 200
$Cl_2C{=}CCl_2$	91 790	72 400	19 400
	98 240	78 700	19 500
	103 000	84 000	19 000

are the in-plane 3p orbitals, which are strongly localized, unlike the chlorine 3pπ orbital. According to the work of Klasson and Manne [K23], the 3pπ pair of electrons are more delocalized onto the ethylenic group and have their ionization potential at 109 400 cm^{-1} (vert.) in chloroethylene. This level should also have associated with it a set of D bands perhaps beginning with that observed at 86 500 cm^{-1} (advert.) in the spectrum of Sood and Watanabe [S45], for it has a term value of 22 900 cm^{-1} (vert.).

Moore's ion-impact energy-loss spectra of the chloroethylenes gave results much like those mentioned for 1,1-difluoroethylene, i.e., a very intense $\pi \to \pi^*$ triplet at $\sim$32 000 cm^{-1} (vert.) followed by a much weaker $\pi \to \pi^*$ singlet at $\sim$56 000 cm^{-1} (vert.) when He$^+$ was used at a projectile, whereas with H$^+$, the transition to the triplet is missing. However, the D-band transitions are quite prominent in the H$^+$ energy-loss spectra, while not appearing at all in the spectra using He$^+$ excitation [M51].

Scott and Russell have reported the spectra of a few mixed chlorofluoroethylenes which make an interesting comparison with the unmixed haloethylenes. In $F_2C{=}CCl_2$ [S24] (Fig. IV.A-23), the $\pi \to$ 3s transition is evident as a weak feature at about 51 200 cm^{-1} (vert.) and has a term value of 28 200 cm^{-1}, which is just between those of $Cl_2C{=}CH_2$ ($\sim$23 000 cm^{-1}) and $F_2C{=}CH_2$ ($\sim$31 500 cm^{-1}), but closer to the latter. It has also been found in other mixed-ligand systems that the fluorines contribute more than their share to the penetration energy, indicating a

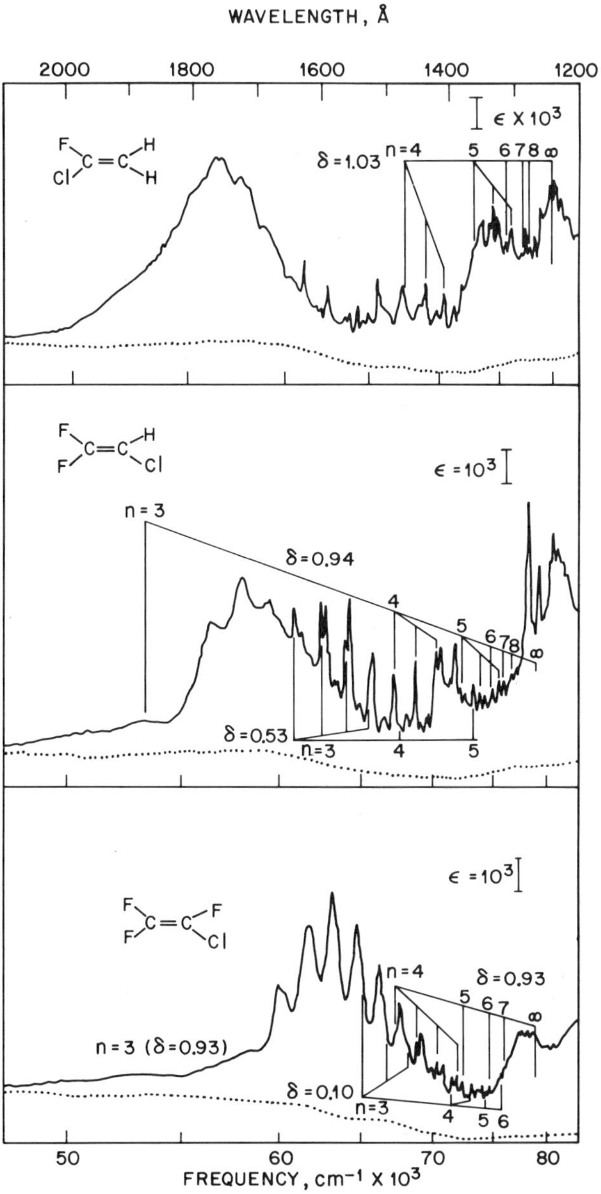

Fig. IV.A-23. Optical absorption spectra of several fluorochloroethylenes in the gas phase [S25]. The dotted lines are the empty-cell baselines.

3s MO which is disproportionately heavy on the fluorine atoms. This pattern is continued in F_2C=CHCl and F_2C=CFCl [S25], in which intermediate $(\pi, 3s)$ values are again observed (Table IV.A-VI). In these mixed halogen compounds, the $\pi \to \pi^*$ excitation frequencies are also intermediate, except that in the fully fluorinated system F_2C=CFCl, the N → V frequency again takes a large step upward, just as in F_2C=CF_2, and even surpasses that of F_2C=CFH. This very unusual effect seems peculiar to the fluoroethylenes.

The $K\beta$ X-ray emission spectra of vinyl chloride and the various dichloroethylenes in which an electron from the valence shell MOs is transferred into the 1s hole on chlorine have been observed [D8, G10, L12, L13]. The spacings and intensities are nicely explained using the observed valence shell photoelectron spectrum and the 3p populations of each valence shell MO at the chlorine atom [K23].

The spectrum of iodoethylene (Fig. IV.A-24), is more like that of ethyl iodide (Section III.B-1) than like that of iodoacetylene (Section IV.D) in the sense that the only absorptions that can be firmly identified are

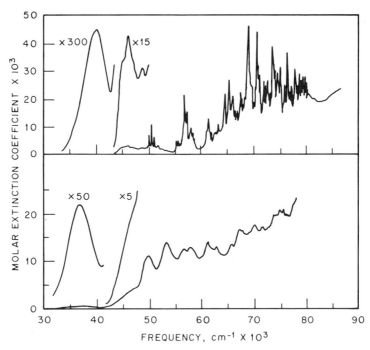

Fig. IV.A-24. Optical absorption spectra of iodoethylene (upper) and allyl iodide (lower) in the vapor phase [B35].

those originating on the iodine atom, as in an alkyl iodide. Its spectrum begins with a weak A band at 39 900 cm^{-1} (vert.) [B35], comparable to its position in ethyl iodide (38 500 cm^{-1} vert.) and iodoacetylene (40 000 cm^{-1} vert.). Another broad band, looking much like a valence shell excitation, comes at 46 190 cm^{-1} (vert.); its term value of 29 000 cm^{-1} is too large for a Rydberg excitation, and so it, too, is probably valence shell. The Rydberg spectrum of iodoethylene begins with two sharp bands at 50 660 and 56 880 cm^{-1} (vert.), which are the B and C bands converging to the lone-pair ionization potentials at 74 980 and 81 100 cm^{-1} (vert.). The spin–orbit splitting due to the 5p^5 configuration in the core is 6100 cm^{-1} in the ion and 6220 cm^{-1} in the (5p, 6s) Rydberg state, which compare rather well with one another but are significantly larger than the values of 4700 cm^{-1} in $C_2H_5I^+$ and 3340 cm^{-1} in HC_2I^+. Contrary to this analysis, Boschi and Salahub feel that the low symmetry quenches the spin–orbit coupling in iodoethylene, and that the splitting of the first Rydberg bands and of the first two ionization potentials is due to the one-electron splitting of the in-plane and out-of-plane lone pairs on the iodine atom. Both effects are probably at work in iodoethylene.

The 5p → 6p Rydberg excitations in iodoethylene (D bands) are expected to have term values of ∼19 000 cm^{-1}, placing then at 56 000 and 62 000 cm^{-1}. Such bands are readily located in the spectrum of Fig. IV.A-24. The C=C pi orbital in iodoethylene is pushed down to 93 000 cm^{-1} (vert.) [B36], so that $\pi \rightarrow$ 3s and $\pi \rightarrow$ 3p transitions are expected at 68 000 and 74 000 cm^{-1}, and may account for the broader bands in these regions. An antiresonance is seen at 75 500 cm^{-1} (Section I.A-2). The $\pi \rightarrow \pi^*$ transition would appear to peak at 65 000 cm^{-1} (vert.) with a much stronger continuum coming at 74 000 cm^{-1}. Boschi and Salahub also point out the presence of 700–900-cm^{-1} vibrational intervals in the 64 000–66 000-cm^{-1} region which could be C=C stretching greatly reduced in the V state, as it is in ethylene itself.

As seen from Table IV.A-VI, an N → V frequency of 65 000 cm^{-1} (vert.) is quite unusual, being exceeded only by that of tetrafluoroethylene (71 600 cm^{-1}). Both this and the ionization potential demonstrate that the iodine atom of iodoethylene has a very strong stabilizing effect on the pi MO.

In allyl iodide, where the iodine and vinyl groups are separated by a methylene group, one expects a more normal B–C spin–orbit splitting interval, and the narrow Rydberg transitions of an alkyl iodide superposed upon a broad ethylenic $\pi \rightarrow \pi^*$ background peaking at its more normal frequency. As seen in Fig. IV.A-24, this is only partially realized.

Beyond the A band at 37 000 cm^{-1} (vert.),† there are three bands at 46 500, 49 900, and 53 480 cm^{-1} (vert.), two of which must be the B–C bands resulting from the spin–orbit coupling within the (5p, 6s) configuration on the iodine atom. Note that the spin–orbit-split ionization potentials come at 75 100 and 78 400 cm^{-1} (vert.) [B36], which agrees with the split between either the first and second bands (3400 cm^{-1}) or between the second and third bands (3580 cm^{-1}). The question of which pair of bands to assign as B–C can be settled using term values, for the choice of the 46 500- and 49 900-cm^{-1} bands as the B–C components results in term values of 28 600 cm^{-1}, whereas the choice of the 49 900- and 53 480-cm^{-1} bands yields term values of 25 100 cm^{-1}. Comparison with the B–C term values in the alkyl iodides (Table III.B-II) convincingly shows that 25 100 cm^{-1} is the more reasonable term value for B–C transitions in an iodide bearing three carbon atoms. Thus the band at 46 500 cm^{-1} is probably a valence shell transition, related to that at 46 190 cm^{-1} in iodoethylene. The D band is recognized by its term value of 18 300 cm^{-1} in allyl iodide as the band at 56 800 cm^{-1} (vert.).

One might guess with little risk that the $\pi \to \pi^*$ absorption in allyl iodide peaks at $\sim$55 000 cm^{-1} (vert.) (Fig. IV.A-24), which is not far from that in propylene (58 000 cm^{-1} vert.). The only surprising feature of the allyl iodide spectrum is that the Rydberg excitations originating at the iodine lone-pair orbitals are so broad and lack vibrational fine structure.

IV.B. Azo and Imine Compounds

The azo group has been relatively neglected compared with the work expended on its isoelectronic counterparts, olefins and ketones. In part, the reason must be that, unlike ethylene and formaldehyde, the parent azo compound, diimide, H—N≡N—H, has only a transient existence [W30]. The *trans*-azo group distinguishes itself by having two equivalent "lone pairs" of electrons in apparently strong interaction. Haselbach *et al.* [H10] report that the ionization potentials from the two "lone-pair" molecular orbitals n_+ and n_- of *trans*-CH$_3$N≡NCH$_3$ are found at 72 400 and 99 200 cm^{-1} (vert.), respectively. Ionization from the pi orbital comes at 95 500 cm^{-1} (vert.). An n_+–n_- split of 20 000–30 000 cm^{-1}

---

† Note that the A bands, being $np \to \sigma^*$(C—X) are dissociating along the C—X line. The fact that the A bands are valence shell conjugates to the $np \to (n+1)$s Rydberg excitations in the haloethylenes and are strongly mixed with them explains why these Rydberg excitations are smooth and structureless, as are the A-band transitions.

IV.B. AZO AND IMINE COMPOUNDS

with the g combination higher has been predicted by MO theory as well [D14, R8, R11].

The $n_+ \to \pi^*$ absorption of the *trans*-azo group has been repeatedly observed in the near-ultraviolet region, and by its low intensity, behavior in different solvents, and according to the calculations, there is no doubt of its assignment. However, the higher transitions of the azo group are not as securely assigned. The $n_+ \to \pi^*$ transition of diimide correlates with the lowest $\sigma \to \pi^*$ ($^1A_g \to {}^1B_{1g}$) band expected in ethylene.

As seen in Fig. IV.B-1, the azoalkanes seem to have a characteristic pattern of three bands in the ultraviolet region below about 70 000 cm^{-1} [R11]. In azomethane, a very weak band at 44 100 cm^{-1} was also observed. Assignment of band I in the 50 000–55 000-cm^{-1} region has been rather hectic. The frequencies, and more quantitatively, the intensities of band I are very close to what one would expect for the symmetry-allowed $n_- \to \pi^*$ excitation, and it was so assigned [R8]. In a later GTO calculation [R11], it was concluded that $n_- \to \pi^*$ would be much higher in energy, and that band I was really an allowed $n_+ \to \sigma^*$ transition, possibly the first member of a Rydberg series. This suggestion can now be tested, for the ionization potentials of both diimide and azomethane have been measured.

The n_+ ionization potential of azomethane is 72 400 cm^{-1} (vert.); if the 54 000-cm^{-1} band in this compound is a Rydberg excitation, then it has a term value of 18 400 cm^{-1}, which would be quite appropriate for an allowed $n_+ \to 3p$ promotion. A $g \to g$ forbidden $n_+ \to 3s$ band would then be expected at about 44 000 cm^{-1}, corresponding to a term value of 28 000 cm^{-1}. This could well account for the weak band found at 44 100 cm^{-1} in azomethane. The allowed $n_+ \to 4p$ member of the series would come at 64 000 cm^{-1}, just where band II is found. The assignment of band II as a transition to 3d instead seems far less likely, because such a transition would be parity forbidden, and its term value of 8400 cm^{-1} is 50% too low for such a promotion.

The regular shift of the spectra to lower frequencies and the constancy of the band I–band II frequency difference would be accounted for if the n_+ ionization potentials are decreasing regularly with increasing size of the alkyl groups, while the term differences remain constant. Since the 54 000-cm^{-1} band of azomethane displays a vibrational progression, the high-pressure effect (Section II.B) can be used to test the Rydberg nature of its upper state. In Fig. IV.B-2, the before- and after-pressurization spectra are compared [R19], and it is seen that the vibrational structure is washed out by the nitrogen perturbing gas, and that there is a general shift of intensity to higher frequencies, as expected for a Rydberg upper state.

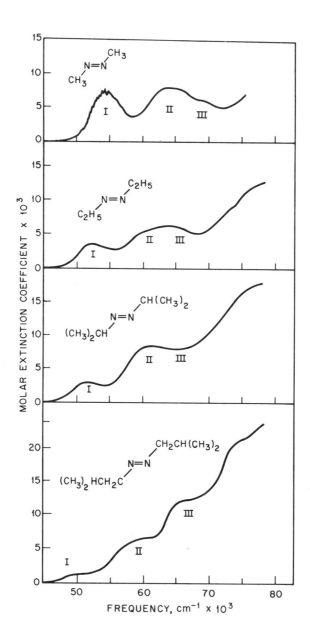

Fig. IV.B-1. Optical absorption spectra of several azoalkanes in the gas phase [R11].

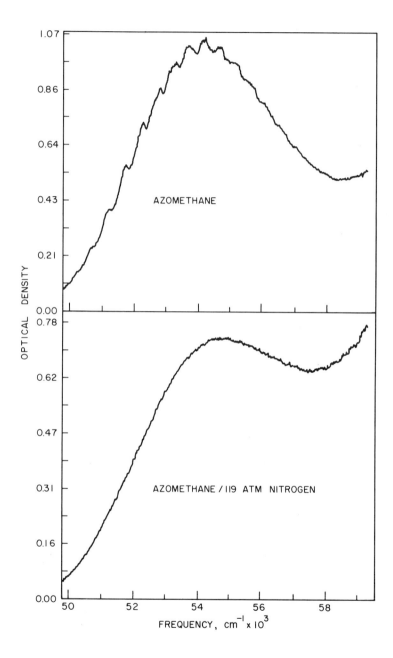

Fig. IV.B-2. Effect of pressurizing the 54 000-cm^{-1} band of azomethane with 119 atm of nitrogen gas [R19].

The vibrational progression in band I of azomethane is regular at 515 cm^{-1} up to 52 500 cm^{-1}, at which point the already diffuse bands become even more indistinct. It is possible that a quantum of the totally symmetric N=N stretch (ν_2', ~1600 cm^{-1}) enters here and forms a new origin for another 515-cm^{-1} progression. The 515-cm^{-1} interval probably originates with ν_{16}, the totally symmetric C—N—N—C angle bending, which has a value of 596 cm^{-1} in the azomethane ground state [W20]. If this is so, then the long, nonvertical vibrational series suggests a large but symmetric change in the C—N—N angles, possibly to a linear upper state. The possibility that there are two electronic origins within the 50 000–53 000-cm^{-1} region is not unexpected, since the $n_+ \to 3p$ transition will show three components due to the aspherical symmetry of the core.

It was noted in the section on ketones (Section IV.C) that the otherwise innocuous alkyl groups seemed able to destroy the spherical symmetry of the lower Rydberg states, with a concomitant redistribution of intensity. It seems probable that a similar phenomenon is operative in the R—N=N—R series, for we see from Fig. IV.B-1 that the $n_+ \to 3p$ Rydberg intensity drops by a factor of over five on going from R = CH$_3$ to R = (CH$_3$)$_2$CHCH$_2$. The missing intensity may be appearing in band III, for its intensity increases in the same series, but it remains to be shown that band III is a Rydberg.

Foner and Hudson [F8] report an ionization potential of 79 400 ± 800 cm^{-1} for diimide (presumably trans). Continuing the argument given above, a weak $n_+ \to 3s$ transition should be found in this molecule at 51 000 cm^{-1}, and the first two of the strong $n_+ \to n p$ bands should appear at about 61 000 and 70 000 cm^{-1}, respectively. Trombetti [T17] has studied the infrared and ultraviolet absorption spectra of H—N=N—H and D—N=N—D in a flow system and concludes from the infrared work that the ground-state molecule is in the planar trans form, with the N—N—H angle equal to 109 ± 1.5°. In addition to an $n_+ \to \pi^*$ band at 28 600 cm^{-1}, Trombetti reports a single, structured band in the 58 000–62 000 cm^{-1} region which displays progressions in the totally symmetric H—N—N—H bending motion ($\nu_3' = 1180$ cm^{-1}, $\nu_3'' = 1286$ cm^{-1}) built upon the origin and upon one quantum of the N=N stretching mode (1874 cm^{-1}). This sounds remarkably like that mentioned earlier for the $n_+ \to 3p$ Rydberg band in azomethane, and it is felt that this allowed band in diimide corresponds to the $n_+ \to 3p$ excitation expected in this region; the region of possible $n_+ \to 3s$ absorption (51 000 cm^{-1}) is unfortunately covered by NH$_3$ absorption in Trombetti's experiment, but the band would be extremely weak in any event. Rotational analysis of the 60 000-cm^{-1} band of diimide led to an out-of-plane polarization, in which case the terminating MO is taken as 3pπ. However, the same analysis led to the conclusion that the ground state of diimide is a spin triplet, rather than the expected singlet.

The foregoing analysis raises the question of where the important $\pi \rightarrow \pi^*$ excitation is to be found in the azoalkanes. Clearly, the only two choices left to us within the framework of the tentative Rydberg assignments given earlier are that the $\pi \rightarrow \pi^*$ excitation is band III (Fig. IV.B-1) or is lost somewhere in the maze of absorption beyond 70 000 cm^{-1}. Since the GTO calculations insist on putting the $\pi \rightarrow \pi^*$ transition of trans-H—N=N—H about 15 000 cm^{-1} higher than that of H$_2$C=CH$_2$, it seems likely that the $\pi \rightarrow \pi^*$ excitation of the azo group can be found between 70 000 and 80 000 cm^{-1}.

The spectrum of trans-F—N=N—F [R11] shows a single strong band ($\epsilon \simeq 2800$) at 66 000 cm^{-1} (vert.) which was originally presumed to be related to band I of the azoalkanes. If this were so, then its term value should be about 22 000 cm^{-1}, as is appropriate for transitions to 3p in fluorinated molecules. Inasmuch as the lowest ionization potential of F—N=N—F comes at 110 000 cm^{-1} (vert.) [B59], the term value for this supposed Rydberg transition is an impossible 44 000 cm^{-1}. It seems more likely that the 66 000-cm^{-1} band of F—N=N—F instead is an allowed n$_+$ $\rightarrow$ σ^* valence shell excitation.

An azo compound of another sort which has been studied is the cyclic compound difluorodiazirine [R13],

$$\begin{array}{c} N \\ \parallel \diagdown \\ \diagup CF_2 \\ \parallel \diagup \\ N \end{array}$$

In the cis-azo compounds, the large n$_+$–n$_-$ split remains with n$_+$ higher, and with the occupied π level coming between them, just as in the trans azoalkanes. The selection rules in the cis system, however, make n$_+$ $\rightarrow$ π^* forbidden and n$_-$ $\rightarrow$ π^* allowed. These two transitions in difluorodiazirine were located at 30 000 and 55 000 cm^{-1}, respectively, with the allowed one having an extinction coefficient of about 100. In the trans azoalkanes, this band is forbidden, and is weaker by a factor of 10–20.

In olefin spectra, the $\pi \rightarrow \pi^*$ band can usually be spotted thanks to its relatively high oscillator strength ($f \simeq 0.3$). In difluorodiazirine, the fact that the π MO is delocalized over three centers, whereas π^* is delocalized over two, acts to reduce considerably the $\pi \rightarrow \pi^*$ oscillator strength. A dipole velocity oscillator strength of 0.15 is calculated for this band of difluorodiazirine, and such a band is observed at 70 000 cm^{-1}. As was the case in cyclopropane (Section III.A-4), one exceptionally strong $\sigma \rightarrow \sigma^*$

($4b_1 \to 5b_1$) band is predicted ($f = 0.419$) for difluorodiazirine, and on the basis of intensity, it is assigned to the strong feature observed at 88 000 cm^{-1}.

The alkyl imines $R_2C{=}NR'$ are isoelectronic with the olefins, azoalkanes, and ketones, but have not been studied at anywhere near the same depth as these classes of compounds. Typical vapor-phase spectra of the alkylated imines are shown in Fig. IV.B-3; as might be expected with such heavily alkylated chromophores, the bands are all broad. As shown in these [S8] and other vapor-phase spectra [R8], the intense band at $\sim$58 000 cm^{-1} (vert.) has a weaker shoulder upon it at about 54 000 cm^{-1} (vert.). In an n-heptane solution of $C_3H_7CH{=}NC_6H_{11}$ [Y2], both of these bands are still present, suggesting that they are both valence shell rather than Rydberg transitions. All workers agree that the feature at 58 000 cm^{-1} ($\epsilon = 7000$–12 000) is the $\pi \to \pi^*$ valence shell transition [R8, S8, Y2] and it seems likely that the 54 000-cm^{-1} band is $n_N \to \sigma^*$. The $n_N \to \pi^*$ transition of the imine group occurs at 41 000 cm^{-1} (vert.).

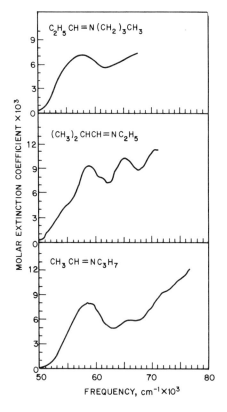

Fig. IV.B-3. Optical absorption spectra of several alkyl imines in the gas phase [S8].

IV.C. Aldehydes and Ketones

The deeper bands beyond 60 000 cm^{-1} have not been assigned yet. In the N-oxide

$$CH_3CH=\overset{\overset{O}{|}}{N}-C_6H_{11}$$

the (π, π^*) upper state acquires considerable charge transfer character and moves out to about 43 000 cm^{-1} (vert.) [Y2]. The photoelectron spectra of several other alkyl imines have been recorded [A4, H10].

IV.C. Aldehydes and Ketones

Ketones and aldehydes (hereafter collectively called ketones) are a most perplexing chromophoric group. On the one hand, ketones repeatedly offer clear examples of phenomena such as long Rydberg series, several series converging to the same ionization potential, and geometric term splitting; in general, the lower members of these Rydberg series are readily identified by their term values. On the other hand, except for the $n_O \to \pi^*$ transition, the valence shell spectra of ketones are poorly understood. The $n_O \to \pi^*$ transition of ketones falls in the 33 000–36 000-cm^{-1} region of the spectrum and, needless to say, no one has ever suggested a different assignment or that it was not a pure valence shell transition. This excitation in ketones has been discussed in detail by Sidman [S35] and will not be mentioned here again.

As in the studies of the other chromophoric groups, that of the ketones is aided considerably by first understanding where the lower Rydberg excitations will come, and what their symmetries are (Section I.C-1). Following such an analysis, one finds that, as in the alcohols, the large difference in penetration energies of the carbon and oxygen atoms in ketones leads to an especially interesting but regular variation of the 3s, 3p, and 3d Rydberg term values with alkylation and/or halogenation. In spite of this regularity, the spectra are sufficiently complex, (Fig. IV.C-1) that one hesitates to state that there is a "ketone spectral pattern" into which all ketone vacuum-ultraviolet spectra can be fit. In fact, in the more heavily alkylated ketones, relatively low-frequency bands appear for which counterparts are difficult to find in the smaller species.

We first discuss the acetaldehyde spectrum as displaying many of the features found in other ketone spectra. According to Walsh [W5], the acetaldehyde spectrum exhibits three long Rydberg series, all converging to 82 500 cm^{-1} (advert.). The quantum defects and term values of the $n = 3$ members of each of the series are: 0.90, 27 500 cm^{-1}; 0.70, 22 300 cm^{-1}; and 0.20, 14 470 cm^{-1}, indicating that the series are ns, np, and

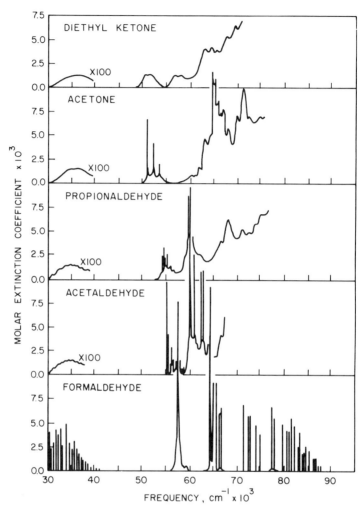

Fig. IV.C-1. Optical absorption spectra of several aldehydes and ketones in the gas phase [B10].

nd, respectively [L7]. Undoubtedly, the originating orbital is n_O, the oxygen lone-pair orbital. The $n_O \to 3s$ band origin of acetaldehyde (Fig. IV.C-2) is at 54 996 cm^{-1} (advert.) and is followed by a complicated vibronic structure which suggested to Walsh that there were *two* electronic transitions in the 55 000–60 000-cm^{-1} region. (As we shall see, the evidence for this is really quite good in the more heavily alkylated ketones.) Vibrational intervals of approximately 1200, 750, and 350 cm^{-1} are the most prominent in this band of acetaldehyde. Assignment

IV.C. ALDEHYDES AND KETONES

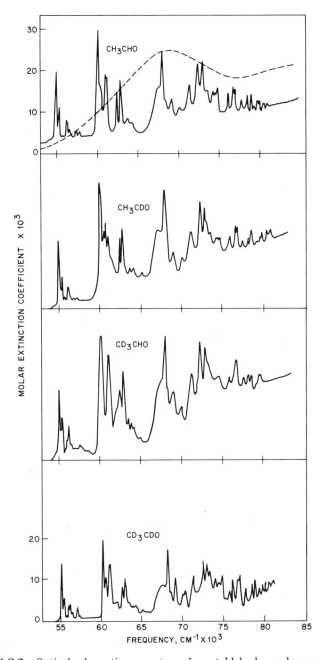

Fig. IV.C-2. Optical absorption spectra of acetaldehyde and several of its deuterated derivatives in the gas phase (solid lines). The spectrum of a solid film of acetaldehyde at 77°K is shown as the dashed curve [L38].

of the 1200-cm^{-1} interval is perplexing since the methyl group deformation, the

deformation, and C=O stretching motions [H20] are all candidates. However, Lucazeau and Sandorfy have studied the corresponding vibronic bands in various deuterated acetaldehydes (Figs. IV.C-2 and IV.C-3) and find the 1200-cm^{-1} intervals in CH_3CHO and CD_3CHO reduced to about 930 cm^{-1} in CH_3CDO and CD_3CDO, and have thereby inferred that the motion involves the

$$\begin{array}{c} C \\ O \diagup \diagdown H \end{array}$$

deformation [L38]. On the other hand, the 300–350-cm^{-1} mode seems indifferent to deuteration at any and all positions, and is thought to be the

$$\begin{array}{c} C \\ C \diagup \diagdown O \end{array}$$

deformation frequency.† Though the $n_O \rightarrow 3s$ excitation of acetaldehyde has an oscillator strength of only 0.037 in the optical spectrum [L1], it is the strongest band in the SF_6-scavenger spectrum [N1].

The strongest band in the acetaldehyde optical spectrum ($f = 0.13$ [L1]) commences at 60 170 cm^{-1} (advert.) and is clearly part of the $n_O \rightarrow 3p$ excitation, as can be seen from its term value (Table IV.C-I). Again the vibronic part of this band is complicated, and Walsh suggests two transitions are responsible. Of the remaining spectrum, almost three dozen additional transitions are assigned as members of the three Rydberg series, the only strong feature not so assigned being a strong doublet at 62 500 cm^{-1} (advert.), which is most likely a component of the $n_O \rightarrow 3p$ manifold or less possibly a valence shell excitation. All of the Rydberg transitions beyond 61 000 cm^{-1} seem to be strangely lacking in any vibronic structure. This is unexpected since the photoelectron band toward which the Rydberg excitations in this region of the spectrum are converging is itself highly structured. It is clear from later spectra [B10, L1, R19] that there is also an underlying continuous band centered at 61 000 cm^{-1} (vert.) in acetaldehyde.

† Since the Rydberg transition under discussion involves the excitation of an electron from the n_O orbital which is not insignificantly C—O antibonding, it is surprising that the C—O vibration is absent from the Rydberg vibronic envelopes of acetaldehyde and numerous other ketones as well.

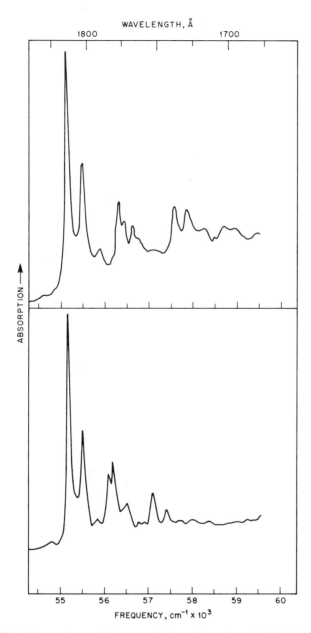

Fig. IV.C-3. Comparison of the $n_O \to 3s$ transitions in acetaldehyde (upper) and acetaldehyde-d_4 (lower), both in the gas phase [L38].

TABLE IV.C-I
RYDBERG TERM VALUES[a] IN THE KETONES

Molecule	Ionization potential	$\phi_i \to 3s$ Frequency	$(\phi_i, 3s)$ Term value	$\phi_i \to 3p$ Frequency	$(\phi_i, 3p)$ Term value	$\phi_i \to 3d$ Frequency	$(\phi_i, 3d)$ Term value
H_2CO	87 787	57 310	30 477	64 267	23 520	71 600	16 187
				65 660	22 127		
CH_3CHO	129 000	100 000	29 000	106 000	23 000	—	—
	82 504	54 996	27 508	60 170	22 334	68 030	14 474
				62 500(?)	20 000(?)		
C_2H_5CHO	80 890	54 702	26 190	59 221	21 670	68 000	12 890
$(CH_3)_2CHCHO$	79 410	53 496	25 910	57 880	21 530	64 880	14 530
				59 762	19 650		
$(CH_3)_2CO$	78 420	51 270	27 150	60 110	18 310	65 250	13 170
	112 000	84 500	27 500			99 000	13 000
	143 000	116 800	26 200				
$CH_3COC_2H_5$	76 890	50 800	26 090	58 800	18 090	—	—
$C_2H_5COC_2H_5$	76 040	51 300	24 740	58 310	17 730	64 200	11 840
$t-C_4H_9COCH_3$	75 660	51 680	23 980	57 390	18 270	62 500	13 160
$\overline{CH_2(CH_2)_2}CO$	77 110	51 710	25 400	58 070	19 040	63 500	13 600
$\overline{CH_2(CH_2)_3}CO$	74 730	50 059	24 670	56 490	18 240	60 950	13 780
$\overline{CH_2(CH_2)_4}CO$	73 780	50 748	23 030	55 992	17 788	60 750	13 030
$H_2C=CHCHO$	81 540	57 200	24 340	60 585	20 955	67 511	14 029
$CH_3CH=CHCHO$	78 800	55 600	23 200	—	—	—	—
$(CH_3)_2C=CHCOCH_3$	72 900	51 300	21 600	—	—	—	—
$OHC—CHO$	85 400	—	—	62 500(?)	22 900	—	—
$CH_3COCOCH_3$	76 870	—	—	57 295	19 575	—	—
F_2CO	109 700	76 000	33 700	—	—	—	—
$(CF_3)_2CO$	97 510	65 000	32 500	—	—	—	—
Cl_2CO	93 200	66 800	26 400	74 000	21 400	—	—
	95 400			76 000	19 400		
	101 600			81 514	20 086		
	109 000			89 040	20 000		
CH_3ClCO	89 850	62 890	26 960	69 440	20 410	—	—

[a] Term values in cm^{-1} (vert.).

Walsh reports that the first five members of the np Rydberg series exhibit a doublet splitting which decreases to zero by the sixth member, and that the early members of the nd series show a similarly decreasing doublet splitting. It is almost certain that this n-dependent splitting of the Rydberg terms is a reflection of the splitting of the degenerate np and nd upper levels by the aspherical ionic core, and it is unfortunate that Walsh did not specifically tabulate the values. The purported splitting of the first band in the ns series must have a different origin, of course, since the 3s orbital is nondegenerate.

As for assignments in the acetaldehyde spectrum, those of the Rydberg bands are most secure; in addition to the suggestive term values (Table IV.C-I), high-pressure experiments on acetaldehyde readily confirm the Rydberg nature of the $n_O \to$ 3s and $n_O \to$ 3p bands [R19], as does the behavior of these bands in the spectrum of a solid film at low temperature (Fig. IV.C-2) and in solution [S12]. Though the 54 996-cm^{-1} band is conceded by Walsh to be the $n = 3$ member of a Rydberg series, he claims it can also be classified as the $n_O \to \sigma^*$(C—O) valence shell excitation calculated to be in this spectral region by McMurry [M20]. However, if such a description were appropriate, one would expect the band envelope to be broad and nonvertical, like the 60 000-cm^{-1} band of water (Fig. III.E-2), in strong contrast to the sharp, vertical excitation observed. Barnes and Simpson instead assign the 54 996-cm^{-1} band of acetaldehyde to a valence shell $n_O' \to \sigma^*$ transition,† reserving the $n_O \to$ 3s assignment for the transition which we assign here as $n_O \to$ 3p [B10]. In a later study, Johnson and Simpson conclude that the 51 000-cm^{-1} band of ketones is polarized in plane and perpendicular to the C=O line, as is appropriate for an $n_O \to \sigma^*$ transition [J8]. Probably the best interpretation of the absorption beginning at about 53 000 cm^{-1} in acetaldehyde is that of Lucazeau and Sandorfy, who point out that there are weak bands preceding the origin at 54 996 cm^{-1} (Fig. IV.C-3) which could well belong to a second transition. In fact, of the 18 vibronic features in the 53 600–58 250-cm^{-1} region, six are assigned by them as $n_O \to$ 3s and 12 to a second transition, which they consider to be an $n_O \to \sigma^*$ valence shell excitation with a highly irregular vibronic pattern. As we continue this study of aldehyde and ketone spectra, several other spectra will be discussed which strongly suggest the presence of overlapped Rydberg and valence shell excitations in the neighborhood of 50 000–55 000 cm^{-1}; still there is no concrete demonstration of this. Lucazeau and Sandorfy similarly feel that there are valence shell $n_O \to \sigma^*$ excitations interleaved with the $n_O \to$ 3p excitation in the 59 000–65 000-cm^{-1} region, but it must be

† The n_O' MO ($5a_1$) in formaldehyde is the second lone pair on oxygen, aligned along the C—O axis, and lies about 40 000 cm^{-1} deeper than the n_O MO [B59].

remembered that this region will appear complex due to the lifting of the degeneracy of the 3p AOs by the molecular field.

Just where the $\pi \to \pi^*$ transition might be in acetaldehyde is another intriguing question. Walsh again assigns an acknowledged Rydberg excitation ($n_0 \to 3p$, 60 170 cm^{-1}) as being alternatively assignable as $\pi \to \pi^*$. This is clearly unacceptable since the two transitions originate with completely different orbitals leading to states of different symmetries, so that they could not be mixed, nor are they alternative descriptions of the same final state. In the spectrum of a solid film of acetaldehyde (Fig. IV.C-2), a broad maximum is observed at 69 000 cm^{-1} (vert.); if it is presumed that this is the $\pi \to \pi^*$ transition and that it has been shifted several thousand cm^{-1} to lower frequencies, as often happens to $\pi \to \pi^*$ bands in the solid phase, then one concludes that the $\pi \to \pi^*$ band of acetaldehyde must be present in the vapor spectrum as an underlying, continuous band centered at perhaps 73 000 cm^{-1} (vert.). Several other indirect lines of reasoning lead to this general frequency for the $\pi \to \pi^*$ transition in ketones. For example, in the sections on the spectra of the amide (Section V.A-1) and carboxylic acid groups (Section V.A-3), it was stated that the $\pi_2 \to \pi_3^*$ transitions in these chromophores were shifted to lower frequencies as compared with the $\pi_1 \to \pi_2^*$ transition in ketones. Since these fall at about 55 000 and 67 000 cm^{-1} (vert.), respectively, a $\pi \to \pi^*$ frequency of more than 70 000 cm^{-1} in ketones is to be expected.

Turning from acetaldehyde to the spectrum of formaldehyde (Figs. IV.C-1 and IV.C-4), Price [P36] reports two Rydberg series in this molecule, with the $n = 3$ members coming at 64 270 and 71 600 cm^{-1} (advert.) and quantum defects of 0.70 and 0.40. The originating orbital for these excitations is again the n_0 ($2b_1$) MO. These transitions are clearly analogous to the second and third series of acetaldehyde, and so are assigned as $n_0 \to n$p and $n_0 \to n$d [A6]. A lower frequency band (57 310 cm^{-1} advert.) is considered by Price to be a possible lower member of the nd series, but from Table IV.C-I and the work of Allison and Walsh [A6], it is seen instead that it is the $n = 3$ member of the expected ns series [L7, W19]. The oscillator strength of the $n_0 \to 3$s band has been measured photographically by Fleming *et al.* [F6] as 0.040, while later photoelectric and electron scattering measurements gave values from 0.03 to 0.038 (Table IV.C-II) [G8, M22, W19]. Fleming *et al.* also find this rather vertical excitation to display single vibrational quanta of 330, 750, 1545, and 2215 cm^{-1}. According to Weiss *et al.* the first two of these intervals are probably due to inversion doubling in a nonplanar ($2b_2$, 3s) upper state [W19]. Note that in the acetaldehyde $n_0 \to 3$s band, very nearly the same two frequencies appear, and the lower of the two was

IV.C. ALDEHYDES AND KETONES

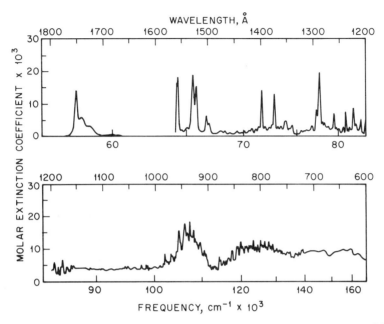

Fig. IV.C-4. Optical absorption spectrum of formaldehyde in the gas phase [M22].

TABLE IV.C-II
OSCILLATOR STRENGTHS OF THE RYDBERG BANDS OF KETONES[a]

Molecule	$n_O \to 3s$	$n_O \to 3p$
H_2CO	0.040	—
CH_3CHO	0.037	0.13
$(CH_3)_2CO$	0.035, 0.046	$\ll 0.035$
$C_2H_5COCH_3$	0.018	0.017
$n\text{-}C_3H_7COCH_3$	0.020	0.016
$iso\text{-}C_3H_7COCH_3$	0.015	0.043
$n\text{-}C_4H_9COCH_3$	0.015	0.018
$iso\text{-}C_4H_9COCH_3$	0.015	0.031
$sec\text{-}C_4H_9COCH_3$	0.023	0.15
$t\text{-}C_4H_9COCH_3$	0.013	0.072
$n\text{-}C_5H_{11}COCH_3$	0.011	0.022
$(C_2H_5)_2CO$	0.010	0.014
$(n\text{-}C_3H_7)_2CO$	0.008	0.012
$(iso\text{-}C_3H_7)_2CO$	0.001	0.018
$(t\text{-}C_4H_9)_2CO$	0.002	0.020
F_2CO	0.15	—

[a] From Reference [I7].

assigned as

deformation. The first photoelectron band of formaldehyde, corresponding to n_O ionization, is also extremely vertical [B2], displaying only one quantum each of ν_1' (2560 cm^{-1}, totally symmetric C—H stretch) and ν_2' (1590 cm^{-1}, C=O stretch), and possibly two quanta of ν_3' (1210 cm^{-1}, totally symmetric C—H deformation). Of the other Rydberg bands, Price reports that only the transitions to 3p and 3d show well-developed vibrational patterns, the former exhibiting 1250- and 1470-cm^{-1} intervals, and the latter exhibiting 1120- and 1260-cm^{-1} intervals. Additionally, the 3p and 4p bands are split by 1370 and 356 cm^{-1}, respectively, by the molecular ion's field [A6, M22], the former yielding a lower $3pb_2$ and an upper $3pa_1$ component. The transition to $3pb_2$ displays intervals of 58 cm^{-1} which are due to the inversion splitting, implying a nonplanar geometry in the upper state. The splitting of the $n_O \to np$ band origins by the molecular core is reminiscent of the splitting in the analogous bands of acetaldehyde, except that they are smaller, and there is not enough data to show that the formaldehyde splittings go to zero with increasing n, as they do in acetaldehyde. Oscillator strengths of several of the higher Rydberg bands of formaldehyde are tabulated by Mentall *et al.* [M22].

As with acetaldehyde, the $n_O \to 3s$ band of formaldehyde also has been assigned as $n_O \to \sigma^*$, valence shell [M20, P31, S3], and as $n_O' \to \pi^*$, valence shell [B10]. McMurry did consider the possibility of a Rydberg assignment, but since there were no further members reported on which to build a series, he decided against it. Barnes and Simpson [B10] assign the $\pi \to \pi^*$ absorption of formaldehyde to one or more of the lines near 73 000 cm^{-1}, realizing that the $n_O \to 3d$ and $n_O \to 4s$ Rydberg transitions were also in this region. Since this value of 73 000 cm^{-1} is quite close to the frequency postulated earlier for the $\pi \to \pi^*$ band in acetaldehyde, we feel it cannot be far wrong, being a little low, if anything. On the other hand, Pople and Sidman [P31] and Sidman [S34] assign the 64 270-cm^{-1} line of formaldehyde as $\pi \to \pi^*$ on the basis of theoretical calculations, but $n_O \to 3p$ seems more likely. In fact, Weiss *et al.* [W19] assign *all* of the sharp features below 90 000 cm^{-1} to Rydberg excitations. Actually, though several semiempirical lines of attack lead one to suspect that the $\pi \to \pi^*$ band of formaldehyde should come near 70 000 cm^{-1}, still it is difficult to find such a discrete absorption in the spectrum with the high oscillator strength expected. Quite possibly, the $\pi \to \pi^*$ band of formaldehyde is a very broad and diffuse one, expressing itself

IV.C. ALDEHYDES AND KETONES

as the continuous absorption in the 70 000–85 000-cm^{-1} region (Fig. IV.C-4). On the other hand, Mentall et al. have recently argued that the $\pi \to \pi^*$ transition is above the $2b_2$ ionization potential (87 710 cm^{-1} advert.), and, being strongly mixed with the continuum extending from the $2b_2 \to npb_2$ limit, is thereby so broadened by rapid autoionization as to be undetectable [M22].

Several transitions beyond the first ionization potential are reported in both the optical and electron scattering spectra of formaldehyde [M22, W19]. The strong features at 100 000–116 000 cm^{-1} have been shown to be strongly autoionizing and are assigned as $5a_1 \to 3p$, going to the third ionization potential ($5a_1$) at 129 000 cm^{-1} (vert.). Its term value of 22 000–24 000 cm^{-1} (vert.), depending upon where one takes the maximum, is acceptable for a 3p-terminating orbital, and the weaker band at $\sim$100 000 cm^{-1} must be the $5a_1 \to 3s$ transition. Note how similar the term values are for the $n = 3$ Rydberg states going to the first and third ionization potentials (Table IV.C-I). Weiss et al. have also tried to identify Rydberg transitions converging upon the second ionization potential ($1b_2$, 117 000 cm^{-1} vert.) but strangely, the $n = 3$ members of the ns, np, and nd series could not be located [W19].

Of the many calculations of the formaldehyde spectrum, that of Whitten and Hackmeyer [W22] is one of the best and most interesting. Using an extensive Gaussian orbital basis and selective configuration interaction on both ground and excited states, they first calculated the singlet and triplet (n_0, π^*) frequencies to within 2400 cm^{-1} of their observed band origins. Since Rydberg AOs were included in the basis set, they also calculated the $n_0 \to 3s$ (60 300 cm^{-1}) and $n_0 \to 3p$ (67 000 cm^{-1}) frequencies in quantitative agreement with the assignments proposed here. An estimate of 90 700 cm^{-1} was also made for the $\pi \to 3s$ Rydberg excitation. Taking the vertical pi ionization potential of Turner et al. [B2, B50] of 117 000 cm^{-1} (vert.), an expected term value of 30 400 cm^{-1} leads to an "experimental" value of $\sim$87 000 cm^{-1} for the $\pi \to 3s$ transition frequency, again in very good agreement with the calculation. Where, then, does the calculation predict the $\pi \to \pi^*$ transition to come? Whitten and Hackmeyer find that the $\pi \to \pi^*$ band of formaldehyde is predicted to come at 91 200 cm^{-1} with an oscillator strength of 0.40, but they do not assign it to any observed feature. Since such calculations tend to overestimate both the $\pi \to \pi^*$ frequency and oscillator strength, we expect the transition below 90 000 cm^{-1}, of moderate strength ($f \simeq 0.2$), and either continuous or showing a long progression of C=O stretching, but not vertical. Such a band remains to be found in formaldehyde. Peyerimhoff and co-workers [B67, P14] report further calculations on formaldehyde of the sort described by Whitten and Hackmeyer. Their results are much like those of the earlier calculation, correctly placing the $n_0 \to \pi^*$, $n_0 \to 3s$, and $n_0 \to 3p$ transitions, placing $\pi \to \pi^*$ at 92 000 cm^{-1}, and additionally predicting the two interesting transitions $n_0' \to \pi^*$ and $n_0 \to \sigma^*$ to come at 69 400 and 84 200 cm^{-1}, respectively. The latter excited state is said to be unbound along the C—O stretching coordinate.

As with the isoelectronic system of ethylene, there would appear to be a similar ambiguity in describing the $3d\pi$ and π^* MOs of formaldehyde and, as in ethylene, the opportunity exists that the π^* MO is strictly valence shell in one configuration and Rydberg in another. Certainly, it is valence shell in the (n_0, π^*) configuration. Whitten [W23] has investigated this problem for the (π, π^*) singlet state of formaldehyde and finds that π^* is very expanded in a low-level calculation that only optimizes π^* in the (π, π^*) configuration, but that extensive mixing with (σ, σ^*) configurations will reduce $\langle \pi^* | x^2 | \pi^* \rangle$ to 3.09 square Bohrs, compared to $\langle \pi | x^2 | \pi \rangle = 1.61$ square Bohrs in the ground state. The strong mixing comes from the fact that the (π, π^*) configuration is highly polar and the (σ, σ^*) configurations act to oppositely polarize the sigma core and thereby increase the binding of the π^* MO. Whitten's conclusion is that the π^* MO in the singlet (π, π^*) configuration is somewhat expanded compared to the π MO, but is much closer to a valence shell orbital than to a Rydberg orbital. This parallels the most recent conclusions on the related state of ethylene (Section IV.A-1).

Extension of the formaldehyde and acetaldehyde interpretations to the dialkyl ketones and cycloalkanones seems straightforward, though there are still some interesting intensity variations among these molecules which remain to be explained. In acetone (Figs. IV.C-1 and IV.C-5), the $n_0 \to 3s$ transition is found to begin at 51 270 cm^{-1} (advert., 27 150 cm^{-1} term) in connection with a short vibrational progression of 1194 ± 5 cm^{-1}, and an oscillator strength between 0.020 [H24] and 0.046 [L1]; Huebner et al. [H35] report an oscillator strength of 0.025 and Ito et al. [I7] report 0.035. A high-pressure experiment convincingly demonstrates the Rydberg nature of its upper state [R15]. The 1200-cm^{-1} vibration in acetone's $(n_0, 3s)$ excited state was first thought to be due to C=O stretching (v_3', 1710 cm^{-1} in the ground state), but Lawson and Duncan [L14] and Ito et al. [I7] find this frequency reduced to 906 ± 5 cm^{-1} in the 51 000-cm^{-1} band of $(CD_3)_2CO$ (Fig. IV.C-7) and assign it instead to v_4', the methyl group deformation motion [1364 cm^{-1} in the $(CH_3)_2CO$ ground state]. The excitation of the methyl group deformation upon excitation of an electron in what can loosely be called "the other end of the molecule" was encountered earlier in the methyl iodide spectrum (Section III.B-1) [W5]. Now in the $(n_0, 3s)$ state, the methyl group may enter into the wave function either through the n_0

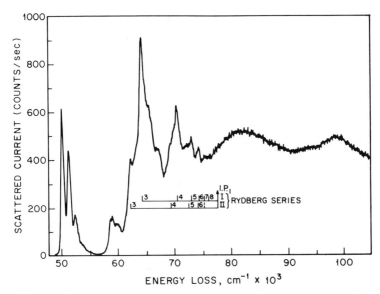

Fig. IV.C-5. Electron-impact energy-loss spectrum of acetone, taken with 100 eV incident energy and at $\theta = 0°$ [H35].

orbital or through the 3s orbital. However, in the n_O photoelectron transition, only the n_O orbital is involved, and since the vibrations in this transition also show the identical isotope shifts, it is clear that the n_O "lone-pair" wave function has density in the methyl groups of acetone.

As shown in Fig. IV.C-5, Huebner et al. [H35] have found two Rydberg series going to the lowest ionization potential with $\delta = 0.03$ and $\delta = 0.315$; they are probably components of the $n_O \to nd$ series. Beyond the first ionization potential of acetone, broad maxima are found at 84 500, 99 000, and 116 800 cm^{-1} (vert.). These bands can be assigned as Rydberg excitations, as shown in Table IV.C-I.

Johnson and Simpson [J8] describe an interesting experiment designed to reveal the polarization of the 51 000-cm^{-1} band of ketones. The experiments were performed with single crystals of the long-chain ketone $C_8H_{17}COC_8H_{17}$, in which the C_α–$C_{\alpha'}$ direction of the molecules is parallel to the crystallographic c axis. The optical density of the 54 000-cm^{-1} band† was recorded using unpolarized light propagating along the c direction, and the crystal was then melted and the optical density redeter-

---

† As shown in Fig. IV.C-6, the 51 000-cm^{-1} band of ketones suffers a large shift to higher frequencies in the condensed phase. Thus the 54 000-cm^{-1} frequency quoted here for the long-chain crystalline ketone would undoubtedly come several thousand cm^{-1} lower for the molecule in the gas phase.

mined. Such randomization experiments showed that in every case the intensity of the 54 000-cm^{-1} band significantly *increased* on melting. Though there are complications due to single-molecule vibronic mixing, hyperchromism, and hypochromism, the simple qualitative result leads to a *c*-axis, in-plane, perpendicular (C_α–$C_{\alpha'}$) polarization for this band, as predicted for an $n_O \to \sigma^*$ assignment, regardless of the nature of σ^*.

One perplexing feature of the acetone spectrum must be mentioned. Though considerations of both term values and the high-pressure effect argue strongly that the 51 270-cm^{-1} band of acetone is a Rydberg excitation, the spectra of neat acetone liquid [L22] and of acetone in several solvents [L22, L23, P25, T20] show a broad, smooth band at about 54 000 cm^{-1} (Fig. IV.C-6). As can be seen from the other spectra in Fig. IV.C-6 and by comparing the gas-phase spectra of references [T20] and [P25], the general effect of a condensed phase on the 51 000-cm^{-1} transition of ketones is to obliterate all vibronic structure, shift the frequency maximum to about 53 000 cm^{-1}, and decrease the maximum molar extinction coefficient by 50% or more. Thus Barnes and Simpson [B10] report that in the spectrum of diethyl ketone, the 50 800-cm^{-1} band in the gas phase has a molar extinction coefficient of 650, whereas this decreases to 60 in isopentane solution. The appearance of the $n_O \to 3s$ band of ketones in condensed phases is contrary to the postulate that Rydberg excitations do not appear in such phases (Section II.C), and two explanations spring readily to mind.

First, it is possible that the Rydberg absorption at approximately 51 000 cm^{-1} in the gas-phase spectrum is obliterated in hexane solution, thereby revealing an underlying $n_O \to \sigma^*$ valence shell excitation. This explanation is especially appealing when one considers molecules such as methyl isopropyl ketone (Fig. IV.C-6), in which the 50 000–54 000-cm^{-1} region shows two pronounced transitions. The presence of two overlapped bands in this same region can easily be imagined in the spectra of other highly alkylated ketones (Figs. IV.C-7 and IV.C-8), and careful inspection of the formaldehyde spectra of Gentieu *et al.* [G8] does show a region of nonzero absorption (59 000–62 000 cm^{-1}) which is distinct from the more intense $n_O \to 3s$ band. Remember, too, that earlier investigators had surmised that the $n_O \to 3s$ region of acetaldehyde consisted of two transitions. The explanation that both a structured Rydberg transition and a weaker, continuous valence shell transition are present at 51 000 cm^{-1} in the gas phase, but only the latter is present in a condensed phase, would be refuted if a condensed-phase spectrum were reported that showed the gas-phase vibronic structure, but no such evidence has been uncovered. On the other hand, the observed polarization of the 54 000-cm^{-1} band in a condensed phase and its diminished intensity with

IV.C. ALDEHYDES AND KETONES

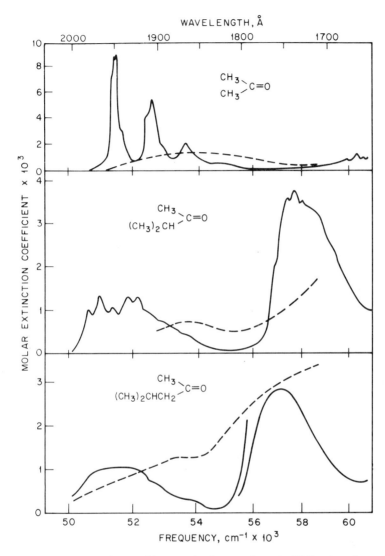

Fig. IV.C-6. Spectra of several ketones in the gas phase (solid lines) and as solutes in paraffin solvents (dashed lines).

respect to the gas-phase spectrum are in agreement with the postulate stated above. Circular dichroism spectra of optically active ketones in the gas phase may be of great value in demonstrating the presence of two bands at 50 000–55 000 cm^{-1}.

A somewhat less tenable argument for the seeming appearance of the

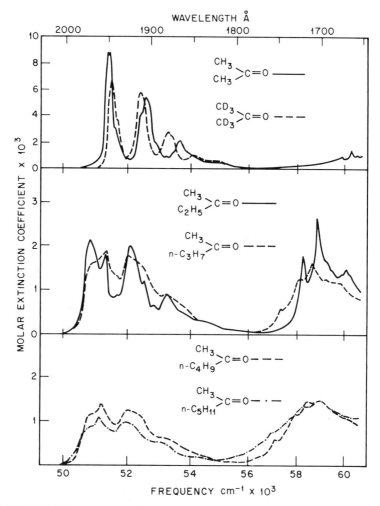

Fig. IV.C-7. The $n_o \to 3s$ and $n_o \to 3p$ regions of several ketones in the gas phase [17].

$n_o \to 3s$ Rydberg excitation of ketones in condensed phases is that the transition observed can no longer be described as $n_o \to 3s$, but instead is a transition to the $n = 2$ Wannier exciton state. As such, it is no longer an excited state of the ketone molecule, but instead belongs to the solvent and solute taken as a supermolecule. A similar explanation has been used already for the appearance of "Rydberg" bands in rare gas and CF_4 matrices doped with organic molecules (Section II.C).

The second ultraviolet transition of the larger ketones can be assigned

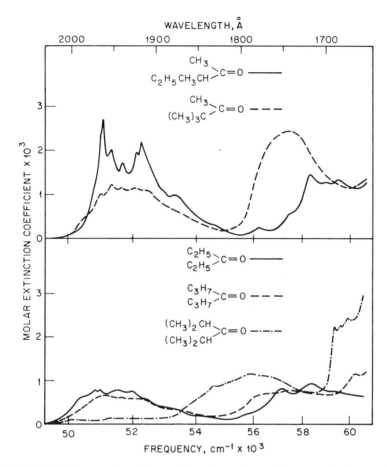

Fig. IV.C-8. The $n_O \to 3s$ and $n_O \to 3p$ regions of several ketones in the gas phase [I7].

as $n_O \to 3p$, just as readily as the first was assigned as $n_O \to 3s$. Again, since the n_O ionization potentials and (n_O, 3p) term values are fairly constant in ketones, so then are the $n_O \to 3p$ frequencies. In the simple ketones, the $n_O \to 3p$ transition comes between 55 000 and 60 000 cm^{-1}, with the monoalkyl ketones (aldehydes) in general coming at the high end [H30] and the cycloketones and dialkyl ketones coming at the intermediate and lower frequencies (Table IV.C-I) [D24, H24, H30, P51, U1]. As with $n_O \to 3s$, the $n_O \to 3p$ transition shows little evidence of the excitation of C=O vibrations in the upper state, indicating that the C=O bond length is essentially unaltered in the transition.

Spectroscopic data on a wide variety of ketones are compiled in Table

IV.C-I, the (n_O, 3s) term values of which are especially interesting. We note first that the CH_2 group of formaldehyde is acting much like the CH_3 group of methanol in reducing the (n_O, 3s) term value of the oxygen atom (36 000 cm^{-1}) to below 31 000 cm^{-1}. Thus, though formaldehyde is the parent compound in the ketone series, its CH_2 group already decreases the (n_O, 3s) term appreciably, so that further alkyl groups will have a much smaller effect here than in water, the parent oxide molecule. Addition of alkyl groups to formaldehyde further decreases the (n_O, 3s) term value, reaching a limiting value of about 23 000 cm^{-1}, just slightly above the alkyl limit as observed in oxides (Fig. III.E-7), amines (Fig. III.D-4), sulfides (Fig. III.F-6), and olefins (Fig. IV.A-13). It is also clear from this table that the regularity of (n_O, 3s) term values will allow one to reliably interpolate ionization potentials after observing just the $n_O \to$ 3s absorption frequency, rather than several consecutive members of a converging Rydberg series. In contrast to the (n_O, 3s) term values, those of the (n_O, 3p) and (n_O, 3d) states are much more constant at 20 000 and 13 000 cm^{-1}, respectively. Substantiation of the claim that the term value in a molecule is independent of the originating MO (Section I.C-1) can be found in Table IV.C-I.

Interestingly, vibrational analyses of the $n_O \to$ 3s bands in cyclobutanone [U1], cyclopentanone [P51], and cyclohexanone [P51] show that the C=O vibrational frequency is absent, and that the transition is very nonvertical in cyclobutanone, but becomes much more vertical in the unstrained cyclopentanone and cycloheptanone ring systems [U1, H30]. The $n_O \to$ 3s bands of methyl ethyl ketone, methyl isopropyl ketone, methyl n-propyl ketone, and methyl sec-butyl ketone all show excitation of a 1200–1300 cm^{-1} vibrational motion which is probably the methyl group deformation, as in acetone, rather than the C=O stretch [D24, I7]. A 470-cm^{-1} progression is also reported for the (n_O, 3s) band of methyl ethyl ketone [I7] and is thought to be due to the

bending motion.

The electronic spectrum of cyclobutanone is very complicated from the point of view of the large number of vibrations which accompany the excitations, but the transitions in general seem readily assignable from their term values. A transition containing at least 40 vibrational bands, many of them hot, is found at 51 710 cm^{-1} (vert.) with an oscillator strength of approximately 0.03 [W21]. Using the photoelectron value of 77 110 cm^{-1} for the vertical n_O ionization potential, a totally reasonable value of 25 400 cm^{-1} is found for its term, showing that the transition is $n_O \to 3s$. This transition and several higher ones as well involve extensive excitation of a 1124-cm^{-1} mode which Whitlock and Duncan assign as C=O stretching, reduced from 1816 cm^{-1} in the ground state. As in many of the other ketones, a more reasonable assignment of this vibration in cyclobutanone would be to a totally symmetric CH$_2$ deformation, which has a frequency of 1499 cm^{-1} in the ground state. A smooth band at $\sim$58 100 cm^{-1} (vert.) having an oscillator strength of about 7×10^{-3} is the $n_O \to 3p$ transition by virtue of its 19 040-cm^{-1} (vert.) term value. A region of strong absorption follows this with a maximum intensity at 63 500 cm^{-1} and broad vibronic bands which were analyzed as those of two overlapping electronic transitions. The term value of 13 600 cm^{-1} (vert.) suggests that this is the $n_O \to 3d$ complex. Following this, a series of doubled bands forms a Rydberg progression going to 75 400 cm^{-1} with $\delta = 1.0555$. It is clearly an ns series, but it is strange that it converges to the adiabatic ionization potential rather than the vertical. Rydberg assignments in the spectra of cyclopentanone and cyclohexanone (Table IV.C-I) parallel those given for cyclobutanone.

There is a most interesting relationship between the $n_O \to 3s$ and $n_O \to 3p$ oscillator strengths in certain series of ketones. In the cyclic ketones, Udvarhazi and El-Sayed [U1] point out that $n_O \to 3s$ is about twice as strong as $n_O \to 3p$ in cyclobutanone, about equal in cyclopentanone, and only about half as strong in cyclohexanone, the sum of the intensities of the two bands remaining approximately constant in the series. Holdsworth and Duncan [H24] and Ito et al. [I7] report a similar reciprocity of $n_O \to 3s$ and $n_O \to 3p$ oscillator strengths in the dialkyl ketones (Table IV.C-II). As with the cyclic ketones, systematic substitution of both methyl groups of acetone by larger alkyl groups results in a dramatic and monotonic diminution of the $n_O \to 3s$ oscillator strength. Thus in acetone, the $n_O \to 3s$ transition is much stronger than $n_O \to 3p$ (which is barely visible, Fig. IV.C-6), whereas in diethyl and di-n-propyl ketone, they are of approximately equal intensity, and in diisopropyl and di-t-butyl ketones, the $n_O \to 3p$ transition is ten or more times more intense than $n_O \to 3s$. The intensity trends in the alkyl methyl ketones parallel those of the corresponding dialkyl ketones; ethyl methyl and n-propyl methyl ketones have $n_O \to 3s$ and $n_O \to 3p$ transitions of approximately equal intensity, whereas in isopropyl methyl and isobutyl methyl ketones, the $n_O \to 3p$ bands are several times stronger than the $n_O \to 3s$ bands (Fig. IV.C-6). Because the intensities of these Rydberg transitions will be dependent upon the geometric disposition of the alkyl groups, the fact that these ketones exist at room temperature in the gas phase as a mixture of rotational isomers [H22, S33] will severely complicate a detailed explanation of the data in Table IV.C-II. An even more important complication involves the proposition that there are $n_O \to \sigma^*$ valence shell transitions underlying both the $n_O \to 3s$ and $n_O \to 3p$ bands, as suggested by Lucazeau and Sandorfy [L38]. In fact, this may be the factor behind a large part of the regular interchange of intensity in the two regions. For example, looking at the spectra in Fig. IV.C-7, it is quite clear that the $n_O \to 3s$ and $n_O \to 3p$ frequencies remain constant in the series of methyl alkyl ketones, but that as intensity is lost in the 51 000-cm^{-1} region, another band seems to make an appearance at about 56 500 cm^{-1}, between the $n_O \to 3s$ and $n_O \to 3p$ bands. It is quite possible that the changing of the alkyl group shifts valence shell $n_O \to \sigma^*$ intensity out of the 51 000-cm^{-1} region and into a second $n_O \to \sigma^*$ band in the 56 500-cm^{-1} region. A similar explanation could also hold for the intensity changes in the dialkyl ketones of Fig. IV.C-8, the second $n_O \to \sigma^*$ band coming at about 55 000 cm^{-1} in these cases.

The observations described here are particularly important for the work of Meyer et al. [M30], who have predicted that in molecules containing both keto and cyclopropyl groups, there will be a cyclopropyl $\to \pi^*(C{=}O)$ charge transfer transition of moderate strength ($f = 0.0$–0.08) lying between the $n_O \to \pi^*$ and $\pi \to \pi^*$ excitations. In their spectral investigation of such compounds as cyclopropyl methyl ketone, they claim to have found the charge transfer band resting between the $n_O \to 3s$ and $n_O \to 3p$ bands ($\sim 55\,600$ cm^{-1} vert.), but it seems that such a band can also appear in materials not containing the cyclopropyl group (cf. Fig. IV.C-9).

It should be noted that the apparent redistribution of $n_O \to 3s$ and $n_O \to 3p$ intensity in the ketones, beginning with acetone, in which $n_O \to 3p$ is extremely weak, does not hold for the less symmetric aldehydes. Thus in acetaldehyde and the higher members of its series (Figs. IV.C-1 and IV.C-9), $n_O \to 3p$ is a strong feature in each, and the relative intensities of $n_O \to 3s$, $3p$, and $3d$ remain constant in the series. This strongly suggests that the mixing in the dialkyl ketones is controlled by symmetry factors.

There is an f-sum rule on the oscillator strengths of the transitions originating at one level, say n_O,

$$\sum_m f_{n_O \to m} = 2, \qquad (IV.6)$$

which can be rewritten as

$$\sum_{vs} f_{n_O \to vs} + \sum_R f_{n_O \to R} = 2, \qquad (IV.7)$$

where the first sum is over valence shell transitions and the second is over Rydberg transitions. If, as seems highly likely, there is a relatively con-

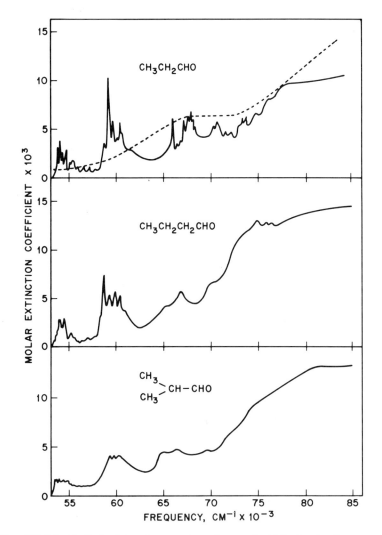

Fig. IV.C-9. Optical absorption spectra of several aldehydes in the gas phase [L38]. The dashed curve is the spectrum of solid propionaldehyde at 77°K.

stant mixing between valence shell states and Rydberg states induced by systematically increasing the size of the alkyl groups on an ultraviolet chromophore such as C=O, then there will be an f-sum rule on the Rydberg transitions

$$\sum_R f_{n_O \to R} = \text{const} < 2 \qquad (IV.8)$$

and the constant will apply equally to all molecules in the series. The bulky alkyl groups, however, will be quite effective in mixing the lower Rydberg orbitals among themselves via a matrix element of the sort $\langle n_O, 3s|V|n_O, 3p\rangle$. Now in the oxygen atom, the 2p → 3p transition is forbidden, whereas 2p → 3s is allowed. Thus in the ketones, the mixing of the 3s and 3p upper orbitals by the alkyl groups will act to intensify the n_O → 3p transition at the expense of the n_O → 3s, just as observed. Of course, if it were only a two-level system, the intensities of the n_O → 3s and n_O → 3p transitions could only approach equality under the strongest perturbation. That n_O → 3p rapidly becomes stronger implies that much of the n_O → 3s intensity is being mixed into Rydberg states higher than $(n_O, 3p)$ and/or that $(n_O, 3p)$ is also gaining intensity from higher (n_O, ns) states. A similar effect where the intensity of the n_+ → 3p Rydberg transition in the azoalkanes diminishes as the pendant alkyl groups increase in size is discussed in Section IV.B.

Dunn has pointed out that of the two atomic transitions p → s and p → d, the latter, having $\Delta l = +1$, will be much more intense than the former, which has $\Delta l = -1$ [D27]. The application of this rule to molecular spectra must be viewed with qualification, for in the cyclic ketones [P51] and in acetone, the n_O → 3d transition ($\Delta l = +1$) is the strongest in the spectrum, but this is not true for either formaldehyde or acetaldehyde (Fig. IV.C-1). Weak or strong, the n_O → 3d transitions will be found in the 60 000–65 000-cm^{-1} region of the spectra of the dialkyl ketones (Table IV.C-I).

Feinleib and Bovey [F3], Johnson [J9], and Schnepp et al. [S18] have reported the circular dichroism spectrum of the optically active ketone (+)-3-methyl cyclopentanone out to 60 000 cm^{-1} and beyond in the vapor phase. Besides the n_O → π^* band at 31 000 cm^{-1} (vert.), electronic origins were found in absorption at 50 200, 56 000, and 60 200 cm^{-1} (Fig. IV.C-10). As explained before, these latter absorptions are n_O → 3s, 3p, and 3d, respectively. The n_O → π^* and n_O → 3d transitions have positive ellipticity, with the Rydberg transition being approximately ten times stronger. Schnepp et al. also found both a weak positive and a weak negative band in the n_O → 3p region which may well be components of the 3p manifold. The splitting is 2000 cm^{-1}. So far, no explanation of the signs of these Rydberg transitions has been offered, nor are there assignments for the bands observed by Johnson in the 60 000–74 000-cm^{-1} region. The circular dichroism spectra of several n_O → 3s bands of optically active ketones are tabulated in reference [K20].

The far-ultraviolet spectra of both F_2CO and Cl_2CO have been reported. In F_2CO, two very complicated weak bands appear at 42 000–56 000 cm^{-1} and at 56 000–62 000 cm^{-1}, with oscillator strengths

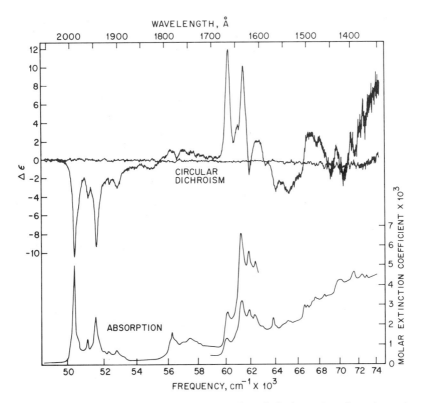

Fig. IV.C-10. Circular dichroism (upper) and optical absorption (lower) spectra of (+)-3-methyl cyclopentanone in the gas phase [J9].

of 3×10^{-4} and 5×10^{-4}, respectively [W32]. The first transition contains over 140 vibronic members, among which a vibrational interval of 950–1020 cm^{-1} (ν_1' the totally symmetric C=O stretch) is very prominent. The second weak band similarly displays a large number of 940–1030-cm^{-1} spacings. As with the first band, vibronic assignments are undoubtedly complicated by excitations originating with vibrationally hot molecules. Two other transitions are found in F$_2$CO; one beginning at 65 597 cm^{-1} (adiab.) consists of four intervals of 780 cm^{-1} (ν_2', the totally symmetric bending frequency) and has an oscillator strength of 1×10^{-3}, whereas the other is continuous, with a maximum at 76 000 cm^{-1} (vert.) and an oscillator strength of 0.15.

Gaussian orbital calculations and the photoelectron spectrum of F$_2$CO both show quite clearly that the n$_O$ orbital (110 100 cm^{-1} vert.) and the C=O pi-bonding MO (112 900 cm^{-1} vert.) are very close in energy with n$_O$ uppermost [B59], whereas in formaldehyde, n$_O$ is about 32 000 cm^{-1} above the pi-bonding MO [W2]. The differences of the MO separations in these molecules provide another example of the "perfluoro effect," in which the substitution of hydrogens by fluorines results in a stabilization of the sigma levels by about 20 000–25 000 cm^{-1}, whereas the pi levels are not shifted at all. In the virtual orbital manifold of F$_2$CO, two low-lying MOs are predicted, π^*, the C=O antibonding MO, and $\sigma^*(9a_1)$, a sigma-antibonding MO with a very high density on oxygen.

Past experience with heavily fluorinated molecules leads us to expect that the $n_O \to 3s$ Rydberg term value in carbonyl fluoride will be in the 30 000–35 000 cm^{-1} range. The first vertical ionization potential of carbonyl fluoride is 109 700 cm^{-1} (vert.), as measured by photoelectron spectroscopy [B59], so that the term values of the four transitions observed at 47 500, 59 200, 67 900, and 76 000 cm^{-1} (vert.) are 62 200, 50 500, 41 800, and 33 700 cm^{-1} (vert.), respectively. Thus only the last two need to be considered as $n_O \to 3s$. On fluorination, we expect the 30 400-cm^{-1} term of formaldehyde to increase, but to remain below that of hexafluoroacetone ($\sim$36 000 cm^{-1}). Of the two choices for F$_2$CO, it seems clear that the 33 700-cm^{-1} term value is to be preferred as that for the (n_O, 3s) Rydberg state. Its oscillator strength is reported as "about 0.15," whereas 0.08 is the upper limit for such Rydberg excitations.

It seems most in keeping with past experience in ketone spectra that the first band at 42 000–56 000 cm^{-1} be assigned as $n_O \to \pi^*$, though the possibility that this band comes instead at 56 000–62 000 cm^{-1} in carbonyl fluoride cannot be definitely ruled out. Working on the assumption that the 42 000-cm^{-1} band is $n_O \to \pi^*$, and that only one such band is possible in F$_2$CO, the very similar valence shell band at 56 000 cm^{-1} must be assigned as either $n_O \to \sigma^*(9a_1)$ or as $\pi(2b_2) \to \sigma^*(9a_1)$. Both of these transitions are predicted to be weakly allowed. The presence of such low-lying bands terminating at σ^* in carbonyl fluoride is related to the presence of similar bands in the fluoroethylenes, Section IV.A-3. The $\pi(2b_2) \to 3s$ Rydberg excitation should come approximately 34 000 cm^{-1} below the $\pi(2b_2)$ ionization potential of 117 800 cm^{-1} (vert.), which places it just before the $n_O \to 3p$ band expected at 89 000 cm^{-1}.

In general, the valence shell transitions below 90 000 cm^{-1} in perfluorinated molecules seem to be much weaker than their perhydro counterparts. Were this true for F$_2$CO, then its $\pi \to \pi^*$ transition might well have an oscillator strength as small as 0.15, in which case the 78 000-cm^{-1} band could be assigned as $\pi \to \pi^*$ rather than $n_O \to 3s$. Lacking any further evidence to support this supposition, one must conclude that once again the $\pi \to \pi^*$ excitation cannot be definitely identified in the spectrum of a ketone. Several bands in the 47 000–56 000 cm^{-1} region originally assigned as part of the $n_O \to \pi^*$ transition of F_2CO [M42] have instead been shown to be the Cameron bands of CO [W33].

A close similarity of the spectra of F_2CO and Cl_2CO is not expected, and one is not disappointed in this regard. La Paglia and Duncan [L6] report six distinct bands in the vacuum-ultraviolet spectrum of Cl_2CO, but no Rydberg series and no vibrational intervals which could be assigned to the C=O stretch in the upper state. The first ultraviolet band, broad and featureless, comes at 65 000 cm^{-1} (vert.) (Fig. IV.C-11), with a vertical term value of 30 400 cm^{-1}, the first ionization potential in phosgene being 95 400 cm^{-1} (vert.) [C10]. Inasmuch as the $(n_O, 3s)$ term value in formaldehyde is 30 400 cm^{-1} and we expect that the substitution of hydrogens by chlorines will reduce this by the same amount as substituting hydrogens by methyl groups, i.e., to the term value of acetone (27 150 cm^{-1} vert.), the 65 000-cm^{-1} band would seem not to be $n_O \to 3s$, by virtue of its larger term value (30 400 cm^{-1}). Also, its molar extinction coefficient of 20 000 at the maximum and its width suggest that it is far too intense for a Rydberg excitation. It is also too intense to be an A band (Section III.B-1), for these have molar extinction coefficients which are usually more than a factor of ten smaller. Thus the most likely

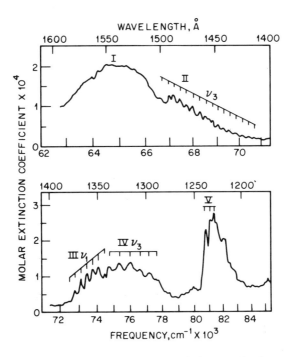

Fig. IV.C-11. Optical absorption spectrum of phosgene in the vapor phase.

assignment of the 65 000-cm^{-1} band on the basis of intensity and frequency is as a $\pi \to \pi^*$ transition. To date, this is the only distinct possibility for a $\pi \to \pi^*$ assignment in a ketone, except for the $\pi \to \pi^*$ assignment in acetyl chloride at 67 570 cm^{-1} (vert.) (Section V.A-3).

The first Rydberg transition in phosgene is more likely the weaker set of structured bands in the 67 000–71 000-cm^{-1} region (Fig. IV.C-11). It is difficult to decide on the vertical frequency of this transition, but the adiabatic value of 66 700 cm^{-1} has a term value of 26 500 cm^{-1} (adiab.) with respect to the adiabatic ionization potential at 93 200 cm^{-1} [C10, T9], just as expected for a transition to 3s. The optical band consists of a 17-member progression of the

bending frequency ν_3', with an apparent origin at 66 707 cm^{-1} and a 267-cm^{-1} interval, according to La Paglia and Duncan. However, the length of this progression is not consonant with the proposition that the excited-state frequency is only 18 cm^{-1} less than that of the ground state, and Herzberg's suggestion that it is ν_2' which is excited ($\nu_2'' = 567$ cm^{-1}) sees more likely. Chadwick et al. [C10] assign the uppermost orbital in phosgene to the oxygen lone pair ($2b_2$ in their notation) and find its ionization is accompanied by the excitation of quanta of 285 cm^{-1} frequency. However, since the oxygen and chlorine lone-pair ionizations are badly overlapped at 88 000–96 000 cm^{-1}, this orbital assignment is somewhat tentative. The corresponding $2b_2 \to 3p$ transitions in phosgene come at 74 000 and 76 000 cm^{-1} (vert.), leading to term values of 21 400 and 19 400 cm^{-1}, respectively. Again there is appreciable excitation of low-frequency vibrations, the frequencies being 307 and 587 cm^{-1}. Herzberg [H20] assigns all of these vibrations to ν_2'.

The most intense band in the spectrum of phosgene is somewhat vertical, with three prominent features at 80 816, 81 185, and 81 514 cm^{-1}. This band (or bands) is strongly reminiscent of the D bands of the alkyl chlorides (Section III.B-1), which were assigned as chlorine $3p \to 4p$. Since a term value of 20 000 cm^{-1} is observed for the D bands, the observed frequency of 81 000 cm^{-1} in Cl_2CO implies a chlorine 3p ionization potential at 101 000 cm^{-1} (vert.). Such a band is found in the photoelectron spectrum at 101 600 cm^{-1} (vert.), so it seems most likely that the transition originates with a chlorine 3p orbital and terminates at a 4p orbital. The 79 000–83 000-cm^{-1} region is also that expected for the $2b_2 \to 3d$ transitions and the transition to 3s from the fourth orbital (a_2) having an ionization potential of 109 000 cm^{-1} (vert.). A sixth band is reported by La Paglia and Duncan to come at 89 040 cm^{-1} (vert.)

IV.C. ALDEHYDES AND KETONES

and it is probably the D band corresponding to the 109 000-cm^{-1} ionization.

The other haloketones require only brief comment. Fluorination of the methyl groups of acetaldehyde and acetone is seen to raise the $(n_O, 3s)$ term value by several thousand cm^{-1}, as expected from the higher term value of the fluorine atom. The sharp feature at 72 140 cm^{-1} in the spectrum of ClCH$_2$CHO (Fig. IV.C-12) is most likely an analog of the alkyl

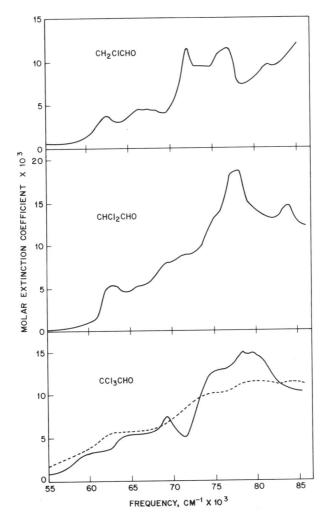

Fig. IV.C-12. Optical absorption spectra of the chlorinated acetaldehydes in the gas phase (solid lines) and as a solid film at 77°K (dashed line) [L38].

chloride D bands (Section III.B-1) which are assigned here as chlorine $3p \to 4p$ and which have term values of 20 000 cm^{-1}. Thus there must be a chlorine 3p ionization potential at 92 200 cm^{-1} (vert.) in this molecule. It is not at all obvious where this band has gone in the dichloromethyl and trichloromethyl acetaldehyde spectra. The 57 000–67 000-cm^{-1} absorption region in the chlorinated acetaldehydes will be complex, because in such systems, both the chlorine $3p \to \sigma^*$(C—Cl) valence shell A bands and the $n_0 \to 3s$ Rydberg excitations fall in this region. In the solid-film spectrum of trichloroacetaldehyde (Fig. IV.C-12), the Rydberg excitations are exorcised, and one component of the A band is observed at 58 000 cm^{-1} (vert.), which is the same frequency at which it appears in chloroform (Fig. III.B-7). The two bands at 75 000 and 80 000 cm^{-1} (vert.) in the solid-film spectrum probably correlate with the σ(C—Cl) $\to$ σ^*(C—Cl) valence shell transitions found at 72 500 and 78 500 cm^{-1} in the spectrum of liquid carbon tetrachloride (Section III.B-2) and the plateau at 65 000 cm^{-1} (vert.) could be the $\pi \to \pi^*$ excitation within the C=O group.

The electronic spectrum of carbonyl cyanide (CN)$_2$CO, has a broad, vibronically structured band centered at 52 700 cm^{-1} (vert.) with $f = 0.07$, and rising absorption from 63 000 to 78 000 cm^{-1} [D25]. This first band closely resembles in intensity and frequency the $n_0 \to 3s$ Rydberg excitation in the alkyl ketones; however, the first ionization potential of carbonyl cyanide is 101 300 cm^{-1} (advert.) [T9], thereby guaranteeing that the band observed in the optical spectrum is valence shell by virtue of its term value. Possibly, it is related to the valence shell excitation found at this frequency in several alkyl nitriles (Section IV.E).

The ketoolefins will be considered next, to be followed by the 1,2-diones. Walsh [W2] reports the vacuum-ultraviolet spectra of acrolein CH$_2$=CHCHO, crotonaldehyde CH$_3$CH=CHCHO, and mesityl oxide

$$(CH_3)_2C=CH\overset{\overset{O}{\|}}{C}CH_3$$

The first two are believed to be trans, whereas mesityl oxide is cis about the C$_2$—C$_3$ bond. The first ultraviolet band of acrolein (strong, broad, 51 600 cm^{-1} vert.), like that of butadiene at 47 800 cm^{-1}, has a term value of 29 900 cm^{-1}, at first suggesting a 3s upper state. However, it seems certain that this band does not correlate with those of the same term value in the simpler ketones [L7] (Table IV.C-I), but instead is the N $\to$ V$_1$ ($\pi_2 \to \pi_3^*$) transition (see Section V.C-1). The molar extinction coefficient of this band of crotonaldehyde in hexane solution (38 000) confirms that it is N $\to$ V$_1$ rather than the lowest Rydberg excitation. In crotonaldehyde, the N $\to$ V$_1$ band moves to 49 200 cm^{-1}

(vert.), and comes at 41 700 cm^{-1} (vert.) in mesityl oxide. The valence shell nature of this band is also confirmed by the large shift to lower frequencies displayed by the 49 200-cm^{-1} band of crotonaldehyde in hexane solution [H12]. In the trans ketoolefins the N → V$_1$ frequency is higher than in the corresponding trans dienes, suggesting that the keto-group $\pi \to \pi^*$ excitation frequency is much higher than that of the corresponding olefin (propylene, 58 000 cm^{-1} vert.), and that the N → V$_1$ transition in the ketoolefins is largely localized within the C=C group. This would also explain why substitution of an alkyl group on the olefinic part of the molecule results in a larger shift of the N → V$_1$ band to lower frequencies (CH$_3$CH=CHCHO, 49 200 cm^{-1} vert.) than when the substitution is on the keto part (CH$_2$=CHCOCH$_3$, 51 300 cm^{-1} vert.) [H30].

Nagakura has studied this N → V$_1$ transition from a different viewpoint, that of the intramolecular charge transfer theory (see Section I.B-2) [N3]. According to this theory, the upper state of the 51 600-cm^{-1} band of acrolein is described as being 65% charge transfer, in which a π electron from the C=C part of the molecule occupies the π^* orbital of the C=O group, and 25% local C=C $\pi \to \pi^*$ excitation. In addition to this band at 50 200 cm^{-1}, two other valence shell transitions at 65 400 and 71 000 cm^{-1} (vert.) are also predicted in this model for acrolein. According to this approach, methylation of the olefin part of acrolein will considerably lower the π-ionization potential of the olefin group, thereby resulting in a lower frequency for the N → V$_1$ band.

In acrolein, three Rydberg series were identified which are analogous to those found in acetaldehyde [W5]. While $n = 3$ members of the nd series ($\delta = 0.15$) and the np series ($\delta = 0.68$) appear at 67 511 and 60 585 cm^{-1} (vert.), respectively, with terms of 14 030 and 20 955 cm^{-1} (vert.), the $n = 3$ member of the ns series ($\delta = 0.95$) was said to be missing. By analogy with the situation in all of the other ketone spectra, the 3s member most likely is the 57 200-cm^{-1} band, though its term value (24 340 cm^{-1}) is somewhat low for a 3s-terminating orbital in a molecule composed of one oxygen and three carbon atoms [compare with the term values in acetone (27 150 cm^{-1}), trimethylene oxide (24 700 cm^{-1}), propionaldehyde (27 000 cm^{-1}), and n-propanol (29 980 cm^{-1})]. In crotonaldehyde, the methyl derivative of acrolein, the transition to 3s has a still lower term value of 23 200 cm^{-1}, which drops again in mesityl oxide to the alkyl limit of 21 600 cm^{-1}.

The lower members of the nd Rydberg series in acrolein are doublets with a splitting that goes to zero as n increases, as in acetaldehyde, and the unassigned doublet band at 62 500 cm^{-1} (advert.) in acetaldehyde also appears in acrolein as a doublet at the same frequency. The two bands at 67 511 cm^{-1} and 68 513 cm^{-1} in acrolein have been assigned by Nagakura as $\pi \to \pi^*$, but the alternate assignments as Rydberg bands seem secure. A region of weak, continuous absorption which is found in acrolein between 59 000 and 65 000 cm^{-1}, and which Walsh points out is analogous to the weak band following the N $\to$ V$_1$ band of propylene (Section IV.A-2), may instead be the higher valence shell transitions calculated by Nagakura.

The close similarity of the acrolein Rydberg series and those in acetaldehyde (excepting the 3s member) strongly suggests that it is an n$_O$ electron which is being excited in the acrolein transitions. Still, the n$_O$ ionization potential of acetaldehyde (82 270 cm^{-1} adiab.) [D11] and the π-ionization potential of propylene (78 160 cm^{-1} adiab.) [D11] are sufficiently close to make one wonder which is lower in acrolein (81 450 cm^{-1} advert.) According to an *ab initio* calculation [S41], the π_2 MO is just above the oxygen lone-pair AO, but the very vertical nature of the acrolein Rydberg transitions implies that the originating orbital is n$_O$ rather than π_2, and that the molecule remains planar in the excited Rydberg states. This conjecture is substantiated by the photoelectron study of acrolein performed by Baker [B8, T21], who found the first ionization potential to be quite vertical and undoubtedly off of the oxygen atom, whereas ionization from π_2 is 6600 cm^{-1} higher. In crotonaldehyde, the separation of n$_O$ and π_2 ionization potentials is only 3870 cm^{-1}, but n$_O$ remains the higher orbital.

The absorption spectra of glyoxal (OHC—CHO) [W4] and biacetyl (CH$_3$CO)$_2$ [E7] are very similar, as one would expect, but the resemblance is in part superficial. There is a weak band beginning at 48 700 cm^{-1} in glyoxal showing six members of a 560-cm^{-1} vibrational progression, whereas in biacetyl, the origin is at 50 647 cm^{-1}, with vibrational intervals which can be interpreted as either 1220 or 610 cm^{-1}, of which the latter seems preferable. What seems to be another set of analogous bands is observed at 57 295 cm^{-1} (advert.) in glyoxal. Since the ionization potential of glyoxal is 8000 cm^{-1} higher than that of biacetyl, it is certain that either these sets of bands are valence shell excitations, or that they have very different assignments in the two molecules, but with a fortuitous coincidence of frequencies. Semiempirical [S54] and *ab initio* [H1, H2, P22] calculations of the electronic structure of glyoxal conclude that the splitting of the two n$_O$ orbitals puts the *gerade* combination higher, with an n$_g$–n$_u$ split of 12 000 cm^{-1} in both glyoxal and biacetyl [C28, T21]. This very large splitting is a classic example of the effect of through-bond interaction splitting AOs which are widely separated and otherwise noninteracting [C28]. Now the transition n$_g \to$ 3s is parity forbidden, and cannot correspond to either of the bands at 50 000 or 57 000 cm^{-1} on the basis of intensity. Walsh reports several other transitions in glyoxal: a broad, strong band at 60 000 cm^{-1} (vert.), a very diffuse band at 62 500 cm^{-1} (vert.), and two fairly strong bands at 73 800 and 75 500 cm^{-1} (advert.). On the basis of term values and intensities in glyoxal, we tentatively propose that the $n_g \to 3s$ band is not observed, that the 60 000-cm^{-1} band corresponds to the $\pi_2 \to \pi_3^*$ ($N \to V_1$) transition of butadiene (Section V.C-1), and that the 62 500-cm^{-1} band has a term value (22 900 cm^{-1}) appropriate to the allowed $n_g \to 3p$ excitation. According to Walsh, the 73 800- and 75 500-cm^{-1} bands are Rydberg excitations; if so, they are either $n_u \to 3s$, allowed with a term value of 24 600 cm^{-1}, or they are $n_g \to 4p$. As for the band at 51 300 cm^{-1}, its term value is far too large for $n_g \to 3s$, its intensity is too low for $\pi_2 \to \pi_3^*$, and so it must be a valence shell $n \to \pi^*$ or $n \to \sigma^*$ transition. The 57 200-cm^{-1} band would appear to be another valence shell excitation, since it clearly is not Rydberg.

The weak band at 51 400 cm^{-1} (vert.) in biacetyl is no doubt analogous to the band at nearly the same frequency in glyoxal. Since this band is observed in the solution spectrum of biacetyl in hexane [L8], it would seem that it is a valence shell excitation, as was concluded for the analogous band in glyoxal. Unfortunately, the solution spectrum of biacetyl [K30] does not extend far enough to give us similar information on the 57 295-cm^{-1} band. Though this would seem to be the analog of the band at 57 200 cm^{-1} in glyoxal, which is valence shell, its term value of 19 575 cm^{-1} (vert.) instead suggests that it is an $n_g \to 3p$ Rydberg transition.

It is interesting that the $\pi_2 \to \pi_3^*$ bands in butadiene and acrolein are found at 47 800 and 51 600 cm^{-1}, respectively, but at 60 000 cm^{-1} in glyoxal. The naive interpretation here is that the $N \to V_1$ transition of acrolein is largely localized in the C=C part of the molecule, and that the $N \to V_1$ band of glyoxal is 12 200 cm^{-1} above that of butadiene because the $\pi \to \pi^*$ transition of formaldehyde is 12 200 cm^{-1} above that of ethylene. Such an argument places the $N \to V_1$ band of formaldehyde at $\sim$72 000 cm^{-1} (vert.), in rough agreement with other empirical estimates. Note here that Walsh's claim that the $N \to V_1$ bands of glyoxal and acrolein can be interpreted equally well as either a Rydberg transition or as a $\pi \to \pi^*$ valence shell transition is inconsistent, for the Rydberg bands in these molecules involve n_O excitation, whereas the valence shell excitation originates with π_2.

Hosoya and Nagakura [H32] have studied the spectrum of tropolone to 62 000 cm^{-1} and found in addition to the well-known bands in the quartz ultraviolet, another at 55 600 cm^{-1} (vert.) of medium intensity and a very strong one peaking beyond 62 000 cm^{-1}. These latter two transitions were assigned as $\pi \to \pi^*$ following a Pariser–Parr–Pople calculation of the spectrum. Optical spectra of the ketonic substances ketene and carbon suboxide are discussed in the section on cumulenes (Section V.D).

IV.D. Acetylenes

Among the classic subjects for study in the vacuum ultraviolet such as ethylene, benzene, etc., acetylene is probably the least understood. This is due in part to the fact that so few derivatives have been studied, and to the fact that there are both multiplet splittings and large distortions away from linearity which complicate the spectrum.

According to an *ab initio* calculation of the electronic structure of acetylene [S41], the ground-state configuration is $(1\sigma_g)^2(1\sigma_u)^2(2\sigma_g)^2(2\sigma_u)^2$-$(3\sigma_g)^2(1\pi_u)^4({}^1\Sigma_g{}^+)$ and the lower singlet excited states are derived from the multiplets of the configuration $\cdots (1\pi_u)^3(1\pi_g)^1$. Though the *ab initio* calculations of the upper states of acetylene are not in quantitative agreement with experiment, they do agree among themselves on the ordering derived from the (π, π^*) configuration [B65, K3]. For example, Kammer found ${}^1\Sigma_u{}^-$ to be lowest with cis- and trans-bent geometries about equally more stable than the linear form. Just slightly above this, ${}^1\Delta_u$ appears, one component of which is bent and the other linear. Far above these two excited states, the strongly allowed transition to ${}^1\Sigma_u{}^+$ is found together with the ${}^1\Sigma_g{}^+ \to {}^1\Pi_g$ transition, derived from the excited configurations $(3\sigma_g)^1(1\pi_g)^1$ and $(1\pi_u)^3(3\sigma_u)^1$.

The acetylene spectrum is shown in Fig. IV.D-1. Experimentally, the lowest valence shell state of the $(1\pi_u)^3(1\pi_g)^1$ configuration, ${}^1\Sigma_u{}^-$, is found in the 40 000–50 000-cm^{-1} region. Ingold and King [I2] and Innes [I3] have performed a beautiful rotational and vibrational analysis of rarely achieved thoroughness to show that the transition is made allowed through the agency of ν_4'' nontotally symmetric vibrations to a trans-bent upper state (${}^1A_u$ in C_{2h}), the polarization being out of plane. The reader is referred to the original papers and to the books of Herzberg [H20] and Murrell [M64] for a more complete discussion of this famous transition. A normal vibration analysis of the ${}^1A_u$ state of acetylene reveals that the C≡C stretching frequency is lower than in the ground state, as expected, but that the C—H wagging frequency has increased [C13].

The spectral calculations, though too high by 15 000 cm^{-1} or more, do predict that the valence shell transition to ${}^1\Delta_u$ should follow closely behind that to ${}^1A_u$. It seems likely that all or part of the absorption in the 50 000–60 000-cm^{-1} region can be assigned to the ${}^1\Sigma_g{}^+ \to {}^1\Delta_u$ transition [M61]. No vibrational analysis has been given for this complex system, and Wilkinson very roughly estimates that the oscillator strength between 52 500 and 64 500 cm^{-1} is 10^{-3}.

The Rydberg term values in acetylene should be of the same size as those in ethylene (Section I.C-1) and this appears to be the case. Price [P37] has described two Rydberg series in acetylene, both converging to 92 000 cm^{-1} (vert.), with origins identified by comparing the spectra of acetylene and acetylene-d_2. Wilkinson later investigated the $n = 3$ members of these series in more detail [W28]. The acetylene term values are 26 000 and 17 500 cm^{-1} (vert.) for the $n = 3$ members of the two series, which compare well with the values of 27 400 and 18 100 cm^{-1} (vert.) for ethylene and 29 500 and 21 700 cm^{-1} (vert.) for ethane. The first Rydberg transition at 65 814 cm^{-1} (vert.) is clearly $1\pi_u \rightarrow$ 3s by its term value, and shows the vibrational excitation of only a few quanta of ν_2', the totally symmetric C≡C stretch (1849 cm^{-1}), as in the ionization of $1\pi_u$ in the photoelectron spectrum (1774 cm^{-1}) [B5]. The second member of this series, $1\pi_u \rightarrow$ 4s, at 80 116 cm^{-1} (vert.), has been studied under high resolution by Herzberg [H19], who finds a simple P, Q, R rotational structure characteristic of a $^1\Sigma_g \rightarrow {}^1\Pi_u$ transition. Hollas and Sutherley studied the Franck–Condon factors for the transition to this state, to the $(1\pi_u, 3s)$ Rydberg state, and to the $^2\Pi_u$ ionic state and conclude that the upper states are all linear with almost identical C≡C and C—H distances [H25]. In these near-ionic and ionic states, the C≡C and C—H distances are about 0.05 and 0.01 Å longer than the respective distances in the ground state. In the hot bands involving the degenerate bending motion ν_4'', Renner–Teller splittings in the (1, 1) sequence band are visible [H19].

The second Rydberg series identified by Price is most likely $1\pi_u \rightarrow$ 3p by its term value, but such an electronic transition is $u \rightarrow u$ forbidden in the centrosymmetric geometry. Alternatively, Greene et al. [G23] assign this series as $1\pi_u \rightarrow$ 3d. While this assignment would be in accord with the apparent allowed character of the bands, it also has a $(\pi_u, 3d)$ term value which is 5000 cm^{-1} larger than that normally found for transitions terminating at 3d. Experimentally, Wilkinson reports the vibronic structure to consist largely of ν_2', but with a few quanta of the nontotally symmetric vibration ν_4'' also appearing, suggesting to him a slight bend in the upper state. However, ν_4'' does not remove the center of symmetry and will not make the transition allowed. Ordinarily, one would be willing to make the claim that the origin of the $1\pi_u \rightarrow$ 3p transition does not appear, and that one quantum of a nontotally symmetric vibration is responsible for the other vibronic features, but in this case, Price has specifically stated that on the basis of a comparison of the acetylene and acetylene-d_2 spectra, the electronic origins of the $1\pi_u \rightarrow n$p series are the strongest features in the spectrum. The $1\pi_u \to 3p$ manifold is further complicated by the splitting of 3p into $3p\pi$ and $3p\sigma$ components, and further, the $(1\pi_u, 3p\pi)$ configuration gives rise to three singlet states, $^1\Sigma_g^+$, $^1\Sigma_g^-$, and $^1\Delta_g$. Since the splitting between these three states is governed by exchange interactions between the $np\pi$ Rydberg orbital and the electrons in the $(1\pi_u)^3$ open shell, their splitting will rapidly tend to zero as n increases, as will the difference between these and the transition to $3p\sigma$. In Price's original work, he does find an interval that goes to zero with increasing n, but he assigns this to a vibrational interval rather than to an electronic splitting.

In keeping with the large number of states expected to result from the $1\pi_u \to 3p$ (or $1\pi_u \to 3d$) promotion, Wilkinson [W28] and Nakayama and Watanabe [N10] find two features close to the $1\pi_u \to 3p$ "origin" which seem not to fit into the vibrational pattern, but are themselves origins for ν_2' progressions. Thus the B, C, and D bands (Fig. IV.D-1) were said by them to be valence shell excitations, but a $1\pi_u \to 3p$ assignment should also be considered for bands B and C. Wilkinson concludes that the upper states in the B and C bands are slightly trans bent. The D band at 80 300 cm^{-1} (vert.) has a term value of 11 700 cm^{-1} (vert.) and so may be a member of the allowed $1\pi_u \to 3d$ complex.

We have already seen that the Rydberg spectrum of acetylene is somewhat anomalous, and those of the various alkyl acetylenes are no less so, as regards both term values and intensities. In methyl acetylene, Watanabe and co-workers [N10, W15] report a Rydberg series converging upon 83 600 cm^{-1} (vert.) and having $\delta = 0.95$ (ns), but with the $n = 3$ member missing. However, the $n = 3$ band is probably part of the weaker absorption centered at 56 800 cm^{-1} (vert.). This contrasts strongly with the situation in acetylene, where the transition to 3s is the most prominent. The strongest Rydberg bands of methyl acetylene form a series having $\delta = 0.57$, as for np upper orbitals. Thus, removing the center of symmetry seems to have shifted the intensity into the np manifold. Another series with $\delta = 0.33$ probably has np-terminating MOs. These same two series also are identified in ethyl acetylene, where all of the bands are considerably broader. Nakayama and Watanabe attempt to assign the strong feature at 69 000 cm^{-1} (vert.) in ethyl acetylene to the $n = 4$ line of an ns series, but it fits better as $\pi_u \to 3d$, with a term value of 13 000 cm^{-1} (vert.). In t-butyl acetylene [K1] (Fig. IV.D-2), the $1\pi_u \to 3d$ transition is most prominent at 67 000 cm^{-1} (vert.) (13 500 cm^{-1} vert. term value), and it seems likely that we have approached the alkyl term limit, with the transitions to 3s and 3p practically degenerate at 58 000 cm^{-1} (vert.). In the corresponding trimethylsilyl acetylene (Fig. IV.D-3), very nearly the same term values are again observed for

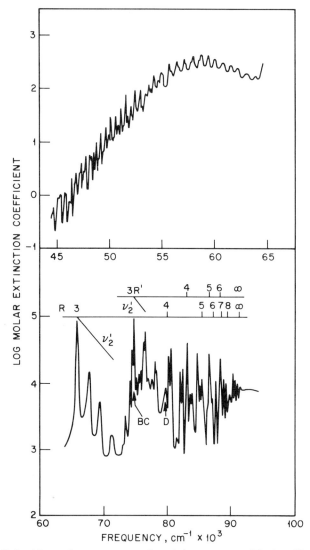

Fig. IV.D-1. Absorption spectrum of acetylene vapor with two Rydberg series delineated [K1].

transitions to 3s, 3p, and 3d. Bowman and Miller have observed the electron energy-loss spectra of acetylene, methyl acetylene, and butyne-1 at low resolution, and have tabulated the relative cross sections of the more prominent bands [B38].

In the monoalkyl acetylenes, the molecules were noncentrosymmetric,

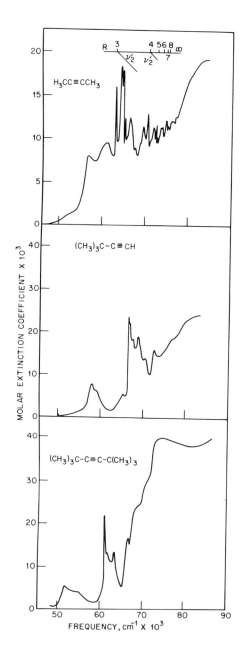

Fig. IV.D-2. Absorption spectra of several alkylated acetylenes in the gas phase [K1].

and only the $\Delta l = \pm 1$ selection rule was in force, whereas in the dialkyl acetylenes, the center of symmetry is reestablished, so that the LaPorte rule functions as well. Among the dialkyl acetylenes (Figs. IV.D.-2 and IV.D-3), the common denominator seems to be a weak lower-energy band at ~55 000 cm^{-1} followed at ~8000 cm^{-1} higher frequency by an intense and very vertical feature which we recognize from its term value and comparison with the monoalkyl spectra as being $1\pi_u \to 3d$. Formally, $1\pi_u \to 3p$ is forbidden and we would normally assume that the weaker band at ~55 000 cm^{-1} is thus $1\pi_u \to 3s$. The term values of these weak bands hover about the alkyl limit of 21 000 cm^{-1}.

The oscillator strength of the vibronically allowed $^1\Sigma_g^+ \to {}^1A_u$ band was originally reported to be of the order 10^{-4}, but Dunn, in a communication to Mulliken [M61], says that the oscillator strength is larger

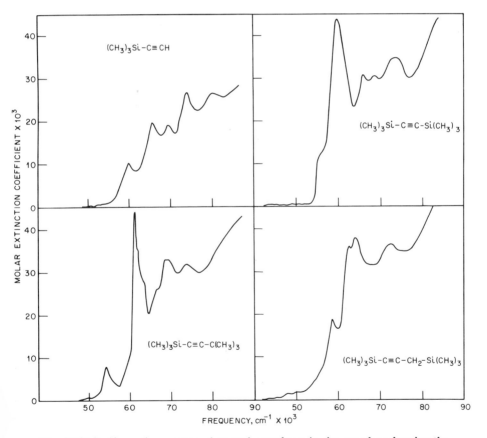

Fig. IV.D-3. Absorption spectra of several acetylenes in the gas phase bearing the trimethylsilyl group [K1].

than this, and that the Franck–Condon maximum is nearer 52 600 cm^{-1}, rather than 45 500 cm^{-1} as previously thought. Using an inverted order of the excited states which places $^1\Sigma_u^+$ at 56 000 cm^{-1} and $^1\Delta_u$ at 74 000 cm^{-1}, Falicov et al. have calculated an oscillator strength of 0.54 for the transition to $^1\Sigma_u^+$, using a correlated wave function [F1].

Moe and Duncan [M45] first determined the oscillator strengths of the (0, 0) bands of both the ns and np Rydberg series members of acetylene photographically, and later studied the oscillator strength of the origin and associated vibrational structure of the $1\pi_u \to 3$s transition [N18], but the results are somewhat unreliable, as suggested by the considerably higher molar extinction coefficients found for acetylene by Nakayama and Watanabe [N10] and Kaiser [K1], using photoelectric detection.

Person and Nicole [P10] list absorption cross sections and photoionization yields in the 88 000–96 000-cm^{-1} regions of acetylene and propylene and their perdeuterated derivatives. Working in the same spectral region, Dibeler and Walker [D15] find that the higher members of the Rydberg series going to vibrationally excited ions are autoionized by the continuum of the (0, 0) ionization, and, in fact, three series with δ values of 0.95, 0.5, and 0.3 were constructed from the autoionization peaks in the photoionization spectrum of acetylene. Similar experiments out to 170 000 cm^{-1} reveal autoionization peaks at 108 000, 114 000, 123 000, and 131 000 cm^{-1} (vert.). These are readily assigned using their term values: the two peaks at 108 000 and 114 000 cm^{-1} have term values of 26 300 and 20 300 cm^{-1} (vert.) with respect to the ionization potential at 134 300 cm^{-1} (vert.) and must be $3\sigma_g \to 3$s and $3\sigma_g \to 3$p, whereas the two peaks at 123 000 and 131 000 cm^{-1} have term values of 28 000 and 20 000 cm^{-1} with respect to the ionization potential at 151 000 cm^{-1} (vert.), and so must be $2\sigma_u \to 3$s and $2\sigma_u \to 3$p. These bands nicely illustrate the independence of the term values on the originating molecular orbitals (Section I.A-1). Going beyond the first ionization potential, several transitions are apparent which may or may not be to the same upper MO. Thus Lassettre et al. [L9] report a weak, sharp band in the electron-impact spectrum at 103 700 cm^{-1} (vert.), while optically, Metzger and Cook find maxima at 108 800, 126 500 and 144 800 cm^{-1} (vert.) [M29]. Collin and Delwiche discuss several other types of experiments which place autoionizing levels at 106 900 and 116 000 cm^{-1} (vert.) [C24]. The band at 144 800 cm^{-1} may be a transition to 3s from the fourth MO, $2\sigma_g$.

None of the higher triplet states of acetylene has been observed optically from the ground state, but two triplet states at 42 000 and 49 000 cm^{-1} (vert.) are clearly seen in the electron-impact spectrum of acetylene when viewed away from $\theta = 0°$ at low impact voltages [T15].

Strangely, although the triplets can be seen at low impact voltages (25 eV) and $\theta = 50°$, under no conditions can the $^1\Sigma_g^+ \to {}^1A_u$ excitation be seen. Lassettre et al. [L9] also failed to see this excitation in electron impact. Another triplet state has been uncovered in acetylene at 65 000 cm^{-1} (vert.), using the trapped-electron technique (Section II.D) [D1].

Some rather interesting effects are observed for acetylene as a solute in a condensed phase. Pysh et al. [P53] studied the spectrum of acetylene as a 1% solution in krypton and argon matrices at 20°K, in which the Rydberg transitions are strongly broadened and shifted. In krypton (Fig. IV.D-4), the two broad bands at 69 900 and 71 800 cm^{-1} were assigned as the origin and one quantum of ν_2' of the $1\pi_u \to 3s$ Rydberg excitation found at 65 700 and 67 600 cm^{-1} in the gas phase. In an argon matrix, these two components are at 72 700 and 74 400 cm^{-1}. The shifts of $\sim$5000 cm^{-1} to higher frequency on going from the gas phase to a rare gas matrix are of the size usually found for Rydberg transitions in a solid that is able to support molecular Rydberg orbitals (Section II.C). In a CF$_4$ matrix, the $1\pi_u \to 3s$ excitation moves still further to higher frequencies, with the strong origin at 74 520 cm^{-1} followed by quanta of ν_2' at 76 340 and 78 125 cm^{-1} [G3].

In both the krypton and argon matrices, there follows another band with origins at 74 100 and 77 400 cm^{-1} (adiab.), respectively, with two quanta of ν_2' (1800 cm^{-1}) attached (Fig. IV.D-4). As discussed by Jortner et al. there is little reason to consider Rydberg orbitals in rare gas matrices with n larger than three, so that $1\pi_u \to 4s$ is not a reasonable assignment for these bands. Alternate possibilities are $1\pi_u \to 3p$, a transition from $1\pi_u$ to the lowest $n = 2$ Wannier exciton, and assignment to one of the valence shell excitations said by Wilkinson [W28] and Nakayama and Watanabe [N10] to be in the region of 74 000 cm^{-1}. Arguments against a valence shell assignment are the near-constancy of the splitting between this band and the origin of the $1\pi_u \to 3s$ Rydberg excitation (4400 ± 300 cm^{-1}) in the gas phase and in condensed phases, and the fact that a value of $\nu_2' = 1800$ cm^{-1} is characteristic of the Rydberg states of acetylene, rather than of (π, π^*) valence shell excited states, in which ν_2' has a value of 1385 cm^{-1} ($^1\Sigma_g^+ \to {}^1A_u$). Thus it seems likely that this second transition in the matrix spectra is also outside the molecular valence shell, but we cannot tell whether it should be described as a molecular excited state or an excited state of the matrix.

A very upsetting and singular fact is uncovered in the crystal spectrum of acetylene in the 50 000–65 000 cm^{-1} region, in which there are at least two transitions (Fig. IV.D-5). In various rare gas matrices as well as in the neat film itself, one sees that the lower-frequency band is completely smeared out in the condensed phase as expected for a Rydberg

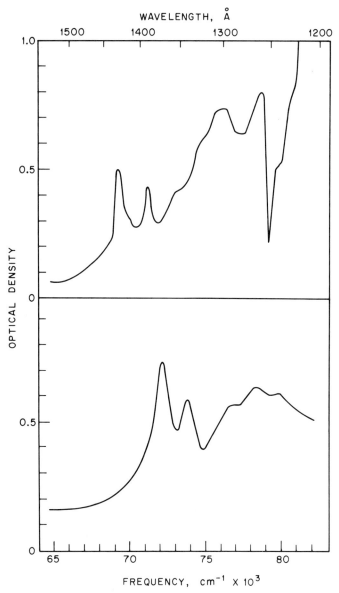

Fig. IV.D-4. Absorption spectra of 1% acetylene doped into krypton (upper) and argon (lower) matrices at approximately 20°K [P53].

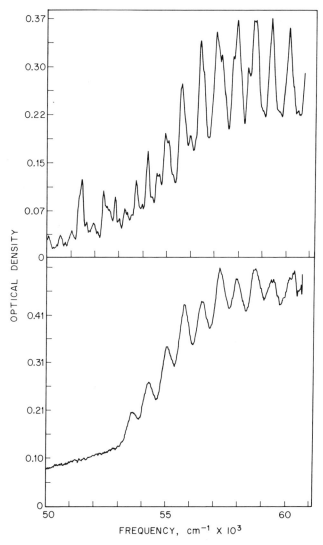

Fig. IV.D-5. Comparison of the spectra of gaseous (298°K, upper) and solid (20°K, lower) acetylene in the 50 000–60 000-cm^{-1} region [R19].

excitation, while the transition at higher frequency is unaffected by the condensed phase, as expected for a valence shell excitation. Now in our scheme of the Rydberg energy levels, there is no room for a Rydberg absorption at such a low frequency; its term value is almost 40 000 cm^{-1} (vert.). One possible explanation is that the band at 53 000 cm^{-1} and below is a valence shell $1\pi_u \to \sigma_g{}^*$ transition which is strongly mixed with the $1\pi_u \to 3s$ Rydberg configuration and so assumes considerable Rydberg character. (See the discussion on Rydberg/valence shell conjugates, Section I.A.-1.)

Klevens and Platt [K30] present the spectra of octyne-1 and octyne-2 as solutes in heptane; each shows a weak band having $\epsilon \simeq 100$ centered at 45 000 cm^{-1} and a much stronger one at about 55 000 cm^{-1}. These probably represent the transitions analogous to $^1\Sigma_g{}^+ \to {}^1A_u$ and $^1\Sigma_g{}^+ \to {}^1\Delta_u$ in acetylene, though the second seems too intense for a formally forbidden excitation.

The spectrum of iodoacetylene is especially interesting since it combines the acetylene and iodine atom chromophoric groups in a common molecule without too great a descent of symmetry (Fig. IV.D-6), and so has several features in common with both acetylene and the alkyl iodides [S4]. The spectrum begins with a weak, structureless band at 40 000 cm^{-1} (vert.), having its counterpart at the same frequency in iodoethylene (Section IV.A-3) and the alkyl iodides (Section III.B-1). The transition in all cases is the A band, the excitation of a 5p lone-pair electron on iodine into the σ^*(C—I) antibonding MO. In iodoacetylene, the A band is followed by a sharply structured feature at 47 642 cm^{-1} (advert.) having a term value of 30 900 cm^{-1} (advert.). Because this term value is about 5000 cm^{-1} larger than that expected for a Rydberg transition to 3s, it is more likely that this is a $\pi \to \pi^*$ excitation analogous to the $^1\Sigma_g{}^+ \to {}^1A_u$ band in acetylene. Unlike the band in acetylene, that in iodoacetylene is quite vertical, suggesting that the upper state may not be bent as it is in acetylene. Rydberg excitations originating with the 5p AOs of the iodine atom commence at 52 326, 55 667, and 59 517 cm^{-1} (advert.). This is the 5p → 6s complex split by spin–orbit coupling, again analogous to the B and C bands of methyl iodide in the same

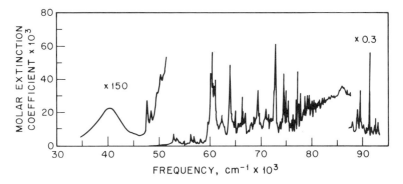

Fig. IV.D-6. Absorption spectrum of iodoacetylene in the gas phase [S4].

region. As in methyl iodide, the first two photoelectron bands of iodoacetylene are strikingly like those of the first two Rydberg excitations. The $^2E_{3/2}$–$^2E_{1/2}$ spin–orbit splitting in iodoacetylene (3340 cm^{-1}) is noticeably smaller than that in methyl iodide (4904 cm^{-1}), an effect due to the delocalization of the 5pπ orbital over the π MOs of the acetylene. This delocalization also makes the Rydberg excitations less vertical. The $\pi^3(^2E_{3/2})$6s and $\pi^3(^2E_{1/2})$6s upper states have the not unusual term values of 26 200 cm^{-1} (advert.), and series having n up to 23 have been constructed by Salahub and Boschi. A very intense transition to 6p at 60 100 cm^{-1} (advert.) has a term value of 18 450 cm^{-1}, appropriately. In the alkyl halides this is called the D band, and must be a very complex affair since the (π^3, np) configuration results in 15 states. There is considerable fine structure beyond the first ionization potential in iodoacetylene, most likely due to Rydberg excitations going to the first ionization potential of the acetylenic group, 97 400 cm^{-1} (advert.).

Price and Walsh [P47] briefly report on the spectrum of divinyl acetylene, a molecule with a pi-electron system resembling that of hexatriene. In this molecule, they found an intense $\pi \rightarrow \pi^*$ absorption centered at about 40 000 cm^{-1}, with pronounced excitation of the central C≡C bond vibration. This is analogous to the N → V$_1$ band of hexatriene (Section V.C-3). Several other weaker bands follow, with the Rydberg excitations beyond 58 800 cm^{-1} converging to an ionization potential of about 84 700 cm^{-1}. The ionization is undoubtedly out of the uppermost pi MO.

IV.E. Nitriles

With the exception of the voluminous work on the CN radical and rather less on HCN, the literature is sadly lacking in data on this interesting chromophoric group, which is isoelectronic with acetylene. In fact, spectral data are available only for acetonitrile [C30], propionitrile [C30], acrylonitrile [M58], dimethyl cyanamide [R2], and cyanogen [B22, P46].

A broad, low band at 60 000 cm^{-1} (vert.) was the first feature identified in the acetonitrile spectrum [H16]. Later work by Cutler [C30] uncovered origins at 77 374, 86 953, and 90 853 cm^{-1} (advert.) which were thought to be members of a Rydberg series converging upon the first ionization potential, which photoelectron spectroscopy shows to be 98 600 cm^{-1} (vert.) [L3, T21]. The band at 60 000 cm^{-1} has a very large term value, and so is most likely valence shell, corresponding either to $n_N \rightarrow \pi^*$ or a forbidden component of $\pi \rightarrow \pi^*$. The 77 374-cm^{-1} band has a term value of 21 230 cm^{-1}, which identifies it as a transition to a 3p orbital.

Since the band is apparently allowed, the transition is most likely to $3p\sigma$, with that to $3p\pi$ coming at higher frequencies. The optical transition displays a C≡N stretch of 2001 cm^{-1} in the upper state (2249 cm^{-1} in the ground state), while 2010 cm^{-1} is found in the ion. Other vibrations of 813, 1239, and 1560 cm^{-1} in the optical spectrum have their counterparts at 810 and 1430 cm^{-1} in the ion. An *ab initio* calculation on acetonitrile [S41] assigns the uppermost occupied MO as $2e$, the C≡N group π MO, and so this is the originating MO for the Rydberg excitations described earlier. A block of continuous absorption centered at about 70 000 cm^{-1} must contain the symmetry-allowed $\pi(2e) \to 3s$ Rydberg band.

Except for a diffuse absorption at 74 500 cm^{-1} (adiab.) there are no features of note in the spectrum of propionitrile [C30]. This is undoubtedly a $\pi \to 3p\sigma$ Rydberg band, as can be seen from the adiabatic ionization potential of 95 500 cm^{-1} [L3] and the term value of 21 000 cm^{-1}. The spectrum of *t*-butyl nitrile shows no absorption maximum out to 62 500 cm^{-1} [Y2].

Onari's spectrum of a solid film of polyacrylonitrile

$$(-CH_2\overset{|}{C}HCN)_x$$

is useful, for it will not exhibit any transitions to the Rydberg states (Section II.C) [O7]. In this material, there is a weak but well-resolved valence shell transition at 50 000 cm^{-1} (vert.) and a stronger shoulder at 62 500 cm^{-1} (vert.). Again, these bands must involve components of the nitrile $\pi \to \pi^*$ or $n_N \to \pi^*$ excitations, but we cannot say which or where. Note that these two types of valence shell excitation in general could be distinguished by comparing the spectra in solutions of pentane and methanol and noting the solvent shift.

The gas-phase spectrum of acrylonitrile, $H_2C=CHCN$, is shown in Fig. IV.E-1 [M58]. The most prominent feature is the $\pi \to \pi^*$ transition of the ethylene group centered at 52 600 cm^{-1} (vert.). The assignment is straightforward, since there is no other transition in the spectrum with an intensity as high as that expected for the $\pi \to \pi^*$ band. The molar extinction coefficient of 4600 at 52 600 cm^{-1} corresponds roughly to an oscillator strength of 0.27, while a value twice as large is reported for acrylonitrile in ethanol [H4]. Mullen and Orloff have analyzed the vibronic structure of the $\pi \to \pi^*$ band and place the origin at 49 260 cm^{-1}, with the remainder of the band built upon the single or simultaneous excitation of quanta of 370, 970, 1680, and 2230 cm^{-1}. Further work on the deuterated species is necessary before these frequencies can be assigned.

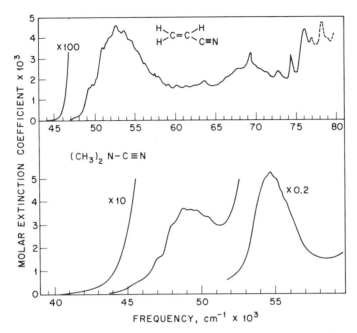

Fig. IV.E-1. Absorption spectra of acrylonitrile [M58] and dimethyl cyanamide [R2], both in the gas phase.

Just to the low-frequency side of the (π, π^*) origin, there appears a weak plateau (47 500 cm^{-1} vert., $\epsilon = 150$) which is regarded as an $n_N \to \pi^*$ transition within the C≡N group [H4, M58]. Vibrational analysis of the 55 000–64 000-cm^{-1} region suggests another electronic origin at 57 970 cm^{-1}.

Several of the bands beyond 63 000 cm^{-1} in acrylonitrile can be assigned as Rydberg bands on the basis of their term values. The ionization potential of acrylonitrile has been measured by photoelectron spectroscopy to be 88 000 cm^{-1} (vert.) [L3], and involves the loss of a pi electron which is almost completely localized in the C=C pi bond. The optical transitions observed at 63 700, 69 300, and 74 400 cm^{-1} (vert.) have term values of 24 300, 18 700, and 13 600 cm^{-1}, respectively, which are very suggestive of 3s-, 3p-, and 3d-terminating Rydberg orbitals. The $n = 4$ members of the ns and np series are found at 76 000 and 78 200 cm^{-1} (vert.), respectively.

The spectrum of dimethyl cyanamide, (CH$_3$)$_2$N—C≡N, has been studied by Rabalais et al. [R2] as a member of the series of three-center cumulenes having 16 valence electrons, and its states elucidated in terms of those of the isoelectronic azide ion. As described in the cumulene discussion (Section IV.D), the single low-lying intense transition in the $D_{\infty h}$ prototype (N_3^-) is $\pi \to \pi^*$, $^1\Sigma_g^+ \to \,^1\Sigma_u^+$. The corresponding transition in dimethyl cyanamide is probably that at 54 600 cm^{-1} (vert.) (Fig. IV.E-1), having $f = 0.32$, and the appropriate term symbols $^1A_2 \to \,^1B_2$. If, instead, this molecule were considered by Nagakura's intramolecular charge transfer theory (Section I.B-2), this intense transition would probably have a large fraction of $(CH_3)_2N \to CN$ charge transfer character, together with a smaller amount of $-C\equiv N$ $\pi \to \pi^*$ local excitation. The weaker bands in the 45 000–51 000-cm^{-1} region are probably derived from the $^1\Sigma_g^+ \to \,^1\Delta_u$ transition, one component of which is formally forbidden in the azide ion, but somewhat allowed in the lower symmetry of dimethyl cyanamide. The forbidden component of $^1\Sigma_g^+ \to \,^1\Delta_u$, the forbidden $^1\Sigma_g^+ \to \,^1\Sigma_u^-$, and the weakly allowed $n_N \to \pi^*$ transitions may also fall in this area. The ultraviolet absorption earlier attributed to $H_2N-C\equiv N$ is said by Rabalais et al. to be due solely to ammonia instead. Since cyanamide has too low a vapor pressure for an optical absorption study at room temperature, and decomposes at higher temperatures, it is an excellent candidate for electron-impact spectroscopy, which can operate with a much smaller sample pressure.

The cyanogen spectrum is a very complicated one and in need of further study. In the region up to 50 000 cm^{-1}, two transitions to triplet states and two to singlet states derived from the $(1\pi_g)^3(2\pi_u)^1$ excited configuration have been identified [B22]. Since the ionization potential of cyanogen is 107 800 cm^{-1} (vert.) [T21], all transitions with frequencies less than about 75 000 cm^{-1} are necessarily valence shell. A very strong band beginning at 60 500 cm^{-1} [B22, P46] is tentatively assigned by Bell et al. as a transition to a $^1\Pi_u$ excited state (probably not $n_g \to 2\pi_u$, which is much too weak). Its complex vibrational structure is a result of Renner instability. Another very strong system is found at 75 800 cm^{-1} (adiab.) in cyanogen. Though its term value is suggestive of a 3s-terminating orbital, this must be a valence shell transition instead, since the transition from $1\pi_g$ to 3s is symmetry forbidden. Another strong, diffuse band occurs at 96 100 cm^{-1} and is followed by several strong, sharp bands, none of which has been assigned, but which are probably Rydberg series converging upon the $1\pi_g$ (107 800 cm^{-1} vert.) and $1\pi_u$ (125 000 cm^{-1} vert.) ionization potentials.

CHAPTER V

Nonaromatic Unsaturates

In the two-center unsaturated systems, the most readily identifiable valence shell excitations are those due to the $n_X \to \pi^*$ and $\pi \to \pi^*$ promotions. As the system is expanded to include more than two centers carrying pi electrons, the effect will be to increase the number of π and π^* MOs that can participate in optical transitions. Since the higher $n_X \to \pi^*$ excitations normally do not have enough intensity to be seen, the practical result is to increase the number of $\pi \to \pi^*$ excitations in the ultraviolet region. Thus in a three-center system such as formic acid, there are two possible $\pi \to \pi^*$ excitations, and in hexatriene there are nine. Of course, to these must be added the various $\pi \leftrightarrow \sigma$ valence shell excitations and the manifolds of Rydberg excitations, making a very complicated picture overall. One practical aid here is that certain of the $\pi \to \pi^*$ excitations will have oscillator strengths far above 0.3, whereas those of the Rydberg excitations do not exceed 0.08 (Section I.A-1). The term value predictions of Section I.C-2 will also be valid for these larger molecules and can be used to sort the spectrum into valence shell and Rydberg excitations, as can the perturbation experiments (Sections II.B and II.C). The opportunities for observing $\pi \leftrightarrow \sigma$ excitations in the larger pi systems are infrequent since $\pi \leftrightarrow \sigma$ excitations are not very intense and can be seen only when the more intense $\pi \to \pi^*$ bands are at much higher frequencies,

V.A. Amides, Acids, Esters, and Acyl Halides

V.A-1. *Amides*

Virtually all of the interest in the optical spectrum of the amide group stems from the fact that it is the basic chromophoric unit in polypeptides, and that the spectra of these biologically important polymers cannot be understood unless the component monomer spectrum is properly assigned. In fact, the polypeptide spectra have been of such overwhelming interest that in certain situations the tables are turned, and the polymer spectra have been interpreted so as to give basic information about the monomeric amide unit (Section V.A-2). Also, because of the application to biopolymers, there has been an active effort toward studying the circular dichroism and optical rotatory dispersion spectra of optically active amides, which adds to our understanding of this group. A further factor of significance to us are the recent all-electron calculations of the electronic structure of the amide group, performed in Gaussian orbital basis sets. In this section, we will consider the spectra of only the simple monomeric amides, whereas dimeric, trimeric, and polymeric amides are discussed in Section V.A-2. Discussion of the less glamorous carboxylic acids, esters, and acyl halides is deferred to Section V.A-3.

At one extreme, the electronic structure of the amide group

$$-C\begin{matrix}\diagup\!\!\!\!\diagup O \\ \diagdown NR_2\end{matrix}$$

can be considered as that of the allyl anion

$$-C\diagup\!\!\!\!\diagup^{C}\diagdown C^{\ominus}- \leftrightarrow -C^{\ominus}\diagup C\diagdown\!\!\!\!\diagdown^{C}-$$

perturbed by the transformation of the terminal CH_2 groups into O and NH_2 groups [P11], and at the opposite extreme, essentially as a ketone somewhat perturbed by charge transfer from the $-NH_2$ group [N3]. Of

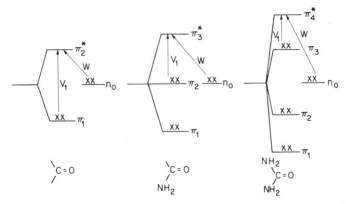

Fig. V.A-1. Valence shell MOs and transitions in the ketones, amides, and urea.

course, given sufficiently large perturbations, the two approaches can be made equivalent, but usually it is found that one or the other is better suited to a *direct* explanation of the evidence. This point will be kept in mind as the spectra and calculations are discussed.

For the moment, we consider the amide group as related directly to the allyl anion, and compare its electronic structure with that of the keto group in Fig. V.A-1. Qualitatively, the introduction of the amino group adjacent to the keto group adds another pi MO to that set, giving a highest filled pi level π_2 very close to that of the nonbonding electron pair on oxygen, n_O, while simultaneously depressing π_1 and raising the empty level π_3^*. Which of the two levels π_2 or n_O is actually higher in a particular amide is an interesting question which is best answered using photoelectron spectroscopy, and will be discussed later.

The results of all-electron calculations of the formamide molecule ground-state wave function confirm the qualitative picture given in Fig. V.A-1 [B11, B12, R7] and are useful not only for answering questions about the ground state, but, just as importantly, for suggesting the orbital natures of several of the lower excited states. Calculation of the formamide ground state in a double-zeta basis gives the energy levels displayed in Fig. V.A-2 and the orbital population analysis of Table V.A-I. A graphic display of the results in this table is given in reference [R7]. Comparison of the rigorously calculated energy levels of formaldehyde and formamide supports the simple picture of Fig. V.A-1. One sees that $n_O(10a')$ and $\pi_2(2a'')$ are very close together in formamide ($\sim$4000 cm^{-1} separation), whereas they are separated by 28 500 cm^{-1} in formalde-

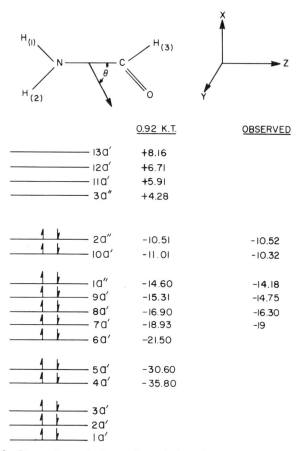

Fig. V.A-2. Comparison of observed vertical ionization potentials in formamide with Koopmans' theorem values (empirically corrected).

hyde, and that π_1 of formaldehyde is higher than π_1 of formamide. The π_2 MO of formamide is calculated to be composed very nearly of equal parts of the $2p\pi$ AOs of oxygen and nitrogen.† It is also to be noted that in the formamide ground state, the n_O orbital $10a'$ has an

---

† Considering the pi electrons only in the ground state, it is seen from Table V.A-II that nitrogen has lost 0.2 of an electron and carbon 0.3 of an electron to the oxygen atom. This is in quantitative agreement with semiempirical calculations of the extent of pi-electron charge transfer in formamide [N3, N4]. However, the *net* charges, pi plus sigma, give a very different picture [B11]. Back-donation of electrons in the sigma MOs leads to a nitrogen atom which is overall *more* negatively charged (−0.758) than the oxygen atom (−0.377), the turnabout coming in

appreciable density on H_3 and somewhat smaller densities on the carbon and nitrogen atoms as well. A similar delocalization of the n_O orbital onto the carbon atom and its adjacent groups is also predicted by calculations on formaldehyde [S41]. We will return to the population analysis again in the discussion of the upper states of formamide.

Concerning the optical spectra of amides, the first point to be made is that there is a characteristic amide-group absorption pattern in the 40 000–80 000 cm^{-1} region. In Fig. V.A-3, the spectra of the diverse amides formamide, fluoroacetamide, and 1-methyl-2-pyrrolidone illustrate the five transitions W, R_1, V_1, R_2, and Q that one can expect in an amide. In certain amides, the band Q may be covered by alkyl-group absorption, as it probably is in 1-methyl-2-pyrrolidone, and the W band is not observed in the absorption spectra of tertiary amides, though its presence can be demonstrated using circular dichroism spectroscopy. As a consequence of the constancy of the amide-group absorption pattern, the calculations of the lower transitions in the formamide spectrum will be relevant to the discussion of the spectra of more complex monoamides. In addition to the five bands mentioned here, the possibility of a sixth is discussed in Section V.A-2.

Judging from Fig. V.A-1, it is to be expected that the $n_O \to \pi_3^*$ transition of amides will be of higher frequency than the $n_O \to \pi_2^*$ band of ketones found at $\sim$36 000 cm^{-1}, but that the $\pi_2 \to \pi_3^*$ transitions of amides will be at lower frequencies than $\pi_1 \to \pi_2^*$ in ketones ($\sim$70 000 cm^{-1}). Interestingly, in urea, the $n_O \to \pi_4^*$ band takes another step to higher frequency, whereas $\pi_3 \to \pi_4^*$ moves again to lower frequency, with the result that in this type of molecule, the $\pi_3 \to \pi_4^*$ transition is expected to precede the $n_O \to \pi_4^*$ transition. In planar systems such as those considered here, the (π, π^*) configurations of increasing energy are given the symbols V_1, V_2, V_3, etc., whereas the ground state is called N. A second transition, $N \to V_2$ ($\pi_1 \to \pi_3^*$), is also possible in the amide group, and is expected to come at about twice the $N \to V_1$ ($\pi_2 \to \pi_3^*$) frequency.

The W band in simple amides and lactams is found near 45 000 cm^{-1} (vert.), and all investigators agree upon an $n_O(10a') \to \pi_3^*(3a'')$ assignment for it. As appropriate for such a transition, the molar extinction

---

large part at the expense of the hydrogen atoms. The final result of the pi-sigma electron flow is that the

$$H-C\overset{\displaystyle O}{\nearrow} \quad \text{and} \quad -NH_2$$

fragments have exactly zero calculated net charge.

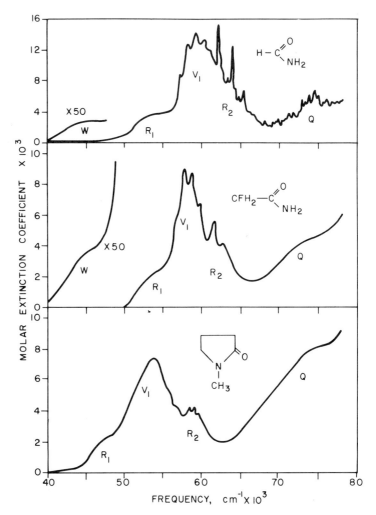

Fig. V.A-3. Optical absorption spectra of various amides displaying the W, R_1, V_1, R_2, Q pattern [B11].

coefficients of the W band in various amides, when corrected for overlapping absorption, are less than 100 ($f = 0.002$–0.004), even though the transition is formally allowed. The reality of this badly overlapped band in amides was confirmed by Litman and Schellman [L30], who pointed out that by going from the usual hydroxylic solvents to nonpolar hydrocarbons, the W band is shifted to lower frequencies whereas the adjacent absorption is shifted to higher frequencies, thereby uncovering the

W-band profile. As an exception to the generalities given above, in the oxamides

$$\begin{array}{c} H_2N \\ \diagdown \\ O \end{array} C-C \begin{array}{c} O \\ \diagup \\ \diagdown NH_2 \end{array}$$

the π^* orbital is delocalized over both amide groups, thus shifting the $n_O \to \pi^*$ band downward to 36 000 cm^{-1}. Similar shifts of the W band may be expected whenever a pi-electron system is placed α to an amide group. The fate of the $n_O \to \pi_3^*$ transition in tertiary amides is revealed by the circular dichroism spectra. Because the $n_O \to \pi_3^*$ rotatory strength and sign make it much more visible in the circular dichroism spectrum than its oscillator strength does in the absorption spectrum, it is readily found in the circular dichroism spectrum of tertiary amides at its expected frequency, near 45 000 cm^{-1}, whereas it is completely buried among stronger bands in the absorption spectrum [R14]. Unlike the $n_O \to \pi_2^*$ band of ketones, the $n_O \to \pi_3^*$ band of the amide group seems always to be smooth and unstructured, even in the gas phase.

The polarized crystal spectrum of myristamide, $CH_3(CH_2)_{12}CONH_2$, has been recorded by Peterson and Simpson (Fig. V.A-4), who interpret the dichroic ratio at 45 500 cm^{-1} as indicating a mixed polarization for

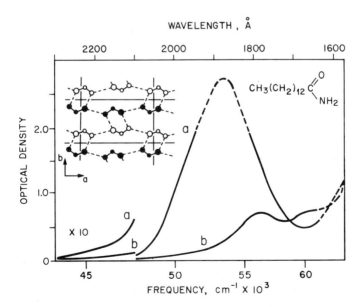

Fig. V.A-4. Polarized absorption spectra along the a and b axes of myristamide. The dashed parts of the curves were measured with relatively lower accuracy [P11].

the $n_O \to \pi_3^*$ band, half in-plane and half out-of-plane [P11]. Group theoretically, the $n_O \to \pi_3^*$ transition of amides is allowed with an out-of-plane polarization, and it is quite possible that the experimental result is complicated by the ever-present overlapping of this band by the stronger in-plane polarized absorption at higher frequencies. This, however, is not to deny the possibility of a transition simultaneously having both allowed and forbidden character and mixed polarization.

As mentioned earlier, the $n_O(10a')$ orbital of formamide is appreciably delocalized throughout the sigma framework, the oxygen contribution to this molecular orbital in the ground state being only 78%. On excitation of an electron from n_O to π_3^*, all of the molecular orbitals reorganize to some extent. If we make the approximation that the space parts of the wave function for the (n_O, π_3^*) open-shell configuration are the same for both the singlet and triplet, with only the spin parts being different, then it is relatively easy to perform the SCF calculation on the triplet configuration and still glean some understanding of the corresponding open-shell singlet. The results of such an "indirect SCF" calculation are given in reference [B11]. It is seen from this work that in the (n_O, π_3^*) excited state, the remaining electron in n_O is now 94% localized on the oxygen atom. Since the π_3^* orbital in the (n_O, π_3^*) configuration is 85% localized on carbon, the $n_O \to \pi_3^*$ transition in formamide is very much an oxygen $\to$ carbon charge transfer transition, and the resulting positive charge on oxygen acts to pull the remaining electron in n_O back onto it. Though deduced from a study of formamide, these conclusions should hold as well for more complex amides.

In the charge transfer model (Section I.B-2), the (n_O, π_3^*) configuration of an amide is localized within the C=O group, and would come at the ketone frequency, 34 500 cm^{-1}, except for the fact that the ground state of the amide is further depressed by about 12 500 cm^{-1} by interaction with the charge transfer configuration [N3]. Thus the W band of amides is expected at about 47 000 cm^{-1}, in good agreement with experiment.

The R_1 band of amides is a relative newcomer to the overall picture of amide spectra, having first been reported in 1966–1967 [B9, B11, B12, K11]. Inasmuch as it does not fit into the simple valence shell scheme of Fig. V.A-1, as a first guess one might assign the R_1 band as a Rydberg excitation. In support of this, in formamide and N,N-dimethyl formamide, the R_1 bands have term values of 29 200 and 24 000 cm^{-1} (vert.), respectively, in the range expected for 3s upper states. Since the R_1 bands are continuous in all amides investigated to date, the big orbit nature of the R_1 upper state cannot be tested experimentally using the high-pressure effect (Section I.B) because the perturbation is too mild to reveal itself in such a smooth band. The R_1 band calls for a more drastic perturbation, i.e., a condensed-phase experiment (Section II.C). In Fig. V.A-5, the gas-phase spectra of formamide and N,N-dimethyl acetamide are compared with their spectra as polycrystalline films at low temperature. In the latter compound, it is seen that in the condensed phase, the V_1 band (52 000 cm^{-1}, vert.) has shifted to lower frequency by about 1000 cm^{-1}, but that R_1 is nowhere to be seen. The spectrum of N,N-dimethyl acetamide was also determined in acetonitrile solution, allowing an intensity calibration of the 52 000-cm^{-1} band. Since the R_1 band of N,N-dimethyl acetamide does not appear in the solution spectrum, and since the intensity of the V_1 band in solution is just that found for this band in the gas phase, it appears that the R_1 band is not sitting beneath it. Similarly, in solid films of formamide, only the shifted V_1 band is apparent in the 50 000–60 000-cm^{-1} region. Thus the 3s Rydberg assignment of the R_1 band of amides is reinforced by its behavior in condensed phases.

The situation in amides is much like that in sulfides, alcohols, and several other classes of compounds in which the Rydberg transitions to 3s are completely devoid of vibrational structure, whereas the transitions

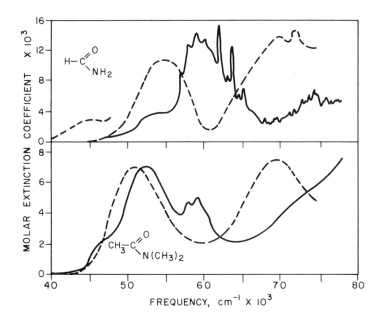

Fig. V.A.-5. Absorption spectra of formamide and N,N-dimethyl acetamide in the gas phase (solid line) and as polycrystalline films at 24.5°K (dashed line) [B12].

to 3p and/or 3d are usually very nicely structured. In the alcohols and water, the lack of vibrational structure comes from the mixing of valence shell O—H antibonding character into the 3s orbital, thus making the potential unbound along the O—H coordinate. Alternatively, we can describe the situation in terms of the mixing of Rydberg and valence shell conjugate configurations. Possibly, an analogous mixing of the 3s orbital and some antibonding valence shell MO is responsible for the lack of vibrational structure in amides. If so, then excitation to the 3s Rydberg state of amides should lead directly to fragmentation.

As is evident from the condensed-phase spectrum of formamide (Fig. V.A-5), the R_2 band in this compound also behaves as is appropriate for a Rydberg excitation. Indeed, Hunt and Simpson [H37] have found the R_2 band of formamide to be the first member of a Rydberg series obeying the formula

$$h\nu = 82\ 566 - [109\ 737/(n - 0.639)^2], \qquad (V.1)$$

with $n = 3, 4, 5$, and 6 having been observed. The 20 400-cm^{-1} term of the $n = 3$ member (R_2), and equally, the size of δ required by the series (0.639), are symptomatic of np upper states. Unlike the R_1 band, the R_2 band in several amides is quite sharp, and in all cases displays the expected asymmetric broadening under high-pressure perturbation [R19]. Tinoco et al. [T14] most recently reassigned R_2 as an n_O' $(9a') \rightarrow \pi_3^*(3a'')$ transition in amides, where n_O' is the oxygen "lone-pair" orbital aligned along the C—O bond. Our work instead leads to the conclusion that R_1 and R_2 are the 3s and 3p members of Rydberg series converging upon the lowest ionization potential of the particular amide in which they are found. The positive identification of the R_2 band as terminating at 3p reinforces our identification of the R_1 band as terminating at 3s. Thus in amides we have an excellent example of one Rydberg series easily identified by its sharp nature (np), but another (ns) missed because it does not conform to the myth that all Rydberg transitions are sharp. For more complete assignments of the R_1 and R_2 transitions, the originating orbitals must be determined. Since the originating orbital for these optical transitions is also the originating orbital for the lowest transition in the photoelectron spectrum, we now turn our attention to this aspect of the amide group's electronic structure.

Further information about the $N \rightarrow R_1$ and $N \rightarrow R_2$ Rydberg transitions of the amide group can be gleaned from the lower-energy parts of the photoelectron spectra of formamide and its two N-methyl derivatives shown in Fig. V.A-6 [B52]. As is apparent from the spectra, and as was suggested earlier [H37], the $\pi_2(2a'')$ and n_O $(10a')$ levels are very close indeed; the application of Koopmans' theorem to the Gaussian

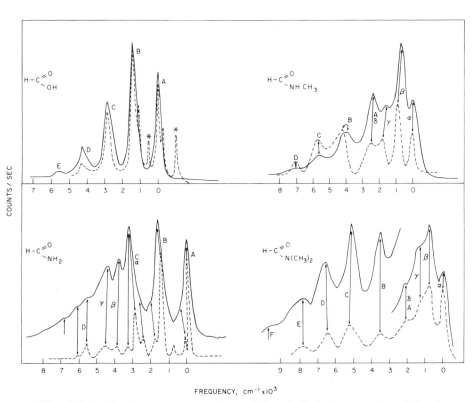

Fig. V.A-6. The lower-energy portions of the photoelectron spectra of formic acid, formamide, and the N-methyl formamides (solid lines), and comparison with certain portions of the Rydberg absorption spectra (dashed lines), all on an arbitrary scale [B52].

orbital calculation (Fig. V.A-2) on formamide also places them as near-degenerate. Looking in detail at the badly overlapped bands in the 72 000–89 000-cm^{-1} region and armed with prior knowledge, one sees that the transition can be decomposed into two bands, one with a characteristic vibrational spacing of 1500–1600 cm^{-1} (labeled with Latin letters), and a second one with a spacing of 600–700 cm^{-1} (labeled with Greek letters). In formic acid (Section V.A-3), where the n_O orbital is considerably above the π_2 MO, only a simple progression of 1460 cm^{-1} is observed in the first band. Arguing by analogy, it is concluded that the 1500–1600-cm^{-1} spacings in the amides are associated with ionization from n_O, whereas ionization from π_2 results in 600–700-cm^{-1} spacings.

Interestingly, careful comparison of the ionization energies using the stated criteria shows that n_O is slightly above π_2 in formamide, but that

TABLE V.A-I
Orbital Population Analysis of the Formamide Ground State

	$8a'$	$9a'$	$10a'(n_O)$	$1a''(\pi_1)$	$2a''(\pi_2)$	$3a''(\pi_3^*)$[a]
$H_{(1)}$	0.0187	0.0791	0.0158	0	0	0
$H_{(2)}$	0.2111	0.0573	0.0147	0	0	0
$H_{(3)}$	0.1683	0.1331	0.2238	0	0	0
C_s	0.0916	0.0190	0.0151	0	0	0
N_s	0.0007	0.0000	0.0056	0	0	0
O_s	0.0876	0.2390	0.0000	0	0	0
C_x	0.1068	0.3661	0.0229	0	0	0
N_x	0.2942	0.1329	0.0009	0	0	0
O_x	0.3296	0.4867	0.4859	0	0	0
C_y	0	0	0	0.6607	0.0334	1.4475
N_y	0	0	0	0.7621	1.0439	0.1340
O_y	0	0	0	0.5772	0.9217	0.4185
C_z	0.4106	0.0391	0.0277	0	0	0
N_z	0.2678	0.0094	0.1101	0	0	0
O_z	0.0145	0.4369	1.0773	0	0	0

[a] This orbital is not occupied in the ground state of formamide.

on N-methylation, n_O is raised by 2500–4000 cm^{-1} whereas π_2 is raised by 5000–11 000 cm^{-1}. Consequently, in both N-methyl formamide and N,N-dimethyl formamide, π_2 is above n_O! Of course, it must be mentioned that the concept of one-electron orbital energies has validity only within the Hartree–Fock approximation, and consideration of correlation effects beyond Hartree–Fock makes the concept of "n_O and π_2 orbital energies" meaningless. That π_2 is so much more susceptible to perturbation by N-methyl groups follows from the fact that 52% of π_2 is centered at the nitrogen atom of the amide group (Table V.A-I) but only 6% of n_O is similarly located [B11]. The promotion of the π_2 orbital above n_O in tertiary amides may be a large contributing factor to the very strong overlap of the $n_O \rightarrow \pi_3^*$ and $\pi_2 \rightarrow \pi_3^*$ transitions in these amides.

The very good agreement between the observed photoelectron band energies and those calculated from molecular orbital theory (Fig. V.A-2) is an important advantage of this theory over the charge transfer model, for which there is nothing as simple and effective as Koopmans' theorem.

Returning now to our analysis of the Rydberg spectra of the amides, we see that the n_O and π_2 orbitals are so close together that the bands previously thought to be R_1 and R_2 in the optical spectrum really must be doubled, R_1 being both $n_O \rightarrow 3s$ and $\pi_2 \rightarrow 3s$, and R_2 being both $n_O \rightarrow 3p$ and $\pi_2 \rightarrow 3p$. In the case of the transition to 3s, the bands are too smooth and broad to reveal the presence of two excitations, but if

TABLE V.A-II
CHARGE DENSITIES IN THE GROUND
STATE OF FORMAMIDE

Atom	Net charge	Pi-electron density
$H_{(1)}$	+0.357	0
$H_{(2)}$	+0.368	0
$H_{(3)}$	+0.152	0
C	+0.258	0.695
N	−0.758	1.806
O	−0.377	1.499

our analysis is correct, the structure of the Rydberg optical bands terminating at 3p may well appear as two overlapping transitions. This it best determined by comparing the optical band envelopes with the first two overlapping bands of the photoelectron spectra (Section II.A). These spectra are compared in Fig. V.A-6, and the match is overall rather good as regards both frequency spacings and relative Franck–Condon factors. Using the analysis of n_O and π_2 band origins in the photoelectron spectra and the close complementarity of these and the optical spectra, one concludes that the origin of the $n_O \rightarrow$ 3p transition in formamide is at 62 160 cm^{-1}, with $\pi_2 \rightarrow$ 3p about 3000 cm^{-1} higher. In N-methyl formamide and N,N-dimethyl formamide, the $n_O \rightarrow$ 3p transitions begin at 62 000 cm^{-1} and 55 300 cm^{-1}, whereas the $\pi_2 \rightarrow$ 3p origins are at 59 900 and 54 300 cm^{-1}, respectively. Of course, because the $\phi_i \rightarrow$ 3p optical transition in an amide will have three nondegenerate components whereas ionization from the ϕ_i MO will give only a single band in the photoelectron spectrum, these bands in the two spectra can look very different. However, the close similarity between optical and photoelectron profiles displayed in Fig. V.A-6 strongly suggests that optical transitions to only one of the components of 3p are appearing in the appropriate spectral region with any strength.

The pertinent experimental quantities for the calculation of the R_1 and R_2 term values in a variety of amide compounds are listed in Table V.A-III. As explained earlier, unless a detailed analysis of the photoelectron spectrum of an amide has been performed, it is not known whether the originating MO in the Rydberg transitions is n_O or π_2 (or both if the optical bands are poorly resolved). In any event, the term values will depend only on the upper MO, and not on whether the transition originates at π_2 or n_O. As explained in some detail in Section I.C-1, the alkylation of the amide group at the carbon and/or the nitrogen atoms

TABLE V.A-III

Ionization Potentials, Absorption Frequencies and Term Values in Amides, Acids, and Esters[a]

	Ionization potential	R_1 Frequency	R_1 Term value	R_2 Frequency	R_2 Term value
HCONH$_2$	83 200	54 000	29 200	63 800	19 400
HCONHCH$_3$	79 600	52 700	26 900	59 900	19 700
HCON(CH$_3$)$_2$	74 600	50 600	24 000	54 300	20 300
CH$_3$CONH$_2$	80 200	51 200	29 000	—	—
CH$_3$CON(CH$_3$)$_2$	73 000	47 500	25 500	—	—
CHC(O)—N(CH$_3$)—C(O)CH	80 600	56 000	24 600	63 000	17 600
CF$_3$CONH$_2$	90 500	54 700	35 800	68 600	21 900
HCOOH	92 840	60 000	32 840	71 000	21 800
	100 900	—	—	80 000	20 900
CH$_3$COOH	87 670	57 100	30 570	68 300	19 370
CH$_3$COOCH$_3$	84 500	58 000	26 500	—	—
CH$_3$COOC$_2$H$_5$	83 600	58 500	25 100	—	—
CHC(O)—O—C(O)CH	88 640	63 100	25 540	68 500	20 140
CF$_3$COOH	96 800	62 400	34 400	77 000	19 800
C$_2$F$_5$COOC$_2$H$_5$	92 300	58 800	33 500	—	—

[a] Vertical values in cm^{-1} listed in all cases.

will have the effect of decreasing the overall penetration of the molecular 3s Rydberg MO, thereby decreasing its term value. In contrast, the 3p term value is much less affected by alkylation and so should remain at $\sim$20 000 cm^{-1} throughout. These expectations are fulfilled (Table V.A-III), for the (n$_0\pi_2$, 3s) term value† of 29 200 cm^{-1} in formamide falls in a somewhat irregular way to 25 500 cm^{-1} in N,N-dimethyl acetamide as it moves toward 21 000 cm^{-1}, the alkyl limit. On the other hand, the (n$_0\pi_2$, 3p) term values remain pegged at about 20 000 cm^{-1} regardless of the extent of alkylation. In general, the effect of fluorination on the (ϕ_i, 3s) term value is an increase toward the perfluoro limit of 36 000 cm^{-1}. This is realized in trifluoroacetamide, where the (n$_0\pi_2$, 3s) term value is 35 800 cm^{-1}, compared with 29 000 cm^{-1} in acetamide itself.

† In the event we are unable to resolve the n$_0 \to \varphi_i$ and $\pi_2 \to \varphi_i$ excitations, the overlapped upper states will be referred to as (n$_0\pi_2$, φ_i).

The R_2 term values in the amides (Table V.A-III) are steady at about 20 000 cm^{-1} regardless of the extent of alkylation, as appropriate for transitions to 3p Rydberg orbitals. This term value is so constant that one could propose that the first ionization potential of any amide can be obtained by simply adding 20 000 cm^{-1} to its R_2-band frequency in the gas-phase spectrum. The one danger in this is illustrated by our earlier work on N,N-dimethyl formamide and N,N-dimethyl acetamide [B52], in which it was decided (before the utility of term values was fully appreciated) that the prominent bands at 61 700 cm^{-1} (vert.) in the former and at 59 000 cm^{-1} (vert.) in the latter were transitions to 3p. Since their term values are 12 900 and 14 000 cm^{-1} (vert.), respectively, it is clear now that the terminating orbitals are 3d, rather than 3p, and that in N,N-dimethyl acetamide, the transition to 3p cannot be resolved at all. A transition to 3d is also the most prominent one in N-methyl pyrrolidone (59 000 cm^{-1}, vert.), whereas that to 3p is barely visible at 55 000 cm^{-1} (vert.) (Fig. V.A-3). Illustrating again the relative insensitivity of the (ϕ_i, 3p) term values, that of trifluoroacetamide is only 2500 cm^{-1} larger than that of formamide.

We now return to the valence shell excited states of the amide group. The V_1 band (53 000–59 000 cm^{-1}) is the most prominent transition in the amides (see, for example, reference [N14]) and corresponds to the singlet–singlet excitation $\pi_2(2a'') \rightarrow \pi_3^*(3a'')$ in the molecular orbital scheme of Fig. V.A-1 [B11, B12, H37, P11, R7, W12]. A second useful approach to the assignment of the amide V_1 band considers the ground state to be composed primarily of the —NR$_2$ and C=O fragments, whereas the upper state is given by the charge transfer configuration (—NR$_2$)$^+$ (C—O)$^-$ [K11, N3, N4, N14, S15]. A mixing of the nonpolar and charge transfer structures then leads to a final ground-state wave function containing approximately 20% of the charge transfer configuration, whereas the V_1 upper state contains 60% of the charge transfer configuration. The locally excited ($\pi_1 \rightarrow \pi_2^*$) configuration within the keto group also contributes substantially to the V_1 state (Fig. I.B-1). Because the frequency of such a charge transfer transition will depend directly upon the ionization potential of the —NR$_2$ group, the frequencies of the N $\rightarrow$ V$_1$ transitions in a number of amides should vary linearly with the ionization potential of the corresponding amine. Furthermore, since only 0.4 of an electron is transferred in the optical transition, the slope of the N $\rightarrow$ V$_1$ frequency versus amine ionization potential curve should be 0.4. Experimentally (Fig. V.A-7), a roughly linear relationship is demonstrated, with a best fit slope of 0.54.

Though the charge transfer and molecular orbital formalisms are outwardly worlds apart, the qualitative and even some quantitative aspects

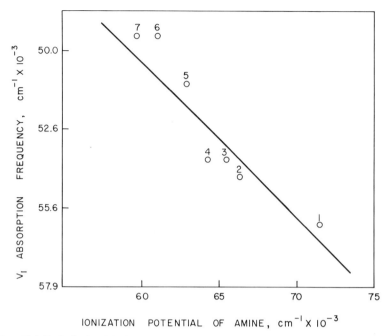

Fig. V.A-7. Relationship between the vertical N → V₁ frequencies in amides and the ionization potentials of the component amines. All compounds are acetamides $CH_3CONR_{(1)}R_{(2)}$: (1) $R_{(1)} = R_{(2)} = H$; (2) $R_{(1)} = H$, $R_{(2)} = CH_3$; (3) $R_{(1)} = H$, $R_{(2)} = C_2H_5$; (4) $R_{(1)} = H$, $R_{(2)} = n\text{-}C_4H_9$; (5) $R_{(1)} = R_{(2)} = CH_3$; (6) $R_{(1)} = R_{(2)} = C_2H_5$; (7) $R_{(1)} = R_{(2)} = n\text{-}C_3H_7$ [S15].

of the two calculations are remarkably alike. Thus in the ground state, 0.21 of a pi electron is transferred from N to C=O in the charge transfer model, and 0.19 of an electron in the molecular orbital calculation. Since the N → V₁ excitation transfers 0.4 of a pi electron from N to C=O in the charge transfer model, but 0.73 pi electrons in the MO model, the latter has more pi-electron charge transfer character than the charge transfer model itself! In the MO model, the charge transfer part of the excitation is 100% nitrogen-to-carbon; in the charge transfer model, it is 70% nitrogen-to-carbon, 30% nitrogen-to-oxygen. The predicted N → V₁ polarization directions in the two models differ by only 8.8°.

It has been found that substituting an alkyl group onto the nitrogen atom of formamide shifts the N → V₁ transition to lower frequencies, whereas substitution of the same group onto the carbon atom shifts it to higher frequencies [N14, S15]. Thus the V₁ band does not cover R₁ in dimethyl acetamide, but in diethyl formamide, the V₁ band comes at a rather low frequency, so low in fact as to cover both the W and the R_1 bands. However, R_1 can still be observed unobstructed in N-methyl formamide and in amides of the type

This response of the $N \rightarrow V_1$ frequency to alkylation is understandable if we recall that a methyl group will promote the generation of a positive charge in the pi system of an adjacent atom, but works against the tendency if the adjacent charge is negative. Since the $N \rightarrow V_1$ promotion is largely a nitrogen-to-carbon charge transfer, the V_1 state of an amide will be stabilized by N-alkylation and destabilized by C-alkylation.

A peculiar intensity effect has been found for the V_1 band of amides. Though all of the amide V_1 bands are of approximately the same width, the molar extinction coefficient of that of formamide (15 000, $f = 0.37$) is almost twice that of the other alkylated amides (~ 8000, $f = 0.23$–0.27). A similar effect is found in the carboxylic acid series (Section V.A-3) and on comparing the spectra of ethylene and cyclopropene (Section IV.A-2). What appears to be happening in the alkylated systems is that we have one pi orbital that does not extend onto the substituent alkyl group for reasons of symmetry, and a second pi orbital that has appreciable alkyl group density. In particular, in cyclopropene, π_2 has the alkyl group contribution but π_3^* does not, whereas in the amides, the node at the carbon atom in π_2 excludes an alkyl group contribution, but this is not so in π_3^*. Because of the selective incorporation of alkyl-group wave function in the pi MOs of these molecules, there results a diminution of the transition density $\pi_2\pi_3^*$ due to a lack of spatial overlap, and hence a lower $\pi_2 \rightarrow \pi_3^*$ oscillator strength than is found in the alkyl-free parent compounds.

The polarization of the amide V_1 band has been deduced by Peterson and Simpson [P11] from the absorption ratios of Fig. V.A-4 and the known orientation of the amide groups in the myristamide unit cell. They find the transition to be in-plane polarized, with $\theta = 17.9 \pm 10°$.† It is also to be noticed in Fig. V.A-4 that there is an obvious transition at 60 000 cm^{-1} in the b polarized crystal spectrum, which Peterson and Simpson assign as the R_2 band. Judging from the contrary behavior of the R_2 band in solution, this assignment seems questionable. This band is discussed more fully in Section V.A-2, where it is tentatively assigned as a valence shell $n_0 \rightarrow \sigma^*$ transition.

† The angle θ is defined in Fig. V.A-2.

TABLE V.A-IV
COMPARISON OF THE OPTICAL SPECTRA PREDICTED FOR HCOX MOLECULES[a]

	$HCONH_2$	$HCOOH$	$HCOF$	$HCOO^-$
$n_O(10a') \to \pi_3^*(3a'')$				
T	51 300	50 600	49 800	55 600
S	55 600	55 300	55 300	59 500
f	0.008	0.007	0.010	0.006
$\pi_2(2a'') \to \pi_3^*(3a'')$				
T	48 900	55 100	54 000	46 900
S	84 700	96 900	105 000	86 900
f	0.422	0.402	0.406	0.161
$n_O'(9a') \to \pi_3^*(3a'')$[b]				
T	85 900	85 400	86 300	56 200
S	90 300	90 300	90 700	59 400
f	2×10^{-5}	0.004	0.0002	0.000
$\pi_2(2a'') \to \sigma^*(11a')$				
T	83 500	94 400	114 000	93 100
S	88 200	97 000	119 000	94 600
f	0.001	0.0002	0.030	0.000
$n_O(10a') \to \sigma^*(11a')$				
T	106 000	97 700	105 000	93 500
S	110 000	104 000	114 000	103 000
f	0.137	0.156	0.221	0.310
$\pi_1(1a'') \to \pi_3^*(3a'')$				
T	77 300	89 900	110 000	81 500
S	113 000	121 000	131 000	115 000
f	0.164	0.121	0.077	0.144

[a] From Reference [B12]. The oscillator strength f is computed using the mixed dipole length–dipole velocity formulation of Hansen [H8]. T and S refer to the excitation frequencies (cm^{-1}) to the triplet and singlet configurations.

[b] The n_O' MO is formed principally of the 2p AO aligned along the C=O bond in formamide, formic acid, and formyl fluoride, but is taken as the second nonbonding MO in formate anion, formed principally of 2p AOs aligned *perpendicular* to the C=O bonds.

The Q band of amides is generally accepted as the second valence shell excitation $N \to V_2$ ($\pi_1 \to \pi_3^*$) [N4, P11]. There seems to be little evidence one can cite either for or against this assignment, except to point out that in the Gaussian orbital calculations on formamide (Table V.A-IV), following the $n_O \to \pi_3^*$ and $\pi_2 \to \pi_3^*$ excitations, the $n_O'(9a') \to \pi_3^*(3a'')$, the $\pi_2(2a'') \to \sigma^*(11a')$, and the $n_O(10a') \to \sigma^*(11a')$ valence shell excitations are all predicted to precede the $\pi_1 \to \pi_3^*$ promotion [B12]. However, only the $n_O \to \sigma^*$ and $\pi_1 \to \pi_3^*$ transitions have predicted oscillator strengths even remotely close to those observed for the Q bands, and as discussed in Section V.A-2, it appears that the $n_O \to \sigma^*$ band comes at about 60 000 cm^{-1}, rather than at the Q-band frequency. Thus it appears by elimination that the Q bands of amides most likely correspond to the $N \to V_2$ transition. Note that in formamide, the Q-band spectral region is also that expected for the $n_O, \pi_2 \to 3d$ Rydberg transitions, and so may be a region of composite valence shell/Rydberg absorption.

In solid amides, a transition is observed at close to the Q-band frequency (Fig. V.A-5), whose intensity is considerably greater than that of the Q band in the gas phase. The new, strong band is also observed in the spectrum of solid acetamide (Fig. V.A-14) [V1] and in the liquid N-methyl acetamide (Fig. V.A-8) [M48]. A similar effect is observed on comparing the gas-phase and thin-film spectra of the substituted benzenes (Section VI.A-2), for which it was suggested that the new, strong band corresponds to a transition terminating at the conduction band of the solid. This apparently is the explanation for this transition in the solid amides as well. In these condensed systems, molecular Rydberg transitions no longer appear, but should be replaced by Wannier excitons which form a Rydberglike series converging to the bottom of the conduction band.

Because genuine phosphorescence from an amide has never been observed [M4], there are no experimental data on the triplet manifold of

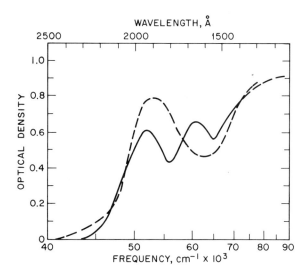

Fig. V.A-8. Absorption spectra of poly-L-alanine in solution (solid line) and of liquid N-methyl acetamide (dashed line) [M48].

this group. However, Maria et al. [M4] argue strongly for a low-lying $\pi_2 \to \pi_3^*$ triplet ($\sim$25 000 cm^{-1}), while the GTO calculations, using a direct SCF of the excited states, places the $\pi_2 \to \pi_3^*$ triplet at 35 000 cm^{-1} and the $n_O \to \pi_3^*$ triplet at 31 000 cm^{-1}.

The electronic spectrum of N-methyl malemide is discussed in Section V.A-3; its R_1 and R_2 term values (Table V.A-III), fall among those of the simpler amides. The spectrum of another amidelike molecule, urea, has been determined using the SF_6-scavenger technique (Section II.D), which revealed a broad, intense peak at 48 000 cm^{-1} (vert.) and a second, weaker band at 65 000 cm^{-1} (vert.) [N1]. In contrast to this result, Rosa and Simpson [R23] report that urea in trimethyl phosphate solution has its intense N $\to$ V$_1$ excitation at 58 400 cm^{-1} (vert.). Optically, we have found strong broad bands at 48 100, 58 800, and 76 000 cm^{-1} in the vapor of N,N,N',N'-tetramethyl urea [R19]. Urea is isoelectronic with the carbonate ion, in which the two uppermost occupied π MOs, π_2 and π_3, are degenerate. This degeneracy will be lifted somewhat in urea, with the result that π_2 and π_3 are separated by $\sim$6000 cm^{-1} according to an *ab initio* Gaussian orbital calculation [S48]. From this, one concludes that the $\pi_3 \to \pi_4^*$ and $\pi_2 \to \pi_4^*$ transitions will be strongly allowed, rather close together, and at about the amide N $\to$ V$_1$ frequency. Thus we assign the first two bands in the ureas as N $\to$ V$_1$ and N $\to$ V$_2$; the $n_O \to \pi_4^*$ transition is not observed.

V.A-2. Polymeric Amides

Once the amide group is incorporated into a dimeric or higher polymeric structure, interesting spectral changes appear, which are being actively studied experimentally and theoretically, but which are only partly understood at present. In the earlier days, when the near-ultraviolet spectrum of the amide group could be taken as consisting of a weak $n_O \to \pi_3^*$ and a strong $\pi_2 \to \pi_3^*$ transition, the application of the exciton theory to dimeric or polymeric helical amides predicted rather large splittings of the V_1 band, but insignificant splittings of the W band [M46]. The difference arises from the fact that the splitting matrix element is directly proportional to the appropriate transition moment integral, which is large for the $\pi_2 \to \pi_3^*$ excitation, but small for $n_O \to \pi_3^*$. As applied to helical polypeptides, the theory seemed to be an immediate success, for in these systems, an unperturbed $n_O \to \pi_3^*$ transition is found at about 45 000 cm^{-1} (vert.), followed by the two $\pi_2 \to \pi_3^*$ exciton-split components polarized parallel (48 000 cm^{-1} vert.) and perpendicular (52 500 cm^{-1} vert.) to the helix axis [B39, G21, H26, J2], just as predicted by Moffitt.

V.A. AMIDES, ACIDS, ESTERS, AND ACYL HALIDES

Similarly, in the rigid amide dimer system of diketopiperazine

$$O=\underset{N}{\overset{N}{\diamondsuit}}=O$$

what appear to be the forbidden and allowed components of the exciton-split V_1 band are found at 50 000 and 53 500 cm^{-1} (vert.) with opposite polarizations in the crystal spectrum (Fig. V.A-9) [K12], while the circular dichroism spectrum of the optically active 3-methyl derivative shows an $n_0 \to \pi_3^*$ transition at about 45 500 cm^{-1} (vert.) followed by two

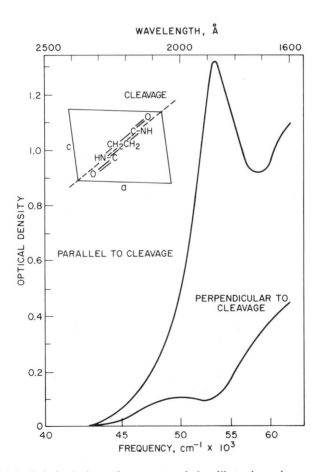

Fig. V.A-9. Polarized absorption spectra of the diketopiperazine crystal parallel and perpendicular to the cleavage plane [K12].

bands of opposite rotatory sign at 49 800 and somewhat above 54 000 cm^{-1} (vert.) in aqueous solution [G25]. The solution spectrum of diketopiperazine in water (to 56 000 cm^{-1}) was also studied by Ham and Platt [H5], who found a single maximum at about 54 000 cm^{-1}, with an extrapolated oscillator strength of 0.19 ± 0.05 per amide group. In the spectra of diglycyl (one amide group) and the open-chain triglycyl (two amide groups), also in water solution, the absorption maxima remain at 54 000 cm^{-1}, with the oscillator strength equal to 0.27 ± 0.07 per amide group. The early dipeptide spectra in solution do not show any exciton splitting [H5].

Recently, the satisfying picture of amide groups splitting under excitonic interaction to produce multiple N → V$_1$ bands in the 50 000-cm^{-1} region has come into question with the realization that the amide R$_1$ band also falls in this region. Thus Barnes and Rhodes [B9] uncovered the R$_1$ band in gaseous monomeric amides, and went from there to assign it as the 45 000 cm^{-1} band of the α-helical polypeptides. Rosenheck et al. [R26], working with poly-L-proline I, Greenfield and Fasman [G25], working with 3-amino-pyrrolidin-2-one and 3-methyl diketopiperazine, and Quadrifoglio and Urry [Q1, Q2, U2], working with poly-L-alanine and poly-L-serine, all admit to the possible presence of R$_1$ in their spectra. Since the R$_1$ band of amides is really two transitions, n$_0$ → 3s and π_2 → 3s, the complications seem doubled. However, we have found repeatedly that the R$_1$ band, being a Rydberg transition, cannot be seen in condensed-phase spectra of the sort described so far. One is doubly sure of this since the band even in the gas phase, is a broad, featureless one. It is likely that the R$_1$ bands will appear in the gas-phase circular dichroism spectra of the cyclic amides studied by Greenfield and Fasman. Nonetheless, at present, all of the experimental evidence is strongly against the assignment of any of the spectral features in condensed-phase spectra around 50 000 cm^{-1} as derived from the R$_1$ excitation.

Our rejection of the appearance of the R$_1$ band of amides in condensed phases is not to say that the spectrum at the lower frequencies will consist only of n$_0$ → π_3^* and π_2 → π_3^* transitions, for there does also seem to be another puzzling valence shell transition in this region. Going beyond the N → V$_1$ transition, Bensing and Pysh [B24] find a characteristic transition near 60 500 cm^{-1} with an oscillator strength of approximately 0.1 in thin films of poly-L-alanine (also seen in Momii and Urry's spectrum [M48]) and poly-L-proline II, while its presence is inferred in the spectra of poly-L-valine and poly-L-proline I. These workers conceded that the 60 500-cm^{-1} band in the gaseous monomer spectrum is the R$_2$ Rydberg transition, but then go on to claim that it persists unchanged in their thin-film polypeptide spectra. Again, there seems to be no experimental evidence in favor of such a situation, and much against it. Indeed, Momii and Urry, who first found this band in poly-L-alanine, specifically point out that it was not present in the spectrum of a liquid film of N-methyl acetamide, the prototype peptide monomer (Fig. V.A-8), and Onari finds it is clearly present in the spectra of poly-L-leucine [O9] but is missing in the spectra of various Nylons [O10]. Thus its presence is real and it is definitely a valence shell transition, but it is conspicuous in only a small number of amide-containing systems.

Actually there is some evidence for the presence of a valence shell excitation at 60 000 cm^{-1} in a few simple amides, though it is not at all apparent in many others. First, after considering the sharp line R_2 absorption in formamide (Fig. V.A-5), one sees that the remaining band envelope could easily contain a second, weaker valence shell transition at 60 000 cm^{-1}. It will be suggested that the amide and carboxylic acid spectra are in one-to-one correspondence (Section V.A-3). Turning to the spectrum of formic acid (Fig. V.A-16), one sees clearly that the sharp line R_2 absorption of this compound is resting upon another broad, but distinct transition at 73 000 cm^{-1} (vert.). Presumably related to this absorption are the obvious bands at $\sim$68 000 cm^{-1} (vert.) in the zwitterionic amino acids (Fig. V.A-17). Finally, the myristamide polarized crystal spectra of Peterson and Simpson (Fig. V.A-4) reveal an obvious transition with b-axis polarization at 60 000 cm^{-1} (vert.), which we infer is valence shell from its appearance in a condensed phase. In all cases where it can be identified, the third valence shell transition of the amide group has an intensity intermediate to those of the first transition ($n_O \to \pi_3^*$) and the second ($\pi_2 \to \pi_3^*$). If one looks for such a band in the diketopiperazine crystal spectrum, possibly it could be assigned to the parallel-polarized feature at about 63 000 cm^{-1} (Fig. V.A-9).

Theoretically, there would seem to be room for such a valence shell transition at 60 000 cm^{-1} in amides. According to the Gaussian orbital calculations [B12], there are three valence shell transitions that are energetically close to the $\pi_2 \to \pi_3^*$ singlet in both formamide and formic acid: $n_O'(9a') \to \pi_3^*(3a'')$, $\pi_2(2a'') \to \sigma^*(11a')$, and $n_O(10a') \to \sigma^*(11a')$ (see Table V.A-IV). Of these, the first two transitions in both formamide and formic acid have predicted oscillator strengths which are smaller than those of the $n_O \to \pi_3^*$ transitions, whereas in both the amide and the acid, the predicted $n_O \to \sigma^*$ oscillator strength is about one-third that of the $\pi_2 \to \pi_3^*$ transition. Since a band weaker than the $n_O \to \pi_3^*$ transition, or even somewhat stronger, could not be seen at all at 60 000 cm^{-1}, whereas the observed intensity is about one-third of the $\pi_2 \to \pi_3^*$ intensity, one can argue for an $n_O \to \sigma^*$ valence shell assignment of the 60 000–70 000-cm^{-1} band on the basis of its intensity. Something of a strike against the $n_O \to \sigma^*$ assignment is the fact that the GTO calculations predict the frequency of the $n_O \to \sigma^*$ band of formic acid to be 6000 cm^{-1} *lower* than that for formamide, whereas it is observed to be about 10 000 cm^{-1} higher. While the absolute accuracy of these calculations was not good, the relative ordering of transitions otherwise seemed to be quite reliable.

The experimentally determined b-axis polarization of the 60 000-cm^{-1} band of myristamide (approximately perpendicular to the N $\cdots$ O line, and in plane) could be a deciding factor in its assignment once the polarizations are calculated for all of the alternatives. Another pertinent bit of evidence is given by Bensing and Pysh [B25], who determined the polarization of the 60 000-cm^{-1} band of poly-L-alanine by partially orienting the sample by needle-shearing during solvent evaporation. The polarized spectra (Fig. V.A-10) are corrected for scattered light and clearly show the 60 000-cm^{-1} band to be polarized along the direction of shear. Using the oriented-gas model and the α-helical structure, the observed absorption is said to be incompatible with both an out-of-plane

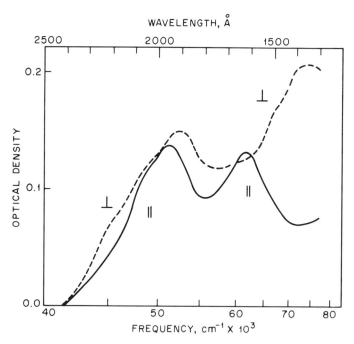

Fig. V.A-10. Polarized absorption spectra of poly-L-alanine with electric vector parallel (solid line) and perpendicular (dashed line) to direction of stroking [B25].

polarization and an in-plane polarization which is perpendicular to the C=O line. Neither of these excludes the suggested $n_O \to \sigma^*$ assignment, which will have a polarization which is in plane and approximately perpendicular to the N $\cdots$ O line. Bensing and Pysh [B25] note experimentally that the sum of the intensities of the 60 000-cm^{-1} band and the V_1 band in peptides is conserved, so that in random-coil polypeptides and in simple amides, V_1 has almost all of the intensity, whereas in helical polypeptides, there is a structurally induced mixing which lends considerable intensity to the 60 000 cm^{-1} band at the expense of the N $\to V_1$ transition.

Applying the foregoing polarization criterion to the crystal spectrum of diketopiperazine (Fig. V.A-9), one finds that the polarization of the band at 53 500 cm^{-1} is consonant with that expected for the $n_O \to \sigma^*$ transition, though the frequency is unusually low; a better guess would be the band at $\sim$63 000 cm^{-1} in the same polarization.

In any event, *three* bands do seem to be present at about 60 000 cm^{-1} in amides, $n_O \to 3p$, $\pi_2 \to 3p$, and, provisionally, $n_O \to \sigma^*$, of which only the third can appear in condensed-phase spectra. Note that though the $n_O \to \sigma^*$ bands of amides, alcohols, and ethers are given the same symbols and fall in the same spectral region, that of the amide group is of the $\sigma \to \sigma^*$ type, whereas in alcohols and ethers, it is of the $\pi \to \sigma^*$ type.

A strong absorption band is also observed at 64 600 cm^{-1} (vert.) in poly-γ-methyl-L-glutamate [M48] and at 64 000 cm^{-1} (vert.) in poly-γ-ethyl-L-glutamate [B39] both of which contain carboxylic ester groups as well as amide groups. Inasmuch as this is just the region of the N $\to V_1$ absorption of esters (Section V.A-3, Fig. V.A-16), a large part, if not all, of this absorption intensity can be safely assigned as coming from the ester groups, with relatively little attributable to the amide "$n_O \to \sigma^*$" transition. Two bands in the neighborhood of 60 000 cm^{-1} are resolved in the CD spectrum of poly-γ-methyl-L-glutamate in hexafluoroisopropanol solution [J10], one as a positive shoulder at 58 100 cm^{-1} (vert.) and the other as a negative peak at 62 500 cm^{-1} (vert.). One of these may be the $n_O \to \sigma^*$ excitation.

One other feature remains to be discussed in the spectra of the polyamides. As reported in Section V.A-1, there is a transition near 75 000 cm^{-1} following the R_2 band of simple amides, labeled Q in Fig. V.A-3. It has an oscillator strength approximately equal to those of the R bands, and will be overlapped by alkyl group absorption in the more heavily alkylated amides such as N,N-dimethyl acetamide (Fig. V.A-5). When these monomeric amides are investigated as solid films at low temperature [B12, V1] or as liquid films at room temperature [M48], the absorption band in the Q region remains, but its intensity increases severalfold relative to the other bands, so that in this phase, it is the strongest in the spectrum (Fig. V.A-5). Such a band is visible as well at 80 000 cm⁻¹ in poly-L-alanine (Fig. V.A-10), but this could be due in part to alkyl sidechain absorption. The assignment of the amide Q band in the gas-phase spectra is quite uncertain, and the present condensed-phase experiments give no help in that direction. However, they are interesting in that this region of increased absorption is also that expected for ionization, and it may be that the 75 000-cm⁻¹ transition in polyamides is an amide → amide intermolecular charge transfer absorption. Alternatively, such an excitation can be viewed as a transition to the conduction band of the solid, which, of course, has no analog in the free-molecule spectrum. This striking difference between the free-molecule and crystal absorptions at higher frequencies has been observed for several different classes of chromophores (see, for example, the situation in the substituted benzenes, Section VI.A-2).

Onari has studied the optical spectra of a large number of polymeric amides. In Nylons of various alkyl-group content (compare Nylon 3 and Nylon 610 in Fig. V.A-11), an $N \rightarrow V_1$ band is prominent at ∼53 000 cm⁻¹ (vert.), followed by a large mass of rather featureless absorption between 60 000 and 80 000 cm⁻¹. The relatively stronger absorption in this latter region in those Nylons with more alkyl groups shows that a large part of this absorption is due to the alkyl groups [O10]. In the polypeptides [O9, O12, O13], Onari finds pretty much the same features, except for the 60 000-cm⁻¹ band, which is present in some, but absent in others. The spectra of poly-L-methionine and of poly-L-serine are much like those of the Nylons, i.e., no —CH₂SCH₃ or —CH₂OH absorptions can be distinguished in these solids. The spectra of several mixed dipeptides are presented in Figs. V.A-12 and V.A-13 and those of tripeptides in Fig. V.A-14. Aside from a by now obvious $\pi_2 \rightarrow \pi_3^*$ band between 50 000 and 55 000 cm⁻¹, it is difficult to identify any other transition in these spectra.

V.A-3. *Acids, Esters, and Acyl Halides*

Since the acid, ester, and acyl halide systems are pi isoelectronic with the amide group, one reasonably expects their spectra to resemble those of the amides. In Fig. V.A-15, the spectra of the isoelectronic molecules trifluoroacetamide, trifluoroacetic acid, and trifluoroacetyl fluoride are compared; the one-to-one correspondence of the amide and carboxylic acid transitions is indisputable, but the acyl fluoride is admittedly less

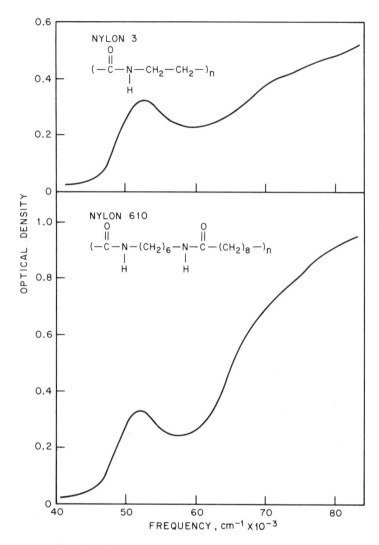

Fig. V.A-11. Absorption spectra of two Nylons [O10].

obviously related [B12]. The close relationship between the carboxylic acid and ester group spectra is evident from Fig. V.A-16. Though it is felt that a convincing argument for the assignment of the carboxylic acid and ester bands can be made solely on the basis of their similarity to those of the amides, in fact this is not necessary, since there is a significant amount of independent evidence which leads to the same conclusions.

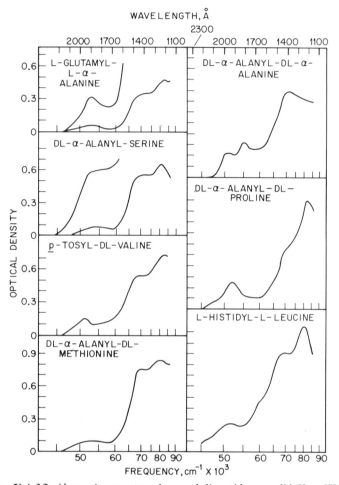

Fig. V.A-12. Absorption spectra of several dipeptides as solid films [V1].

This is presented later. The reader should study the section on the amide-group spectrum (Section V.A-1) in parallel with the present one.†

† The calculated charges in the ground states of formic acid and formyl fluoride are of some interest [B12]. As does the $-NH_2$ group of formamide, the $-OH$ group in formic acid assumes a positive charge in the pi system, but is overall negatively charged due to a back-donation of sigma electrons. Also, as one goes from fluoride to acid to amide, the pi-electron densities tend to become equal, indicating, as Barnes and Simpson have commented [B10], that this is the direction in which allylic character is expected to increase. Similarly, viewed within the charge transfer theory [N4], the larger amount of charge transfer in the ground state of formamide (21%) compared with that in acetic acid (13%) shows that amides are closer to allylic resonance than are the carboxylic acids.

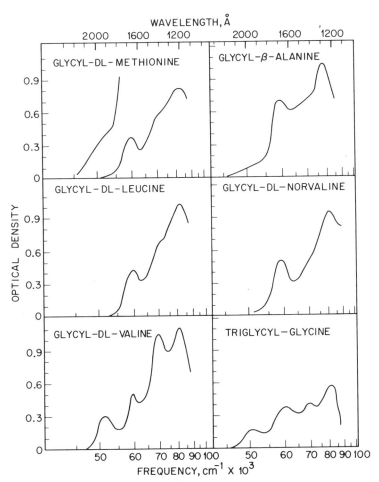

Fig. V.A-13. Absorption spectra of several dipeptides and of triglycylglycine as solid films [V1].

Because our present interpretation of the carboxylic acid and ester spectra is at variance with much of the earlier work, it must be considered tentative, though it is by far the least complicated scheme so far. In the following discussion, the correspondence of the carboxyl and amide bands will be presumed, so that the designations W, R_1, V_1, R_2, Q used for the labeling of the amide transitions can be used here as well. As with the amides, in certain highly alkylated acids and esters, the R_2 band may be too weak to see, whereas the transition to 3d is strong.

In earlier studies, Nagakura and co-workers [N5] observed the vapor spectra of formic acid, acetic acid, and ethyl acetate in the 52 000–65 000-

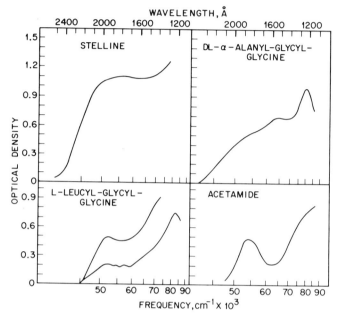

Fig. V.A-14. Absorption spectra of several tripeptides and of acetamide as solid films [V1].

cm^{-1} region, in all three instances obtaining the smooth profile of the band here called R_1. They, however, presumed it to be the V_1 band, and using the intramolecular charge transfer theory, proceeded to calculate a frequency of 62 200 cm^{-1} for this band in formic acid. Our view is that the V_1 band of formic acid really comes at 67 000 cm^{-1} (vert.) (Fig. V.A-16), but that even with this adjustment, the agreement with experiment still can be considered as quite good. In a similar but more extensive study, Barnes and Simpson [B10] have recorded the spectra of aldehydes, ketones, acids, esters, and amides, and propose several interesting correlations among these. In the carboxylic acid and ester group spectra (Fig. V.A-16) [B10, S12], they assign the band in the 55 000–61 000-cm^{-1} region (R_1) as a valence shell $n_o' \rightarrow \pi_3^*$ excitation, where n_o' is the second lone-pair orbital on the keto oxygen atom (orbital $9a'$, Fig. V.A-2), aligned with the C=O axis. They have also assigned the structured 69 000-cm^{-1} band of acetic acid as V_1, based upon a correlation with the similarly structured V_1 band of formic acid. Though Barnes and Simpson also argue for the correlation of amide and acid spectra, they did not observe the R_1 band of the amide group, and so arrived at a correspondence of bands different from that proposed here.

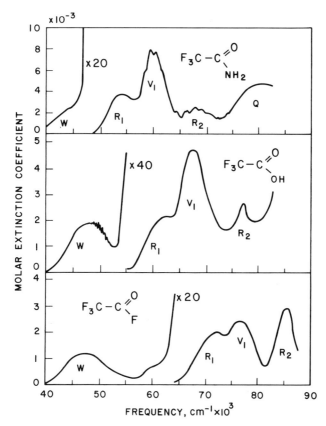

Fig. V.A-15. Absorption spectra in the gas phase of trifluoroacetamide, trifluoroacetic acid, and trifluoroacetyl fluoride, showing the W, R_1, V_1, R_2, Q sequences in each [B12].

Price and Evans [P41] studied the formic acid spectrum in detail, though they did not make detailed assignments of the bands. The most prominent aspect of their analysis was the delineation of a Rydberg series having the R_2 band (71 000 cm^{-1}, adiab.; Fig. V.A-16) as its $n = 3$ member, and $\delta = 0.60$. In the members of this series, rather long progressions with vibrational spacings of 1450–1500 cm^{-1} (C=O stretching) are observed, each of which is an origin for one quantum of several vibrations having frequencies in the 600–1000-cm^{-1} range. This Rydberg series is assigned as originating with the n_O orbital on oxygen, and it is further implied that the 67 000-cm^{-1} band is the first member of another Rydberg series. One other feature of interest is mentioned by them. They report a group of bands around 80 000 cm^{-1} which are "different in character

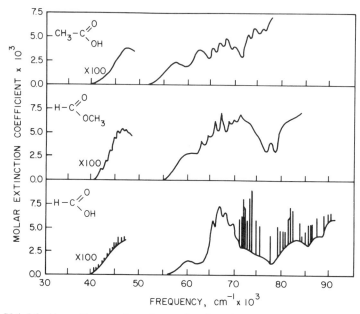

Fig. V.A-16. Absorption spectra of several carboxylic acids and of methyl formate, all as vapors [B10].

from the surrounding bands and do not fit into the analysis," and a second such group at 89 400 cm⁻¹.

Our aim is to place the first five bands of the acid, ester, and acyl fluoride spectra into the mold previously cast for the amide group (Section V.A-1). The logic of such a correlation is supplied by *ab initio* calculations on these systems, which show them to be closely related spectroscopically (Table V.A-IV) [B12]. It is evident from Table V.A-IV that the virtual orbital calculations of excitation energy are not in particularly good agreement with experiment, being in general too high. However, the trends in excitation energy would seem to have more meaning, and the oscillator strengths f calculated using the mixed dipole length–dipole velocity formula [H8] are rather good guides for spectral assignment. The orbital numbering in Table V.A-IV follows that of Fig. V.A-2. Unlike the situation in formamide, the n_O orbital is considerably above the π_2 MO in formic acid, and will no doubt remain so in the higher carboxylic acids. However in the alkyl esters, π_2 will be more destabilized than n_O, and so the two levels will be much closer together, necessitating a detailed photoelectron study to decide their ordering.

Agreement is universal for the assignment of the W band at **47 000** cm⁻¹ in the acids and esters as $n_O \rightarrow \pi_3{}^*$, and no more need be said about it. The notion that the second band of acids and esters (R_1) at about 60 000 cm^{-1} is a valence shell excitation is refuted by the matrix spectrum of trifluoroacetic acid [B12], which shows a broadened and red-shifted V_1 band, but no second band [R19]. Also, the R_1 band in formic acid has an oscillator strength 20 times larger than that calculated for the $n_O' \rightarrow \pi_3^*$ excitation (Table V.A-IV) [B12]. The Rydberg nature of the 57 000-cm^{-1} band of acetic acid (Fig. V.A-16) explains a seeming anomaly. In n-heptane solution, acetic acid shows a clean $n_O \rightarrow \pi_3^*$ band at 48 800 cm^{-1} and then rising absorption which reaches $\epsilon = 100$ at 57 000 cm^{-1} [P25, R28], whereas in the gas phase, $\epsilon = 2500$ at 57 000 cm^{-1}. The very large difference is readily explained in terms of an $n_O \rightarrow 3s$ Rydberg absorption in the gas phase at 57 000 cm^{-1} which is completely smeared out in n-heptane solution (Section II.C). As with the amides, the R_1 band in the acids and esters is completely lacking in vibronic structure, whereas R_2 is usually sharp.

The proposed correlation of the transitions in Fig. V.A-15 is strengthened by the constancy of the separation (14 000 cm^{-1}) of the bands assigned as R_1 and R_2 in these compounds, a constancy which follows from the constancy of 3s and 3p Rydberg term values. This difference of 14 000-cm^{-1} in the frequencies of the $n_O \rightarrow 3s$ and $n_O \rightarrow 3p$ transitions in the CF$_3$COX series is significantly larger than that found in other ketones and amides ($\sim$10 000 cm^{-1} or less) and requires some comment. In line with the view promulgated in this book, the (n_O, 3s) term value is strongly affected by nonchromophoric substituents, in particular, increasing when fluorine atoms are added to the system (Section I.C-1), whereas the (n_O, 3p) term is much less affected, being less penetrating. Consequently, the effect of CF$_3$ groups will be to increase the frequency separation between $n_O \rightarrow 3s$ and $n_O \rightarrow 3p$ transitions by moving the (n_O, 3s) configuration to higher term values. As seen from Table V.A-III, the explanation proposed for the molecules in the trifluoro series results in very reasonable term values for trifluoroacetamide and trifluoroacetic acid, and can be checked further for the acyl fluoride once its ionization potential has been determined.

The photoelectron spectra once again contribute to our understanding of the optical spectra of acids. In formic acid, the 1500-cm^{-1} vibrational progression of the R_2 optical band appears in the first photoelectron band (Fig. V.A-6), with an almost perfect frequency and intensity match. Since this ionization is unambiguously calculated to arise from ionization of an n_O lone-pair electron [B52], and since the δ value of the Rydberg series is characteristic of np upper states, the R_2 band of formic acid is therefore $n_O \rightarrow 3p$. This parallels the assignment of the R_2 band in amides, and we generalize it to the acids and esters. It seems inescapable that if R_2 is $n_O \to 3p$ in acids and esters, then R_1 must be $n_O \to 3s$, as it is in amides. The extraneous bands found by Price and Evans at 80 000 and 89 400 cm^{-1} (vert.) in formic acid have term values of 20 000 and 10 600 cm^{-1} with respect to ionization from the π_2 molecular orbital (100 900 cm^{-1} vert.) [B52] and could well be assigned as $\pi_2 \to 3p$ and $\pi_2 \to 4p$, respectively. Working backward from here, we would expect the $\pi_2 \to 3s$ band to have a term value of 29 000 cm^{-1}, which would put it at about 71 000 cm^{-1}. Referring to Fig. V.A-16, we see that in the optical spectrum of formic acid, there are sharp, Rydberglike features at 71 000–72 000 cm^{-1} which do not appear in the n_O photoelectron band, and hence are not part of the $n_O \to 3p$ optical transition which comes in the same region. We therefore suggest tentatively that they are part of an overlapping $\pi_2 \to 3s$ Rydberg excitation. As an alternative explanation, it could be that the "extraneous" bands in the 71 000–72 000-cm^{-1} region of formic acid are components of the 3p manifold, the degeneracy of which is broken by the electrostatic field of the ionic core. Such a splitting would also occur in the higher np members, but would be missing completely in the corresponding photoelectron band (Section II.A). This latter explanation is to be preferred since transitions to 3s in acids are not structured, in general.

Having identified R_1 and R_2 in the acid and ester spectra, the assignment of the 67 000-cm^{-1} band as V_1, $\pi_2 \to \pi_3^*$ seems straightforward. The extinction coefficient of this band is highest in formic acid (7500), but is less than 5000 in the alkylated acids [B12]. A parallel effect was noted in formamide and the alkyl amides (Section V.A-1), and was explained as due to the preferential participation of the carbon-bonded alkyl groups in the π_3^* MO, thereby reducing the $\pi_2\pi_3^*$ transition density. In the CF_3 series (Fig. V.A-15), the $N \to V_1$ extinction coefficient goes from 8000 in the amide to 4800 in the acid and 2500 in the acyl fluoride. The calculations in the parent HCOX series predict only a slight decrease of the $N \to V_1$ oscillator strength on going from the amide to the fluoride (Table V.A-IV). In the $N \to V_1$ transition of formic acid, a long vibronic progression of 1470–1500 cm^{-1} spacing is observed [P41] and assigned as C=O stretching (1770 cm^{-1} in the ground state); in the same band of methyl acetate, the vibrational frequency is 1450 cm^{-1} [B10].

In formic acid, the R_2 absorption band is clearly seen to be resting upon another band, centered at 73 000 cm^{-1}. A corresponding broad band has been identified in the amide and polypeptide spectra (Sections V.A-1 and V.A-2) as most likely being the $n_O \to \sigma^*(11a')$ transition on the basis of its intensity and frequency. According to Table V.A-IV, such an $n_O \to \sigma^*(11a')$ transition in formic acid is expected to fall about 6000 cm^{-1} to the high-frequency side of the $\pi_2 \to \pi_3^*$ transition, with an oscillator strength approximately one-third that of the $N \to V_1$ absorption. These characteristics agree quite well with the 73 000-cm^{-1} band observed in formic acid, whereas all other energetically reasonable choices for this band have computed oscillator strengths which are smaller than the observed by at least a factor of 50. This band, which in general is badly overlapped by $N \to V_1$ and $N \to R_2$ absorption, could be much more prominent in the circular dichroism spectrum.

Since the compounds are readily available, one can do an interesting term value study of the carboxylic esters. For example, we know that the $(n_O, 3s)$ term values of RCO_2R for large alkyl R groups will be about 22 000 cm^{-1}, whereas their value rises to 36 000 cm^{-1} if the R groups are perfluorinated. Consider now the case of the hermaphrodite $C_2F_5CO_2C_2H_5$. If the Rydberg 3s orbital in this molecule is centered on the C_2H_5 group, the excited-state term value will be close to 22 000 cm^{-1}, but somewhat above it, since C_2H_5 is still rather far from the alkyl limit. If, on the other hand, the orbital is centered on the C_2F_5 half of the molecule, a term of about 35 000 cm^{-1} is expected, while less localized distributions of the 3s orbital will yield intermediate term values. Appropriatey, in $C_2H_5CO_2C_2H_5$, we find an $(n_O, 3s)$ term value of 25 700 cm^{-1}, but in $C_2F_5CO_2C_2H_5$, the term value is 33 500 cm^{-1}, strongly suggesting that the 3s Rydberg orbital is confined to the C_2F_5 half of the molecule (Table V.A-III).

Onari's [O7] spectrum of the polymeric ester polymethyl methacrylate

$$\left(-CH_2-\underset{\underset{COOCH_3}{|}}{\overset{\overset{CH_3}{|}}{C}}- \right)_n$$

is quite unlike that expected for the ester group in a condensed phase. A weak transition at 46 000 cm^{-1} is no doubt the $n_O \to \pi_3^*$ band of the ester group, but there then follows a moderately intense band at 50 000 cm^{-1} (vert.) and a strong one at 54 300 cm^{-1} (vert.). These last two bands are puzzling inasmuch as the second valence shell transition in methyl acetate vapor is found beyond 59 000 cm^{-1} (vert.) [S12]. The possibilities must be considered that large exciton interactions are grossly shifting the carboxylate excited states in this polymer and/or that the $n_O \to \sigma^* (11a')$ transition discussed earlier is appearing in the polymer at a relatively low frequency compared to its frequency in formic acid.

Platt et al. [K30, R28] have made a spectrographic study of the fatty acids in n-heptane solution, obtaining transmission to 58 000 cm^{-1}. In the nonconjugated olefinic acids, such as linoleic acid, $CH_3(CH_2)_4CH= CHCH_2CH=CH(CH_2)_7COOH$, what appears to be the carboxylic

$n_O \to \pi_3^*$ transition is found at 44 000 cm^{-1} (vert.) and a strong band (ϵ = 20 000) is centered at 51 000 cm^{-1}. This strong band is undoubtedly due to the $\pi \to \pi^*$ excitations of the unconjugated olefinic groups (Section IV.A-2), but like the $n_O \to \pi_3^*$ transition, it, too, is at a lower frequency than one would expect. In crotonic acid, trans-CH$_3$HC=CHCOOH, the conjugation between the C=C and C=O groups puts the strong N $\to$ V$_1$ ($\pi \to \pi^*$) band at 48 500 cm^{-1}. This transition is probably closely related to the $\pi_2 \to \pi_3^*$ band of butadiene (Section V.C-1). There are no R$_1$ bands to contend with here, since they would not be present in solution spectra.

Experimental data on the absorption spectra of carboxylate anions are sparse indeed due to the necessary inconvenience of solvent absorption, and, of course, without experiments, there are few calculations. Aqueous solutions of alkali formates show an $n_O \to \pi_3^*$ absorption at 50 000 cm^{-1} (vert.), about 4000 cm^{-1} to the high-frequency side of that for formic acid [B12, J6, L23]. Investigation of a solution of sodium formate in hexafluoroacetone hemihydrate showed that following the $n_O \to \pi_3^*$ band, the absorption rises monotonically from 50 000–62 000 cm^{-1}; the $\pi_2 \to \pi_3^*$ maximum must be just beyond this upper frequency. Though only investigated to 53 000 cm^{-1}, the acetate ion spectrum seems much like that of the formate ion. In the polarized crystal spectrum of sodium formate [C12], the absorption in the 42 000–46 000-cm^{-1} region is polarized along the O $\cdots$ O line, whereas the allowed $n_O \to \pi_3^*$ transition is predicted to be polarized out of plane. It seems more likely that the absorption in this region is part of the broad $\pi_2 \to \pi_3^*$ absorption band, which will have O $\cdots$ O polarization, but there is also the possibility that the observed absorption is due to the *forbidden* $n_O \to \pi_3^*$ transition made allowed by vibronic mixing with the $\pi_2 \to \pi_3^*$ configuration.

Theoretically, Gaussian-type orbital calculations on the formate ion (in the gas phase, of course) predict that the two n_O orbitals on the two equivalent oxygen atoms mix so as to give two $n_O \to \pi_3^*$ transitions, one allowed and one forbidden, separated by only 160 cm^{-1}, and just above that predicted for formic acid [B12, P12]. Further, the calculations place the formate ion $\pi_2 \to \pi_3^*$ band at the frequency found for it in formamide, rather than formic acid, with a mixed oscillator strength of only 0.16, less than half that predicted for formic acid.

Spectra of several carboxylate anions are given by Vinogradov and Dodonova [V1], who studied the aliphatic amino acids in their zwitterionic forms in solid films (Fig. V.A-17). The strong band at ~59 000 cm^{-1} (vert.) in all of these spectra is most likely the N $\to$ V$_1$ band of the —CO$_2^-$ group. This band is found at 62 500 cm^{-1} (vert.) in glycine, and its frequency drops as the size of the alkyl groups grows. These frequencies are undoubtedly characteristic of the solid phase and could be con-

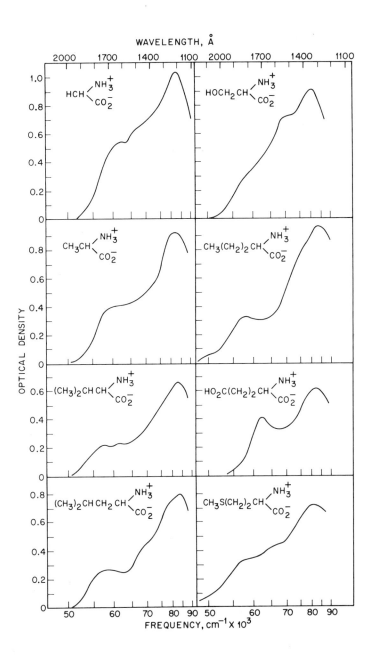

Fig. V.A-17. Absorption spectra of several amino acids in their zwitterionic forms, as solids [V1].

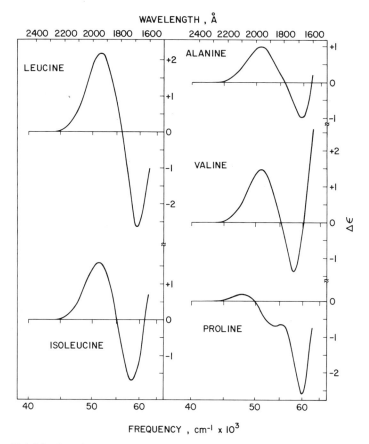

Fig. V.A-18. Circular dichroism spectra of several amino acids in hexafluoroisopropanol solution [S43].

siderably different in the gas phase. There is evidence of a second band, at 68 000 cm⁻¹ (vert.), which would also seem to be due to the carboxylate anion, the two possible assignments being either $\pi_1 \to \pi_3^*$ or $n_0 \to \sigma^*$, the latter of which is preferred. The 82 000-cm⁻¹ band is most likely a combination of alkyl-group absorption and transitions into the conduction bands of the solids. The absorption and circular dichroism spectra of several of these amino acids in hexafluoroisopropanol solution also have been recorded to 63 000 cm⁻¹ (Fig. V.A-18) and the peak positions differ little from those observed by absorption in the solid films. In solution, the circular dichroism spectrum shows a clear $n_0 \to \pi_3^*$ band at 50 000 cm⁻¹ (vert., positive rotatory sign), the presence of which can only be inferred in the absorption spectra. The $\pi_2 \to \pi_3^*$ transitions appear as negatively rotating bands in each of the amino acids at 58 800–60 200 cm^{-1} (vert.). As seems to be the case in almost all of the chromophores studied in Section V.A, the $n_0 \to \pi_3^*$ and $\pi_2 \to \pi_3^*$ bands have nearly equal but opposite rotatory strengths.

In the oxalates, the π_3^* group orbitals in the two halves of the molecules combine in symmetric and antisymmetric combinations with a large splitting, whereas the corresponding effect with the π_2 and n_0 orbitals is minimal due to a much smaller overlap. The net result is that $n_0 \to \pi^*$ and $\pi \to \pi^*$ in oxalates should be found at about 10 000 cm^{-1} lower frequency than in formic acid. The only compound of this sort for which we have vacuum-ultraviolet data is the diester, dimethyl oxalate (Fig. V.A-19) [M5]. Though this molecule is trans-planar in the crystal, this is not necessarily so in the gas phase, and we assume, along with Maria and McGlynn, that the absorber in Fig. V.A-19 is nonplanar. Though

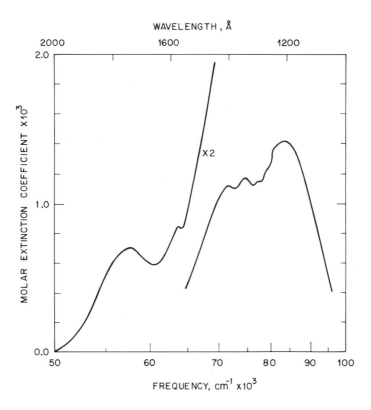

Fig. V.A-19. Absorption spectrum of dimethyl oxalate in the gas phase [M5].

we are no longer certain of which MO is uppermost or even what the first ionization potential of dimethyl oxalate is, we still expect to see the first transitions to the 3s and 3p Rydberg orbitals separated by ~ 6000 cm^{-1}. This is nicely fulfilled by the two bands at 57 500 and 63 000 cm^{-1} (vert.), with the first having an oscillator strength of 0.012. If this interpretation is correct, then the first ionization potential of dimethyl oxalate is 83 500 cm^{-1} (vert.). The mass of stronger absorption which follows at 65 000 cm^{-1} probably consists of $n_O \rightarrow \sigma^*$ and/or $\pi \rightarrow \pi^*$ transitions. In either case, note that the $\pi \rightarrow \pi^*$ transition seems to have shifted to higher frequency with respect to its frequency in the monomer (Fig. V.A-16). One way of looking at this uses the independent systems approach (Section III.A-3) in which the two $O \cdot \cdot \cdot O$ polarized $\pi - \pi^*$ transitions in the two halves of the molecule are pictured as nearly parallel, interacting oscillators. For this geometric configuration, the $\pi \rightarrow \pi^*$ component shifted to lower frequency compared with the monomer will be weak (forbidden in the planar molecule), whereas the larger part of the $\pi \rightarrow \pi^*$ oscillator strength will appear in the component shifted to higher frequency. However, this strongly allowed $\pi \rightarrow \pi^*$ band will have an extinction coefficient much larger than the value of 1200–1400 reported for the bands in the 70 000–80 000 cm^{-1} region. Perhaps the intensities of these bands should be checked.

Besides the spectrum of CF$_3$COF (Fig. V.A-15), the only other acyl halide spectrum in the literature is that of acetyl chloride [L41, W4]. In this compound, there is an $n_O \rightarrow \pi_3^*$ transition at 42 640 cm^{-1} (vert.) followed by what appears to be a pattern of R$_1$, V$_1$, and R$_2$ bands at 62 890, 67 570, and 69 440 cm^{-1} (vert.), respectively. In a related compound, phosgene (COCl$_2$), the $\pi \rightarrow \pi^*$ transition is tentatively assigned to an intense band at 65 000 cm^{-1} (vert.), which is not too different from that in CH$_3$COCl. Another band at 71 400 cm^{-1} (vert.) in acetyl chloride is probably the D band of the chlorine lone-pair electrons (Section III.B-1), or possibly the R$_2$- and D-band assignments should be reversed. The assignment of the R$_1$ and R$_2$ bands is substantiated by their term values of 26 800 and 20 300 cm^{-1} (vert.), respectively. Again we see that the chlorine atom has a penetration effect very nearly equal to that of a methyl group, for the $(n_O, 3s)$ and $(n_O, 3p)$ term values are very nearly equal in the series Cl$_2$CO, CH$_3$COCl, (CH$_3$)$_2$CO.

V.B. Oxides of Nitrogen

The polyatomic molecules of this neglected group are especially interesting since their electronic structures and spectra seem to parallel those of the analogous carbon–oxygen systems such as ketones, amides, acids, urea, carbonates, etc. We begin our discussion with the simplest oxides of nitrogen, the derivatives of the —N=O chromophoric group, which may be considered as isoelectronic with the corresponding aldehydes

$$\begin{array}{c} \text{H} \\ | \\ -\text{C}=\text{O} \end{array}$$

In the spectrum of 1-chloro-1-nitrosocyclohexane, Tanaka et al. [T5] find the $n_O \to \pi^*$ transition at 13 300 cm^{-1} (vert.) and two very sharp features at 50 700 and 55 900 cm^{-1} superposed upon a rising absorption with a maximum near 62 000 cm^{-1} ($\epsilon > 2600$). The two sharp features look very much like Rydberg excitations, and though the term values of these excitations cannot be evaluated since the ionization potential for this molecule is not known, the separation of 5200 cm^{-1} between the bands is just that expected between transitions terminating at 3s and 3p Rydberg orbitals in a molecule of this size. If this is so, then the ionization potential for 1-chloro-1-nitrosocyclohexane will be close to 74 500 cm^{-1} (advert.), and ionization originates with the lone-pair electrons on the oxygen atom; ionization from the lone-pair electrons on chlorine will come at 88 000 cm^{-1} (see, for example, Section III.B-1), with an intense D band expected at $\sim 66 000$ cm^{-1}. The investigators of this spectrum concluded that the broad band centered near 62 000 cm^{-1} is the $\pi \to \pi^*$ (N $\to$ V$_1$) transition of the nitroso group.

A preliminary study of the related nitroso compound CF$_3$NO has been made using photoelectron and electron-impact spectroscopies [R19]. In this molecule, a weak $n_O \to \pi^*$ band is found at 11 300 cm^{-1} (vert.), followed by four bands at 51 000 (moderate), 60 800 (weak), 71 700 (strong), and 84 000 cm^{-1} (vert., very strong). With respect to the first ionization potential at 88 900 cm^{-1} (vert.), the first three of these bands have vertical term values of 37 900, 28 100, and 17 700 cm^{-1}; since the transition to 3s will have a term value of 34 000 cm^{-1} and that to 3p will be 21 000 cm^{-1} (Section I.C-2), it seems unlikely that any of these bands are Rydberg excitations. The large, featureless lump of absorption at 84 000 cm^{-1} is due to the B- and D-band Rydberg absorptions of the CF$_3$ group (Section III.B-3), and the valence shell $\pi \to \pi^*$ band is most likely that at 71 700 cm^{-1}. The bands between 51 000 and 61 000 cm^{-1} are likely to involve $n_O \to \sigma^*$ promotions.

In the cases described here, we can take the chromophoric group as —N=O, with a pendant R group, chlorocyclohexyl or trifluoromethyl. Such systems are isoelectronic with the keto group discussed in Section V.A-1. In that section, it was shown that in those cases where the R group bears formally nonbonding pi electrons, as when R is a halide, —OH, or —N(CH$_3$)$_2$, for example, the n$_O \rightarrow \pi^*$ transition is moved to higher frequencies, whereas N $\rightarrow$ V$_1$ is moved to lower frequencies. Thus, comparing ketones to carboxylic acids, the n$_O \rightarrow \pi^*$ frequencies in these two chromophores are 36 000 and 47 000 cm^{-1}, whereas the N $\rightarrow$ V$_1$ frequencies are $\sim$72 000 and 65 000 cm^{-1}, respectively. Using a molecular orbital description, the interactions responsible for these shifts are displayed in Fig. V.A-1. In such three-center pi-electron systems, the two transitions $\pi_2 \rightarrow \pi_3^*$ and $\pi_1 \rightarrow \pi_3^*$ can alternately be described using the intramolecular charge transfer theory of Nagakura [N3] in which the N $\rightarrow$ V$_1$ transition involves the charge transfer of an electron from the R group into the π^* MO of the C=O group, and N $\rightarrow$ V$_2$ is largely a local $\pi \rightarrow \pi^*$ excitation within the C=O group (Section I.B-2). The MO and charge transfer descriptions seem to be equivalent as regards their predictions.

It is interesting now to see if the spectral shifts of the ketone absorptions induced by halide, hydroxyl, and amine groups are also present when these groups are attached to the nitroso chromophore. Note, however, that because the $\pi \rightarrow \pi^*$ transition of the nitroso group is at substantially lower frequency than that of the keto group, the corresponding $\pi \rightarrow \pi^*$ bands in the substituted nitrosyl derivatives will fall at lower frequencies than in the analogous keto derivatives. In nitrosyl chloride, Cl—NO, the n$_O \rightarrow \pi_3^*$ transition is located at about 17 000 cm^{-1} (vert.) [G16], and it is followed by two broad bands centered at 50 000 and 67 500 cm^{-1} [G16, L19, P4]. At least three other transitions having molar extinction coefficients between 0.1 and 30 appear in the 17 000–35 000-cm^{-1} region, and must in part represent chlorine 3p $\rightarrow \sigma^*$ (N—Cl) A bands (Section III.B-1). According to the charge transfer calculations of Tanaka *et al.*, the strong (ϵ = 2500) band at 50 000 cm^{-1} in nitrosyl chloride is well described as a chlorine 3p $\rightarrow \pi^*$ (N=O) excitation, where π^* is largely localized within the N=O group (N $\rightarrow$ V$_1$). Though the assignments quoted here must be taken as tentative, they are in full qualitative accord with expectations, i.e., upon adding the chlorine atom to the nitrosyl group, the n$_O \rightarrow \pi^*$ transition moves to higher frequency (+2700 cm^{-1}), whereas the lowest $\pi \rightarrow \pi^*$ excitation (N $\rightarrow$ V$_1$) moves to lower frequency ($-$10 000 cm^{-1}). The comparable ketonic molecules would be acetaldehyde and acetyl chloride, for which the n$_O \rightarrow \pi^*$ shift is +1300 cm^{-1}, but unfortunately the $\pi \rightarrow \pi^*$ frequency for neither of these molecules has been positively identified.

In methyl nitrite, CH$_3$ONO, we have an analog of the carboxylic ester methyl formate, and the spectra seem to be similar. The n$_O \rightarrow \pi_3^*$ transition of methyl nitrite falls at 26 000 cm^{-1} (vert.), at considerably higher frequencies than is found in nitrosoalkanes.† This is followed by a second transition at 47 200 cm^{-1} having a molar extinction coefficient of 1050 [T5]. Since the ionization potential of ethyl nitrite is 85 000 cm^{-1} [D10], one estimates that of methyl nitrite to be beyond 88 000 cm^{-1} (vert.), in which case the vertical term value of the 47 200-cm^{-1} band in the latter molecule is far too large to allow a Rydberg assignment. In line with this, the 47 200-cm^{-1} band of methyl nitrite is interpreted by Tanaka et al. as being largely a charge transfer transition from the pi electrons of the OCH$_3$ group into π^* of the N=O group. Alternatively, the assignment can be rephrased as $\pi_2 \to \pi_3^*$ (N $\to$ V$_1$) in MO language. Here one has a very large shift of the N $\to$ V$_1$ band to lower frequency on going from the nitrosoalkanes (62 000 cm^{-1}) to the alkyl nitrites (47 000 cm^{-1}), paralleling that found in the ketone–carboxylic ester systems. The comparison fails, however, on the point of intensity, for the observed N $\to$ V$_1$ oscillator strength of 0.052 in the nitrite is only about one-fifth that observed for the same band in carboxylic esters, and perhaps should be remeasured. Moreover, the 88 000 cm^{-1} (adiab.) ionization potential of methyl nitrite will place its first Rydberg excitation at about 60 000 cm^{-1} (adiab.), far to the high-frequency side of the $\pi_2 \to \pi_3^*$ excitation, in contrast to the situation in carboxylic esters, where the transition to R$_1$ precedes that to V$_1$. As was the case with the carboxylate and amide groups, the ordering of the n$_O$ and π_2 MOs is not immediately obvious in nitrites, nitrates, etc., and would make a good subject for a photoelectron spectroscopy study. In nitromethane, Dewar et al. find the two lowest adiabatic ionization potentials to be 90 600 and 95 100 cm^{-1}, but make no assignments [D10].

The corresponding amidelike oxide of nitrogen is represented by N,N-dimethyl nitrosamine, (CH$_3$)$_2$N—N=O, the spectrum of which shows an n$_O \to \pi_3^*$ transition at 36 500 cm^{-1} (vert.) and two stronger bands at 44 100 (ϵ = 5200) and 48 800 cm^{-1} (vert., ϵ = 2350). Again using the charge transfer model, Tanaka et al. conclude that the 44 100-cm^{-1} band involves an electron transfer from the dimethylamino group to π^* of the nitroso group; once again the observed oscillator strength (f = 0.15) is noticeably lower than that of the $\pi_2 \to \pi_3^*$ band of the corresponding amide, dimethyl formamide (50 000 cm^{-1}; f = 0.24) [H37]. However, as in the ketone–amide series, charge transfer to the —N=O group from the —N(CH$_3$)$_2$ radical results in the lowest-frequency $\pi_2 \to \pi_3^*$ transition. If the 48 810-cm^{-1} band is n$_O \to$ 3s or $\pi_2 \to$ 3s, then the ionization potential of N,N-dimethyl nitrosamine is approximately 73 000 cm^{-1} (vert.).

† A parallel situation exists in the ketones and carboxylic esters, where the frequency of the n$_O \to \pi^*$ transition is ~10 000 cm^{-1} higher in the ester than in the ketone.

McEwen has investigated theoretically the spectrum of the related molecule H_2N—NO by a semiempirical MO technique, and finds the $\pi_2 \to \pi_3^*$ excitation, which is very much of an $NH_2 \to \pi^*(N$—$O)$ charge transfer, to come at 45 600 cm^{-1} with an oscillator strength of 0.18 [M19]. This agrees nicely with the data and interpretation of the 44 100-cm^{-1} band of N,N-dimethyl nitrosamine as $\pi_2 \to \pi_3^*$.

The analogies discussed here can be extended to the nitro group —NO_2 as the basic chromophore. Nagakura [N2] and Loos et al. [L33] report a smooth, strong band ($\epsilon = 5000$, $f = 0.16$) centered at 50 500 cm^{-1} in the gas-phase spectrum of nitromethane, CH_3NO_2. According to pi-electron calculations by Tanaka [T2] (reported in Nagakura's paper) and McEwen [M18], the transition from the highest filled pi MO of the nitro group ϕ_2 to the lowest empty one ϕ_3^*, where

$$\phi_2 = 0.7071(\chi_{O_1} - \chi_{O_2}) \qquad \text{(V.2)}$$
$$\phi_3^* = 0.7009\chi_N - (0.7133/\sqrt{2})(\chi_{O_1} + \chi_{O_2}) \qquad \text{(V.3)}$$

in an obvious notation, has a calculated oscillator strength of 0.38, and no doubt corresponds to the observed band at 50 500 cm^{-1}.† From the form of the wave functions given here, it is readily seen that the $\phi_2 \to \phi_3^*$ transition of the nitro group is closely analogous to the lowest $\pi \to \pi^*$ excitations of the nitrate and nitrite ions at 50 000 cm^{-1} (Section VII.A). Since the vertical ionization potential of nitromethane is 91 300 cm^{-1} [R3], the lowest Rydberg excitation in this molecule should not appear below 60 000 cm^{-1}, or below 55 000 cm^{-1} in the higher nitroalkanes. What is apparently an $n_O \to \pi_3^*$ transition is observed at 37 000 cm^{-1} in nitromethane [L33].

If the alkyl group of nitromethane is replaced by one having nonbonding pi electrons, then the spectrum may be expected to show charge transfer as well as local excitations. Kaya et al. [K10] studied both ethyl nitrate C_2H_5O—NO_2 and nitramide NH_2—NO_2 with this idea in mind. Ethyl nitrate vapor displays a single broad band at 52 600 cm^{-1} (vert., $\epsilon = 6800$; $f = 0.23$), whereas what appears to be the same transition in nitramide comes at 50 500 cm^{-1} (vert., $\epsilon \simeq 6000$; $f = 0.23$) in n-heptane solution. It is clear both from the appearance of these bands in the solu-

† In the carboxylic acids and amides, it was found that the $\pi_2 \to \pi_3^*$ oscillator strength was considerably larger in those compounds bearing a hydrogen atom on the carbon of the three-center chromophore as compared to the alkylated chromophores. The explanation involves a contribution of the alkyl group to π_3^* which is missing from π_2 due to symmetry, with a concomitant decrease of the $\pi_2 \to \pi_3^*$ transition density. A similar effect is expected in the H—NO_2, R—NO_2 series, and may account for the discrepancy between the f values calculated for HNO_2 and observed for CH_3NO_2.

tion spectra (Section II.C) and from their oscillator strengths that they are valence shell excitations rather than Rydberg. Both of these transitions were interpreted as being intramolecular charge transfer excitations, or equivalently as $\pi_3 \to \pi_4^*$, as shown in Fig. V.A-1. Since the π_2–π_3 split should be small in these compounds (they are degenerate in the nitrate ion), the $\pi_2 \to \pi_4^*$ transition should closely follow the $\pi_3 \to \pi_4^*$ band, and, in fact, Kaya et al. claim that $N \to V_1$ and $N \to V_2$ are superposed in both ethyl nitrate and nitramide. Mullen and Orloff report that pentaerythritol tetranitrate, $C(CH_2O-NO_2)_4$, also has an intense band at 51 700 cm^{-1} (vert., $\epsilon = 20\,400$ in acetonitrile solution) which they calculate to consist of the two allowed $\pi \to \pi^*$ transitions of the NO_3 group [M59]. In hexahydro-1,3,5-trinitro-s-triazine, $(-CH_2NO_2N-)_3$, the splitting of the two lowest $\pi \to \pi^*$ transitions is much more obvious, the peaks coming at 42 400 cm^{-1} ($\epsilon = 11\,000$) and 51 100 cm^{-1} ($\epsilon = 16\,400$) [O14]. In the alkali metal nitrates, the corresponding $\pi_2, \pi_3 \to \pi_4^*$ degenerate excitation is found at 50 000 cm^{-1} (vert.) (Section VII.B).

This problem assumes another dimension of complexity in the nitroethylenes [L33], where not only do the $\pi_2 \to \pi_4^*$ and $\pi_3 \to \pi_4^*$ excitations of the nitro group fall near 50 000 cm^{-1}, but the ethylenic $\pi \to \pi^*$ local excitation is also a possibility in this region. Experimentally, transitions are observed at 41 300 ($f = 0.012$) and 49 400 cm^{-1} (vert., $f = 0.304$) in gaseous nitroethylene, whereas in the unconjugated system 3-nitro-propene-1, only one strong band is found, at 46 900 cm^{-1} (vert., $f = 0.211$). Between 34 000 and 38 000 cm^{-1}, $n_O \to \pi^*$ bands are observed in all compounds. Of the three possibilities, $\pi_2 \to \pi_3^*$ localized in the NO_2 group, $\pi \to \pi^*$ localized in the ethylenic group, and the ethylene $\to NO_2$ charge transfer transition, the first two never have an oscillator strength as low a 0.01, thus ruling them out as assignments for the band at 41 300 cm^{-1} in nitroethylene. This band could be a Rydberg transition terminating at 3s, though even if this were true, the originating orbital is still in doubt. On the other hand, Pariser–Parr–Pople calculations on these systems do predict a relatively low oscillator strength ($f = 0.017$) for the charge transfer transition, followed by a rather strong ($f = 0.41$) $\pi \to \pi^*$ transition localized within the $-NO_2$ group [L33]. Alternatively, on the basis of CNDO-CI calculations, it has been proposed that the 41 300-cm^{-1} band of nitroethylene is instead the second $n_O \to \pi^*$ transition [L17], in analogy with the results of a calculation on nitromethane [T13]. However, it would seem that the observed intensity is too high for this, and a charge transfer assignment is to be preferred. It is interesting to note how the MO descriptions of these lower states parallel closely the charge transfer local-excitation descriptions given to similar molecules by Nagakura and co-workers.

The intense band in the 47 000–50 000-cm^{-1} region of the conjugated nitroolefins has an undoubted analog in the spectra of the conjugated carboxyl olefins, crotonic acid (trans-CH_3CH=$CHCO_2H$) and 2-heptadecenoic acid [R28], both of which show a very strong, broad $\pi \rightarrow \pi^*$ transition centered at about 47 000 cm^{-1} ($\epsilon \simeq 16\,000$ in n-heptane solution), with a suggestion of a second weaker band at about 42 000 cm^{-1} (Section V.A-3).

The spectra of the inorganic nitrate and nitrite anions are closely related to those of the organic nitrogen oxides discussed in this section, and are described more fully in Section VII.B.

V.C. Dienes and Higher Polyenes

V.C-1. *Dienes*

In this and the following subsections we discuss consecutively the spectra of conjugated all-carbon dienes, as typified by butadiene, nonconjugated dienes such as 1,4-cyclohexadiene, and dienes having a heteroatom carrying pπ lone-pair electrons, as in furan. The key papers on the spectra of the conjugated dienes describe the experimental work of Price and Walsh [P43, P44] and Carr et al. [C7] and the theoretical interpretations of Mulliken [M60]. Before embarking on any discussion of the excited electronic states of butadiene and its derivatives, we must first consider the probable geometries of the molecules in a gaseous sample at room temperature. There has been much discussion in the past regarding the cis–trans isomerization about the central single bond of butadiene, and the relative amounts of each isomer at equilibrium; the estimates range from less than 1% to over 50% cis isomer in the mixture. In the most recent study, the negative result of a search for microwave absorption in butadiene gas places the cis percentage below 1% [L10]. We shall take the butadiene molecule as planar, centrosymmetric, and trans for our purposes. The other dienes will also be taken as trans-planar, unless there are steric or spectral reasons for thinking otherwise.

The two ethylenic groups of a conjugated diene may be thought of as interacting to produce two filled 2p pi MOs, π_1 and π_2, and two empty pi MOs, π_3^* and π_4^*. The pi-electron configuration in the ground state is then $\pi_1^2\pi_2^2$, called N, and the lowest-energy excited singlet state, V_1, has the configuration $\pi_1^2\pi_2^1\pi_3^1$. Though it is generally assumed that the highest filled sigma MO, $7a_g$, is below π_1, this might not be so, and they are very close together in any case [B59, B63, D12]. Experimentally, the N $\rightarrow$ V_1 ($\pi_2 1b_g \rightarrow \pi_3^* 2a_u$) transition in butadiene is centered at **47 800**

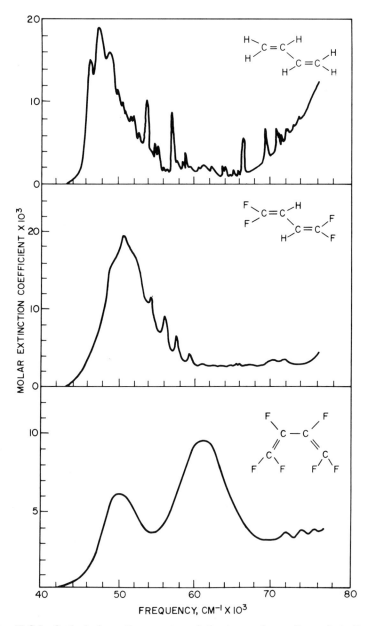

Fig. V.C-1. Optical absorption spectra of the trans planar dienes butadiene and 1,1,4,4-tetrafluorobutadiene, and the cisoid twisted diene hexafluorobutadiene [B53].

cm⁻¹ (Fig. V.C-1), and successive substitution of its hydrogens by methyl groups shifts the $N \to V_1$ band to lower frequencies by about 1000 cm⁻¹ per methyl group. The $N \to V_1$ transition in butadiene is allowed ($f = 0.4$), and appears with even higher oscillator strengths in the alkyl-substituted butadienes [J1]. However, the vibrational structure of the $N \to V_1$ band uniformly consists of a short progression of about 1400 cm⁻¹, the totally symmetric C=C stretch, reduced by about 200 cm⁻¹ from the ground-state value. In a theoretical calculation, Shih et al. [S32] predict not only the excitation of C—C stretching in the $N \to V_1$ band of butadiene, but significant twisting of the two CH_2 groups, as in the $N \to V$ transition of ethylene (Section IV.A-1).

Interestingly, the $N \to V_1$ band of butadiene has a term value of 25 400 cm⁻¹, which, by our work in the preceding sections, can be taken as an indication of a 3s terminating orbital. This is clearly not so, and serves to point out that the term value by itself is necessary but not sufficient evidence for a Rydberg upper state. That the $N \to V_1$ band is totally valence shell rather than Rydberg is clearly shown by the crystal spectra of butadiene and its derivatives (Fig. V.C-2), in which the $N \to V_1$ band is solvent-shifted to lower frequencies as expected, but displays none of the more drastic solvent effects found for Rydberg transitions in condensed phases [R19].† There is no experimental measurement of the polarization of the $N \to V_1$ transition in a trans diene, but it is probably along the line connecting the centers of the olefin groups.

The next two lowest configurations of *trans*-butadiene, $\pi_1^1\pi_2^2\pi_3^1$ and $\pi_1^2\pi_2^1\pi_4^1$, have the same orbital symmetry (a_g in C_{2h}) and are very nearly degenerate. Their interaction results in two mixed-configuration states, transitions to both of which are parity forbidden from the ground state. The very successful Pariser–Parr calculations of Allinger and Miller [A5] predict that after configuration interaction, the $N \to V_2$ transition of *trans*-butadiene comes at 56 000 cm⁻¹, with $N \to V_3$ at 59 000 cm⁻¹. Experimentally, in the trans dienes, a region of weak, continuous absorption is observed centered at 60 000 cm⁻¹ in the gas-phase spectra, upon which are superposed many sharp, Rydberglike features (Fig. V.C-1). The reality of this underlying absorption is reinforced by the spectrum of polycrystalline butadiene (Fig. V.C-2), in which the Rydberg absorption does not appear, but which clearly shows the valence shell transition at

† Apropos the furor over the diffuseness of the V state of ethylene (Section IV.A-1), Shih et al. [S32] have raised the same complication in the butadiene spectrum. Their extended calculations place the $^1B_u(V_1)$ state below the $^1A_g(V_2)$ state and predict that the π^* MO of the V_1 state has considerable 3p character to it. As with ethylene, this supposed diffuseness in the upper state of butadiene is not at all evident on comparing the spectra of gaseous and solid samples (Fig. V.C-2).

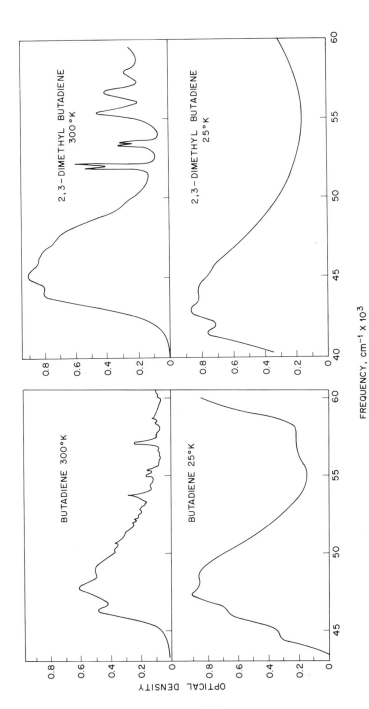

Fig. V.C-2. Optical absorption spectra of butadiene and 2,3-dimethyl butadiene in the gas phase at 300°K (upper) and as solid films at 24°K (lower) [R19].

60 000 cm^{-1} (vert.). Thus it seems fairly safe to assume that the valence shell N → V$_2$ and N → V$_3$ bands of the trans dienes are found in the 55 000–65 000 cm^{-1} region, submerged beneath the more prominent Rydberg excitations [M60].†

In opposition to the simple picture just given, Schulten and Karplus [S21] have recently discussed in detail the possibility mentioned earlier by others [B63] that the intense N → V$_1$ band of butadiene is *not* the lowest singlet–singlet excitation. They find instead that the interaction between the configurations that form the V$_2$, V$_3$ complex is so large that the lower component (V$_2$,^{1}A$_g$) actually lies below the V$_1$ configuration. Viewed as coupled excitons, the low-lying V$_2$ state is formed from the two ethylenic groups simultaneously excited to their triplet $\pi \to \pi^*$ states, and then coupled to form an overall singlet state. Dumbacher [D23] finds the V$_2$ transition to be below V$_1$ at all angles of twist from *trans*- to *cis*-butadiene. Experimental data on this interesting transition are still lacking.

In a cis diene, a conformation achieved by ring closure, the π MOs will differ from those in the trans configuration by virtue of an altered 1–4 interaction, possible nonplanarity, and ring-strain effects. In general the cis cyclic dienes have their N → V$_1$ bands at somewhat lower frequencies (43 100 cm^{-1} vert. in cyclopentadiene, 40 300 cm^{-1} vert. in 1,3-cyclohexadiene, compared with 46 700 cm^{-1} vert. in *trans*-2,4-hexadiene), and most importantly, the N → V$_1$ oscillator strength suffers a severe reduction to about one-third that of its trans counterpart [H14, J1]. In those dienes in which the two double bonds are parallel rather than just cis, the N → V$_1$ oscillator strength is predicted to be zero.

The presence of the V$_2$, V$_3$ complex in the 60 000–65 000-cm^{-1} region is amply confirmed in the spectra of the cis dienes, (Fig. V.C-3), in which the N → V$_2$, V$_3$ bands appear with a high oscillator strength.‡ In fact, it has been suggested that the weak absorption observed at 60 000 cm^{-1} in butadiene may be due to a very small amount of the cis isomer in equilibrium with the trans. An interesting aspect of the N → V$_1$ and N → V$_2$, V$_3$ bands of the butadienes is their relative oscillator strength as a function of the dihedral angle θ between the planes of the ethylenic

---

† Because the $7a_g$ MO is so close to $\pi_1(1a_u)$, one might expect that the $\sigma(7a_g) \to \pi_3^*(2b_u)$ transition in butadiene might fall close to the N → V$_2$ pi-electron excitation. Actually, both bands might be contributing to the absorption at 60 000 cm^{-1}.

‡ If, in fact, the absorption of V$_2$ precedes that to V$_1$, then the absorption in the 60 000–65 000-cm^{-1} regions of cis dienes should be assigned more simply as N → V$_3$. Until this is definitely shown to be the case, we will continue to speak of the V$_2$, V$_3$ complex as if the two bands are still quasidegenerate.

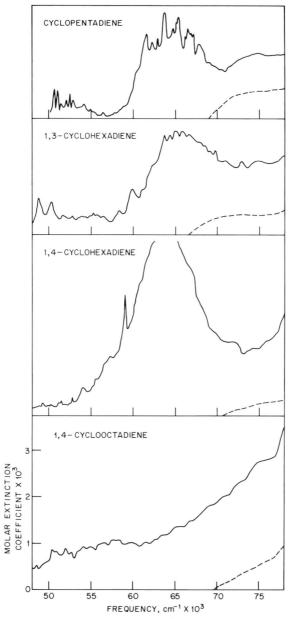

Fig. V.C-3. Optical absorption spectra of some conjugated and some nonconjugated cyclic dienes [D2].

groups. Allinger and Miller [A5] calculate that on twisting the molecule from the cis geometry ($\theta = 0°$) to the trans ($\theta = 180°$), the $N \to V_1$ oscillator strength decreases to zero at $\theta = 90°$ and then increases again to a local maximum at $\theta = 180°$. The $N \to V_2, V_3$ oscillator strength, on the other hand, shows a monotonic increase as θ goes from $180°$ to $0°$. As a result, the $N \to V_1$ transition is much stronger in and around the trans configuration, whereas $N \to V_2, V_3$ is about twice as strong as $N \to V_1$ in and around the cis configuration. Since the Allinger and Miller calculations predict $N \to V_2$ to be considerably weaker than $N \to V_3$ for all values of θ, if the $N \to V_2$ band is found at all, it will be as a step on the low-frequency wing of the $N \to V_3$ band.

There are numerous examples of the strong effect geometry has on the $(N \to V_1)/(N \to V_2, V_3)$ intensity ratios in conjugated dienes [B42, B53, W34]. Thus in cyclopentadiene and 1,3-cyclohexadiene, for which there are only qualitative measurements, it is found that the $N \to V_2, V_3$ bands at 65 000 cm^{-1} (vert.) (Fig. V.C-3) are much stronger than the $N \to V_1$ bands located at 43 100 and 40 300 cm^{-1} (vert.). In hexafluorobutadiene (Fig. V.C-1), the $N \to V_1$ band at 50 700 cm^{-1} is less intense than the $N \to V_2, V_3$ band at 61 000 cm^{-1}, indicating that the ground state structure of this molecule is certainly not trans planar and may be cisoid, though not necessarily planar [B53]. Because the predictions of the independent-oscillator model (Section III.A-3) are intimately related to the molecular geometry, it was applied to the twisted-butadiene intensity problem, with the result

$$\frac{f(N \to V_1)}{f(N \to V_3)} = \frac{5 - 3 \cos \theta}{3 + 3 \cos \theta}. \qquad \text{(V.4)}$$

Combining this equation with the experimental oscillator strength ratio of hexafluorobutadiene leads to a "measurement" of $\theta = 42°$ in the cisoid conformation. In regard to this, electron diffraction work on hexafluorobutadiene has confirmed this nonplanar cisoid geometry [C11], and the diminution of the pi-electron overlap in the twisted molecule is readily apparent in its photoelectron spectrum [B53]. 1,1,4,4-Tetrafluorobutadiene, on the other hand, has the normal trans planar spectrum (Fig. V.C-1), consistent with its known centrosymmetric structure. Hexachlorobutadiene, having a nonplanar cisoid structure like that of the hexafluoro compound, also should have an $N \to V_2, V_3$ transition which is stronger than its $N \to V_1$ band. Braude reports that the $N \to V_1$ band of hexachlorobutadiene is beyond 48 000 cm^{-1}, with an extinction coefficient less than 10% of that of butadiene [B43]. He also reports that both 1,1,3-trimethyl butadiene and 1,1-dimethyl-3-chlorobutadiene have $N \to V_1$ intensities which are only 34% of that butadiene. Presumably they are twisted, with rather intense $N \rightarrow V_2$, V_3 transitions. A similar inversion of intensity due to twisting occurs in 2,3-di-t-butyl butadiene [W34]. In butadienes bearing trimethyl silyl substituents in the 1,4 positions, the spectrum is "normal," i.e., that expected for a trans planar arrangement, whereas with substituents in the 2,3 positions, the spectrum is that of a noncentrosymmetric diene [B32]. See Section III.G for details.

As a cis diene is twisted away from its planar configuration, the frequencies of the $N \rightarrow V_1$ and $N \rightarrow V_2$, V_3 transitions are expected to approach one another. Thus it is found in planar dienes such as cyclopentadiene that the V_1–V_3 split is 22 000 cm^{-1}, whereas in the twisted system hexafluorobutadiene, this is reduced to 12 000 cm^{-1}. In 1,4-cyclohexadiene, where the double bonds are almost totally uncoupled (not by twisting, but by the intervention of a methylene group), the V_1–V_3 splitting collapses to give a single broad absorption feature at 64 000 cm^{-1} (vert.) (Fig. V.C-3).

The transition to the ^1A_g state proposed to lie below V_1 is forbidden in the trans dienes, but is formally allowed in the cis isomers. It is interesting to note in this regard that in cyclopentadiene, Pickett et al. find a weak structured band displaying four quanta of a 770-cm^{-1} vibration resting upon the broad, structureless $N \rightarrow V_1$ band [P17, S13]. It is clearly not part of the $N \rightarrow V_1$ excitation, and it is clearly not a Rydberg transition to 3s since it appears unperturbed in the hexane-solution spectrum (Section II.C) and its term value with respect to the ionization potential of 69 100 cm^{-1} (advert.) is about 7000 cm^{-1} larger than expected for a Rydberg (Section I.C-2). Tentatively, it would appear that this band is the $\pi \rightarrow \pi^*$ excitation predicted to precede $N \rightarrow V_1$ in dienes [S21]. A similar band could not be found in the spectra of 1,3-cyclohexadiene or 1,3-cyclooctadiene [R19].

The final $\pi \rightarrow \pi^*$ transition, $\pi_1{}^2\pi_2{}^2 \rightarrow \pi_1{}^1\pi_2{}^2\pi_4{}^1$, is an allowed one which Allinger and Miller predict to come at 75 000 cm^{-1} in $trans$-butadiene, but with a very low oscillator strength for a formally allowed transition ($f \simeq 10^{-4}$). While a peak is observed at 78 000 cm^{-1} (vert.) in butadiene and at 74 000 cm^{-1} (vert.) in isoprene [P43], these are much stronger transitions than predicted, and might very well involve sigma orbitals in one or both states, especially since π_1 and $\sigma 7a_g$ are so close together.

The spectra of the trans dienes are rich in Rydberg transitions. In butadiene itself, there are two Rydberg members at 50 560 cm^{-1} and 53 650 cm^{-1} (advert.) [C7], the analogs of which also appear in the alkylated dienes; all such bands show a 1500–1600-cm^{-1} vibrational progression, the upper-state C=C stretch. By their term values of 22 600 and 19 600 cm^{-1} (advert.), it would appear that these bands in butadiene are

$\pi_2 1b_g \to 3p$ transitions split by the nonspherical core. The Rydberg nature of these bands in butadiene and 2,3-dimethyl butadiene is readily demonstrated by their failure to appear in the neat-film spectra (Fig. V.C-2), and additional examples can be found by comparing the vapor and solution spectra of dienes reported by Platt and co-workers [J1, K30]. In liquid 1,3-cyclohexadiene [S46], the transitions to V_1 and V_3 appear at close to their vapor-phase frequencies, but with another clear band coming between them at 51 000 cm^{-1} (vert.). However, as the other diene spectra show, this band in the liquid cannot be due to the Rydberg absorption, which comes at the same frequency in the gas phase, but instead must be due to another valence shell transition which underlies the Rydberg bands. The band at 80 000 cm^{-1} (vert.) in liquid 1,3-cyclohexadiene is also present at this frequency in benzene, cyclohexene, and cyclohexane.

The $\pi_2 1b_g \to 3s$ Rydberg excitation in butadiene is symmetry forbidden, and will have a term value about equal to that of transitions terminating at 3s in butane, placing it at $\sim$48 000 cm^{-1} (advert.). It probably corresponds to the weak but definite structure seen in the absorption curve in the 50 000–51 000-cm^{-1} region (Fig. V.C-1). Price and Walsh [P43] identify another Rydberg system beginning at 57 000 cm^{-1}, and can fit two series converging upon the same ionization potential from among the strong bands beyond 66 000 cm^{-1}. The first series has $n = 3$, 4, ..., 8, with $\delta = 0.10$, and the second has $n = 4, 5, ..., 8$, with $\delta = 0.50$, indicating nd and np series, respectively. The 53 650-cm^{-1} Rydberg band is the $n = 3$ member of the second series (55 540 cm^{-1} calculated). As with the ns series, the transitions to members of the nd series from the $1b_g$ MO are parity forbidden and must appear through the agency of nontotally symmetric vibrations. Because the butadiene molecular ion is so far from being spherical (or even linear), the lower terms of np, nd, and nf series will undoubtedly show splittings which will decrease to zero as n increases to infinity [L24]. In fact, in the two Rydberg series mentioned above, such a decreasing splitting in successive terms is observed.

A calculation of the molecular structure of the butadiene positive ion by Hutchinson [H39] predicts a symmetric structure in which the two terminal C—C bonds expand by 0.05 Å while the central bond contracts by the same amount. Presumably, the same structure would apply to the core of the butadiene Rydberg states.

The Rydberg spectra of isoprene and 1,3-dimethyl butadiene are very similar to that of butadiene except for the shift to lower frequencies, reflecting their lower ionization potentials. Chloroprene, CH$_3$CH=CCl—CH=CH$_2$, in addition to the above two series, shows another strong one beginning at 70 000 cm^{-1}, which corresponds to the chlorine lone-pair excitation. Its terms value of 22 000 cm^{-1} shows that it is analogous to the D band of methyl chloride (Section III.B-1), and terminates in a chlorine 4p orbital. Price and Walsh infer that it is a chlorine 3pπ orbital that is being excited, but recent photoelectron spectra of molecules containing chlorine adjacent to C=C pi orbitals [B3, B28] show that there is an appreciable split (10 500 cm^{-1}) between the pi lone pair and the lone pair perpendicular to it, and these workers suggest that it is the perpendicular pair which has the lower ionization potential, in agreement with later theoretical work [K23].

Sugden and Walsh [S51] claim to have assembled a third Rydberg series in butadiene converging upon an ionization potential 2500 cm^{-1} lower than that given by the other two series. Inasmuch as Dewar and Worley [D9] searched specifically for such an ionization potential using photoelectron spectroscopy and failed to find it, one must doubt the reality of this third series.

Comparison of the Rydberg spectra of butadiene and 1,1,4,4-tetrafluorobutadiene is especially interesting. Since the perfluoro effect (Section II.A) is operative in this pair, the first pi-ionization potential of the fluoro derivative (75 000 cm^{-1} vert.) [B53] is very nearly equal to that in butadiene (73 200 cm^{-1} advert.). However, the presence of four fluorine atoms in the molecule acts to dramatically increase the penetration energy in the $(\pi_2, 3s)$ Rydberg state, leading to an increased term value, which in turn places the $\pi_2 \to 3s$ transition considerably lower in the fluoro compound, 46 500 cm^{-1} (vert.) versus 52 000 cm^{-1} (vert.) in butadiene (Fig. V.C-1). On the other hand, since the ionization potentials are almost equal and the $(\pi_2, 3p)$ term value is much less sensitive to fluorination, the $\pi_2 \to 3p$ transitions commence at 53 000–54 000 cm^{-1} in both butadiene and tetrafluorobutadiene.

The cyclic cis dienes are similarly rich in Rydberg series absorptions. The N $\to$ V$_1$ band of cyclopentadiene has superposed upon it two sharp sets of bands beginning at 38 880 and 45 800 cm^{-1} [P17, S13]. The first of these has a 31 000-cm^{-1} term value which is much too large for a Rydberg excitation, and so it may in fact be the lowest $\pi \to \pi^*$ excitation (see earlier discussion), but the second has a term value of 24 000 cm^{-1} (advert.), which is appropriate for a transition to 3s in a molecule as large as cyclopentadiene. However, the transitions may or may not be real, since Price and Walsh do not mention them at all in their work on the same compound. Price and Walsh do identify a short, four-membered series beyond 63 000 cm^{-1} in cyclopentadiene which converges upon 69 550 cm^{-1}. Most recently, Derrick et al. [D7] have determined the π_2 ionization potential of cyclopentadiene to be 69 080 cm^{-1} (vert.), and instead proposed a series having $\delta = 0.45$. The first member of this series at 50 380 cm^{-1} has a term value of 18 700 cm^{-1} (3p) and an electronic splitting of 450 cm^{-1}. The vibronic structures of the first band in the photoelectron spectrum and in the optical spectrum at 50 380 cm^{-1} are very similar.† The situations in 1,3-cyclohexadiene [P44] and 1,3-cyclooctadiene are much the same as that in cyclopentadiene, except that the two weaker band systems are not observed in the N → V$_1$ region; high-pressure and crystal spectra of these two dienes have shown conclusively that the bands at ~ 50 000 cm^{-1} are Rydberg excitations [R19].

In cyclopentadiene and 1,3-cyclohexadiene, the allowed N → V$_3$ band displays considerable vibronic structure [P17, P44], the most prominent features of which are the 1450 and 480 cm^{-1} separations in cyclopentadiene and the 1400 and 100 cm^{-1} spacings in 1,3-cyclohexadiene. The larger vibrational interval is assigned to the C=C stretching mode, whereas the smaller is thought to be due to the CH$_2$-group rocking motion.

One puzzling feature of the diene optical spectra is that the terms definitely identified by Price and Walsh as fitting to Rydberg series are all claimed to be absolutely vertical, without vibrational fine structure. It is of course reasonable that the bonding to nonbonding $\pi_2 \to n$p transitions are more vertical than the bonding to antibonding $\pi_2 \to \pi_3^*$, but since the π_2 ionization in the photoelectron spectrum of butadiene displays an obvious progression of 1500 cm^{-1}, as well as intervals of 1200 and 520 cm^{-1} [B53], it does seem strange that the Rydberg transitions converging on this ionization potential are really without vibrational structure.

In the unconjugated dienes, such as 1,4-cyclohexadiene or norbornadiene

there is a diminished but still nonzero overlap between the two olefinic groups as well as a through-bond interaction, both of which act to split the degeneracy of the two filled π MOs, π_1 and π_2. An alternative approach to the spectra of such molecules considers the C=C groups to have

† Byrne and Ross [B72] state the general rule that the lowest singlet–singlet electronic transition in a molecule will have the sharpest vibronic structure, but then quote the case of cyclopentadiene as an apparent exception to the rule since the 51 000-cm^{-1} bands are much sharper than those at 43 000-cm^{-1}. The explanation here is that the 51 000-cm^{-1} bands are Rydberg excitations ($\pi_2 \to$ 3p) and are so weakly coupled to the valence shell manifold that they are not relaxed rapidly by the lower V$_1$ state.

zero overlap, but to be interacting through an electrostatic resonance force (Section III.A-3). In this independent-oscillator picture of the nonconjugated dienes, the two parallel oscillators interact so as to yield one excited state to which transitions are forbidden from the ground state, and a higher one to which transitions are twice as strong as in the single oscillator transition [W14]. A qualitatively similar result is obtained from a delocalized MO picture, for the $N \rightarrow V_1$ transition is predicted to be lowest but forbidden, whereas the $N \rightarrow V_2$ band is allowed, but displaced about 8000 cm^{-1} higher [H15, W24].

The low-lying band centered at 47 400 cm^{-1} (vert.) in the norbornadiene spectrum (Fig. V.C-4) was long thought to be the forbidden $\pi_2 \rightarrow \pi_3^*$ transition predicted theoretically to fall at this frequency. However, a vibrational analysis of this band [R10] later showed that there was intensity at the origin, that the long vibrational progression was a totally symmetric one, and that, therefore, the transition is an electronically allowed one, possibly the lowest Rydberg excitation from π_2. Supporting evidence for a Rydberg assignment comes from the norbornadiene photoelectron spectrum, in which one finds a first band which has vibrational intervals and Franck–Condon factors [B61] almost identical to those of the optical transition under discussion. Moreover, the optical absorption is just the type of sharp, easily accessible transition on which the high-pressure effect (Section II.B) works best. The result of this experiment on norbornadiene is rather interesting (Fig. V.C-4), for the high pressure of perturbing helium gas readily washes out the 384-cm^{-1} vibrational progression, thus showing the big-orbit nature of its upper state, but in doing so, uncovers a second transition having a 1200-cm^{-1} vibrational interval [R15]. The same effect is observed on going from the gas phase into the low-temperature crystal (Fig. V.C-4) [R10] and the sharp features of this underlying valence shell transition are identifiable in the gas-phase spectrum as well, once their presence is appreciated.

Thus these simple experiments lead to the conclusion that a valence shell band does come at about 47 000 cm^{-1} (vert.) in norbornadiene, as calculated for $\pi_2 \rightarrow \pi_3^*$, but is normally covered by an allowed Rydberg excitation composed of a long progression in a low-frequency vibration (384 cm^{-1}). Since $\pi_2 \rightarrow 3s$ is an allowed transition in norbornadiene, we assign the 47 400 cm^{-1} transition to this, noting that the seven carbon atoms in the molecule will lead to a (π_2, 3s) term value rather close to the alkane limit. Thus it is no surprise that we find the term value to be 22 800 cm^{-1} (vert.). The band centered at 54 000 cm^{-1} (Fig. V.C-5) seems a likely candidate for the $\pi_2 \rightarrow 3p$ transition, and $\pi_2 \rightarrow 4s$ should come just beyond this. The strongly allowed $N \rightarrow V_2$ transition has not been positively identified as yet, but could come at 55 000 cm^{-1} (vert.).

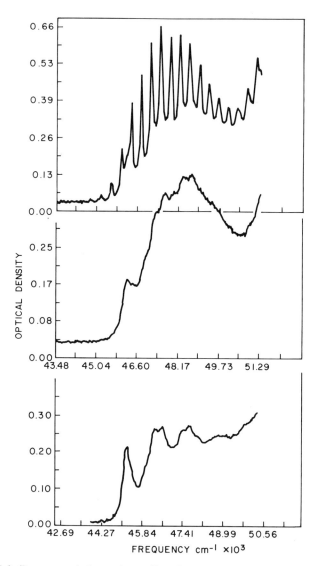

Fig. V.C-4. Response of the norbornadiene band at 47400 cm^{-1} (vert.) in the gas phase (upper) to 136 atm of He gas (middle), and the spectrum as a thin film at 24°K (lower) [R15].

Another aspect of the electronic structure of such nonconjugated systems of particular interest is the extent of interaction between the two C=C bonds [D14, H23]. Rephrased, using Koopmans' theorem within the framework of the molecular orbital theory, we ask what the difference

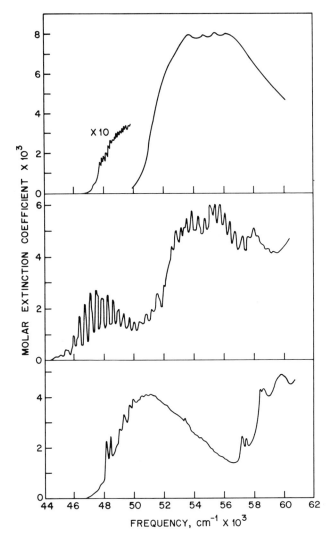

Fig. V.C-5. Comparison of the optical absorption spectra of cyclopentene (upper), norbornadiene (middle), and norbornene (lower), all in the gas phase at 300°K [R10].

is between the ionization potentials of the π_1 and π_2 MOs. Comparison of the photoelectron spectra of norbornane, norbornene, and norbornadiene convincingly demonstrates that this splitting in the latter molecule is 6860 cm^{-1} [B31, B61]; in 1,4-cyclohexadiene, it is 8100 cm^{-1} [B31]. For comparison, the π_1–π_2 splitting in butadiene is 24 000 cm^{-1}. Thus the interaction in these "nonconjugated" dienes, while smaller than those in the conjugated dienes, is nonetheless appreciable. As a consequence of the relatively small splitting of the π_1 and π_2 orbitals, the optical spectra of such unconjugated dienes will contain badly overlapping Rydberg series originating at these two levels, making the identification of any but the lowest member most difficult. However, the 6860-cm^{-1} splitting of the π_1 and π_2 MOs in norbornadiene immediately leads one to expect the $\pi_1 \to 3s$ transition (allowed with "out-of-plane" polarization) in norbornadiene to come 6860 cm^{-1} beyond the $\pi_2 \to 3s$ transition at 47 400 cm^{-1} (vert.), and we assign the band at 53 900 cm^{-1} (vert.) accordingly. The $\pi_1 \to 3s$ transition is probably sitting upon the strongly allowed $N \to V_2$ valence shell excitation at 55 000 cm^{-1} (vert.). The reality of a valence shell assignment for the 55 000-cm^{-1} band is supported by the matrix spectrum, which shows this band shorn of its Rydberg overcoat [R10].

The fact that the $N \to V_1$ bands of both butadiene and norbornadiene come at 47 000–48 000 cm^{-1} is at first unexpected since the pi-electron interaction is so much larger in the former. The explanation rests in the fact that the appropriate monomer absorption for butadiene comes at 58 000 cm^{-1} (propylene), whereas the norbornadiene monomer, norbornene, has its $N \to V_1$ at 51 000 cm^{-1} (Fig. V.C-5). Thus the π_1–π_2 splitting is about three times larger in butadiene than in norbornadiene even though their $N \to V_1$ transitions come at the same frequency.

It should be pointed out before we leave this molecule that the terminating Rydberg orbitals cover the molecular core like a glove so that the basic tenet of all of the interacting-oscillator theories applied to norbornadiene, no overlap of oscillator wave functions [S1], is grossly violated for these transitions. On the other hand, the V_1–V_2 split of 7600 cm^{-1} is very close to that calculated (8600 cm^{-1}) by the nonoverlapping interaction of two parallel ethylene oscillators separated by 2.37 Å each having $f = 0.35$. Moreover, the observed frequencies are nicely split about the $N \to V_1$ frequency of 51 000 cm^{-1}, as predicted by the simple interacting-oscillators model.

V.C-2. *Heterocyclic Dienes*

The heterocyclic dienes considered in this work consist of two unsaturated olefin groups in conjugation with a heteroatom bearing lone-pair electrons, such as nitrogen, oxygen, and sulfur. The olefinic groups need not be in direct conjugation themselves, but when they are, then the spectra bear a strong resemblance to those of the corresponding simple cis or trans dienes.

In their planar configurations, the heterocyclic dienes support six pi electrons, like benzene, but of course the lower symmetry breaks the benzene degeneracies. Thus the $1e_{1g}$ orbitals of benzene become the $1a_2$ and $2b_1$ occupied MOs of the heterocycles (π_2 and π_3), while the $1e_{2u}$ benzene virtual orbitals become the π_4^* and π_5^* virtual MOs of the heterocycles. Almost all of the transitions so far identified in the heterocyclic dienes involve transitions from $1a_2$ and $2b_1$ into the π^* orbitals (N → V_n) or into ns, np, and nd Rydberg orbitals.

Though the experimental frequencies of the furan absorption bands have been determined several times with unanimous agreement, their interpretation is much less certain. Price and Walsh [P44] and Pickett [P16] found the spectrum of furan to look much like those of cyclopentadiene, 1,3-cyclohexadiene, and thiophene. The first strong feature is a broad band centered at $\sim$48 700 cm^{-1} (vert.) which displays several quanta of symmetric C═C stretching ($\sim$1200 cm^{-1}) (Fig. V.C-6). To the low-frequency side of this N → V_1 transition, Pickett also finds a few quanta of 700–800 cm^{-1}, which may belong to a second transition. Beginning at 52 230 cm^{-1} in furan, there are a large number of bands of a complex nature. Of these bands, Price and Walsh were able to construct three Rydberg series, but Watanabe and Nakayama [W16] later showed from the photoionization spectrum that the purported ionization potentials did not agree with their measurements, and proceeded to reassign the spectra. These revised series of Watanabe and Nakayama were in turn revised by Derrick et al. [D4], who claim two series going to the first ionization potential (71 630 cm^{-1} vert.) with $\delta = 0.55$ and 0.04, and a third series with $\delta = 0.82$ going to the second ionization potential at 83 120 cm^{-1} (vert.).

Photoelectron studies of furan [D4, E2] assign the first two ionizations to loss of electrons from π MOs in the diene part of the molecule. According to an *ab initio* calculation, the upper one of these has $1a_2$ symmetry and the lower has $2b_1$ symmetry [S36]. Now the third Rydberg series converging upon the second ionization potential has strong origins and consists of allowed transitions. This is consistent with a $2b_1$ originating MO and an ns (a_1) terminating MO, as implied by the large value (0.82) of the quantum defect. On the other hand, transitions from $1a_2$ to ns are symmetry forbidden, and will be so weak that the only chance for observation is at $n = 3$. We can expect that the ($1a_2$, 3s) term value in furan will be much the size of the ($2b_1$, 3s) term in the same molecule (23 600 cm^{-1} vert.), or of the (a'', 3s) term values in tetrahydrofuran (24 600 cm^{-1} vert.) or diethyl ether (23 900 cm^{-1} vert.), i.e., the $1a_2 \to 3s$ transition should come at $\sim$48 000 cm^{-1} in furan. This is just the frequency at which Pickett found evidence for a weak excitation separate

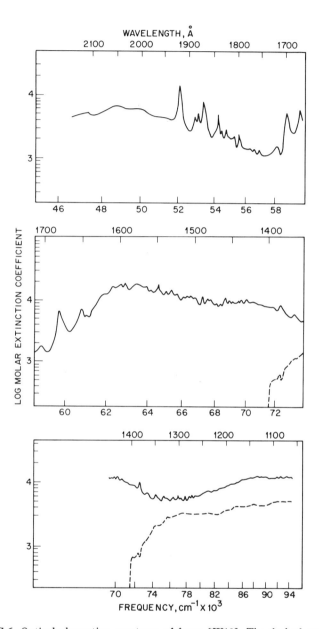

Fig. V.C-6. Optical absorption spectrum of furan [W16]. The dashed curve is that of the photoionization spectrum.

from the $N \to V_1$ band, and we tentatively assign it as $1a_2 \to 3s$. A less likely possibility is that this weak band is the $N \to V_2$ ($\pi \to \pi^*$) excitation predicted to come at unusually low frequencies in the dienes, and possibly seen in this spectral region in cyclopentadiene (Section V.C-1).

The first band of the $1a_2 \to 3p$ Rydberg series in furan has a strong origin at 52 230 cm^{-1} (a term value of 19 400 cm^{-1} advert.), followed by over 20 vibronic components, most of which have been assigned by Pickett as involving progressions of v_1' (1395 cm^{-1}) in combination with one or two quanta of v_2' (848 cm^{-1}), v_3' (1068 cm^{-1}), and v_4'. A spacing of 465 cm^{-1} may be due to a hot band, a difference of vibrational quanta, or an electronic splitting within the 3p manifold. The C=C stretch is appended to each of the five members of the np series, and very similar vibrational intervals are observed in the first band of the furan photoelectron spectrum [D4]. Similarly, the structure of the $2b_1 \to ns$ bands is much like that of the $2b_1$ photoelectron band. The $1a_2 \to nd$ series appropriately begins at 58 700 cm^{-1}, yielding a term value of 12 900 cm^{-1} (vert) for the $n = 3$ member.

Several strong features in the spectrum of furan beyond the first ionization potential, which are evident in the optical and photoionization spectra [P44, W16], can be assigned as Rydberg transitions going to higher ionization potentials. The strong band at 72 200 cm^{-1} appears prominently in both the absorption and photoionization spectra, and is assigned as $2b_2 \to 4s$ by Derrick et al. These workers also assign a doublet band with components at 91 610 and 92 410 cm^{-1} (vert.) as associated with the third ionization potential, which shows a doublet structure with components at 103 900 and 104 700 cm^{-1} (vert.). The term value here is 12 800 cm^{-1}, thereby assigning the band as $9a_1 \to 3d$. Another band at 88 000 cm^{-1} in the optical spectrum has a term value of 23 100 cm^{-1} with respect to the ionization potential at 111 100 cm^{-1} (vert.), and so can be assigned as $6b_2 \to 3s$.

The f values over certain spectral regions and the absorption coefficients of furan are given in the papers of Watanabe and Nakayama [W16] and Pickett et al. [P19].

Being a conjugated cis diene, several $N \to V$ transitions of moderate intensity are expected in furan, with $N \to V_3$ more intense than $N \to V_1$. In furan, we can tentatively assign these two bands to the continuua centered at 48 700 cm^{-1} and 62 500 cm^{-1} (vert.). One sees the effect of oxygen here, for in its absence (as in cyclopentadiene), the V_1–V_3 split is 22 000 cm^{-1}, but as the fifth position contributes a pair of pi electrons, thus making the pi system more "round," the V_1 and V_3 states tend to become degenerate, the splitting being reduced to 13 800 cm^{-1} in furan.

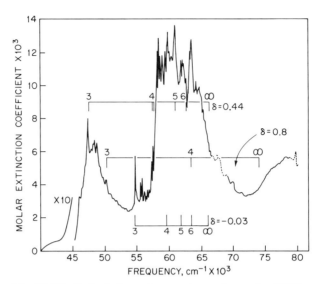

Fig. V.C-7. Optical absorption spectrum of pyrrole [M57].

The $N \to V_1$ band of furan in hexane solution has a molar extinction coefficient of 4970 at the maximum [H29].

As one might expect, the optical spectrum of pyrrole, (Fig. V.C-7) is much like that of furan [M57, P44] and the similarity extends to the photoelectron spectra as well [D7, E2]. In their recent study of pyrrole in the gas phase, Mullen and Orloff [M57] found a band at 42 500 cm^{-1} with an extinction coefficient of about 40 in a sample of very high purity, and argued that this was the $N \to V_1$ band of pyrrole. However, Horváth and Kiss [H29] also studied this spectrum in hexane solution, looking for impurity bands, and did not report the 42 500-cm^{-1} band. Such behavior is more characteristic of a Rydberg transition, and we shall return to this point later. The more commonly accepted $N \to V_1$ ($\pi_3 \to \pi_4^*$) band is found in the 45 000–53 000-cm^{-1} region [P44] as a rather featureless lump, while $N \to V_3$ ($\pi_2 \to \pi_4^*$ according to reference [S44]) appears as a strong feature of the same outline at 57 000–67 000 cm^{-1}. In hexane solution, the $N \to V_1$ band of pyrrole is about 50% more intense than that of furan [H29, K30]. Superposed upon these featureless valence shell bands in the gas phase are several obvious Rydberg excitations, most of which Derrick et al. [D6] have placed in series going to the first ionization potential ($1a_2$, 66 200 cm^{-1} vert.) or the second ($2b_1$, 74 190 cm^{-1}, vert.). Their assignments are shown in Fig. V.C-7. Klevens and Platt [K30] report the spectra of pyrrole and furan in heptane solution, in which the $N \to V_1$ and $N \to V_3$ bands are clearly seen, but no trace of the sharp Rydberg bands is evident, as expected for big-orbit states in a condensed phase (Section II.C).

As in furan, the Rydberg series to the second ionization potential in pyrrole is an ns series with a $(2b_1, 3s)$ term value of 23 900 cm^{-1} (vert.). Moreover, experience shows that the same term value can be expected for $(1a_2, 3s)$, placing it as a symmetry-forbidden band at 42 300 cm^{-1} (vert.), which is just the frequency at which Mullen and Orloff found the weak feature in the gas-phase spectrum. Thus we assign this band as $1a_2 \to 3s$ rather than $\pi_3 \to \pi_4{}^*$ as did Mullen and Orloff. The np and nd series in pyrrole converging upon the first ionization potential have the regular term values. A sharp dip in the absorption at 63 000 cm^{-1} (Fig. V.C-7) is suggestive of an antiresonance (Section I.A-2).

The $1a_2 \to 3p$ transition beginning at 47 320 cm^{-1} in pyrrole and at 47 277 cm^{-1} in N-d_1-pyrrole is quite vertical and so shows only a few vibrational quanta [M31]. To the origin in pyrrole (N-d_1-pyrrole) is added one quantum of 1037 cm^{-1} (869 cm^{-1}) and two of 1469 cm^{-1} (1054 cm^{-1}). The ratio of $1037/869 = 1.40$ identifies that vibration as being the N—H stretch, while the $1469/1054 = 1.19$ ratio of the second vibration suggests a large contribution of N—H motion of some sort. This is all rather surprising since the $1a_2$ MO is largely in the C=C parts of the molecule. The $n = 4$ member of the $1a_2 \to n$p series is split by 145 cm^{-1} [D6], with each component showing a vibrational progression much like that of the $1a_2$ band in the photoelectron spectrum [D6, P21, S13].

The structure of the $1a_2 \to 3d$ band of pyrrole at 54 670 cm^{-1} (vert.) was studied by Pickett et al. [P21], who identified excited-state quanta of 870, 1070, 1425, and 1510 cm^{-1}, and possibly 730 cm^{-1}, the first three of which are close to values found in the $1a_2 \to 3p$ band of furan.

The spectra of furan and pyrrole are of special spectroscopic interest as cautionary examples to theoreticians who calculate electronic spectra. In at least four separate papers, semiempirical pi-electron calculations have been juggled and twisted so as to fit the experimental spectra, which are in fact almost totally outside of the pi-electron valence shell. These examples of Rydberg excitations assigned to levels calculated in a valence-shell basis are in no way unique, and represent a huge waste of effort which hopefully can be avoided in the future.

Two highly structured bands in the spectrum of N-methyl pyrrole with origins at 42 000 and 52 730 cm^{-1} have been studied by Milazzo [M32, M36, M38]; they are most likely the $1a_2 \to 3s$ and $1a_2 \to 3d$ Rydberg transitions, as one sees by comparison with the pyrrole spectrum. N-Methyl pyrrole also offers an apparent example of something which is quite rare in polyatomic molecules, fluorescence from a Rydberg state.

Beginning at ~42 000 cm^{-1}, Milazzo has found a weak series of absorption bands ($\epsilon \simeq 40$) which are most likely the $1a_2 \to 3s$ Rydberg excitation. Moreover, when N-methyl pyrrole is excited by an electrical discharge in a Schüler tube, a fluorescence is observed with several bands in common with the absorption, but lacking a common origin [M34, M35, M37]. If our assignment by analogy with the pyrrole spectrum is correct, then the fluorescence is from the ($1a_2$, 3s) excited Rydberg state.

For a long time, the only spectrum of thiophene available was a published photograph of Price and Walsh's original photographic plate [P44]; the spectral curve of Di Lonardo et al. [D17] has appeared only recently (Fig. V.C-8). Working from the published photograph, Derrick et al. [D5] have once again succeeded in identifying three Rydberg series which are obviously closely related to those found by them in furan and pyrrole, i.e., two series going to the first ionization potential with

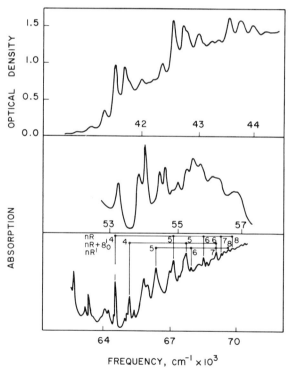

Fig. V.C-8. Optical absorption spectrum of thiophene (upper) and of thiophene-d_4 (middle and lower) [D17].

quantum defects of 0.40 (np) and 0.03 (nd) and a third series going to the second ionization potential with $\delta = 0.82$ (ns). While the two ionization potentials in question here involve pi MOs, it is not clear from the semiempirical calculations whether the $2b_1$ MO is slightly below or slightly above the $1a_2$ MO. This question has been studied using the He II effect in the photoelectron spectrum [R16] and shows clearly that $1a_2$ (71 540 cm^{-1} vert.) lies above $2b_1$ (76 770 cm^{-1} vert.). Thus the orbital ordering in thiophene is like that in the other heterocyclic dienes. As discussed for furan, the allowedness of the various Rydberg series also places $1a_2$ above $2b_1$, as do the *ab initio* theoretical calculations of Gelius *et al.* [G6, G7]. According to the calculations, the $1a_2$ MO is totally within the butadiene part of the thiophene pi-electron system, whereas $2b_1$ is the sulfur lone-pair pi orbital.

Price and Walsh report two very long progressions of 965 cm^{-1}, separated by 340 cm^{-1}, in the 41 500–47 500 cm^{-1} region of thiophene [P44]. The 340-cm^{-1} interval is most likely a difference frequency, while the 965-cm^{-1} quanta are assigned as totally symmetric C=C stretching in the upper state. Milazzo [M33] and Di Lonardo *et al.* [D17] have also studied the vibronic structure of this band. The term value here (30 000 cm^{-1} vert., 27 000 cm^{-1} adiab.) is too large for a Rydberg excitation in a molecule of this size, and so we must be observing the $N \rightarrow V_1$ transition. Indeed, spectra in paraffin solvents show the $N \rightarrow V_1$ frequency of thiophene to be about 5000 cm^{-1} lower than those of furan and pyrrole [H29, K30]. The $N \rightarrow V_3$ band is similarly low at 56 000 cm^{-1} in heptane solution [K30].

The Rydberg transitions in thiophene as assigned by Derrick *et al.* [D5] seem to be slightly unusual, suggesting that perhaps the origins are one quantum or so removed from those quoted by them. Thus, as the assignments now stand, the ($1a_2$, 3p) term value is 22 700 cm^{-1} (vert.), whereas something less than 20 000 cm^{-1} is expected, the transition to 3d is masked by heavy diffuse absorption, and the ($1b_2$, 3s) term is 22 900 cm^{-1} (vert.), whereas a value closer to 24 000 cm^{-1} is expected.† Di Lonardo *et al.* have questioned some of the finer points of Derrick's Rydberg analysis, challenging the validity of the series going to the sec-

---

† Inasmuch as the thiophene molecule contains a sulfur atom, one would normally begin numbering the ns series at $n = 4$ rather than at $n = 3$. However, since the molecular Rydberg orbital is felt to have considerable 3s contributions from the carbon atoms and since the 3s and 4s term values are indistinguishable anyway (Section I.A-1), it is arbitrary as to how one picks the principal quantum numbers and quantum defects, and we choose to start with $n = 3$ so as to maintain the resemblance with the corresponding states in furan and pyrrole.

ond ionization potential and pointing out that the vibrations tied to the origin at 53 270 cm^{-1} show an isotopic shift upon deuteration which is much larger than those shown by all of the other Rydberg excitations, which leads them to believe that this band is not Rydberg. The controversy could be settled using the high-pressure effect (Section II.B). We also point out that, as in the alkyl sulfides (Section III.F), the spectrum of thiophene will be complicated by $1b_2 \rightarrow \sigma^*$(S—C) valence shell excitations at relatively low frequencies. According to the term values, it is clear that the $1a_2 \rightarrow 3s$ promotion corresponds to the band at 48 300 cm^{-1} (advert.), but this then leaves the highly structured band at 41 600 cm^{-1} (advert.) as the $N \rightarrow V_1$ transition, whereas $N \rightarrow V_1$ in dienes is usually only slightly structured.

The spectrum of selenophene shows two broad, overlapping bands at 40 750 and 43 000 cm^{-1}, and what appears to be a Rydberg excitation at 47 800 cm^{-1} (adiab.). This spectrum does not look much like those of the other heterocyclic dienes, possibly due to an inversion of the $1a_2$ and $2b_1$ MOs, and to the possibility that the molecule is nonplanar [T16]. Disagreements as to the presence of weak, sharp bands near 37 000 and 41 000 cm^{-1} and of the vibrational assignment of the sharp bands near 48 000 cm^{-1} are voiced by Trombetti and Zauli [T16] on one side, and by Milazzo on the other [M40].

To this point, we have considered the heterocyclic dienes, the spectra of which were largely those of a cis diene perturbed by the heteroatom. In systems where the two vinyl groups are not conjugated, this pattern of levels no longer holds, though there is an apparent interaction of the two vinyl groups through the heteroatom. In dioxene

the first strong band is at 55 000 cm^{-1} (vert.) and is largely the C=C $N \rightarrow V_1$ promotion, with several quanta of C=C stretch excited. Two weaker valence shell excitations precede this band [P18]. The spectrum of divinyl ether shows a broad maximum at 49 180 cm^{-1} ($f = 0.45$), which is obviously a pi-electron transition $\pi_3 \rightarrow \pi_4^*$ which essentially changes a C—O pi antibond into a C—C antibond, and so is related to the 55 000-cm^{-1} band of dioxene [H9] and the 48 700-cm^{-1} band of furan. The allowed component of the $N \rightarrow V_1$ interactions in dioxadiene would seem to come at ~53 000 cm^{-1} (vert.) with any one of several weak, lower-frequency bands being the forbidden component. It is difficult to assign bands here since there are eight pi electrons and no ionization potentials have been determined. Pickett and Sheffield note that vibronic separations of 550 and 220 cm^{-1} are very prominent in the dioxadiene spectrum [P18].

V.C-3. Higher Polyenes

The only conjugated polyene beyond butadiene that has been studied in the vacuum-ultraviolet region is 1,3,5-hexatriene, investigated by Price and Walsh [P47]. The N → V$_1$ ($\pi_3 \to \pi_4^*$) band of this molecule is located in the 39 700–45 700-cm^{-1} region (41 500 cm^{-1} vert.), and displays two progressions of 1615 cm^{-1} each, separated by 1230 cm^{-1}. Rather than assigning the 1230-cm^{-1} separation to a vibrational interval, Price and Walsh have attributed it instead to the presence of two conformationally different molecules in the gas at room temperature. As in all of the molecules of this type, the 1615-cm^{-1} separation is again thought to be C=C stretching, but with a frequency suprisingly close to that expected for the ground state (~1650 cm^{-1}).

Two other band systems were identified in 1,3,5-hexatriene. Beginning at 52 700 cm^{-1}, there are both sharp and diffuse bands among which 330-cm^{-1} intervals are discerned (C=C twisting?), superposed upon a strong, underlying, continuous absorption having a maximum at 57 000 cm^{-1}. Which, if any, of these sharp features are Rydberg absorptions remains to be tested. Four members of a Rydberg series are identified beyond 60 600 cm^{-1}, converging upon the lowest ionization potential, 66 800 cm^{-1}, a value since confirmed by photoelectron spectroscopy. The reality of the continuous absorption at 57 000 cm^{-1} is questioned by Price and Walsh, since its relative intensity varied from experiment to experiment. This band appears prominently at 57 700 cm^{-1} (vert.) in the trapped-electron spectrum (Section II.D) of Knoop [K32] along with bands at 41 300 cm^{-1} (vert.) ($^1A_g \to {}^1B_u$) and 64 000 cm^{-1} (vert.).

Electron energy-loss spectra of 1,3,5-cycloheptatriene and 1,3,5,7-cyclooctatetraene have been reported out to 85 000 cm^{-1}, but the spectra beyond 50 000 cm^{-1} are not very clear cut, and it is difficult to even count the number of transitions [K31].

In the diene discussion (Section V.C-1), the possibility was briefly mentioned that the strongly allowed N → V$_1$ band of butadiene was not the lowest singlet excited state, but that a forbidden $\pi \to \pi^*$ transition ($^1A_g \to {}^1A_g$) may come lower. The theoretical work of Schulten and Karplus [S21] predicts that this forbidden transition will also precede the intense $N \to V_1$ bands of hexatriene and octatetraene, and possibly those of the larger polyenes as well. These interesting transitions have yet to be found.

Seliskar and McGlynn [S27, S28, S29] have studied the interesting series of heterocyclic polyenes, 2,2-dimethyl-3-cyclopentene-1,3-dione (**I**),

I **II** **III**

N-methyl maleimide (**II**), and maleic anhydride (**III**), and compared their spectra (Fig. V.C-9). In the region below 43 000 cm^{-1}, there are one or more $n_O \to \pi^*$ excitations. With the insulating —C(CH$_3$)$_2$ group in the ring, the pi part of the molecule is isoelectronic with hexatriene and has its $\pi_3 \to \pi_4^*$ absorption maximum at 47 700 cm^{-1} (vert.), whereas that of hexatriene is at 41 500 cm^{-1}. It is a common circumstance that substitution of C=O for C=C in polyenes raises the $N \to V_1$ frequency. The corresponding band in maleic anhydride is found at 49 800 cm^{-1} (vert.). However, in N-methyl maleimide, the authors report that the band at 44 620 cm^{-1} (advert.) is the second $\pi \to \pi^*$ excitation, the first coming at 35 000 cm^{-1} (vert.), on the basis of its molar extinction coefficient of 750. Because it seems unreasonable that the exchange of the —O— group for —NCH$_3$— should shift the $N \to V_1$ transition by —14 800 cm^{-1} (compare the $N \to V_1$ frequencies of acids and amides, Section V.A), it is felt more likely that it is the 44 620-cm^{-1} band instead that is the $N \to V_1$ transition.

In order to unravel these complicated spectra, we take the usual route of first locating the Rydberg excitations, using the photoelectron spectra and anticipated Rydberg term values. Once this separation into Rydberg and valence shell excited states is accomplished, the valence shell excitations can be further assigned using MO arguments. We have determined the ionization potentials of the anhydride and N-methyl imide by photoelectron spectroscopy [R19] and find the first ionization potential is 80 580 cm^{-1} (vert.) for N-methyl maleimide and 88 640 cm^{-1} (vert.) for the anhydride. This large difference in ionization potentials leads one at first to think that it is a pi MO which is involved in the ionization rather than a lone pair. However, consideration of the lone pair ionizations in simple acids and amides [B52], refutes this. Now the term value for the band at 44 620 cm^{-1} in the imide (36 000 cm^{-1} vert.) is really too large for it to be considered as a Rydberg transition to 3s. In order to more completely secure a valence shell assignment, we also investigated

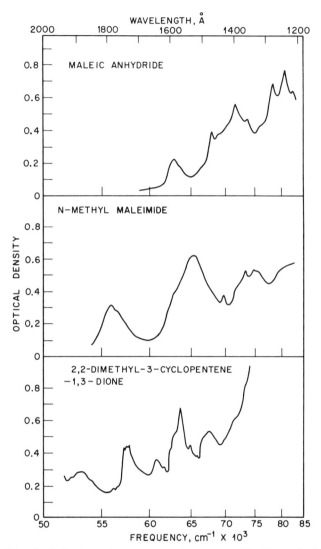

Fig. V.C-9. Optical absorption spectra of several pi-isoelectronic heterocyclic polyenes in the gas phase [S27, S28, S29].

the high-pressure effect on the 44 620-cm^{-1} band (Fig. V.C-10), finding that the (0, 0) and a few higher vibronic bands remain sharp under perturbation, but that several others display the broadening to higher frequencies characteristic of Rydberg transitions! Our suggestion here is that certain members of the $\pi \rightarrow \pi^*$ vibronic band are strongly mixed with the $n_O \rightarrow 3s$ $(\sigma \rightarrow \sigma^*)$ Rydberg excitation. Seliskar and McGlynn state that

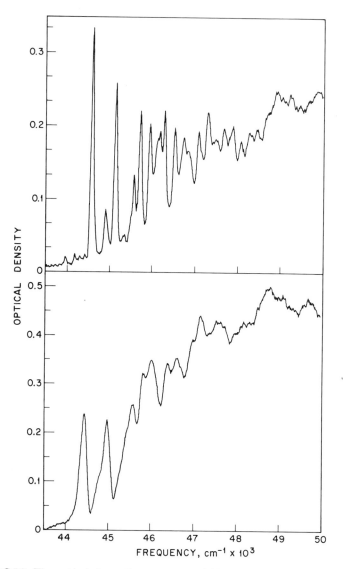

Fig. V.C-10. The optical absorption spectrum of N-methyl maleimide vapor before (upper) and after (lower) pressurizing with 146 atm of nitrogen [R19].

all vibronic structure is washed out of the 44 620-cm^{-1} band of N-methyl maleimide in solution. The 49 800-cm^{-1} band of the anhydride also showed evidence for appreciable Rydberg character; however, its term value of 38 800 cm^{-1} (vert.) suggests that the upper state is basically valence shell, in spite of the strong Rydberg admixture. In all three of the molecules, there is more or less evidence for two overlapped transitions near 50 000 cm^{-1}.

Considering that N-methyl maleimide has five carbon atoms in it, one would guess that the lowest transition to the 3s Rydberg orbital would have a term value somewhat less than that of N,N-dimethyl acetamide (25 500 cm^{-1} vert.). This argument leads to the assignment of the band at 56 000 cm^{-1} (vert.) as terminating at 3s, since its term value is 24 600 cm^{-1} (vert.). The transition $n_O \rightarrow 3s$ is symmetry allowed, as is the band at 56 000 cm^{-1} (ϵ_{max} = 10 000; f = 0.34). All of our earlier experience with Rydberg excitations (Section I.A-1) suggests that such an $n_O \rightarrow 3s$ Rydberg excitation in N-methyl maleimide should have an oscillator strength no larger than 0.08, and so if our assignment of the 56 000-cm^{-1} band is correct, there must be a second, overlapping valence shell band of appreciable strength at the same frequency. The corresponding transition to 3p appears as the weak shoulder at 63 000 cm^{-1} (vert.) with a term value of 17 600 cm^{-1} (vert.), and that to 3d comes at 70 000 cm^{-1} (vert.), the term value being 10 600 cm^{-1} (vert.). This leaves the very strong feature at 65 500 cm^{-1} (vert.) as another $\pi \rightarrow \pi^*$ transition. Now, in an all-trans polyene, it is $N \rightarrow V_1$ which is strongest and the higher frequency $N \rightarrow V_n$ are considerably weaker, whereas if the elongated system is bent as if cyclized, the $N \rightarrow V_1$ is weak and the stronger band appears among the higher-frequency $N \rightarrow V_n$. This would seem to be happening here, for in hexatriene it is $N \rightarrow V_1$ which is strongest, but in N-methyl maleimide it is $N \rightarrow V_2$. The spectrum of N-ethyl maleimide is said to strongly resemble that of N-methyl maleimide.

Now the photoelectron spectrum gives the first ionization potential of maleic anhydride as 88 640 cm^{-1} (vert.), which means that the corresponding transition to 3s should fall near 64 000 cm^{-1}, considering that O replaces N—CH$_3$ of the imide. In this way, we come to assign part of the strong band at 63 100 cm^{-1} (vert.) as terminating at 3s, analogous to the band at 56 000 cm^{-1} in the imide. The transition to 3p (68 500 cm^{-1}) has a term value of 20 100 cm^{-1} (vert.) with respect to the n_O-ionization potential, and the transition to 3d comes at 78 500 cm^{-1} (vert.) with a 10 100-cm^{-1} term. The strong $N \rightarrow V$ band is that peaked at 72 000 cm^{-1} (vert.). The Rydberg term values deduced for N-methyl maleimide and maleic anhydride are listed in Table V.A-III along with those of other amides and carboxylic acids, and it is seen that they fit in nicely with the simpler compounds of that type.

There are no ionization potential data on the cyclopentene-dione, but the bands at 53 200, 57 500, and 65 000 cm^{-1} (vert.) have the proper 3s, 3p, and 3d terms for a first ionization potential at 76 000 cm^{-1} (vert.).

On this basis, the strong N → V band would correspond to the peak at 63 500 cm^{-1}.

V.D. The Cumulenes

The polyatomic molecules with cumulated double bonds are gathered together here, without regard for their component groups. Beginning with allene H$_2$C=C=CH$_2$, there is a long isoelectronic series, including ketene H$_2$C=C=O, diazomethane H$_2$C=N=N, hydrazoic acid H—N=N=N, and isocyanic acid H—N=C=O, which makes a very interesting comparative study; though the spectral observations and interpretations are still incomplete, Rabalais et al. [R2] have made a significant contribution toward correlating the states of these molecules. To these polyatomic systems one could add isoelectronic triatomics such as O=C=O and N=N=O, but we ignore these since they are adequately discussed elsewhere [H20, R2]. Similarly, the cumulated triatomics such as NO$_2$, SO$_2$, CS$_2$, etc., are covered in detail by Herzberg, except for recent photoelectron spectral data [B51, B56], and will not be considered here. Carbon suboxide O=C=C=C=O, is another cumulated molecule of much current interest, but companion systems such as H$_2$C=C=C=C=CH$_2$ or S=C=C=C=S have not been investigated in the vacuum-ultraviolet region.

Before going into the details of the specific molecular spectra, let us pause to consider briefly the electronic structure of the 16-valence-electron cumulene prototype, N$_3$$^-$ [M16]. In the linear symmetric azide ion, the highest filled orbital has symmetry π_g and the lowest vacant orbital has π_u symmetry. The excited configuration $(1\pi_g)^3(2\pi_u)^1$ will lead to the three upper states $^1\Sigma_u^+$, $^1\Sigma_u^-$, and $^1\Delta_u$, and excitation from the $^1\Sigma_g^+$ ground state will be allowed only to the $^1\Sigma_u^+$ component. In isoelectronic molecules of lower symmetry such as the cumulenes considered here, the $^1\Delta_u$ state will be split into two components, and transitions to the formerly forbidden states may become weakly allowed, though probably not as strong as that to $^1\Sigma_u^+$, which comes at higher frequencies than the others. Valence shell excitations may also be found which correlate with the lowest $\sigma \rightarrow \pi^*$ ($^1\Sigma_g^+ \rightarrow {}^1\Pi_u$) transition of the azide ion.

In order to illustrate the azide ion transitions, let us first focus on the spectrum of hydrazoic acid and the alkyl azides. In these systems, the R and H groups are off the N—N line and so reduce the symmetry to C$_s$. The absorption spectrum of hydrazoic acid has been published by both McDonald et al. [M16] and Okabe [O3], and they are in agreement with respect to all major features (Fig. V.D-1), except for the absolute intensities. The weak band in HN$_3$ at 37 900 cm^{-1} (vert., $f = 6 \times 10^{-4}$)

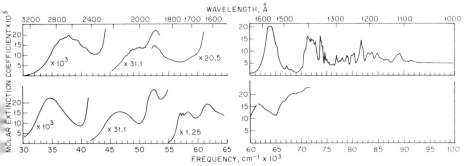

Fig. V.D-1. Optical absorption spectra in the gas phase of hydrazoic acid (upper) and n-amyl azide (lower) [M16].

has been assigned by McDonald et al. as a valence shell $\pi \to \pi^*$ transition derived from the forbidden $^1\Sigma_g^+ \to {}^1\Sigma_u^-$ transition of the azide ion. However, Closson and Gray [C23] prefer to assign it as one component of the $^1\Sigma_g^+ \to {}^1\Delta_u$ promotion in the linear system. The second component of this doubly degenerate excitation is assigned by them to the band at 50 000 cm^{-1} (vert., $f = 9 \times 10^{-3}$). McDonald et al. [M16] also assign the 50 000-cm^{-1} band of alkyl azides as derived from the $^1\Sigma_g^+ \to {}^1\Delta_u$ transition of the azide ion, but assign its second component to the band at 52 900 cm^{-1} (vert., $f = 1.5 \times 10^{-2}$). In an alkyl derivative such as n-amyl azide in the gas phase (Fig. V.D-1), the corresponding weak, low-frequency $\pi \to \pi^*$ bands are observed at 34 800, 46 700, and 52 400 cm^{-1} (vert.) [M16]. Though the precise assignment of this triplet of weak bands in the azides is not yet clear, they are no doubt derived from the forbidden azide ion transitions, $^1\Sigma_g^+ \to {}^1\Sigma_u^-$ and $^1\Sigma_g^+ \to {}^1\Delta_u$. All three of these bands in HN$_3$ are rich in vibrational structure: The first shows intervals of N—N—N asymmetric stretching (1475 cm^{-1}, $\nu_2'' = 2140$ cm^{-1}) and a bending mode (645 cm^{-1}, $\nu_6'' = 672$ cm^{-1}); the second shows intervals of N—N—N symmetric stretching (779 cm^{-1}, $\nu_3'' = 1274$ cm^{-1}) and a bending mode (428 cm^{-1}), which also appears in the third band.

Several other valence shell excitations in hydrazoic acid were identified by comparison with the states of the azide ion. The two bands at 58 800 cm^{-1} and 64 000 cm^{-1} (vert.) are said to be the symmetry-split components of a valence shell $^1\Sigma_g^+ \to {}^1\Pi_u$ transition arising from the allowed $\sigma_g \to \pi_u^*$ promotion in hydrazoic acid, while in n-amyl azide, the two components have more equal intensity, with their maxima at 56 800 and 61 700 cm^{-1} (vert.), according to McDonald et al. These workers then go on to assign the second strong feature in the hydrazoic acid spectrum at 71 300 cm^{-1} (vert.) as derived from the strongest band of the azide ion, $^1\Sigma_g^+ \to {}^1\Sigma_u^+$. However, since the 64 000- and 71 300-cm^{-1} bands both have oscillator strengths of 0.3 [R2], one cannot be sure which is to be correlated with the transition to the $^1\Sigma_u^+$ state. The polarizations of the two bands will differ, with the $\Sigma \to \Sigma$ transition being polarized along the N—N—N line, and the $\Sigma \to \Pi$ being polarized perpendicular to it.

Among the strong valence shell excitations in hydrazoic acid, the first three or four members of four Rydberg series were identified: 65 230 cm^{-1} (3s, $\delta = 1.00$, 27 400 cm^{-1} term), 73 855 cm^{-1} (3p, $\delta = 0.50$, 18 500 cm^{-1} term), and 77 040 cm^{-1} (3d, $\delta = 0.20$, 15 500 cm^{-1} term), all of which converge to the first ionization potential, 92 600 cm^{-1}, and a second series beginning at 74 765 cm^{-1} (3s, $\delta = 1.00$, 27 100 cm^{-1} term) converging upon a second ionization potential of 101 850 cm^{-1}. In all cases, it is seen that the quantum defects and term values are quite reasonable for the proposed upper states. Unfortunately, the photoelectron spectroscopic work of Eland [E3] does not support the Rydberg limits deduced by McDonald et al., for he finds the first ionization potential of hydrazoic acid to be very vertical at 86 630 cm^{-1} (advert.) and the second to be 98 400 cm^{-1} (vert.). Thus these Rydberg series, which look so reasonable, are in fact, completely erroneous! To begin with, the uppermost filled MO in hydrazoic acid is a nonbonding π MO and should look very nearly like its counterpart in the azide anion, i.e., a nodal pattern similar to a 3d AO. The photoelectron-band envelope [E3] confirms the nonbonding nature of this orbital. Note, however, that 3d $\to$ ns is forbidden and therefore the ns series going to the first ionization potential will not appear. However, in n-amyl azide, the s and d symmetries will be very badly upset and the ns Rydberg series should appear. This is our tentative explanation for the otherwise magical appearance of the moderately intense band at 57 500 cm^{-1} (vert.) in n-amyl azide. The np series will appear in hydrazoic acid, beginning with $n = 3$ members at 65 500 and 67 200 cm^{-1} (advert.). These terms are 21 100 and 19 400 cm^{-1} (advert.). Higher members of the series can be readily identified from the spectrum of Fig. V.D-1 and the term table of Appendix A.

The second lowest occupied MO of hydrazoic acid is rather more localized on the nitrogen atom bearing the hydrogen atom and is more like the sigma lone pair in an aldimine. This and the uppermost occupied MO form the doubly degenerate nonbonding $1\pi_g$ set of the azide ion. Transitions from this lower orbital will be allowed to ns, np, and nd upper Rydberg orbitals, and we tentatively assign the transition to 3s to part of the strong absorption at 71 300 cm^{-1}. This block of absorption must contain other transitions as well since that to 3s will not have an oscillator strength larger than 0.08 (Section I.A-1), but 0.3 is observed.

Turning from hydrazoic acid to isocyanic acid, HNCO, we would expect to find very nearly the same pattern of absorption bands in the two isoelectronic compounds. As seen in Fig. V.D-2, this is only partially realized. In the low-frequency region, three weak bands appear (50 000, 52 600, and 59 900 cm^{-1} vert.) [O5, R1], which would seem entirely analogous to the three weak bands of hydrazoic acid in the same region resulting from the forbidden components of the $1\pi_g \to 2\pi_u^*$ valence shell excitation. A fourth weak band is found at 64 100 cm^{-1} (vert.) in isocyanic acid, but probably does not come from the $\pi \to \pi^*$ excitation since only three such weak bands are expected.

The sharper, much stronger features at 72 930 cm^{-1} (advert.) and the structured band at 78 000 cm^{-1} (vert.) in isocyanic acid at first would seem to be related to the stronger features of hydrazoic acid in the same region. However, comparison with the photoelectron spectrum of isocyanic acid (Fig. V.D-2) [E3] again shows that these two features look almost exactly like the first two ionizations from the split $1\pi_g$ manifold, as regards both frequency spacings and Franck–Condon factors, and so both must be Rydberg excitations. Their term values of 20 660 and 21 900 cm^{-1} (vert.) clearly show that they are components of the $1\pi_g \to$ 3p excitation. The corresponding band to 4p starts at 82 900 cm^{-1} (advert.), and that to 3d is prominent at 80 500 cm^{-1} (vert.) with a 13 100-cm^{-1} term value. Working backward, we can now guess that the weak band at 64 100 cm^{-1} (29 500 cm^{-1} term value) is the Rydberg transition to 3s, made somewhat more allowed in isocyanic acid than in hydrazoic acid by the lower end-to-end symmetry in the pi system. As expected, the 3s term value in isocyanic acid (29 500 cm^{-1}) is very close to that observed for formamide (29 200 cm^{-1}), a molecule also having but one C, N, and O atom.

The big question now is where has the strongly allowed $^1\Sigma_g^+ \to\ ^1\Sigma_u^+$ excitation gone? Since the forbidden $\pi \to \pi^*$ bands of HCNO are 10 000–15 000 cm^{-1} higher than in HN$_3$, the same exhaltation would place $^1\Sigma_g^+ \to\ ^1\Sigma_u^+$ at about 80 000 cm^{-1} in HNCO; still there is no sign of it out to 84 000 cm^{-1}. Presumably it is beyond this.

In an alkyl isocyanate like C$_2$H$_5$NCO [R1], virtually all of the bands described for HNCO are observed (Fig. V.D-2), but at lower frequencies. Again there is no strong peak below 80 000 cm^{-1} corresponding to the expected $^1\Sigma_g^+ \to\ ^1\Sigma_u^+$ transition. Since the first excitation to 3p in ethyl cyanate comes at 63 800 cm^{-1} (vert.), its first ionization potential will be $\sim$83 500 cm^{-1} (vert.). That of methyl isocyanate is 85 340 cm^{-1} (vert.) [E3]. Rabalais et al. argue that the peak at 70 800 cm^{-1} (vert.) is the long-sought $^1\Sigma_g^+ \to\ ^1\Sigma_u^+$ excitation, but it appears to us to be the second transition to 3p, converging upon an ionization potential of $\sim$90 500 cm^{-1} (vert.). Two of the three weak $1\pi_g \to 2\pi_u$ valence shell excitations in HNCS have been observed at 40 800 and 50 800 cm^{-1} (vert.) [M17].

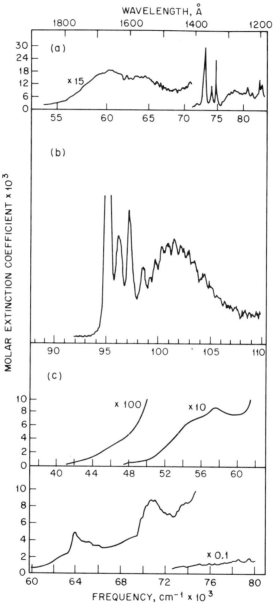

Fig. V.D-2. (a) The absorption spectrum of isocyanic acid in the gas phase [O5]; (b) the first two photoelectron bands of isocyanic acid [E3]; and (c) the absorption spectrum of ethyl isocyanate in the gas phase [R1].

The alkyl cyanates (RCNO) are isoelectronic with the alkyl isocyanates (RNCO) described earlier, and should have similar spectra. As with the isocyanates, the cyanates also have the alkyl groups displaced from the CNO line, but the states of the chromophoric part of the molecule can still be profitably described using the symmetry labels of the $C_{\infty v}$ point group. In the only alkyl cyanate measured so far, $(CH_3)_3C$—C≡N—O, there are two weak bands at 45 950 cm^{-1} (vert., $f = 0.016$) and 53 730 cm^{-1} (vert., $f = 0.022$) which would seem to be the forbidden transitions $^1\Sigma^+ \to {}^1\Sigma^-$ and $^1\Sigma^+ \to {}^1\Delta$. Following these, there is the allowed $\pi \to \pi^*$ transition ($^1\Sigma^+ \to {}^1\Sigma^+$) at 57 930 cm^{-1} (advert.), with $f = 0.206$ [Y2]. Interestingly, in the bent nitrones

$$R_2C=\overset{\overset{O}{|}}{N}-R$$

the azide ion pattern is completely obliterated.

Turning next to allene, $H_2C=C=CH_2$, the photoelectron spectrum [B7, T21] shows that the first ionization potential is a complicated affair, and so, too, should be the Rydberg transitions leading up to it. The least tightly bound electron of allene is in a doubly degenerate pi MO (2e), and its loss puts the molecular ion in a ^2E state which is Jahn–Teller unstable. Consequently, a broad, double-peaked band is observed in the photoelectron spectrum, with the lower vertical ionization potential at 80 600 cm^{-1} and the upper one at 85 500 cm^{-1} (vert.). The first peak shows a regular vibrational series with 720 cm^{-1} mean spacing, which Baker and Turner [B7] assign as v_3', the totally symmetric C—C—C stretch. Following the double peak, the next ionization (also out of an e MO) begins at 113 700 cm^{-1} (adiab.). Haselbach has performed semiempirical calculations to determine the Jahn–Teller-distorted equilibrium geometries of allene positive ion in the lower ^2E state, and finds the lowest configuration to have the two H_2C planes at a dihedral angle of 38° [H11]; consequently, he instead interprets the 720-cm^{-1} vibration as a torsional excitation. The second lowest configuration has a 90° dihedral angle, but unequal C=C bond lengths of 1.264 and 1.371 Å. Since the lowest empty MO in allene is π^*3e [S41], the lowest of the allene valence shell excited states will be generated from the $(\pi 2e, \pi^*3e)$ configurations, just as in the azide ion.

As promised by the photoelectron spectrum, the optical spectrum of allene is most complex (Fig. V.D-3), but Sutcliffe and Walsh [S52, S53] have made a brave attempt at unraveling it. Following on their heels, other groups have also taken a turn at interpreting this complex spectrum [I9, R2]. Again using the correlation with the states of the azide ion,

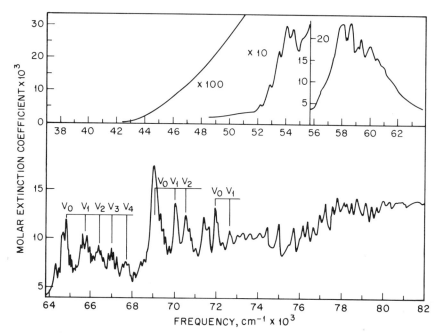

Fig. V.D-3. The gas-phase absorption spectrum of allene [R2].

Rabalais et al. [R2] tentatively identify the $^1\Sigma_g^+ \to {}^1\Pi_g$ ($^1A_1 \to {}^1E$ in D_{2d}) excitation with the weak ($f = 0.03$), structured band at 54 000 cm^{-1} (vert.) (Fig. V.D-3), sporting a ~650–550 cm^{-1} vibrational progression. Our view instead is that this is the $\pi 2e \to 3s$ Rydberg excitation with a vibronic envelope looking very much like that in the photoelectron spectrum. As a transition terminating at 3s, it has a reasonable term value of 26 600 cm^{-1} (vert.), not too different than those of other three-carbon chromophores, i.e., $\pi \to 3s$ of propylene (25 400 cm^{-1} vert.; Section IV.A-2), $\pi \to 3s$ of methyl acetylene (26 800 cm^{-1} vert.; Section IV.D), or $2b_2 \to 3s$ of propane (27 900 cm^{-1} vert.; Section III.A-3).

As Rabalais et al. point out, on the basis of frequency, the 650-cm^{-1} progression in the 54 000-cm^{-1} band of allene is most likely either the torsion mode b_1 (820 cm^{-1} in the ground state) or the skeletal bending mode e (838 cm^{-1} in the ground state). Note now that the ^1E excited state is Jahn–Teller unstable, and that for such a state in the D_{2d} point group, the b_1 vibration can be active in reducing the symmetry, but the e vibration will not [H20]. Thus we have a clear situation in which a nontotally symmetric torsion appears as a progression in single quanta, which is a positive indicator of a Jahn–Teller influence, as Liehr points out [L27, L28].† Outwardly, the situation in the twisted ($\pi 2e$, 3s) Rydberg state of allene resembles that in the twisted (πb_{2u}, 3s) Rydberg state of ethylene (Section IV.A-1); however, in ethylene, the Jahn–Teller effect is not operative and so the torsion must appear as double quanta rather than as single as in allene.

Beginning at 55 000 cm^{-1} in allene and stretching to about 64 000 cm^{-1}, there is a very strong absorption, $f = 0.34$, structured with many vibrational quanta of 650 ± 150 cm^{-1} [R2]. The absorption pattern is complex, and Sutcliffe and Walsh readily admit to the possibility of two transitions in this region. From this broad, strong band and many sharper ones at higher frequencies, Sutcliffe and Walsh have constructed no less than seven Rydberg series, all converging to the common ionization potential, 82 200 cm^{-1}. Since the photoelectron spectrum gives the lowest ionization potential as 80 600 cm^{-1}, it is immediately clear that the original identification of the series members as to n and δ will have to be revised as dictated by the true ionization limit. Rabalais et al. have confirmed the earlier Rydberg assignments, but again their limit does not agree with the more reliable ionization potential obtained directly by photoelectron spectroscopy. Most recently, Iverson and Russell [I9] have reanalyzed the optical data and composed two Rydberg series converging upon 80 788 cm^{-1} (advert.), in much better agreement with the photoelectron value.

Rabalais et al. assign the strong feature at 58 000 cm^{-1} (vert.) as two overlapping transitions, one the $^1\Sigma_g^+(^1A_1) \rightarrow {}^1\Sigma_u^+(^1B_2)$ valence shell band derived from the ($\pi 2e, \pi^* 3e$) configuration and the other the $\pi 2e \rightarrow 3s$ Rydberg excitation. We totally agree with the first of these and feel that the largest part of the oscillator strength of this band comes from the $\pi \rightarrow \pi^*$ excitation. However, we feel instead that there are *two* other Rydberg excitations in this same region: (i) the transition from $\pi 2e \rightarrow 3p$ at $\sim$59 000 cm^{-1} (vert.) converging upon the first ionization potential at 80 600 cm^{-1} (advert.) with a term value of 21 600 cm^{-1} (vert.), and (ii) the transition from $\pi 2e \rightarrow 3s$, also at $\sim$59 000 cm^{-1} (vert.), but converging upon the *second* ionization potential at 85 500 cm^{-1} (vert.) with a term value of 26 500 cm^{-1} (vert.). Since this second transition to 3s leaves the ionic core in the geometry of the upper Jahn–Teller component, we suspect that the antisymmetric C=C stretch ($\nu_6'' = 1980$ cm^{-1}) will appear as a progression of single quanta in this transition. The very com-

---

† The isotope shift on this vibration in allene-d_4 would seem to argue for an e vibration instead. We observe vibrational intervals of 560, 428, 420, 370, and 320 cm^{-1} [R19], which show an anharmonicity even larger than that in allene-h_4. Using only the first member of the progression for comparison with allene-h_4, the frequency ratio in the excited state is 1.15; for $\nu_{10}(e)$ in the ground state, the ratio is 1.26, whereas for $\nu_4(b_1)$, it is 1.41 [L34].

plicated nature of the transition at 69 150 cm⁻¹ (advert.), which we assign as $\pi 2e \to 3p$ converging upon the first ionization potential, might possibly be due to the effects of l-uncoupling [F2]. The corresponding transition to 3p from the other Jahn–Teller component is found at 64 800 cm⁻¹ (advert., 20 700 cm⁻¹ term) and is similarly complicated, perhaps for the same reason. However, in this case, the band is also overlapped by a transition to 3d [I9].

According to the most recent *ab initio* calculation of the valence shell spectrum of allene [S9], the strong transition to $^1B_2$ is preceded by forbidden transitions to $^1A_2$, $^1B_1$, and $^1A_1$ states, all of which are derived from the lowest (π, π^*) configuration. These forbidden bands no doubt appear as the weak absorption between 43 000 and 52 000 cm⁻¹ in allene (Fig. V.D-3) and between 36 000 and 50 000 cm⁻¹ in ethyl allene [C6].

It is interesting to consider the spectrum of tetramethyl allene in order to assess the effect of alkylation on the term values. In general, the effect is to drive the 3s term value toward 22 000 cm⁻¹ and the 3p term value toward ~19 000 cm⁻¹. Our determination of the photoelectron spectrum of this compound gives a doubled first band, with Jahn–Teller components at 68 760 and 72 700 cm⁻¹ (vert.), just as in allene [R19]. The optical spectrum of Scott and Russell [S26] has a weak feature at 46 300 cm⁻¹ (vert.) which we think is the $e \to 3s$ Rydberg excitation because it has a term value of 22 460 cm⁻¹ (vert.). Following this, there is a much stronger transition ($\epsilon = 12\,500$) centered at 51 150 cm⁻¹. Its term value of 17 610 cm⁻¹ (vert.) is rather low and so its assignment as $e \to 3p$ is tentative. However, the intensities of these two bands would seem to be behaving in the usual fashion for 3s and 3p Rydbergs on alkylation, i.e., even though symmetry allowed, the transition to 3s becomes vanishingly weak while that to 3p becomes strong as the chromophore is alkylated. The intense excitation at 59 140 cm⁻¹ (vert.) in tetramethyl allene ($f \simeq 0.9$) houses the allowed $^1A_1 \to {}^1B_2$ ($\pi \to \pi^*$) excitation as well as several Rydberg excitations. Unlike the situation in ethylene and butadiene, in which methylation moves the long-axis allowed $\pi \to \pi^*$ band rapidly to lower frequencies, that of tetramethyl allene (59 140 cm⁻¹ vert.) is actually somewhat *higher* than that of allene (58 000 cm⁻¹ vert.). The suggested splitting of this strong band in allene and tetramethyl allene is more clearly realized in 1,1-dimethyl allene, in which distinct transitions are centered at 57 900 and 63 130 cm⁻¹ (vert.) [S26]. The spectra of other alkylated allenes are given by Jones and Taylor [J12].

The Rydberg spectrum of ketene, $H_2C=C=O$, presents us with an apparently anomalous Rydberg series. Since the photoelectron spectrum of ketene shows that the first two ionizations at 77 700 cm⁻¹ (advert.) and 114 600 cm⁻¹ (vert.) are separated by over 35 000 cm⁻¹ [B7, T21], all

Rydberg absorptions below 80 000 cm^{-1} will originate at the highest filled level in the molecule. According to the Gaussian orbital calculation on ketene [S41], the two lowest ionization potentials come at 74 200 and 113 700 cm^{-1} (vert.), with the first being out of a $2b_2$ pi orbital which is mostly C=C bonding, but which has a large component of C=O antibonding character as well. In line with this, the first photoelectron band of ketene is accompanied by several quanta each of 2140 and 1020 cm^{-1}, the ν_2' out-of-phase and ν_2' in-phase C=C=O stretches, respectively.

In the optical spectrum (Fig. V.D-4) [P49, R2], there is a weak band at 46 900 cm^{-1} (advert.), having a term value of 30 800 cm^{-1} (advert.), which is quite appropriate for a $2b_2 \rightarrow 3s$ excitation. For comparison, the oxygen lone pair $\rightarrow 3s$ term values in other systems having one oxygen and two carbon atoms are 27 100 cm^{-1} (acetaldehyde, Section IV.C), 27 500 cm^{-1} (dimethyl ether, Section III.E-3), and 30 700 cm^{-1} (ethyl alcohol, Section III.E-2). Of course, the $\pi \rightarrow 3s$ term value in ketene is larger than that in allene since the penetration at the oxygen atom is significantly larger than that at a carbon atom (Section I.A-1). As with the first Rydberg band of allene, the 46 900-cm^{-1} band of ketene in the optical spectrum has a frequency spacing (1040 cm^{-1}) and Franck–Condon factors that look much like those of the first band in the photoelectron spectrum [T21]. The $2b_2 \rightarrow 3s$ oscillator strength is 0.01 [R2].

Price *et al.* [P49] also report a very nice Rydberg series having $n = 3$ at 54 680 cm^{-1} (advert.), stretching to $n = 8$, and converging upon 77 500 cm^{-1}. The first member of this series has a term of 22 800 cm^{-1}, as expected for a 3p upper orbital, but the quantum defect for this series is

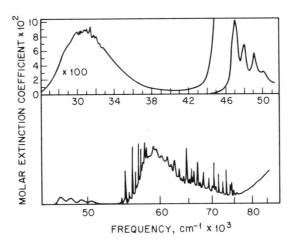

Fig. V.D-4. The gas-phase absorption spectrum of ketene [P49, R2].

1.07, as is appropriate for an ns series! If, instead, we accept the quantum defect as indicating that the 54 680-cm^{-1} band has a 3s upper orbital, then its term value is approximately 7000 cm^{-1} below that expected for $2b_2 \to 3s$. Furthermore, in this assignment, the 46 900-cm^{-1} band would then have to be assigned as valence shell. This is the anomaly referred to earlier. The solution to this problem rests in the following reassignment of the bands. The series having $\delta = 1.07$ has ns orbitals, but the $n = 3$ member is at 46 900 cm^{-1}, rather than 54 680 cm^{-1}. On the other hand, the 54 680-cm^{-1} band is not a member of the ns series at all, but instead is the $n = 3$ member of the np series, having a term value of 22 800 cm^{-1}. This value is somewhat high, but not unacceptably so for a transition to 3p.

We have already seen that the $(2b_2, 3s)$ and $(2b_2, 3p)$ term values in ketene are about 2000 cm^{-1} larger than otherwise might have been expected. Perhaps it is no surprise then that the $(2b_2, 3d)$ term value is similarly high. It seems that the $n = 3$ member of the nd series appears at 61 350 cm^{-1} (advert.) (Fig. V.D-4), with a term value of 16 150 cm^{-1} (advert.). The only other possibility for a Rydberg assignment at this frequency in ketene is that the $(2b_2, 3p)$ manifold is split by the aspherical core, putting 3p term values at 16 000 and 22 000 cm^{-1}. A similar anomaly is found in the spectra of several of the other cumulenes discussed later.

In their original work on the ketene spectrum, Price et al. mention parenthetically that the continuum with maximum intensity at 59 000 cm^{-1} (vert.) may be that of a water impurity rather than that of ketene. Braun et al. [B45] confirm this suspicion, but find another broad continuum centered at 56 500 cm^{-1} (vert.) having $\epsilon \simeq 7400$ ($f \simeq 0.3$). This is most likely the allowed $\pi \to \pi^*$ excitation corresponding to that at 53 900 cm^{-1} in allene and at 61 700 cm^{-1} in ethylene. The absorption profile of ketene given by Braun et al. is very suggestive of antiresonance interactions (Section I.A-2) between Rydberg transitions and the underlying continuum in the 54 000–57 000-cm^{-1} region. The weak bands observed in the ketene spectrum at frequencies below 40 000 cm^{-1} [N17] no doubt are the analogs of the forbidden $^1\Sigma_g^+ \to {}^1\Sigma_u^-$ and/or $^1\Sigma_g^+ \to {}^1\Delta_u$ transitions of the azide ion.

Merer [M23] has made an extensive study of the deeper states of diazomethane, $H_2C{=}N{=}N$, and its deutero analog. Eight members of a Rydberg series having $\delta = 0.10$ and converging to 72 585 cm^{-1} were found, with all members being very vertical [H18]. The 13 100-cm^{-1} term of the $n = 3$ series member, together with the δ value, undeniably argue for nd upper states in the Rydberg series. The early members of the series are split into three components by the aspherical symmetry of the ionic core. The $n = 3$ and 4 members of a fragmentary np series having

$\delta = 0.67$ were also identified and the core splitting was again observed in the transitions to 3p and 4p. Strong, unassigned features are found at 56 870 and 57 300 cm^{-1}, the first of which has a term value of 15 710 cm^{-1} (advert.). This is related to the puzzling bands in allene and ketene, which also have this intermediate term value; they are components of either the 3p or 3d manifold.

In the nd Rydberg series of diazomethane, only ν_2' vibrations appear with certainty, and very weakly at that. That the transitions converging upon the lowest ionization potential of diazomethane are so vertical and consequently are to excited states with very nearly the ground-state geometry is strong evidence that the transitions originate with an MO that is quite nonbonding. Looking at the theoretical calculations [A7, S41], one finds the highest-filled MO, $2b_2$, to be a pi MO which is large on the terminal atoms, but virtually zero at the central (nitrogen) atom, looking much like the nonbonding MO π_2 in the carboxylic acids (Section IV.A-3). In line with this nonbonding character, the ν_2 vibration suffers only the smallest reduction in frequency upon excitation to the Rydberg orbitals. In the excitations to np Rydberg orbitals, progressions of ν_6', the out-of-plane bending, are excited, but the states are nonetheless said to be planar.

The very low ionization potential of diazomethane (72 580 cm^{-1}) places the $n_N \rightarrow 3s$ Rydberg band at about 44 000 cm^{-1}, at which frequency (43 500 cm^{-1} vert.) a weak, continuous band has been reported.

Carbon suboxide, O=C=C=C=O, is the longest of our cumulenes, and its spectrum has been studied in some detail recently. In the vacuum ultraviolet (Fig. V.D-5), there is a structured band centered at 56 200 cm^{-1} (vert.), apparently resting upon a continuous absorption centered at the same frequency. This structured band show 14 quanta of 385 cm^{-1} spacing and has an oscillator strength of 0.08 [B21, K13, R20]. Beyond this, there is an extremely intense band centered at 63 300 cm^{-1} (vert.), with an oscillator strength of 1.5 and an apparent shoulder at 66 600 cm^{-1} (vert.). A series of rather diffuse bands follows with decreasing intensity and spacing, much like Rydberg bands converging upon an ionization potential. In fact, Roebber et al. identified an ns Rydberg series having $\delta = 1.00$ and converging to 85 500 cm^{-1} (advert.) [K13, R20], just the value found by photoionization [K13] and photoelectron experiments [B6, G5, T21]. However, they propose that the $n = 3$ member of this series is the very strong band at 63 300 cm^{-1}, for which an anomalously low term value of 22 200 cm^{-1} is calculated. Since high-pressure experiments show convincingly that the 56 200-cm^{-1} band of carbon suboxide is a Rydberg [R21], it seems much more natural to assign it, with its term of 29 300 cm^{-1} and 0.08 oscillator strength, as the 3s

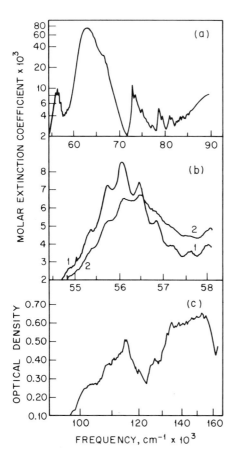

Fig. V.D-5. The absorption spectrum of carbon suboxide in the gas phase [R20, R21]. In (b), the curves labeled 1 and 2 are the spectra before and after pressurizing the sample with 102 atm of argon.

member, and the very strong band at 63 300 cm^{-1} as a valence shell transition. In fact, all of the evidence to date suggests that the Rydberg excitation to 3s in carbon suboxide should have an oscillator strength of no more than 0.16 (Section I.A-1), a value ten times smaller than that found for the band at 63 300 cm^{-1}.

A few members of an np series having $\delta = 0.76$ and a regular first term of 19 000 cm^{-1} were also identified in the carbon suboxide spectrum. Vibrational structure was observed in the transitions to the 4s, 5s, and 6s orbitals consisting of multiple quanta of the totally symmetric C=C stretch ($\nu_2' \simeq 700$ cm^{-1}; $\nu_2'' = 830$ cm^{-1}) and fewer quanta of the totally symmetric C=O stretch ($\nu_1' = 2100$–2200 cm^{-1}; $\nu_1'' = 2200$ cm^{-1}). As in the optical spectrum, Baker and Turner report the first band in the photoelectron spectrum of C_3O_2 to be quite vertical, with a few quanta of 660 and 1950 cm^{-1} evident [B6].

As regards the valence shell of carbon suboxide, the pi-electron configuration in the linear $^1\Sigma_g^+$ ground state is $(1\pi_u)^4(1\pi_g)^4(2\pi_u)^4(2\pi_g)^0(3\pi_u)^0$. What is apparently the lowest singlet configuration, $(1\pi_u)^4(1\pi_g)^4(2\pi_u)^3$-$(2\pi_g)^1$, leads to $^1\Delta_u$, $^1\Sigma_u^+$, and $^1\Sigma_u^-$ states. Of these, transitions to $^1\Delta_u$ and $^1\Sigma_u^-$ are lowest and forbidden, and probably correspond to the weak absorption found near 36 000 cm^{-1}, whereas the strongly allowed transition to $^1\Sigma_u^+$ is most likely that at 63 300 cm^{-1}, though Bell et al. [B21] prefer to assign it to the 56 200-cm^{-1} band. Roebber et al. also show that the $(2\pi_u)^3(3\pi_u)^1$ and $(1\pi_g)^3(2\pi_g)^1$ configurations lead to low-lying $^1\Sigma_g^+$, $^1\Sigma_g^-$, and $^1\Delta_g$ states, and any one of these could be responsible for the continuous absorption beneath the 56 200-cm^{-1} band. Since $2\pi_u$ is the highest filled MO in carbon suboxide, it is necessarily the originating MO for the lowest-frequency ns and np Rydberg series described earlier.

Operating with differential pumping, Roebber et al. [R20] also investigated the C_3O_2 absorption in the 100 000–160 000-cm^{-1} region. Several of the bands in this high-energy region (Fig. V.D-5) can be assigned to Rydberg excitations, using the known ionization potentials and the inversion symmetry of the various occupied MOs [G5, S41]. Thus the $1\pi_g$ ionization potential at 119 800 cm^{-1} (vert.) is preceded by an allowed $\pi_g \to$ 3p optical excitation at 102 600 cm^{-1} (vert., 17 200 cm^{-1} term value). The same optical absorption also fits as $1\pi_u \to$ 3s (25 000 cm^{-1} term) with the $1\pi_u$ ionization potential at 127 600 cm^{-1} (vert.). Excitation from $1\pi_u$ is also allowed to 3d, and this excitation is found at 114 300 cm^{-1} (vert., 13 300 cm^{-1} term value). Rydberg excitations from the $5\sigma_u$ MO (139 400 cm^{-1} ionization potential) to the 3s and 3d orbitals will be allowed, and correspond to the transitions observed at 114 300 cm^{-1} (25 100 cm^{-1} term value) and 126 700 cm^{-1} (12 700 cm^{-1} term value), respectively. Finally, the $6\sigma_g$ MO has an ionization potential of 155 700 cm^{-1}, and the allowed excitation from it to 3p is observed at 137 000 cm^{-1} (vert., 18 700 cm^{-1} term value).

The absorption system at 56 200 cm^{-1} in carbon suboxide is a most interesting one. First of all, we assign it as the lowest member of the ns Rydberg series, but it is vibronically very different from both the higher members of the series and the corresponding photoelectron band, for these are strongest at (0, 0) whereas the 56 200-cm^{-1} band is maximal at the $v' = 8$ member, and they have no frequencies in common.†
Second, carbon suboxide in the ground state has a very low-frequency

---

† Actually, there are a number of examples of molecules in which the transition to 3s is utterly different from the remaining Rydberg spectrum (see, for example, the water spectrum, Section III.E-1). It is thought that this is due to the mixing of the (ϕ_i, 3s) Rydberg configuration with the conjugate valence shell configuration (ϕ_i, σ^*).

bending mode about the central carbon atom ($v_7'' = 63$ cm^{-1}) which will be highly excited at room temperature, and its frequency is expected to increase considerably in certain excited states. In fact, Bell et al. [B21] conclude that v_7' in the upper state of the 56 200-cm^{-1} transition is approximately 450 cm^{-1}, and that all of the vibronic structure in this band is due to $nv_7'' \to nv_7'$ sequences rather than to a vibrational progression. According to this explanation, all bands but the (0, 0) are hot bands and would disappear in a sufficiently cold gas. If the structure instead is a vibrational progression and the upper state is linear, then the vibrations must be totally symmetric stretching motions (or double quanta of bending motions, which seems slightly unlikely if the upper state really is linear), whereas if the upper state is bent, then low-frequency bending motions may appear in a progression. Of course, each member of such a vibrational progression will be a composite of the v_7 sequences, which must occur in any event. The solution to this interesting puzzle undoubtedly rests with the effect of temperature on the vibronic intensities, an experiment recently reported by Roebber [R21]. He finds that the relative Franck–Condon factors in this band do not change on going from 290 to 195°K, thus ruling out the sequence explanation with certainty. However, he denies the more natural assignment of an allowed 3s Rydberg excitation and prefers to call it instead a forbidden valence shell excitation. Since the 56 200-cm^{-1} band of carbon suboxide broadens significantly under perturbation by a second high-pressure gas (Fig. V.D-5), he presumes that the vibronic intensity results from mixing with an allowed Rydberg excitation.

The vacuum-ultraviolet spectrum of cyanoazide, NC—N$_3$, is presented without comment in reference [O4]. The sharp features at 53 200 and 62 300 cm^{-1} (vert.) have the proper splitting for an assignment to 3s and 3p upper orbitals. If this is so, then the first ionization potential of this compound is 83 200 cm^{-1} (vert.). The strong $^1\Sigma_g^+ \to {}^1\Sigma_u^+$ transition of hydrazoic acid at 64 000 cm^{-1} would seem to be shifted upward to 74 600 cm^{-1} (vert.) in the cyano derivative.

CHAPTER VI

Aromatic Compounds

VI.A. Phenyl Compounds

On going from the straight-chain polyenes to their planar, cyclic counterparts, the major spectroscopic features are still found to be the $\pi \to \pi^*$ excitations, but with certain small differences. Thus, in the more rigid cyclic systems, the $\pi \to \pi^*$ excitations are vibronically more highly structured than in the open chains. Also, it is the first $\pi \to \pi^*$ excitation ($N \to V_1$) which is the most intense in the open chains, whereas in the ring and fused-ring systems, the lower $\pi \to \pi^*$ bands are weak and the higher bands have the largest part of the oscillator strength. This redistribution of intensity holds not only for benzene, where the symmetry is high, but for molecules such as phenanthrene, where it is not. As in all systems, the $\pi \leftrightarrow \sigma$ excitations are again difficult to identify in the aromatic compounds, due to their low oscillator strengths. Rydberg excitations are most prominent in benzene itself, but in the higher aromatics, they appear less significantly as weak bumps resting upon a background of more intense valence shell excitation. Since the lowest π-ionization potential drops to $\sim 60\,000$ cm^{-1} or so in the larger aromatics, a typical 3s term value of 21 000 cm^{-1} places the first Rydberg excitations in such compounds at $\sim 40\,000$ cm^{-1}.

VI.A-1. *Benzene*

Because the benzene molecule is one of the cornerstones of organic chemistry, it is no wonder that its electronic spectrum has been subject to repeated and detailed studies under a wide variety of conditions. An excellent account of progress in unraveling the intricacies of its spectrum, replete with the historical development, is given in the review by Dunn [D27]. In spite of its high symmetry, the benzene molecule is nonetheless of sufficient size as to make accurate calculations of its electronic structure presently "impossible." The uppermost orbital of benzene is the doubly degenerate $1e_{1g}$ pi MO (called π_2, π_3); of this there can be no doubt. However, the relation of the lower pi MO, π_1 ($1a_{2u}$), to the uppermost sigma MO $3e_{2g}$ has been a problem of great concern [A14, A15, E5, L29, S7, T21], though the answer seems to be sigma above pi, thanks to the study of Åsbrink *et al.* [A15]. The photoelectron spectrum does show these two orbitals to be accidentally near-degenerate, so it is understandable that the approximate calculations occasionally give one ordering and occasionally the reverse.

One-electron transitions from the highest filled pi MOs ($1e_{1g}$) to the lowest empty pi MOs ($1e_{2u}$) lead to three excited singlet states $^1B_{2u}$, $^1B_{1u}$, and $^1E_{1u}$, which have been located experimentally in the vicinities of 40 000, 48 000, and 57 000 cm^{-1} (Fig. VI.A-1) [N16]. A transition to a

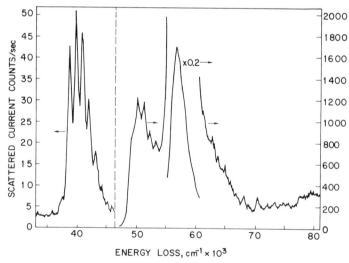

Fig. VI.A-1. Electron-impact energy-loss spectrum of benzene vapor, recorded with a 40-eV impact energy at $\theta = 0°$ [L9].

fourth state, $^1E_{2g}$, is also predicted by pi-electron theory to be in the vicinity of the transition to $^1E_{1u}$, but its frequency has not been determined directly. For a discussion of the $^1A_{1g} \rightarrow {}^1B_{2u}$ transition, the reader is referred to the accounts of Callomon et al. [C1] and Herzberg [H20]. Because of its ill-defined vibronic structure and close proximity to the very strong band just to its high-frequency side, the assignment of the absorption in the 50 000-cm^{-1} region of benzene is somewhat insecure. While most argue that the upper state is $^1B_{1u}$, others argue less convincingly that the excited state is $^1E_{2g}$ [P6, S39], or possibly a Rydberg transition [N11]. As will be discussed, the first of these alternate suggestions can now be shown to be incorrect, while there may be a grain of truth to the second, though not in the manner originally thought.

It is understandable that the transition at 50 000 cm^{-1} in benzene has received far less attention than that at 38 000 cm^{-1}, since it is much more diffuse and is overlapped by the strong band at 56 000 cm^{-1}. The theoretical descriptions of this upper state show a clear demarcation along the following lines. In single configuration calculations of the ASMO–Pariser–Parr type, the state lying between the $^1B_{2u}$ and $^1E_{1u}$ states is found to be $^1B_{1u}$, whereas on extending the calculations to include configuration interaction, or equivalently, in a valence bond calculation that includes polar structures, the intermediate state has symmetry $^1E_{2g}$. Though one would rightly be tempted to place more confidence in the results of the higher-order configuration interaction calculations, the experimental evidence points instead to the $^1B_{1u}$ assignment.

The experimental determination of the symmetry of the 50 000-cm^{-1} band of benzene rests largely on the interpretation of the vibronic structure of the band. Thus, if the upper state is $^1B_{1u}$, the transition from the ground state is forbidden for electric-dipole radiation, but the state can be mixed vibronically with the adjacent $^1E_{1u}$ state via e_{2g} vibrations in order to gain an in-plane-polarized intensity. On the other hand, the electronically forbidden transition to $^1E_{2g}$ will be mixed with $^1E_{1u}$ by vibrations of symmetry e_{1u}, b_{1u}, and/or b_{2u}. Appended to the symmetry-allowing vibrations quoted above, there may be a progression of totally symmetric vibrations (a_{1g}) as well. In arguing for an $^1E_{2g}$ upper state, Dunn and Ingold [D26] conclude that the vibronic structure of the transitions to $^1B_{1u}$ should closely resemble that to $^1B_{2u}$ (perturbing vibration e_{2g}), whereas the experimental vapor-phase spectrum argues against such a similarity between the bands at 38 000 and 50 000 cm^{-1}.

Experimentally, the facts are these. The vapor-phase spectrum of benzene shows five or six features in the 49 000–54 000-cm^{-1} region (Fig. VI.A-1). Due to sequence congestion, hot bands, and radiationless decay to lower states, the bands are rather broad at room temperature, with

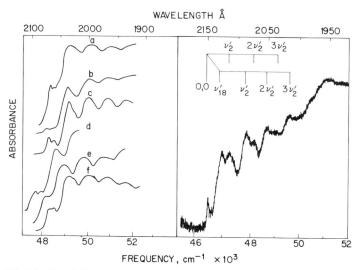

Fig. VI.A-2. (Left) Optical absorption spectrum of the 50 000-cm^{-1} band of benzene in several transparent matrices; (a) C$_6$H$_6$ in argon, (b) C$_6$H$_6$ in krypton, (c) C$_6$D$_6$ in krypton, (d) C$_6$H$_3$D$_3$ in xenon, (e) C$_6$H$_6$ in nitrogen, and (f) C$_6$D$_6$ in nitrogen. All of the samples were deposited and measured at 20°K, except (b) and (c), which were deposited at 40°K and measured at 20°K [K8]. (Right) The optical absorption spectrum of a solid film of benzene approximately 300 Å thick, annealed and measured at ~10°K [P1]. All of the spectra in this figure are uncorrected for the nonuniform spectral output of the lamps.

many signs of smaller splittings. These bands sharpen considerably in low-temperature matrices to reveal five bands with an average spacing of 920 ± 50 cm^{-1} in C$_6$H$_6$ and 850 ± 50 cm^{-1} in C$_6$D$_6$ [K5, K8]. Additionally, there are site splittings in the matrix spectra (Ar, Kr, N$_2$, Xe) which amount to a few hundred cm^{-1}. In the pure crystal [B48, P1], the spectrum is shifted into the 46 500–50 000-cm^{-1} region and shows doubling of the bands and a first band which is approximately one-third as wide as those that follow (Fig. VI.A-2). In spite of the perturbation offered by a condensed phase, the electronic origin does not appear in any of the matrix spectra. Instead, Katz et al. [K5, K8] argue that the first feature in the matrix spectrum (48 340 cm^{-1}, C$_6$D$_6$ in Kr) corresponds to the excitation of one quantum of $\nu_{18}'(e_{2g})$. Following this, there are four quanta of the totally symmetric vibration $\nu_2'(a_{1g})$. (See reference [H17] for a description of these vibrations.) A second progression begins with one quantum of $\nu_{16}'(e_{2g})$ (approximately 1500 cm^{-1}), which happens to fall very close to $(0, 0) + \nu_{18}' + \nu_2'$ (approximately 600 + 900 cm^{-1}). Due to this coincidence, the progression of ν_2' vibrations attached to the false origin at $(0,0) + \nu_{16}'$ coincides with that beginning at $(0,0) + \nu_{18}' + \nu_2'$. Brith *et al.* [B48] also feel that they are observing two progressions in the spectrum of crystalline benzene (Fig. VI.A-2), but cannot decide whether this splitting into two totally symmetric progressions is due to nontotally symmetric vibrations or to crystal-field effects. Pantos and Hamilton [P1] also have studied the $^1B_{1u}$ region of solid benzene, arriving at a slightly different interpretation (Fig. VI.A-2). Under proper conditions of deposition, an electronic origin was observed (46 510 cm^{-1}) and progressions of ν_2' are built upon this (weak) and upon one quantum of an e_{2g} vibration (strong), which Pantos and Hamilton assign as ν_{18}' rather than ν_{16}'. Absorption at the origin is induced by the crystal field. These analyses of the vibronic structure in the 45 000–50 000-cm^{-1} region of benzene are unanimous in the implication of single quanta of e_{2g} vibrations as making the transition allowed, in accord with a $^1B_{1u}$ upper electronic state. In the methylated benzenes of lower symmetry, discussed in the following section, the vibronic analysis confirms the $^1B_{1u}$ assignment of the 50 000-cm^{-1} transition.

It is clear that the apparent splitting of the 49 400-cm^{-1} band in the vapor spectrum is missing in the low-temperature spectra and so must be due to a hot band. Katz *et al.* specifically assign it as a (1, 1) sequence band due to excitation of the e_{1u} vibration, which has a frequency of 404 cm^{-1} in the ground state.

The vibrational structure of the 50 000-cm^{-1} transition of benzene has also been observed by the electron-impact technique [L9], with results which are yet to be explained. As seen in Fig. VI.A-3, the relative Franck–Condon factors within the band are changing as the scattering angle is varied from $\theta = 0$ to 8°. The spectra, however, are independent of impact voltage over the range 40–100 eV. On the basis of the relative changes of the Franck–Condon factors, it was suggested that there are actually *two* overlapping transitions near 50 000 cm^{-1}, for the Franck–Condon factors are otherwise constant within a single vibronic transition, as illustrated by the angular data on the $n_N \rightarrow 3s$ band of ammonia (Fig. VI.A-3). Working in the same spectral region, Doering [D20] has shown that the intensity ratio of the peaks at 50 000 and 57 000 cm^{-1} in benzene is independent of scattering angle from 9 to 75° at 20 eV incident energy, and questions the existence of two states at 50 000 cm^{-1}.

One peculiarity of the Rydberg spectrum of benzene is that though transitions to both ns and nd upper MOs are electronically $g \leftrightarrow g$ forbidden, those to nd are readily observed, whereas those to ns have not been seen. A reasonable value of 26 000 cm^{-1} for the $(1e_{1g}, 3s)$ term value would place the absorption at $\sim$48 000 cm^{-1}, i.e., just under the transition to $^1B_{1u}$. Now the condensed-phase spectra show that the vibronic

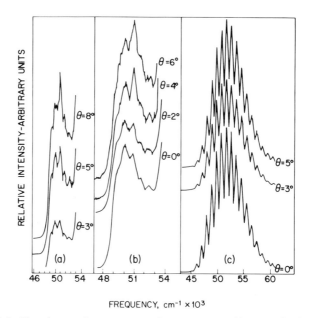

Fig. VI.A-3. The electron-impact energy-loss spectrum of benzene in the 50 000-cm^{-1} region, recorded with an electron impact energy of (a) 50 eV and (b) 90 eV, at several different scattering angles. In contrast, note the $n_N \to 3s$ transition of ammonia (c) taken at an impact energy of 50 eV at several angles [L9].

structure is definitely valence shell; however, the angular dependence (if genuine) of the gas-phase electron-impact spectrum in the 50 000-cm^{-1} region is characteristic of two overlapping transitions. Perhaps we have a broad, smooth $\pi(1e_{1g}) \to 3s$ Rydberg transition (allowed for electric-quadrupole radiation) partially underlying a structured valence shell transition to $^1B_{1u}$.

Honig et al. have suggested that the symmetry of the 50 000-cm^{-1} band of benzene could be deduced from the two-photon spectrum, for they calculate that the transition to $^1E_{2g}$ has a molecular cross section of $\sim 10^{-48}$–10^{-47} cm^4 sec/photon, whereas that to $^1B_{1u}$ is $\sim 10^{-52}$–10^{-51} cm^4 sec/photon [H27]. Following this suggestion, Monson and McClain [M49] performed the experiment, and finding no two-photon absorption, placed an upper limit of $\sim 1 \times 10^{-51}$ cm^4 sec/photon on the cross section of the two-photon transition, thus supporting the $^1B_{1u}$ assignment.

The position of the forbidden $^1A_{1g} \to ^1E_{2g}$ transition in benzene has been a long-standing problem without a convincing solution. Theoretical work is quite ambiguous as to where $^1E_{2g}$ falls among the other (π, π^*) states [K4, R25, T12]. Experimentally, Morris and Angus have uncovered an interesting bit of evidence which suggests to them that they have found the $^1A_{1g} \to {}^1E_{2g}$ band of benzene. Studying the excitation spectrum of benzene luminescence in rare gas matrices, weak structure was observed in the 45 750–46 800-cm^{-1} region (xenon host), whereas the transition to $^1B_{1u}$ has an origin estimated as 47 400 cm^{-1}. After eliminating other possibilities, Morris and Angus finally conclude that this is the $^1A_{1g} \to {}^1E_{2g}$ transition, falling 1000–2000 cm^{-1} below the transition to $^1B_{1u}$. They relate their finding to the peculiar behavior of Lassettre's electron-impact spectrum in this region.

Since the transition to the $^1E_{1u}$ state of benzene is allowed from the ground state with an oscillator strength predicted to be near one, there is little risk in assigning the intense feature centered at 57 000 cm^{-1} to it (Fig. VI.A-1) [N16]. Vibrational analysis of this band in the gas phase is hampered by an overlapping Rydberg transition. However, Katz et al. [K5] find that in the matrix spectra, the $\pi \to \pi^*$ transition is shifted to lower frequencies, thereby separating it from the Rydberg transition, which shifts considerably to higher frequencies. Under such conditions, five quanta of $\nu_2'(a_{1g})$ are observed to be excited, with an average spacing of 920 ± 50 cm^{-1} in benzene and 850 ± 50 cm^{-1} in benzene-d_6. Note that the $\nu_2'(a_{1g})$ mode occurs with just these frequencies in the $^1B_{2u}$ and $^1B_{1u}$ states as well. In the transition to $^1E_{1u}$, the vibrations form a simple, totally symmetric progression, with no sign of the e_{2g} vibrations, the Jahn–Teller-active modes. Indeed, theory shows that there is an accidental cancellation of such vibronic coupling terms in the $^1E_{1u}$ state [K8]. The vibronic linewidths in the transition to the $^1E_{1u}$ state are approximately ten times larger than those in the transition to $^1B_{2u}$, due to radiationless relaxation of the $^1E_{1u}$ state by lower states [B44].

Birks [B30] points out that transitions that are in the vacuum-ultraviolet region and that are forbidden from the ground state S_0 occasionally can be observed if the molecule is first brought to its lowest excited singlet S_1 or triplet T_1 state, and then the absorption $S_1 \to S_n$ or $T_1 \to T_n$ observed in the quartz ultraviolet. Thus for benzene in a hydrocarbon glass, Godfrey and Porter [G15] report an allowed $T_1 \to T_n$ band at 41 600 cm^{-1} (vert.). Since T_1 in benzene is 29 700 cm^{-1} above S_0, we therefore know that the $S_0 \to T_n$ transition comes at 71 300 cm^{-1}, a region of sharp-line Rydberg absorption superposed upon a rising background absorption (Fig. VI.A-1). In the gas phase, the corresponding $T_1 \to T_n$ transition of benzene is found at 43 200 cm^{-1} (vert.) with an oscillator strength of 0.06 [B71]; since T_1 is known to be $^3B_{1u}$, Burton and Hunziker feel the transition is an allowed one terminating at $^3E_{2g}$, but with an oscillator strength severely depressed by the effect of strong configuration interaction. This argument leads to a $^1A_{1g} \to {}^3E_{2g}$ assignment for the absorption at 71 300 cm^{-1}. The uppermost triplet in benzene observed by electron impact is at 45 200 cm^{-1} (vert.) [D20].

Bonneau et al. [B34] have pumped benzene to S$_1$ (^{1}B$_{2u}$, 38 400 cm^{-1} vert.) and then observed the S$_1 \rightarrow$ S$_n$ absorptions in the nanosecond interval. They observed a 20 400-cm^{-1} band, which is therefore 58 800 cm^{-1} (vert.) above S$_0$. Pariser [P3] has predicted that the fourth lowest excited singlet state of benzene is ^{1}E$_{2g}$ and though ^{1}A$_{1g} \rightarrow$ ^{1}E$_{2g}$ is forbidden, ^{1}B$_{2u} \rightarrow$ ^{1}E$_{2g}$ is allowed in the excited-state manifold and could correspond to the band found by Bonneau et al. If this is the case, then the forbidden transition from the ground state to ^{1}E$_{2g}$ would correlate with the weak tail observed in the electron-impact and optical spectra at 58 000 cm^{-1} (Fig. VI.A-1). Also, if this is the proper assignment for the ^{1}E$_{2g}$ state, then there is no longer any doubt about the transition at 50 000 cm^{-1} in benzene having the ^{1}B$_{1u}$ upper state, and the interpretation of the luminescence excitation spectrum by Morris and Angus must be in error.

The beautiful work initiated by Price and Wood on the Rydberg spectrum of benzene [P38] has since been amplified by the high-resolution study of Wilkinson [W26]. Price and Wood first described a series in benzene which would have its $n = 3$ member at 55 881 cm^{-1}, δ = 0.46, and a limit of 74 587 cm^{-1} (Fig. VI.A-4). Its term value of 18 706 cm^{-1} would identify the terminating orbital as 3p. Each member of this series, called nR, shows a set of sequence hot bands, a strong origin, successive quanta of 974 and 915 cm^{-1} which are the totally symmetric breathing motion ν_2', and a strongly anharmonic progression of high intensity beginning with a 695-cm^{-1} quantum (Fig. VI.A-5). This latter is assigned as $2\nu_{18}'(e_{2g})$. Origins for all members of the series are readily found by comparing the spectra of C$_6$H$_6$ with C$_6$D$_6$, for only the origins show a constant isotope shift of 35 cm^{-1}. One also sees combination bands excited involving ν_2', ν_{18}', and ν_{20}'. According to Liehr and Moffitt [L25], the nR series terminates at $np\pi$ (a_{2u}) orbitals, so that the transitions are $1e_{1g} \rightarrow na_{2u}$ (^{1}A$_{1g} \rightarrow$ ^{1}E$_{1u}$) and are electronically allowed with in-plane polarization. However, this leads to the unusual result that the sharp 3R Rydberg and the strong, broad, underlying valence shell transition at 57 000 cm^{-1} have the same upper-state symmetry (^{1}E$_{1u}$), yet do not show any signs of mixing. Perhaps it is more reasonable that the ^{1}A$_{1g} \rightarrow$ 3R transition instead terminates at 3pσ (e_{1u}), giving an out-of-plane allowed ^{1}A$_{1g} \rightarrow$ ^{1}A$_{2u}$ component. This is an interesting problem well suited to a magnetic circular dichroism experiment.

With respect to this problem of the assignment of the Rydberg excitation at 55 881 cm^{-1} in benzene, Scheps et al. present an interesting argument [S16]. They propose the generality that an antiresonance interaction between a sharp-line and an underlying quasicontinuum will be

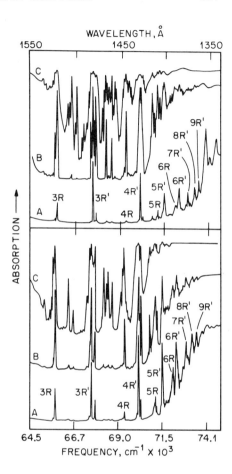

Fig. VI.A-4. Detailed assignments of the Rydberg band origins in the vapor spectra of benzene-h_6 (upper) and benzene-d_6 (lower) [W26].

manifest only if the transition moment of the discrete transition has a nonzero projection on the transition moment of the continuous transition (Section I.A-2). Looking in detail at the band in question in benzene, they conclude that there is no antiresonance interaction at work, and that since the continuum has an in-plane transition moment ($^1A_{1g} \to {}^1E_{1u}$), the Rydberg transition must be polarized out-of-plane ($^1A_{1g} \to {}^1A_{2u}$).

Price and Wood report a second Rydberg series in benzene also going to the first ionization potential which shows a strong origin for each value of n, as well as a pair of nearby origins, as judged from the small isotope shift. Wilkinson succeeded in placing each of the components of each triplet into series called nR' ($\delta = 0.16$), nR'' ($\delta = 0.11$), and nR''' ($\delta = 0.05$). As n increases, the spacing between nR', nR'', and nR''' de-

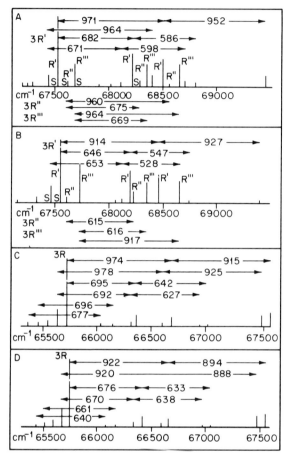

Fig. VI.A-5. Schematic splitting patterns among the 3R, 3R′, 3R″, and 3R‴ Rydberg orbitals of benzene-h_6 (A and C) and benzene-d_6 (B and D). S indicates superposed bands [W26].

creases, and is essentially zero at $n = 8$. The nR′ series is the most intense, followed by nR‴, whereas nR″ is relatively weak. As in the nR series, the primed series also show the excitation of ν_2' and what is assigned by Wilkinson as $2\nu_{18}'$. The spectra of these $n = 3$ Rydberg complexes (Fig. VI.A-4) are shown diagrammatically in Fig. VI.A-5. Now the quantum defects of the primed series would seem to be good evidence for components of nd upper orbitals split by the ionic core. Note, however, that such transitions are $g \rightarrow g$ forbidden, yet the isotope effects clearly show we are looking at electronic origins. One possible solution is that the bands are made allowed by one quantum of a nontotally symmetric vibration that shows a small increase in excited-state frequency in C_6D_6 as compared with its frequency in C_6H_6.†

Intensities of the various bands of benzene are of some interest, and have been measured both directly and indirectly in the vapor and condensed phases. These measurements are hampered somewhat by the overlap of the $^1A_{1g} \rightarrow {}^1B_{1u}$ and $^1A_{1g} \rightarrow {}^1E_{1u}$ vibronic envelopes in the region of 52 000 cm^{-1}. For the transition to $^1B_{1u}$, Hammond and Price integrated the photoelectrically determined absorption from 49 000 to 54 000 cm^{-1} after correcting it for overlap with the transition to $^1E_{1u}$, and found an oscillator strength of 0.094 ($\epsilon = 6800$ at the absorption maximum) [H7]. Using the less accurate photographic technique, Pickett et al. report a value of 0.12 for the same band [P20], while Platt and Klevens obtained 0.1 ± 0.02 for the $^1B_{1u}$ band of benzene dissolved in heptane [P26]. The oscillator strength to the $^1E_{1u}$ state is much less certain than that to $^1B_{1u}$. Hammond and Price again give the most reliable value, $f = 0.88$, obtained photoelectrically and corrected for overlapping absorption in the 52 000–62 000-cm^{-1} region, whereas Pickett et al. again report a value about 30% higher ($f = 1.23$). The reflection spectrum of liquid benzene in the vacuum ultraviolet has been analyzed to yield an oscillator strength of 1.25 ± 0.15 for the transition to $^1E_{1u}$ [W29], while in heptane solution, the same transition is reported to have an oscillator strength of 0.69 [P26]. This relatively low value for the $^1A_{1g} \rightarrow {}^1E_{1u}$ oscillator strength in solution finds support in the work of Potts [P33], who reports a value of 0.60 ± 0.07 for benzene in a paraffin solution at room temperature and 0.63 ± 0.07 when frozen to a glass at 77°K. Under the same conditions, the oscillator strength to $^1B_{1u}$ was 0.10 ± 0.01. It appears that there is a systematic error in the concentration–path length measurements of Pickett et al., for the ratio of their $^1B_{1u}$ and $^1E_{1u}$ oscillator strengths (0.098) is close to that of Hammond and Price (0.107) and that determined from the electron-impact spectrum (0.099) [S38], yet the absolute values seem somewhat too high. Goto [G18] reports quantitative absorption data for the 65 000–95 000-cm^{-1} region, while Bunch et al. [B70]

† Note that the e_{2g} MO of benzene is a "3d" orbital in the sense of having two perpendicular nodal planes. From this, transitions will be allowed to $n p$ and $n f$ Rydberg orbitals, but not to $n s$ and $n d$. Regarding the R', R'', and R''' series in benzene, it is possible that they are components of $n f$ orbitals, for which three allowed components are predicted group theoretically, and for which δ values close to zero are expected. These $n f$ series, referred to fleetingly by Liehr and Moffitt [L25] and Gilbert et al. [G13], if genuine, would be the first examples of $n f$ series identified in a polyatomic molecule. However, it should be noted that such series will have term values of 6940 cm^{-1} for their first members, appropriate to $n = 4$, $\delta = 0$, whereas the reported series begin with $n = 3$, $\delta \simeq 0$.

have similar data extending to 230 000 cm^{-1}. The absorption cross sections and photoionization efficiencies of benzene and benzene-d_6 have been carefully measured in the 74 000–94 000-cm^{-1} region by Person [P8].

The oscillator strength of the forbidden $^1A_{1g} \to {}^1B_{1u}$ transition in benzene ($f = 0.1$) is remarkably high due to its near-degeneracy with the very intense $^1A_{1g} \to {}^1E_{1u}$ band, from which it borrows vibronically. The high intensity of the transition to the $^1E_{1u}$ state is another example of how cyclizing a molecule shifts the intensity into higher-frequency transitions, for in the open-chain analog, hexatriene, the first $\pi \to \pi^*$ band is the most intense, rather than the third as in benzene.

As pointed out by Inagaki [I1], the valence shell absorption spectra of liquid and solid benzene to 60 000 cm^{-1} look very much alike, but are distinctly different from that of the vapor [R22]. Thus, in the solid and liquid, the transition to $^1B_{1u}$ maintains its intensity, but is at about 3000 cm^{-1} lower frequency, while that to $^1E_{1u}$ is shifted down by ~5000 cm^{-1} and broadens considerably to the high-frequency side with a concomitant decrease of the maximum molar extinction coefficient from 79 000 to 25 000. A peak at 58 000 cm^{-1} in the energy loss function $-\text{Im}(1/\epsilon)$ has been calculated from reflection data on liquid benzene and assigned as a plasma oscillation involving all of the pi electrons in the molecule; it is said to be similar to a plasma oscillation near 56 000 cm^{-1} in graphite, and distinct from the transition to $^1E_{1u}$ that appears as a shoulder at 6400 cm^{-1} lower frequency in $-\text{Im}(1/\epsilon)$ [W29]. Since the same shoulder appears in the optical absorption spectrum of a thin liquid film of benzene [I1, S46, S47], it is probably not a collective excitation (Section I.A-3), but instead is an exciton component of the intense $^1A_{1g} \to {}^1E_{1u}$ valence shell excitation.

Katz and co-workers [K5, K7] have studied the vacuum-ultraviolet spectrum of benzene in various solid rare gas matrices with the hope of revealing the fate of Rydberg absorptions in condensed phases of high electronic mobility. Using host/guest ratios of ~100, they found "extra" weak, broad lines at 57 500 and 60 350 cm^{-1} in an Xe matrix, which are shifted to 59 030 and 62 230 cm^{-1} in Kr; in an Ar matrix, only a single line at 61 880 cm^{-1} was observed. Their interpretation of the isotope shifts and vibronic structures of these bands led them to postulate that the lower-frequency component is the 55 881 cm^{-1} Rydberg excitation of the gas phase shifted several thousand cm^{-1} to higher frequencies (Section II.C). Under these experimental conditions, the higher molecular Rydberg states do not exist as such, but in their place, Wannier excitons appear which converge upon the bottom of the rare gas conduction band. Katz *et al.* assign the second feature at higher frequency to the lowest such Wannier exciton. This optical work was repeated by Angus and Morris

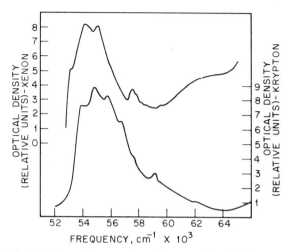

Fig. VI.A-6. Optical absorption spectra of benzene doped into xenon (upper) and krypton (lower) matrices (1:200), deposited at 40° and measured at 20°K [A10].

[A10], who found spectra of the sort shown in Fig. VI.A-6. The low-frequency bands reported by Katz et al. are observed, and possibly the high-frequency one as well in Kr, but not in Xe. None of the bands in question is seen in the spectrum of benzene in a methyl cyclohexane glass at 77°K [G20]. Following a photoemission study of the doped solids, Angus and Morris concluded that all of these features are most likely Wannier excitons, and, as such, are unrelated to the free-molecule Rydberg states. The so-called "antiresonances" first reported by Pysh et al. [P53] in benzene-doped rare gas crystals are not found in the more recent studies, and are now thought to be artifactual.

The benzene spectrum beyond the first ionization potential has been studied optically, by the inelastic electron energy-loss technique and by the SF_6-scavenger technique (Section II.D.). In the electron excitation spectra [H33, L9], the spectral resolution was low, and only the envelopes of groups of Rydberg excitations were observed (Fig. VI.A-1). Similarly, the resolution in the optical work of Yoshino et al. [Y12] was not adequate to resolve the immense amount of fine structure in the benzene spectrum. The best optical spectrum of benzene beyond the first ionization potential is that of Koch and Otto [K37] obtained using synchrotron radiation (Figs. VI.A-7 and VI.A-8). Using the ionization potentials determined by photoelectron spectroscopy and also the vibronic envelopes of these ionization processes, these authors were able to assign the earlier members of Rydberg series leading up to the second through fifth ioniza-

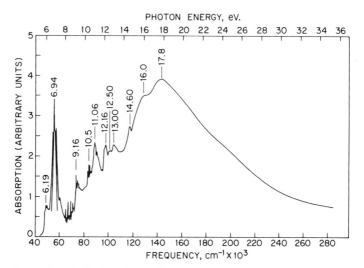

Fig. VI.A-7. Optical absorption spectrum of benzene vapor [K37].

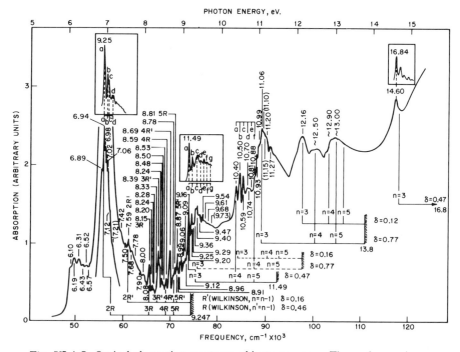

Fig. VI.A-8. Optical absorption spectrum of benzene vapor. The peak energies are given in eV, and the insets show the photoelectron band envelopes toward which the Rydberg excitations are converging [K37].

tion potentials. Our interpretation differs but slightly from those of Koch and Otto (Fig. VI.A-8) and Jonsson and Lindholm [J14].

The sharp line at 73 880 cm^{-1} (advert.) has a term value of 18 800 cm^{-1} with respect to the $3e_{2g}$ σ-ionization potential at 92 680 cm^{-1} (advert.) and can be assigned as the allowed $3e_{2g} \rightarrow 3p(a_{2u}, e_{1u})$ Rydberg excitation. In support of this, Koch and Otto point out the close similarity of the optical and photoelectron band envelopes (Fig. VI.A-8). The $n = 4$ member of this series is within the complex of bands at 84 000 cm^{-1}. The deepest pi MO, a_{2u}, has its ionization potential at 99 050 cm^{-1} (vert.) and transitions from it to 3s are allowed, the expected term value being 25 000 cm^{-1}. Since the a_{2u} ionization process is structureless, we are led to expect a broad $a_{2u} \rightarrow$ 3s excitation centered at 74 000 cm^{-1}. This transition probably accounts for the band underlying the structured $3e_{2g} \rightarrow$ 3p excitation at 74 000 cm^{-1}. Transitions from the fourth lowest level, σ (e_{1u}, ionization potential at 111 300 cm^{-1} advert.), to 3s and 3d are allowed by symmetry, but only that to 3d at 98 100 cm^{-1} can be positively identified. The strong feature at 89 200 cm^{-1} (vert.) in the optical spectrum has a term value of 22 100 cm^{-1} with respect to the $3e_{1u}$ ionization potential, which is rather too large for a transition to 3p and rather too small for a transition to 3s, the assignment preferred by Koch and Otto. There is another ionization potential in benzene at about 119 000 cm^{-1} (vert.), but it seems too distant to have given rise to the band at 89 200 cm^{-1}. Finally, the band at 116 900 cm^{-1} (advert.) has a term value of 18 900 cm^{-1} with respect to the $3a_{1g}$ ionization potential at 135 800 cm^{-1} (advert.), and most likely is the $3a_{1g} \rightarrow 3p(a_{2u}, e_{1u})$ transition. Strangely, though we have a myriad of obvious Rydberg excitations in benzene, none can be assigned with any confidence as terminating at 3s.

The intense overlapping continua with apparent maxima at 137 000 cm^{-1} (vert.) in benzene are apparently related to the continua in the same region reported for the alkanes (Section III.A) and the alcohols (Section III.E-2), and most likely originate with the C—H sigma-bonding MOs. In the electron-impact energy-loss spectrum of solid benzene, this broad region broadens even more, and Otto and Lynch have resolved it into equally intense peaks centered at 151 000 and 189 000 cm^{-1} (vert.) [O15]. Though the latter peak meets the first requirement of a collective excitation, being intense in the electron-impact spectrum, but missing optically (Section I.A-3), it should be remembered that the optical and electron-impact spectra were run on two different phases of benzene.

VI.A-2. Alkyl Benzenes

As one might reasonably expect, the effects of alkyl groups on the benzene spectrum are not large, so that in almost every case, the pattern

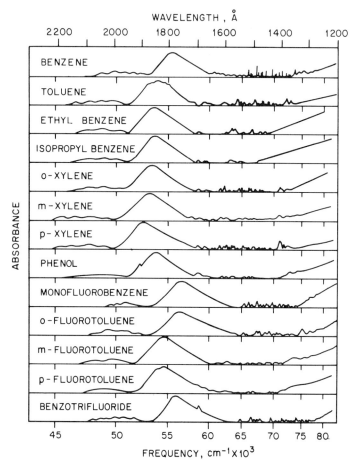

Fig. VI.A-9. Gas-phase absorption spectra of several monosubstituted and disubstituted benzene derivatives [H6].

of $^1B_{2u}$, $^1B_{1u}$, and $^1E_{1u}$ valence shell upper states followed by Rydberg transitions can be readily discerned in the gas-phase spectra (Fig. VI.A-9). Of course, in the lower symmetries of the alkyl benzenes, the state symbols will not be those appropriate to the D_{6h} point group, but in order to stress the close relationship of these spectra to that of benzene, the benzene notation still will be used, even though technically incorrect. In general, the effects of added alkyl groups are to shift the benzene bands to lower frequencies, this being largest for transitions to $^1E_{1u}$ and smallest for transitions to $^1B_{2u}$, and to broaden the bands, especially the Rydberg transitions. As in benzene, the valence shell bands are sharpest in transitions to $^1B_{2u}$ and broadest in transitions to $^1E_{1u}$.

Potts first found that the valence shell spectra of benzene and its alkylated derivatives in the vacuum ultraviolet are much sharper in paraffin matrices at 77°K than at room temperature in the gas phase [P33], probably due to the repression of hot bands, while Katz *et al.* found the bands sharper still in krypton matrices at 20°K [K9]. An interesting feature of this work is that whereas the transition to $^1B_{1u}$ is electronically forbidden in benzene and no (0, 0) component is observed even in matrices, a strong (0, 0) appears in the analogous bands of toluene (46 300 cm^{-1}) and p-xylene (45 290 cm^{-1}) (Fig. VI.A-10). This observa-

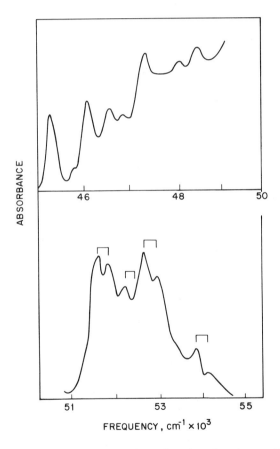

Fig. VI.A-10. Absorption spectrum of p-xylene in a krypton matrix deposited at 40° and measured at 20°K [K9]. The electronic splitting of 350 cm^{-1} in the 52 000-cm^{-1} band is indicated by brackets.

tion secures the $^1B_{1u}$ assignment for the 48 000-cm^{-1} band of benzenoid compounds, since the transition to $^1E_{2g}$ is still electronically forbidden in p-xylene. The vibronic analysis of Katz et al. for p-xylene is based largely upon the excitation of the totally symmetric vibrations ν_2' (700 cm^{-1}) and ν_{15}' (1120 cm^{-1}).

The $^1A_{1g} \rightarrow {}^1B_{1u}$ excitation in the hexamethyl benzene crystal comes at $\sim$45 000 cm^{-1} (vert.) in the crystal spectrum, and was found to have virtually all of its intensity polarized in plane [N11], as expected if the transition were borrowing intensity from the in-plane-polarized transitions to $^1E_{1u}$. At the time, the experimental result was said to rule out a Rydberg assignment with out-of-plane polarization, but of course such a band would not be seen in the crystal at essentially the gas-phase frequency (Section II.C).

The transition to the $^1E_{1u}$ state of the alkyl benzenes in a krypton matrix also shows vibrational structure which in toluene, toluene-d_8, and m-xylene is just a single progression in the totally symmetric vibration ν_2' built upon the (0, 0). In the corresponding transition of p-xylene, both ν_2' and ν_{15}' appear and each prominent line has a satellite of equal intensity displaced by 350 $\pm$ 50 cm^{-1} (Fig. VI.A-10). That this is due to site splitting was discounted by comparison with the spectrum in the 37 000-cm^{-1} region, which showed the site occupancy ratios to be very different (50:5:1) from the intensity ratio of the 350-cm^{-1} split components. It was concluded that the splitting is electronic and indicates that the $^1B_{2u}$ and $^1B_{3u}$ components of $^1E_{1u}$ in the lower symmetry are no longer degenerate. Note how much more structure is discernible in the matrix spectrum (Fig. VI.A-10) than in the vapor spectrum of p-xylene (Fig. VI.A-9).

An exceptionally interesting band in the spectrum of hexamethyl benzene has been reported by Nelson and Simpson [N11]. Between the transitions to $^1B_{2u}$ and $^1B_{1u}$, a band with a frequency of 42 500 cm^{-1} (vert.) and a maximum extinction coefficient of $\sim$2400 is found in the gas-phase spectrum of this molecule. An analogous band is absent in benzene and all other alkylated benzenes reported so far. A hint that this "extra" band might be a Rydberg transition comes from the observation that the band is not found in either paraffin solution or in the pure-crystal spectrum. In these circumstances, however, it could be covered by stronger, shifted absorption. If it is a Rydberg band, then its low frequency would demand a 3s terminating orbital. The vertical ionization potential of hexamethyl benzene is 65 600 cm^{-1} [R19], so that the 42 500 cm^{-1} vertical frequency of the "extra" band results in a term value of 23 100 cm^{-1}, in good agreement with our expection for a highly alkylated chromophore. Since the $\pi(e_{1g}) \rightarrow 3s$ transition is parity forbidden in hexamethyl benzene as in benzene, it is surprising that it was first found in the highly symmetric compound rather than in the less symmetric derivatives. The term value found here does suggest that the $(\pi_3, 3s)$ term values of benzenoid compounds will follow those of nonaromatic and saturated compounds once they are identified.

In a recent study of the luminescence excitation spectrum of benzene in rare gas matrices, Morris and Angus [M54] detected a weak feature between the $^1B_{2u}$ and $^1B_{1u}$ excitations which they eventually concluded was due to the excitation of the elusive $^1E_{2g}$ state. Apropos of this, they also then assigned the 42 500-cm^{-1} band of gas-phase hexamethyl benzene, coming as it does between the excitation to $^1B_{2u}$ and $^1B_{1u}$, to the $^1A_{1g} \to {}^1E_{2g}$ transition. Our feeling is that it is clear from the frequency of the benzene transition in the matrix that it is valence shell (its gas-phase frequency would be about 42 000 cm^{-1} if it were a Rydberg, giving it an improbably large term value of 32 500 cm^{-1}), and is not related to the "extra" band in hexamethyl benzene, the frequency and matrix behavior of which strongly argue for a Rydberg upper state.

One sees from the spectra of Fig. VI.A-9 that in the region of 70 000–80 000 cm^{-1}, there is absorption which in large part could be assigned to the alkyl groups. Note, however, that even in hexamethyl benzene [N11], it does not rival the intensity to $^1E_{1u}$. With this in mind, it is interesting to view the spectra of the alkyl benzenes as pure films at 77°K; in Fig. VI.A-11 [V2] it is seen that there is present a huge block of absorption beginning at $\sim 67\,000$ cm^{-1} which dwarfs the $^1A_{1g} \to {}^1E_{1u}$ absorption, and is not present in the gas-phase spectra. An analogous situation was found in the spectra of solid amides (Section V.A-1), and was attributed to transitions into the conduction band of the molecular solid. This would also seem to be the case here.

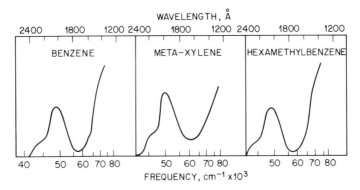

Fig. VI.A-11. Optical absorption spectra of benzene and its alkyl derivatives as solid films at 77°K [V2].

TABLE VI.A-I

SPECTRAL DATA ON THE $^1A_{1g} \to {}^1B_{1u}$ AND $^1A_{1g} \to {}^1E_{1u}$ TRANSITIONS OF THE ALKYL- AND HALOBENZENES IN n-HEPTANE SOLUTION[a]

Molecule	$^1B_{1u}$ Frequency (cm^{-1} vert.)	$^1B_{1u}$ Oscillator strength[b]	$^1E_{1u}$ Frequency (cm^{-1} vert.)	$^1E_{1u}$ Oscillator strength[c]
Benzene	49 000	0.10 (51 300)	54 500	0.79
Toluene	47 800	0.12 (50 000)	53 000	1.09
Ethyl benzene	48 000	0.11 (50 000)	52 900	1.11
n-Butyl benzene	47 600	0.11 (50 000)	52 900	1.04
sec-Butyl benzene	47 000	0.18 (50 200)	53 100	1.37
t-Butyl benzene	47 900	0.16 (50 400)	53 200	1.57
o-Xylene	47 400	0.11 (49 000)	52 500	1.22
m-Xylene	47 700	0.18 (49 200)	51 800	1.46
p-Xylene	47 200	0.14 (49 200)	52 000	1.23
Mesitylene	46 600	0.12 (47 400)	50 300	0.89
Hexamethyl benzene	45 300	$\sim$0.1	49 400	1.03[d]
Hexaethyl benzene	43 400	—	47 200	>0.7[e]
Fluorobenzene	48 100[f]	0.15 (52 500)	55 000	0.90
Benzotrifluoride	47 200[f]	0.12 (51 500)	54 300	1.22
o-Fluorotoluene	47 300[f]	0.11 (50 000)	53 400	1.26
m-Fluorotoluene	46 300[f]	0.13 (50 300)	53 300	1.09
p-Fluorotoluene	46 500[f]	0.11 (50 000)	54 000	1.19
Perfluorotoluene	50 500[f]	0.13 (52 000)	56 200	1.04
Chlorobenzene	46 600	—	52 700	1.1[g]
o-Dichlorobenzene	45 700	—	51 200	$\sim$0.6[g]
m-Dichlorobenzene	45 800	—	51 000	$\sim$0.65[g]
p-Dichlorobenzene	44 400	—	51 800	$\sim$0.7[g]
1,3,5-Trichlorobenzene	45 000	—	49 400	1.0[g]
Hexachlorobenzene	42 500	—	46 000	1.6[g]
Bromobenzene	46 500	—	52 400	$\sim$0.65[g]
Iodobenzene	—	—	51 400	1.6[g]

[a] From References [K28, K30, P26, P27].
[b] Oscillator strength determined by integration from the onset of absorption to the frequency given in parentheses.
[c] Oscillator strength obtained by integrating both $^1B_{1u}$ and $^1E_{1u}$ bands together.
[d] Estimated from a molar extinction coefficient of 51 000 [U3].
[e] $^1E_{1u}$ seems split into two components, and $f = 0.7$ applies only to the first of these.
[f] Onset of absorption, rather than vertical value.
[g] Estimated from molar extinction coefficients at the $^1E_{1u}$ maxima.

Being electronically allowed, the $^1A_{1g} \to {}^1E_{1u}$ oscillator strengths are of some interest in the alkyl benzenes. In order to make a consistent comparison, it is best to consider the heptane-solution data of Platt and Klevens [K30, P26, P27] (Table VI.A-I). In Table VI.A-I, the oscillator strength listed for the transition to $^1E_{1u}$ is actually that measured for the $^1E_{1u}$, $^1B_{1u}$ combination; inasmuch as the transition to $^1B_{1u}$ borrows its intensity from that to $^1E_{1u}$, the listed values may be considered as the experimental strengths to $^1E_{1u}$ before vibronic mixing. First, one sees that there is a rough correlation between the oscillator strengths of the two transitions, with that to $^1B_{1u}$ being 10–13% that of the total in all cases. For transitions to $^1E_{1u}$, alkyl groups have a surprisingly strong effect on the oscillator strengths, especially when placed so as to lower the D_{6h} symmetry of the parent molecule. The frequencies are much less sensitive to alkylation. Though the general tendency is for alkylation to increase the oscillator strength to $^1E_{1u}$, the number of alkyl substitutents is less important than their relative placements on the ring, for the oscillator strengths to $^1E_{1u}$ in mesitylene and hexamethyl benzene are actually lower than those for any monoalkyl or dialkyl compound of lower symmetry in our list. These results imply simply that the transition moment to $^1E_{1u}$ has a component along the C—C (ring-substituent) axis, such that these additional sources of transition moment sum to zero vectorially in compounds of symmetry D_{3h} or D_{6h}, but not in D_{2d}.

Rydberg spectra in the alkyl benzenes are less sharp than that of benzene, but Hammond et al. [H6] have found Rydberg series which have nd upper orbitals starting at either $n = 3$ or $n = 4$. Specifically, in benzene, an nd series has $\delta = 0.03$; in toluene, there is a series with $\delta = 0.05$, while in ethyl benzene, $\delta = 0.10$, and in o-xylene, $\delta = 0.08$. A second series in toluene has $\delta = 0.50$, and probably terminates at np MOs [P48].

Using laser excitation, Burton and Hunziker [B71] have excited toluene vapor to its lowest triplet state T_1, and then observed a $T_1 \to T_n$ excitation at 41 700 cm^{-1} (vert.) having $f = 0.05$. By analogy with the corresponding absorption in benzene (Section VI.A-1), this is assigned as $^3B_{1u} \to {}^3E_{2g}$. The corresponding $S_0 \to T_n$ transition must come at about 71 000 cm^{-1}.

Some rather peculiar spectra of solid thin films of toluene and mesitylene have been obtained by Lewis et al. [L21] by the technique of electron-impact energy-loss spectroscopy applied in reflection rather than transmission as usually done. They find energy-loss peaks in the reflected beam which correspond with known optical transitions in these molecules and additionally observe sharp peaks at the gas-phase ionization potentials out to 100 000 cm^{-1}. This latter feature is missing in conventional gas-phase energy-loss spectra.

The absorption spectrum of 2-phenyl-3,3-dimethyl butane in n-heptane solution shows the familiar transitions to $^1B_{2u}$, $^1B_{1u}$, and $^1E_{1u}$ at 38 750, 47 900, and 53 100 cm^{-1} (vert.). In addition, the circular dichroism of this optically active material has been recorded [S5] and shows three positive CD bands at the absorption frequencies given above, with $\Delta\epsilon$ values in the relative ratio 1:10:100, and with no sign of any fourth transition out to 54 000 cm^{-1}. The transition to $^1B_{1u}$ is much more highly structured in the CD spectrum than in absorption.

VI.A-3. Halobenzenes

Perturbations of the benzenoid pi spectrum by the introduction of fluorine, chlorine, and bromine atoms about the ring seem to be no stronger than those introduced by alkyl groups, i.e., in these compounds, the original benzene spectrum is readily identified, despite small frequency shifts and relatively larger intensity alterations. In the case of iodobenzene, the perturbation is more severe, however, and the benzenoid portion of the spectrum is less obvious (Section VI.A-5).

We first discuss the spectra determined in solution at room temperature [K30], for the Rydberg transitions will be obliterated under this condition (Section II.C), leaving only the valence shell excitations. In the compilation of Klevens and Platt [K30] (Table VI.A-I), the spectra of the chlorobenzenes in n-heptane solution are recorded, and show the usual pattern of $^1B_{2u}$, $^1B_{1u}$, and $^1E_{1u}$ bands (Fig. VI.A-12). The spectrum of hexachlorobenzene is anomalous in this respect, for it shows an additional strong band at 38 000 cm^{-1}, which at first sight would appear to be related to the "extra" band at 42 500 cm^{-1} (vert.) in hexamethyl benzene (Section VI.A-2). However, that band in hexamethyl benzene is probably a Rydberg excitation, whereas that in hexachlorobenzene is valence shell and more likely related to the A bands of the chloromethanes (Section III.B-1). The A bands in halogen-containing compounds result from the promotion of a halogen lone-pair electron into the antibonding sigma MO formed between the halogen atom and the adjacent carbon atom. Related low-frequency bands should appear in the highly brominated and iodinated benzenes, though they are unaccountably missing in chlorobenzene, the dichlorobenzenes, and trichlorobenzene.

On successive addition of chlorine atoms to the benzene ring, the absorption maximum of the transition to $^1E_{1u}$ in solution shifts to lower frequencies by about -1600 cm^{-1} per chlorine atom (Table VI.A-I). The $^1B_{1u}$ band also moves to lower frequencies, but its vertical frequency in several compounds is difficult to quote due to overlapping absorption. For fluorobenzenes in solution [K24, K28], the $^1E_{1u}$ vertical frequencies show a shift of $+500$ cm^{-1} per fluorine atom, whereas in the gas phase [G13], the shift per atom is more like -1000 cm^{-1}, while at the same time, the $^1B_{1u}$ transition is rather static. In n-heptane solution, the $^1E_{1u}$ bands of bromobenzene and iodobenzene shift by about -2000 to -3000

VI.A. PHENYL COMPOUNDS

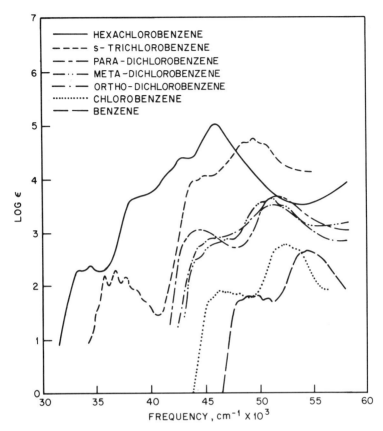

Fig. VI.A-12. Optical absorption spectra of various chlorobenzenes in n-heptane solution at room temperature [K30].

cm^{-1} with respect to benzene (Fig. VI.A-13). Interestingly, in none of these compounds of whatever low symmetry is there any resolvable splitting of the doubly degenerate $^1E_{1u}$ upper state in the solution spectra, even though the photoelectron spectra show large ($\sim$8000 cm^{-1}) splittings of the originating e_{1g} pi MOs [T21].

Though the pattern of energy levels throughout the halobenzenes remains remarkably benzenelike, the intensities show much larger variations (Table VI.A-I). The variations seem to be random, for the $^1B_{1u}$ band does not gain intensity in the lower-symmetry compounds, nor does it gain intensity when the transition to $^1E_{1u}$ intensifies, as it does in the alkyl benzenes. Most striking is the extreme enhancement of the intensities to $^1B_{1u}$ and $^1E_{1u}$ in hexachlorobenzene. In the series of monosubstituted

VI. AROMATIC COMPOUNDS

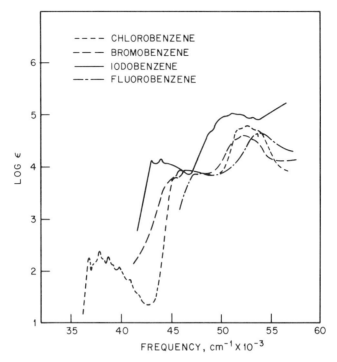

Fig. VI.A-13. Optical spectra of the monohalobenzenes in n-heptane solution at room temperature [K30].

benzenes, the iodo compound looks grossly out of line, with an extinction coefficient of 40 000 reported at the $^1B_{1u}$ maximum and 90 000 at the $^1E_{1u}$ maximum. No such anomaly appears in the gas-phase spectrum of iodobenzene [K16]. If these intensities in iodobenzene and hexachlorobenzene are correct, it seems probable that they are due to intense halogen $n\text{p} \rightarrow \pi^*$ charge transfer transitions underlying the benzenoid bands in the 50 000-cm^{-1} region. The oscillator strengths to the $^1E_{1u}$ states of the dichlorobenzenes are anomalously low, and perhaps should be remeasured.

In contrast to the solution spectra, a forest of sharp Rydberg bands fills the region between 55 000 and 70 000 cm^{-1} in the gas-phase spectra of the halobenzenes. The fluorobenzenes have received special attention recently (Fig. VI.A-14). Hammond et al. [H6] first identified two Rydberg series in fluorobenzene converging to the first ionization potential. One series was given a $\delta = 1.05$, and the $n = 3$ member was placed at $\sim 50\,000$ cm^{-1}. Since the band at 50 000 cm^{-1} is better assigned as valence

VI.A. PHENYL COMPOUNDS 233

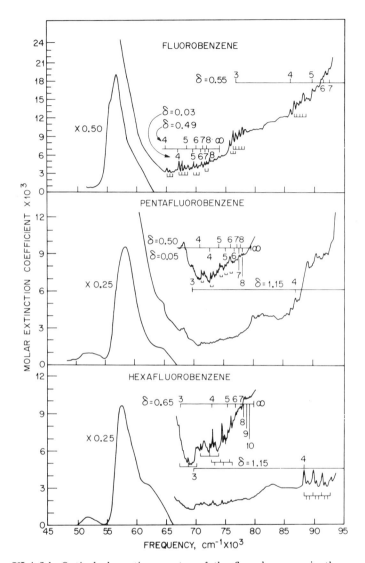

Fig. VI.A-14. Optical absorption spectra of the fluorobenzenes in the gas phase, with several Rydberg series and vibrational progressions delineated [S40].

shell $^1B_{1u}$, it is more reasonable to place the $n = 3$ member at 61 200 cm^{-1} and reduce δ to 0.05. Thus we have an nd series with a $(\pi_3, 3d)$ term value of 13 000 cm^{-1}. A second series with δ = 0.5 (np) also converges to the first ionization potential; Gilbert and Sandorfy report a sharp band at 55 309 cm^{-1} as $n = 3$ [G11], but Smith and Raymonda do not report this band in their analysis [S40]. The most recent work has revised the quantum defects to 0.03 and 0.49 for these two series in fluorobenzene [G11, S40].

Gilbert and Sandorfy claim two other Rydberg series in fluorobenzene going to the second ionization potential, but Smith and Raymonda deny this, claiming instead that this absorption is part of a series going to the third ionization potential with $\delta = 0.55$ ($n = 3, \ldots, 7$), accompanied by 490- and ~ 600-cm^{-1} vibrations. Being superposed on the continuum of the first ionization potential, the Rydbergs of the third series are undoubtedly autoionized and therefore appear somewhat broadened.

The Rydberg spectrum of hexafluorobenzene is especially interesting. According to Smith and Raymonda, a long nd series having $\delta = 0.05$ can be identified going to the first ionization potential accompanied by 1550- and 487-cm^{-1} vibrations. However, such an $e_{1g} \to n$d excitation is parity forbidden and so the series must be allowed through the intervention of an odd-parity vibration. A second ionization potential begins at 88 275 cm^{-1} and has $\delta = 1.15$. Since the second highest MO in hexafluorobenzene is $\pi(a_{2u})$ [B59], the series to ns upper orbitals (as implied by the large value of δ) is electronically allowed, whereas that to np is forbidden. This second series is very interesting in that it is the first ns series observed in a substituted benzene. The (πa_{2u}, 3s) term value of 31 000 cm^{-1} (adiab.) is appropriately large for a highly fluorinated substance.

In pentafluorobenzene, the Rydberg bands are more diffuse than in either fluorobenzene or hexafluorobenzene. The usual two series with $\delta = 0.50$ and 0.05 are observed here converging on the first ionization potential, together with the early members of a series going to the second ionization potential with $\delta = 1.15$. This latter absorption is clearly related to the series with the same δ value observed in hexafluorobenzene. Analyses similar to those given here have been reported for 1,3,5-trifluorobenzene by Gilbert et al. [G12]. In this molecule, two Rydberg series going to the first ionization potential (e'', 77 765 cm^{-1} advert.) were identified, with δ values of 0.06 and 0.47 starting at $n = 3$. The profile of the $e'' \to 3$p optical transition closely resembles that of the e'' photoelectron band.

One sees in the fluorosubstituted benzene spectra (Fig. VI.A-14) that there is a further valence shell transition on the high-frequency wing of the transition to the $^1E_{1u}$ state, 60 000–64 000 cm^{-1}, which is not nearly as obvious in benzene itself. Two possibilities present themselves for an assignment. Arguing by analogy with the fluoroethylenes (Section IV.A-3), one could argue that the perfluoro effect acts to lower the σ^* manifold with respect to π^*, so that a $\pi \to \sigma^*$ transition is much lower in the fluorobenzenes than in benzene itself. It seems unlikely that this is the explanation, however, since the band is at nearly the same frequency in both fluorobenzene and hexafluorobenzene. A similar argument may be used against an A-band assignment [F lone pair $\rightarrow \sigma^*$(C—F)] for the band in the 60 000–65 000-cm^{-1} region. Better is the assignment to the $^1E_{2g}$ state, thought to be near 60 000 cm^{-1} in benzene (Section VI.A-1). If this is correct, then there must be a very strong vibronic coupling between this state and $^1E_{1u}$, for the band is quite strong in hexafluorobenzene even though it is parity forbidden.

Gilbert et al. [G13] have also found that the intensity of the massive absorption around 80 000 cm^{-1} in benzene is progressively decreased as the molecule is fluorinated, and suggest that this is the region of $\sigma \rightarrow \sigma^*$ absorption, which in general moves to higher frequency upon fluorination.

As was the case with fluorobenzene, Hammond et al. [H6] also found a long Rydberg series in benzotrifluoride (Fig. VI.A-9), to which they assigned $\delta = 1.05$ with the $n = 3$ member at about 48 000 cm^{-1}. It is more likely that the 48 000-cm^{-1} band is the transition to the valence shell $^1B_{1u}$ state, and that the n values in the series should be decreased by one, with $\delta = 0.05$, i.e., a δ value characteristic of an nd series. An np series was also identified in benzotrifluoride having $\delta = 0.50$ and a term value of 23 000 cm^{-1} for the $n = 3$ member. An almost identical series can also be seen in p-fluorotoluene (Fig. VI.A-9).

We see from these spectra of the alkyl and fluorobenzenes a remarkable constancy both in the valence shell and Rydberg spectra. In all cases, the valence shell transitions originate at the e_{2g} MOs, and terminating at e_{2u}, generate the familiar pattern of $^1B_{2u}$, $^1B_{1u}$, and $^1E_{1u}$ excited states. Rydberg excitations originating at e_{2g} are not observed to ns, but an extended np series is visible in almost all of the molecules ($\delta \simeq 0.5$) and another with $\delta \simeq 0.0$ has nd-terminating MOs. This latter series in benzene itself is further resolved into three electronic components (the R′, R″, and R‴ series of Wilkinson), but these are not seen in its derivatives. Though transitions to 3s in benzenoid compounds are not seen generally from the upper pi MO (e_{2g}), in the fluorobenzenes they are seen originating from the lower pi MO (a_{2u}), and probably could be traced in the alkyl benzenes with ease, since the (a_{2u}) ionization potentials are known and the term values easily estimated (Eq. I.27).

Like the fluorobenzenes, the chlorobenzenes and bromobenzenes display only a slightly perturbed valence shell benzene spectrum shifted to lower frequencies. The chlorobenzenes show the characteristic D band found in the spectra of almost all chlorine-containing molecules; however, according to Price and Walsh [P48], the Rydberg transitions originating at the benzene pi MOs are not seen. In chlorobenzene and o-dichlorobenzene, the D bands come at 68 500 and 69 000 cm^{-1}, respectively. Since the chlorine 3pπ ionization potential in chlorobenzene is 91 300 cm^{-1} (vert.), the D-band term value in this compound is 22 900 cm^{-1}. Similar bands at similar frequencies are reported in the chloroethylenes (Section IV.A-3), in chloroprene (Section V.C), and in the alkyl chlorides (Section III.B-2).

The absorptions of bromobenzene and iodobenzene are again like those of benzene, except that the Rydberg transitions originating with the halogen lone pairs are even more prominent than in chlorobenzene. In bromobenzene, the B and C bands (4p → 5s) come at 48 800 and 50 700 cm^{-1} and the D band falls at 62 500 cm^{-1} [P48]. The B–C splitting in a cylindrically symmetric bromide such as methyl bromide is due to spin–orbit coupling in the ionic core configuration (· · · 4p^5)$^+$, but in bromobenzene, there is an additional factor in the splitting due to the nonequivalence of the in-plane and out-of-plane 4p lone-pair AOs. In methyl bromide, where only spin–orbit splitting is a factor, the B–C separation is 3145 cm^{-1}, decreasing to 2540 cm^{-1} in the ion. In the bromobenzene positive ion, the splitting of the bromine lone pairs is 4590 cm^{-1} [T21]. Since the D band of bromobenzene has terms of 27 500 and 23 000 cm^{-1} with respect to the ionization potentials at 90 000 and 85 500 cm^{-1}, respectively, it is more likely that the D band is converging upon the lower of these two ionization potentials, and that there is another D band at about 67 000 cm^{-1}. According to Turner et al. [T21], the lower of the two bromine ionization potentials corresponds to loss of an electron from the 4p lone-pair AO that is in the plane of the benzene ring.

Sergeev et al. [S31] have studied the photoionization spectrum of bromobenzene and found two series of autoionizing lines converging to an ionization potention of 82 500 cm^{-1}, with quantum defects of 0.5 and ~0.0. However, there is no ionization potential in bromobenzene at 82 400 cm^{-1} according to the photoelectron spectrum, and the autoionizing frequencies are much too close together to be members of a series converging to a higher genuine ionization potential. It is more likely that the line at 77 100 cm^{-1} is a Rydberg origin [4p(lone pair) → 4d; 12 900 cm^{-1} term value with respect to the ionization potential at 90 000 cm^{-1}], but that the other features are either vibronic structure on this origin or upon other origins that are components of the 4p → 4d complex.

Interestingly, whereas no splitting of the $^1E_{1u}$ state can be seen in the gas-phase spectrum of toluene, a splitting of ~1400 cm^{-1} is readily observed in the spectrum of α,α,α-trichlorotoluene [K18]. However, this may be vibrational rather than electronic. In the gas-phase spectrum of chlorobenzene, Kimura and Nagakura [K16] note that the transition to $^1E_{1u}$ is twice as broad as that in benzene, and they separate it into two components split by about 1300 cm^{-1}. There is an obvious splitting of 2800 cm^{-1} in the $^1A_{1g} \to {}^1E_{1u}$ band of bromobenzene.

Kimura and Nagakura have investigated the excited states of the halobenzenes using their intramolecular charge transfer approach (Section I.B-2) [K16]. They conclude that in the fluoro, chloro, and bromo compounds, the charge transfer effects are only weak perturbations on the benzene local excitations, but in iodobenzene, the two iodine $(5p\pi, \pi^*)$ charge transfer configurations mix very strongly with the $^1B_{1u}$ and $^1E_{1u}$ benzene configurations and truly complicate the spectrum (Fig. VI.A-13). According to the calculations, the bands at 44 000 and 50 000 cm^{-1} are largely charge transfer, with the more benzenoid transitions following at 53 000 cm^{-1} and beyond. On the other hand, Price and Walsh [P48] assign the 44 000-cm^{-1} band as a benzenoid $^1B_{1u}$ transition, and the 50 000-cm^{-1} band as a Rydberg B band analogous to the 48 000-cm^{-1} band of methyl iodide; the expected D bands in iodobenzene are observed in the 57 000–59 000-cm^{-1} region, and are analogous to the band at 59 000 cm^{-1} in methyl iodide.

VI.A-4. *Azabenzenes*

The two most interesting aspects of the azabenzene spectra are the resemblance or lack of same to the benzene spectrum, and the spectral consequences of the introduction of one or more lone-pair orbitals on the nitrogen atoms. The first aspect has been studied repeatedly in the pi-electron approximation, while the second expresses itself as the problem of the relative ordering of the lone-pair and pi MOs. The Rydberg spectra in the vacuum ultraviolet are of great use in this latter respect when combined with the results of photoelectron spectroscopy. Lindholm and his co-workers especially have used this technique to good advantage in interpreting the spectra of the azabenzenes.

The ground-state vibrations in the azabenzenes are closely related as regards both the frequencies and types of displacements, and so it is convenient to use a common descriptive label in all molecules of whatever formal symmetry. In this, we follow the example of Innes and co-workers [I4, P5], who have labeled the most conspicuous vibrations in the pyridine and pyrazine spectra as in Fig. VI.A-15. One can readily imagine the displacements for the corresponding vibrations in systems of different symmetries.

As was demonstrated in the earlier sections, the benzene spectrum is remarkably resistant to substitutive perturbations, and can be recognized virtually unchanged in many formally different classes of compounds.

238　　　　　　　VI. AROMATIC COMPOUNDS

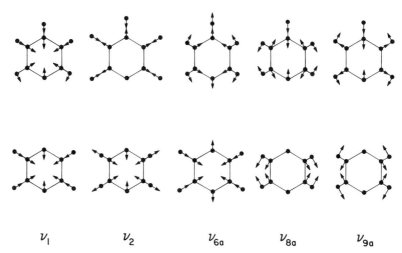

Fig. VI.A-15. The totally symmetric normal modes for pyridine (upper) and pyrazine (lower) that appear most frequently in the spectra of all azabenzenes [I4, P5].

Thus it is no surprise that the spectrum of pyridine (Fig. VI.A-16) so closely resembles that of benzene. One sees first the equivalent of the transition to $^1B_{2u}$ (benzene notation) at 40 300 cm^{-1} (vert.) in pyridine, but with a greatly increased intensity thanks to the lower symmetry. A weak $n_N \to \pi^*$ transition lies just under the low-frequency wing of this band. Following this are the two stronger transitions to $^1B_{1u}$ and $^1E_{1u}$

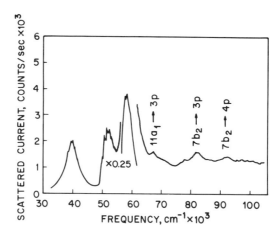

Fig. VI.A-16. The electron-impact energy-loss spectrum of pyridine vapor as measured with 100-eV electrons scattered at $\theta = 0°$ [J13].

(in the benzene notation). The former comes at 52 000 cm^{-1} (vert.) with an oscillator strength of 0.1 in the gas phase [P21] but 0.2 in heptane solution [K30]. It is interesting to note that though the transition to $^1B_{1u}$ is vibronic in benzene but is formally electronically allowed in pyridine, the oscillator strengths in the gas phase are equal in the two compounds. It seems likely that there is vibronic borrowing for this transition in pyridine even though it is electronically allowed. The transition to $^1E_{1u}$ in pyridine (58 300 cm^{-1} vert. in the gas phase, 56 800 cm^{-1} vert. in heptane solution) has an oscillator strength of 1.3 [P21, K30]. Though the lower symmetry of pyridine formally splits the benzene $^1E_{1u}$ state into the components $^1A_1$ and $^1B_2$, no such splitting can be experimentally demonstrated. Indeed, theoretical calculations ([M21], for example) predict that the $^1A_1$–$^1B_2$ splitting will be less than 2000 cm^{-1}. A few intervals of ~900 cm^{-1} are visible in this band, but could easily be quanta of ν_1' (Fig. VI.A-15), the same vibration which is conspicuous in the transition to $^1E_{1u}$ in benzene and the alkyl benzenes. (Note that this ring-breathing vibration in benzene is called ν_2'.)

According to the photoelectron studies of Gleiter et al., the uppermost filled MO in pyridine is the lone-pair orbital on nitrogen n_N (a_1) with an ionization potential of 77 400 cm^{-1} (vert.), 75 000 cm^{-1} (adiab.) [G14]. Just 1200 cm^{-1} beyond that is the first pi-electron ionization potential $\pi_3(a_2)$, while $\pi_2(b_1)$ comes at 84 700 cm^{-1} (vert.). Note, however, that there have been many arguments over the ordering of the levels in the azabenzenes, and, for example, Jonsson et al. [J13] prefer π_3, π_2, n_N for pyridine. However, we shall hold to the ordering of Gleiter et al. as the most consistently convincing. One interesting aspect of pyridine and the diazabenzenes is that in all of them, the $^2E_{2g}$ ionic state of benzene is split into components separated by up to 12 000 cm^{-1} according to the photoelectron spectra, yet the electronic transitions to the $^1E_{1u}$ states show no signs of splitting, except for a possible splitting in pyrimidine.

An allowed $n_N \rightarrow 3s$ transition in pyridine would be expected at ~50 000 cm^{-1} (vert.), but none has been seen yet. This parallels the situation in benzene and the other diazabenzenes, where there is again no positive identification of the lowest transition to the 3s MO. Transitions to 3p, however, are prominent in benzene and the diazabenzenes, and are no less so in pyridine, though not apparent in the low-resolution spectrum of Fig. VI.A-16. El-Sayed points out the presence of both a sharp and a diffuse band system resting upon the $^1E_{1u}$ band of pyridine; the diffuse system has an origin at 56 400 cm^{-1} (adiab.) accompanied by several quanta of ν_1' (950 cm^{-1}) [E6]. The term value of 18 600 cm^{-1} (adiab.) characterizes it as a symmetry-allowed $n_N \rightarrow 3p$ Rydberg transition. A second, sharper progression of bands in the same region is another component of the $n_N \to 3p$ manifold, but El-Sayed instead suggests that it is the long-sought valence shell transition to $^1E_{2g}$, which is formally allowed in pyridine. A pressure-effect experiment (Section II.B) would settle this problem.

Doering and Moore [D21] have studied the ion-impact energy-loss spectrum of pyridine at low resolution with both H⁺ (3.0 keV) and He⁺ (2.8 keV) ions for excitation. With the latter, a singlet–triplet excitation at 33 000 cm⁻¹ (vert.) was observed, but the Rydberg excitations following that to $^1E_{1u}$ were missing, whereas with H⁺ excitation, the singlet–triplet band is missing, but the Rydberg bands beyond 65 000 cm⁻¹ are quite intense. A similar discrimination against Rydberg excitations using He⁺ excitation was also found for the substituted ethylenes (Section IV.A) and suggests its possible use as a tool for distinguishing Rydberg and valence shell upper states. In the H⁺ ion-impact spectrum, the broad Rydberg excitations come at 68 000 and 88 000 cm⁻¹ (vert.), whereas in the trapped-electron spectrum of pyridine (Section II.D), bands are observed at 72 000 and 84 000 cm⁻¹ (vert.) [P24]. Since relative intensities in the trapped-electron spectrum can be very different from those observed using nonthreshold impact techniques (in the trapped-electron spectrum of pyridine, the transition to $^1B_{1u}$ is considerably more intense than that to $^1E_{1u}$, for example), one cannot be completely certain that the Rydberg bands in the ion-impact and trapped-electron spectra are the same excitations. Neither of them agrees very closely with the excitation frequencies obtained using the electron-impact energy-loss technique (Fig. VI.A-16).

El-Sayed *et al.* [E5] have also claimed several Rydberg series going to ionization potentials of 74 700, 83 100, and 93 000 cm⁻¹ in pyridine, the last of which is certainly spurious, as the photoelectron spectrum shows [T21]. Only a few broad bands beyond the first ionization potential have been characterized in the electron-impact energy-loss spectrum [J13]. These are identified by their term values in Fig. VI.A-16.

In pyrazine (1,4-diazabenzene), the benzenelike transition to $^1B_{2u}$ is found at about 41 000 cm⁻¹ (vert.), as in benzene itself, but with an oscillator strength (0.1) much increased over that of benzene [K30]. Four absorption systems beyond 45 000 cm⁻¹ have been delineated so far in pyrazine (Figs. VI.A-17 and VI.A-18). Following the intense transition to $^1B_{2u}$, a much weaker transition is seen in the vapor spectrum at 49 500 cm⁻¹ (vert.) which is structureless and seems not to appear in the solution spectrum [K30] or in the lower-resolution electron-impact spectrum [F13]. Since there is no analogous transition in the pi-electron spectrum of benzene, and since the weak pyrazine band does have a term value of 28 100 cm⁻¹ (vert.), it can tentatively be assigned as a

VI.A. PHENYL COMPOUNDS

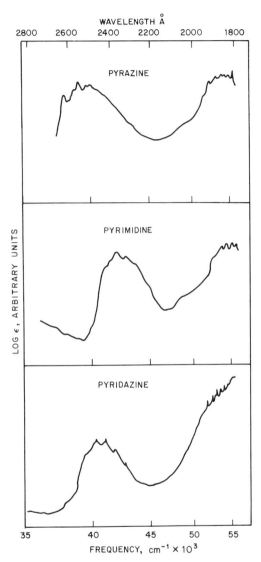

Fig. VI.A-17. Absorption spectra of the three diazabenzenes in the vapor phase [P5].

transition to 3s that is electronically forbidden.† Gleiter et al. [G14] and Fridh et al. [F13], in their photoelectron studies of pyrazine, do conclude

† However, see the later comments concerning the same transition in pyrimidine and pyridazine. The evidence is considerably stronger for such a transition to 3s in s-triazine.

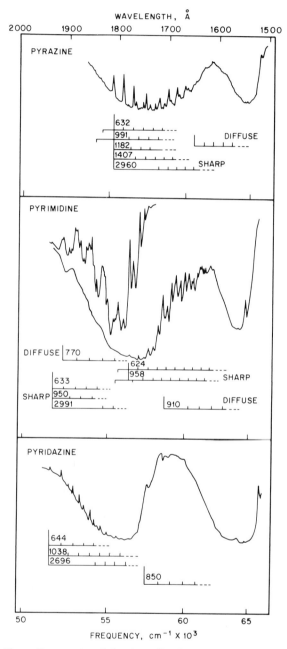

Fig. VI.A-18. Absorption spectra of the three diazabenzenes in the vapor phase [P5].

that a lone-pair a_g MO lies uppermost in this molecule, but a more sophisticated *ab initio* calculation [H3] places an occupied b_{1g} pi MO slightly above the $6a_g$ MO. In either case, the transition to 3s is forbidden, as expected, but transitions to all three components of the 3p manifold will be allowed from $6a_g$, whereas transitions to only two of these are allowed from b_{1g}.

Another strong benzenoid transition is centered at 54 000 cm^{-1} (vert.) in pyrazine [E6]. Klevens and Platt [K30] estimate an oscillator strength of 0.145 for this band in the solution spectrum, and assign it as analogous to the $^1B_{1u}$ transition of benzene (50 000 cm^{-1} vert.; $f = 0.12$) (Section VI.A-1). Parkin and Innes [P5] have analyzed the vibronic structure of this band (the vibrations also appear in the solution spectrum) and find that it consists of the totally symmetric vibrations ν_1' and ν_2' radiating from an origin of moderate intensity. Actually, since the symmetry of pyrazine is the same as that of *p*-xylene, the demonstration in the latter molecule of the electronic allowedness of the 50 000-cm^{-1} band and from that the assignment of the upper state as related to $^1B_{1u}$ rather than $^1E_{2g}$ (Section VI.A-2) is equally valid for the corresponding band in pyrazine.

An obvious Rydberg transition in pyrazine has its origin at 55 154 cm^{-1} (adiab.), with maximum absorption at 55 786 cm^{-1} (vert.) [P5, S17]. Since Gleiter *et al.* [G14] give a first ionization potential of 77 600 cm^{-1} (vert.), one obtains a term value of 21 810 cm^{-1} (vert.), which identifies the transition as terminating at 3p. Fridh *et al.* and Scheps *et al.* also favor this assignment. In both the pyrazine-h_4 and -d_4 spectra, only the totally symmetric modes ν_1', ν_2', ν_{6a}', ν_{8a}', and ν_{9a}' are excited (Fig. VI.A-15), with 600-cm^{-1} progressions of ν_{6a}' being prominent. This motion distorts the molecule along the N—N line. Scheps *et al.* [S17] have placed this band of pyrazine as the $n = 3$ member of an $np\pi$ series having $\delta = 0.50$ and converging to the first ionization potential. The linewidths in this region of the spectrum suggest a mixing of the Rydberg configuration and the underlying valence shell state [S17].

Parkin and Innes also suggest that there is a second system with an origin near 54 000 cm^{-1} since there are several bands which do not fit into the analysis of the previously discussed transition. Of course, this is to be expected, since the threefold degeneracy of the 3p manifold will be lifted in pyrazine. The situation is much like that near 57 000 cm^{-1} in pyridine. Fridh *et al.* [F13] have analyzed this band system with electron-impact spectroscopy and have come to the same conclusions regarding the $np\pi$ series, except that they deny the excitation of ν_2'. Additionally, their vibronic analysis uncovers another of the components to 3p, with an origin at 56 885 cm^{-1} (adiab.). Unfortunately, in the analysis of Scheps et al., this line is instead assigned as a vibronic component of the $\delta = 0.50$ series, but these authors uncover another series in which the $n = 3$ band has its origin at 54 413 cm^{-1} (adiab.), the series having $\delta = 0.59$. If there are truly three origins in the transitions to 3p, then this unambiguously assigns the symmetry of the originating MO as a_g, whereas only two origins are expected for excitations from b_{1g}. For the moment, it appears that the latter is the correct choice, and we will operate on this premise.

Parkin and Innes attempted to perform a contour analysis of the rotational envelope of the (0, 0) band at 55 288 cm^{-1} in pyrazine-d_4, and concluded that the polarization was in plane, leading to either a $^1B_{1u}$ or $^1B_{2u}$ upper state. In a later report, Innes et al. [I4] quote a short-axis, in-plane polarization ($^1A_g \rightarrow {}^1B_{2u}$) for this Rydberg excitation. This is the polarization to be expected for a transition which is $\pi(b_{1g}) \rightarrow 3p\pi(b_{3u})$. According to Fridh et al., the second overlapping Rydberg transition to 3p is $^1A_g \rightarrow {}^1A_u$, but such a forbidden excitation is contrary to the intense origin claimed by them.

The analog of the strongly allowed transition to $^1E_{1u}$ in benzene has an oscillator strength of about 1.0 in pyrazine and a gas-phase absorption maximum at 61 900 cm^{-1} (vert.). As in benzene and pyridine, the transition is accompanied by several quanta of the totally symmetric ring-breathing vibration, but shows no signs of splitting into two components. Of course, in benzene, such a splitting would be due to Jahn–Teller effects, which might be small as suggested by theory, but in pyrazine, the symmetry of the molecule breaks the degeneracy, so that in the ground state, Gleiter et al. find the two components of the e_{1g} MO of benzene to be separated by about 15 000 cm^{-1}. Still, no splitting is observed for the $\pi \rightarrow \pi^*$ band originating at the e_{1g} components (unless it is very large) and in fact, a semiempirical calculation by McWeeny and Peacock [M21] predicts that the two components ($^1B_{2u}$ and $^1B_{1u}$) will be accidentally near-degenerate. Unfortunately, in the more sophisticated calculation of Hackmeyer and Whitten, only the energy of the $^1B_{2u}$ component was calculated [H3].

A doublet of sharp bands appearing weakly at 64 500 cm^{-1} (Fig. VI.A-17) with a term value of 13 100 cm^{-1} (vert.) is a component of $\pi(b_{1g}) \rightarrow 3d$, which is $g \rightarrow g$ forbidden. Other apparent Rydberg origins are reported at 65 746 and 69 504 cm^{-1} (adiab.) [I4], with the first of these being much too strong to be part of the $\pi(b_{1g}) \rightarrow 3d$ complex; as suggested by Innes et al. [I4], these bands are the higher members of the Rydberg series having $n = 3$ at 55 154 cm^{-1} (adiab.). Fridh et al. and Scheps et al. also accept this assignment.

Possible Rydberg excitations at higher frequencies converging to higher ionization potentials are also identified in the electron-impact spectrum of pyrazine. A strong band at 78 641 cm^{-1} (advert.) has a term value of 15 730 cm^{-1} (advert.) with respect to the $1b_{2g}\pi$ ionization potential at 94 370 cm^{-1}, and is probably $1b_{2g} \to 3p$, though its term value is unusually low for a 3p terminating orbital.

The absorption spectrum of pyrimidine (1,3-diazabenzene) (Figs. VI.A-17 and VI.A-18) is remarkably like that of pyrazine, in spite of the fact that the center of symmetry is no longer present. However, since all of the $\pi \to \pi^*$ transitions are already allowed in the D_{2h} symmetry of pyrazine (except the benzenoid $^1A_g \to {}^1E_{2g}$), they would be expected at about the same frequencies and intensities in the lower C_{2v} symmetries of pyrimidine and pyridazine. Thus in pyrimidine, the strong transition to $^1B_{2u}$ (using the benzene notation) is found at 42 000 cm^{-1} (vert.) with $f = 0.052$ in solution [K30]; an even stronger transition to $^1B_{1u}$ comes at 55 000 cm^{-1} (vert.) with $f = 0.16$ in solution, and the two transitions to the components of $^1E_{1u}$ appear at 62 000 cm^{-1} (vert.) with $f \simeq 1$ [E6, P5]. In all of these $\pi \to \pi^*$ excitations of pyrimidine, the 950-cm^{-1} vibration ν_1' is prominent. The electronic spectrum of the pyrimidine derivative uracil (Chapter VIII) has been interpreted as showing a large splitting of the benzenoid $^1E_{1u}$ state, but none is apparent in pyrimidine itself.

Because of the near degeneracy of the lone pair and π MOs in the diazabenzenes, there is a considerable problem in general in establishing the orbital ordering, and pyrimidine is no exception. In the past, virtually every possible permutation of the two lone-pair and two pi MOs has been proposed. Recently, this problem was tackled with the perfluoro effect [B59, B60], i.e., when the molecule is fluorinated, the lone-pair ionization potentials are stabilized by 15 000–25 000 cm^{-1}, whereas the pi MOs show no shift. According to this test, one sees immediately (Fig. VI.A-19) that the first ionization potential of pyrimidine involves the lone-pair orbital. Gleiter et al. [G14] and Åsbrink et al. [A16] recently came to the same conclusion from a different direction. The ionization potential of this uppermost b_2 lone-pair MO is 78 500 cm^{-1} (vert.).

As in the spectrum of pyrazine, a weak band is also found between the strong $^1A_g \to {}^1B_{2u}$ and $^1A_g \to {}^1B_{1u}$ excitations in pyrimidine. Its frequency, 50 000 cm^{-1} (vert.), gives it a term of 28 100 cm^{-1} (vert.), but unlike the case in pyrazine, the transition from the highest occupied MO ($1b_2$) to $3s(a_1)$ is allowed electronically. Yet the intensity of this band is no higher in pyrimidine, according to the data of Parkin and Innes [P5]. Thus it must be admitted that this band might be an $n_N \to \pi^*$ transition, in spite of its Rydberglike term value. Its appearance at 47 700

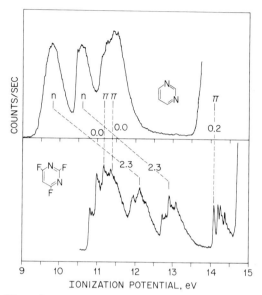

Fig. VI.A-19. Photoelectron spectra of pyrimidine and its trifluoro derivative in which the lone-pair ionizations shift 18 000 cm^{-1} to higher energy while the π MOs are unshifted [R19].

cm^{-1} (vert.) in the solution spectrum (methyl cyclohexane) argues strongly against a Rydberg assignment [C16].

Under high resolution (Fig. VI.A-17), the broad vibronic structure of the transition to $^1B_{1u}$ is seen to be intermixed with a second transition consisting of a complex of sharp lines with an origin at 51 759 cm^{-1}. The term value of this sharp-line absorption (24 100 cm^{-1} adiab.) is about 3000 cm^{-1} smaller than that expected for the $1b_2 \rightarrow 3s$ transition and about 4000 cm^{-1} larger than that expected for $1b_2 \rightarrow 3p$. However, since it has a vibronic envelope much like that of the first photoelectron band [A16], it is a Rydberg excitation, most likely terminating at 3s. This transition has ν_{6a}' as its most prominent totally symmetric vibration. A considerably stronger Rydberg system begins at 56 271 cm^{-1} (adiab.), and again only totally symmetric vibrations appear with very long progressions in ν_{6a}' (620 cm^{-1}). Its term value of 19 500 cm^{-1} (adiab.) is sufficient to identify it as a member of the 3p manifold. A sharp single line at 64 500 cm^{-1} (vert.) has a term value of 13 600 cm^{-1} (vert.) and is the $1b_2 \rightarrow 3d$ transition, which appears prominently in benzenoid compounds regardless of symmetry.

Loustauneau and Nouchi [L36] found the transition from the lowest triplet state of pyrimidine to the next higher one to begin at 29 000 cm^{-1}.

Since the lowest triplet is 29 000 cm^{-1} above the ground state, the newly found upper triplet lies 58 000 cm^{-1} (adiab.) above the ground state.

The valence shell spectrum of pyridazine (1,2-diazabenzene) is like those of the other diazines, and requires little further comment. According to Gleiter et al. [G14] and Åsbrink et al. [A17], the highest occupied MO in pyridazine is again a lone-pair orbital ($1b_2$), as in pyrimidine. On the high-frequency side of the transition to $^1B_{1u}$, sharp Rydberg bands are observed, with an origin at 50 865 cm^{-1} (adiab.) and the ever-present progressions in the ν_{6a}' vibration of ~650 cm^{-1} frequency. Since the ionization potential of pyridazine is 70 220 cm^{-1} (adiab.) [G14], the Rydberg transition has a term value of 19 360 cm^{-1} (adiab.), which we assign as $1b_2 \to 3p\pi$, as judged by comparison with the term values in the other diazabenzenes. Two Rydberglike bands at 67 000 and 64 000 cm^{-1} (vert.) are too high for excitation from n($1b_2$) to 3d, but fit nicely as excitations from $\pi_3(1a_2)$ to 3p, the $1a_2$ level having an ionization potential of 84 550 cm^{-1} (advert.) [G14]. The corresponding $1a_2 \to 3s$ transition was identified by Åsbrink et al. [A17] as a diffuse progression beginning at 57 300 cm^{-1} (advert.) and resting upon a $\pi \to \pi^*$ band. Its term value is 27 200 cm^{-1} (advert.). In the trapped-electron spectrum of pyridazine [P24], the Rydberg spectrum appears as a broad peak centered at ~68 000 cm^{-1} (vert.). An identical peak is found as well in the spectra of pyrimidine, pyrazine, and s-triazine.

A triplet–triplet absorption experiment on pyridazine uncovered a transition with origin at about 25 000 cm^{-1} [L36]. Since the lowest triplet is also 25 000 cm^{-1} above the ground state, the unidentified upper triplet is about 50 000 cm^{-1} (adiab.) above the ground state.

Brinen et al. [B47] have published a brief account of the spectra of s-triazine and its derivatives in the 50 000–60 000-cm^{-1} region. The notable feature uncovered was a set of very strong, sharp lines superposed upon the low-frequency side of a broad, continuous absorption. The sharp bands in s-triazine begin at 55 782 cm^{-1} (adiab.) with a maximum intensity at 57 400 cm^{-1} (vert.). In s-triazine-d_3, the adiabatic value is 55 903 cm^{-1}, whereas in trimethyl-s-triazine, it has shifted to 51 190 cm^{-1}; intermediate values were obtained for the monomethyl and dimethyl derivatives. Vibrational assignments were proposed in which multiple quanta of 950 cm^{-1} (s-triazine itself) are joined to the true origin and to false origins having 660- and 730-cm^{-1} displacements. These authors argue that the sharp profile of the (0, 0) band is that of a parallel-type transition, and suggest a $^1A_1' \to {}^1A_2''$ Rydberg assignment, but Innes et al. [I4] argue instead that the profile is more like that expected for an in-plane-polarized $^1A_1' \to {}^1E'$ transition.

Photoelectron studies of s-triazine place the ionization potential at

84 300 cm^{-1} (vert.) [B60, G14], giving a term value of 26 900 cm^{-1} (vert.) to the absorption band of Brinen *et al.* The large size of this term value shows that it is not related to the 3p Rydberg excitations of benzene and the diazines found in the vicinity of 55 000–60 000 cm^{-1}. Since the term value of 26 900 cm^{-1} is indicative of a 3s-terminating MO, but $e''(\pi_2,\pi_3) \to 3s$ is electronically forbidden, it must be that the originating MO instead is of e' (sigma) symmetry, as is the upper component of the lone-pair orbitals on the nitrogen atoms. Indeed, it is concluded from the perfluoro effect [B60] and from semiempirical calculations [F12, G14] that the uppermost MO in s-triazine is $6e'$. The $^1E'$ upper state is Jahn–Teller unstable, and this may account for the appearance of the nontotally symmetric vibrations of 660 and 730 cm^{-1}. The 660-cm^{-1} vibration is the ubiquitous ν_{6a}', which is totally symmetric in the diazines, but belongs to e' in s-triazine; it is the Jahn–Teller-active mode for an E$'$ state. The 950-cm^{-1} vibration in this band has its counterpart in the first photoelectron band, where its frequency is 970 cm^{-1}, and can be assigned as either $\nu_1'(a_1')$ or $\nu_{12}'(a_1')$, both of which are ring-breathing modes [L4]. Another ring-breathing mode comes at 340 cm^{-1} in the ground state (ν_{14}'', e''), and may be the source of the 90-cm^{-1} splitting observed in many of the bands. Since the (0, 0) of the $n(e') \to 3s$ band of trimethyl-s-triazine comes at 51 190 cm^{-1}, the alkyl limit term value of 21 000 cm^{-1} leads to the prediction of 72 200 cm^{-1} (adiab.) for its first ionization potential. In s-triazine itself, another complex of sharp bands corresponding to the $e' \to 3p\sigma, 3p\pi$ transitions are expected near 64 000 cm^{-1} (vert.), where they would be very badly overlapped by the strong $\pi \to \pi^*$ transition to $^1E_{1u}$. Again, it should be pointed out that $\phi_i \to 3s$ transitions in benzene and its derivatives are rare, and if these bands at 55 000–60 000 cm^{-1} in s-triazine are correctly assigned as $6e' \to 3s$ [F12], it is a most unusual situation to find them so highly prominent.

The electron-impact spectrum of s-triazine has been reported [F12] at a resolution much poorer than that of the optical study, but of course, over a much larger range. This study reveals the strong $^1A_1' \to {}^1E'$ transition centered at 62 700 cm^{-1} (vert.), followed by several weaker features which are probably Rydberg excitations. The benzenoid transition to $^1B_{1u}$ (in the benzene notation) appears as an intense shoulder at 55 600 cm^{-1} (vert.), upon which the $6e' \to 3s$ Rydberg transition rests. In the higher frequency region between 70 000 and 100 000 cm^{-1}, Fridh *et al.* find numerous transitions originating with the deeper MOs, and terminating at 3s, 3p, 3d, etc.

Fridh *et al.* [F14] have assigned the pi-electron spectrum of s-tetrazine as consisting of $^1B_{2u}$ (40 300 cm^{-1} vert.), $^1B_{1u}$ (57 300 cm^{-1} vert.), and the two components of $^1E_{1u}$ at 61 300 cm^{-1} (vert.; $^1B_{1u}$) and 66 900 cm^{-1}

VI.A. PHENYL COMPOUNDS

(vert.; $^1B_{2u}$). Since the lone-pair ionization potentials are spread between 78 000 and 107 000 cm^{-1} [G14] and intimately penetrate the occupied pi manifold, there is good reason to expect that the $\pi \to \pi^*$ and $n_N \to \pi^*$ excitations will be badly overlapped, as will $\pi \to R$ and $n_N \to R$ Rydberg excitations. There is a weak feature in the electron-impact spectrum of s-tetrazine at 52 400 cm^{-1} (vert.) which at first sight appears to be a Rydberg transition, since it has a term value of 25 800 cm^{-1} with respect to the first ionization potential (b_{3g}, 78 200 cm^{-1} vert.), implying a $b_{3g} \to$ 3s assignment. However, the possibility of an $n_N \to \pi^*$ assignment cannot be ruled out on the basis of the current evidence, and in fact, it seems more likely that this band of s-tetrazine is related to the bands at $\sim$50 000 cm^{-1} in the diazines and s-triazine, which are probably $n_N \to \pi^*$ excitations. A choice between valence shell and Rydberg upper states could be made once the spectrum of s-tetrazine is determined in paraffin solution (Section II.C). Fridh et al. assign higher Rydberg transitions as originating from deeper MOs in the molecule and terminating at 3s and 3p.

VI.A-5. Substituent Effects in Benzene

In the previous sections, the spectra of several types of substituted benzenes were discussed in which the substituent shifted the frequencies of the benzenoid bands and altered their intensities somewhat, but added no new features below 70 000 cm^{-1}. Thus in molecules like toluene, fluorobenzene, and phenol, the first three excited states are clearly derived from the $^1B_{2u}$, $^1B_{1u}$, and $^1E_{1u}$ states of benzene (Fig. VI.A-20) and related Rydberg transitions fill the 60 000–70 000-cm^{-1} gap. In contrast, we now consider substituted benzenes in which the substituent can introduce lowlying bands of two kinds, local excitations within the substituent, and charge transfer transitions between the ring and the substituent (or between substituents), at frequencies below 65 000 cm^{-1}.

In the first group of compounds mentioned, i.e., those bearing alkyl, hydroxy, fluoro, and chloro groups, the perturbation on the benzene ring is so mild that the spectrum, though somewhat shifted, is easily recognized as still being in one-to-one correspondence with that of benzene up to 70 000 cm^{-1} at least. Spectrally, a stronger perturbation results when the substituent itself has a local excitation or can participate in the formation of charge transfer configurations that have frequencies higher than that of the excitation to $^1E_{1u}$. In such a case, the benzenoid pattern is intact, but closely followed by other bands not characteristic of benzene. If the transitions involving the substituent (local excitation, charge transfer) fall among the benzenoid bands, then their presence and exten-

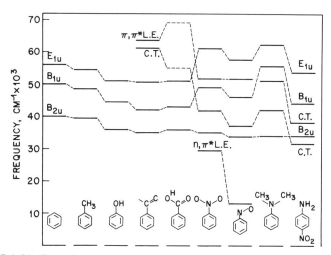

Fig. VI.A-20. Perturbations of the benzene spectrum as stronger substituents are added to the ring. The labels C.T. and L.E. signify charge transfer and local excitations, respectively.

sive mixing will destroy the characteristic pattern, making identification in terms of benzene transitions difficult as well as ambiguous.

A very neat and compact scheme has been devised for handling such composite systems theoretically (Section I.B-2). It is essentially a perturbative mixing in which the local states of the benzene ring ($^1A_{1g}$, $^1B_{2u}$, $^1B_{1u}$, $^1E_{1u}$) and of the substituent (usually $\pi \rightarrow \pi^*$ excitations) are considered as zeroth-order basis functions along with the ring-to-substituent and substituent-to-ring charge transfer configurations. The unperturbed energies are simply taken as the experimental values, except for the charge transfer configurations, where the usual approximation is made that the energy is equal to the ionization potential of the donor plus the electron affinity of the acceptor diminished by their Coulombic attraction. The mixing elements are calculated by MO theory, and the resulting energy matrix then leads to excitation energies, oscillator strengths, and polarizations.

As an intermediate situation, consider the spectra of styrene and benzoic acid. The optical spectrum of styrene vapor has been presented in several works [K17, W6, Y12], with that of Yoshino et al. [Y12] extending furthest, to 167 000 cm^{-1}. The very intense feature at 50 800 cm^{-1} (vert., $\epsilon = 68\,000$) is obviously related to the $^1A_{1g} \rightarrow {}^1E_{1u}$ band of benzene. Since the width of this band at half-height is only 5500 cm^{-1}, there does not seem to be any splitting of the degeneracy due to the lower symmetry. The band at 42 000 cm^{-1} (vert.) has $f = 0.24$, and would seem to be the transition to $^1B_{1u}$, while that to $^1B_{2u}$ is observed at 35 000 cm^{-1} (vert.) (Fig. VI.A-20). In benzoic acid, these three benzenoid bands are found at 51 000, 43 000 and 36 000 cm^{-1} (vert.) (Fig. VI.A-20) [H31, T3].

Unlike the first three compounds listed in Fig. VI.A-20, styrene has two additional levels to be considered, the olefinic local excitation, which appears as a band centered at 63 500 cm^{-1} (vert.), and a ring-to-olefin charge transfer thought to come at 61 500 cm^{-1} (vert.). A similar set of bands is anticipated in benzoic acid, as shown by the dashed lines in Fig. VI.A-20. Now the point here is that in these compounds of intermediate perturbation, the perturbing levels are *above* the benzenoid set, and, though mixing with them, do not change in any qualitative way the benzenoid appearance of the spectrum below 55 000 cm^{-1}.

Several other phenyl compounds have spectra of this intermediate type. As might be expected, the spectra of benzaldehyde and styrene are quite similar, except for the presence of an $n_0 \to \pi^*$ local excitation in the former [K17, W3, W6]. In benzaldehyde, the $^1E_{1u}$ band comes at 51 500 cm^{-1} (vert.) with $\epsilon \simeq 50 000$. The band of mixed $^1B_{1u}$/CT character appears at 42 900 cm^{-1} (vert.). In acetophenone, the frequencies are much the same, and in each there is the weak shoulder at 56 000 cm^{-1} (vert.) following the excitation to $^1E_{1u}$. Walsh reports the spectra of phenyl isocyanate and phenyl acetylene to be very much like that of benzene except for the shift to lower frequencies [W6]. In phenyl isocyanate, a strong, diffuse band at 57 100 cm^{-1} is identified as a Rydberg transition; if it terminates at 3p, then the first ionization potential of phenyl isocyanate will be 77 000 cm^{-1}. A similar Rydberg band is said to appear at 60 600 cm^{-1} in phenyl acetylene. The acetaniline spectrum has a low-lying benzenoid pattern, the shift being about -5000 cm^{-1} [T3]. Solution spectra of many of these molecules to 57 500 cm^{-1} are presented by Klevens and Platt [K30].

A considerably different spectrum develops once the locally excited and charge transfer configurations get within the benzene manifold, as in nitrobenzene, nitrosobenzene, etc. (Fig. VI.A-20). Let us consider nitrobenzene in some detail, since it is a particularly good example [N6]. In zeroth order, we have the $^1B_{2u}$ (38 000 cm^{-1} vert.), $^1B_{1u}$ (50 000 cm^{-1} vert.), and $^1E_{1u}$ (56 000 cm^{-1} vert.) excited states of the benzene fragment, and the $n_0 \to \pi_3^*$ (37 000 cm^{-1} vert.) and $\pi_2 \to \pi_3^*$ (50 500 cm^{-1} vert.) excitations of the nitro group, as typified by the spectrum of nitromethane (Section V.B). Additionally, there will be the benzene $\to$ nitro charge transfer configuration to consider. The experimental spectrum is shown in Fig. VI.A-21. The configuration interaction calculation offers

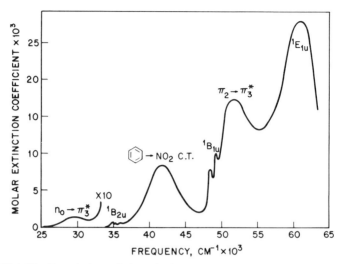

Fig. VI.A-21. Description of the nitrobenzene absorption spectrum in terms of charge transfer and local excitations [N6].

the following explanation, beginning at the low-frequency end. The weak band at about 30 000 cm⁻¹ (vert.) is the nitro group $n_0 \to \pi_3^*$ excitation, down from 37 000 cm⁻¹ in the nitroalkanes. Another weak excitation is apparent at 35 700 cm⁻¹ (vert.) which the calculation assigns as an almost equal mixture of benzene $^1B_{2u}$ excitation and charge transfer character. The stronger band ($f = 0.17$) at 41 700 cm⁻¹ (vert.) in the gas-phase spectrum is almost totally charge transfer in character, benzene → nitro. As expected for a transition to a highly polar upper state, the frequency of this excitation is strongly dependent upon solvent polarity, being 37 000 cm⁻¹ (vert.) in aqueous solution. Two badly overlapped excitations appear in the region near 50 000 cm⁻¹, a weaker band at 49 500 cm⁻¹ (vert.) and a stronger one at 51 800 cm⁻¹ (vert.). The calculations again offer an explanation: The weaker but structured band is the excitation to the benzenoid $^1B_{1u}$ state, whereas the stronger band is the localized nitro group excitation $\pi_2 \to \pi_3^*$. This finally leaves the very intense ($f = 0.87$) feature at 61 000 cm⁻¹ (vert.) as the benzenoid excitation to $^1E_{1u}$, again with the degeneracy essentially unsplit. The upward shift of the $^1E_{1u}$ level is due to the fact that the charge transfer configuration is below it and pushes it upward, whereas in compounds with weak substitutents, the order is reversed, and the $^1E_{1u}$ frequency is pushed downward. Consideration of the nonpolar molecule nitromesitylene and of the oscillator strengths of the various excitations only serves to reinforce the conclusions briefly sketched here. Molecular orbital studies of the nitrobenzene spectrum are discussed in references [M9, S37]. In the CNDO work of Sieiro and Fernandez-Alonso [S37], the first two bands are calculated to be $n \to \pi^*$ and $^1A_{1g} \to {}^1B_{2u}$ as in Fig. VI.A-21, but then the band at 41 700 cm^{-1} is assigned as benzenoid $^1A_{1g} \to {}^1B_{1u}$ rather than as intramolecular charge transfer, and the bands at 51 800 and 61 000 cm^{-1} are assigned as local benzene ($^1A_{1g} \to {}^1E_{1u}$) and nitro-group excitations ($\pi_2 \to \pi_3^*$), rather than the reverse as in the figure. It seems more likely that the stronger of the two bands has the $^1E_{1u}$ upper state.

Judging from the large number of bands below 60 000 cm^{-1} in nitrosobenzene and the relatively high frequency of the $^1E_{1u}$ band (62 000 cm^{-1} vert.), this must be another molecule in which the charge transfer and locally excited configurations are down among the benzenoid transitions. The configuration interaction calculation confirms this [T1] and offers the correlation of levels depicted in Fig. VI.A-20.

It is interesting that in benzoic acid the charge transfer band must be above $^1E_{1u}$ (Fig. VI.A-20), but in *trans*-cinnamic acid (C_6H_5—CH=CH—CO_2H), it is at 36 000 cm^{-1} (vert.), bringing it into the class of strongly perturbed phenyl compounds along with nitrobenzene and nitrosobenzene [T3]. In heptane solution, the $^1E_{1u}$ band of *trans*-cinnamic acid is far above 55 000 cm^{-1} (vert.), whereas the corresponding band is at 51 000 cm^{-1} in benzoic acid, illustrating again how the charge transfer configuration pushes the $^1E_{1u}$ higher if it is low, but lower if it is high.

Turning from electron-withdrawing substituents to strongly electron-donating substituents, such as found in the anilines [K15], we have a similar situation in that there are the two charge transfer configurations resulting from promotion of a lone-pair electron from the amine group to either of the near-degenerate, empty, benzene pi MOs, but in this case there seem to be no locally excited configurations (see Section III.D-1 for a discussion of amine spectra). Thus the complicated spectrum of N,N-dimethyl aniline is readily explained as having these two charge transfer bands interleaved between the $^1B_{2u}$ and $^1B_{1u}$ states of benzene (Fig. VI.A-20), with the mixing again elevating the $^1E_{1u}$ state to 62 500 cm^{-1} (vert.). The same pattern of bands is evident in aniline and several other of the alkylated anilines [K15].

The spectral consequences of decoupling the amine group from the ring have been studied by Klevens and Platt [K25], who observed the spectra of dialkyl anilines bearing bulky substituents in the o and o' positions. They found that as the —NR$_2$ group was twisted out of the benzene plane, the intensities of the 42 000- and 51 000-cm^{-1} charge transfer bands fell dramatically, and that their vertical frequencies also decreased. With this depression of the charge transfer frequencies, the transition to $^1B_{1u}$

also moved to lower frequencies but maintained its intensity so that its frequency approached more closely that of benzene itself. This solution study did not penetrate sufficiently deep to surmount the $^1E_{1u}$ maximum, but its frequency shift is no doubt following that of the $^1B_{1u}$ band to lower frequencies.

The frequencies of the charge transfer bands are lowest in disubstituted compounds of the sort p-nitroaniline. In this molecule, the lowest —$NH_2 \rightarrow$ —NO_2 charge transfer configuration even falls below the $^1B_{2u}$ configuration, and so the spectrum begins with a strong, broad feature, rather than a weak, structured band [T3]. Since the charge transfer bands are so low in p-nitroaniline, the mixing with $^1E_{1u}$ is diminished and this band has a relatively low frequency again (58 500 cm^{-1} vert.) as compared with the other strongly perturbed benzene spectra in Fig. VI.A-20.

As substituents of intermediate strength are appended to benzene, the first ionization potential stays at about the same frequency or increases somewhat, while at the same time, the strong transition to $^1E_{1u}$ moves to lower frequencies. Since the $^1E_{1u}$ and $\pi_2 \rightarrow 3p$ Rydberg bands are badly overlapped (at 57 000 cm^{-1} vert.) in benzene, the substitution has the effect of moving the $^1E_{1u}$ band out from under the $\pi_2 \rightarrow 3p$ Rydberg excitation, leaving it relatively unmolested on the high-frequency edge of the $^1E_{1u}$ band. In the vapor-phase spectra [K17], weak bands are found on the high-frequency wing of the $^1E_{1u}$ band in styrene (55 500 cm^{-1} vert.), in benzaldehyde (56 200 cm^{-1} vert.), and in benzonitrile (59 000 cm^{-1} vert.), as anticipated for Rydberg excitations to 3p. Unfortunately, the solution data of Klevens and Platt for these compounds [K30] do not extend far enough to test the Rydberg nature of these weak bands. Looking at the term values instead, ionization potential data are available only for benzaldehyde (79 000 cm^{-1} vert.) and benzonitrile (80 800 cm^{-1} vert.) [B3]; with respect to these values, the optical bands have term values of 22 800 and 19 800 cm^{-1} (vert.), respectively. These would seem to be close enough to 20 000 cm^{-1}, considering the error in estimating the vertical optical frequencies, to allow one to argue tentatively for $\phi_i \rightarrow 3p$ Rydberg assignments.

Walsh reports a strong, doubled band at 59 340 cm^{-1} in benzaldehyde [W3] and a similar set of bands beginning at 60 000 cm^{-1} in benzonitrile [W6], both of which, he suggests, are Rydberg excitations. However, since Kimura and Nagakura [K17] do not report these bands in their spectra of the same systems, this point should be reinvestigated. That Walsh also found the corresponding bands in other phenyl compounds and assigned them as Rydberg prompts the argument that Kimura and Nagakura somehow missed these Rydberg bands, and that the broad

VI.A. PHENYL COMPOUNDS

bands they saw at lower frequencies are possibly the valence shell $\pi \rightarrow \pi^*$ promotions to the $^1E_{2g}$ state, expected to come just beyond $^1E_{1u}$ [D22].

With proper substitution, the first ionization potential of the benzene ring can be lowered so as to bring the corresponding Rydberg excitations into better view. This is apparently the situation in N,N-dimethyl aniline and N,N-diethyl aniline (Fig. VI.A-22), where several bands that appear

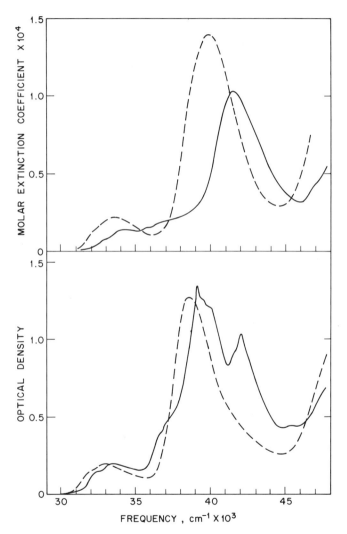

Fig. VI.A-22. Absorption spectra of N,N-dimethyl aniline (upper) and N,N-diethyl aniline (lower) in the gas phase (solid lines) and in solution in n-heptane (dashed lines) [T18].

in the gas-phase spectra are not present in the solution spectra (n-heptane), thereby suggesting Rydberg upper states (Section II.C). Thus Tsubomura and Sakata point out that in N,N-dimethyl aniline, there is a band at 37 800 cm^{-1} (vert.) which is absent in the solution spectrum [T18]. The first ionization potential of this compound being 60 400 cm^{-1} (vert.) [T21], the Rydberg excitation has a term value of 22 600 cm^{-1}, identifying it as $\phi_i \rightarrow 3s$. As such, it is related to the band at 42 500 cm^{-1} (vert.) in hexamethyl benzene (Section VI.A-2). In N,N-diethyl aniline, there are no ionization potential data available, yet one can see from the spectra that transitions to 3s and 3p occur at 37 000 and 39 000 cm^{-1} (vert.), and others, probably originating with deeper MOs, come at 42 300 and 45 300 cm^{-1} (vert.).

The electronic spectrum of polystyrene is not very different from that of styrene itself, but nonetheless has become the focus of attention with regard to a very intriguing theoretical problem. In addition to $^1B_{2u}$ and $^1B_{1u}$ excitations in optical transmission experiments, polystyrene shows a clear excitation to $^1E_{1u}$ at 51 500 cm^{-1} (vert.) in the solid, but at 52 900 cm^{-1} in solution [B62, K29, O7, P7]. Though the frequency shift is not large on going from solution to the solid (1400 cm^{-1}), the molar extinction coefficient does change more drastically, going from 60 000 in solution to half that in the solid [B6, P7]. Atactic and isotactic polystyrene behave in the same way in this respect. This type of hypochromism of a strong band on going from a solution to a solid is frequently caused by an excitonic coupling between the monomeric units in their excited states. According to Partridge [P7], the coupling is between benzene rings on adjacent chains. Following the transition to $^1E_{1u}$, the absorption spectrum shows a shoulder at $\sim$77 000 cm^{-1} (vert.) leading to an immense block of absorption peaked at 128 000 cm^{-1} (vert.). In electron transmission experiments, the peak appears at $\sim$170 000 cm^{-1} (vert.) and is followed by another plateau at $\sim$320 000 cm^{-1} (vert.) [L11]. A number of polystyrenes in which the phenyl groups bear methyl or chloro substituents have also been investigated by Onari [O7] who finds them to look much like polystyrene itself out to 83 000 cm^{-1}.

Partridge feels that the very strong absorption peaking in the 130 000-cm^{-1} region in polystyrene is related to similar bands at 144 000 cm^{-1} (vert.) in benzene (see Fig. VI.A-7) and at 125 000 cm^{-1} (vert.) in ethane (see Fig. III.A-3), and in fact finds the polystyrene curve to fit semiquantitatively to the sum of the benzene and ethane absorptions. He concludes that these are single-particle excitations resulting in photoionization and photodissocation continua. A contrary view is expressed by Carter et al. [C8], who determined the complex dielectric constant of polystyrene by studying its reflectivity, and conclude that the

130 000-cm^{-1} band is a "plasmon" excitation (Section I.A-3) involving the simultaneous coherent excitation of the 40 valence electrons per monomeric unit. LaVilla and Mendlowitz [L11] use the plasmon assignment for not only the 130 000-cm^{-1} band of polystyrene, which they observed at 170 000 cm^{-1} by electron transmission energy-loss spectroscopy, but for the 55 000-cm^{-1} band, which they call an E_{1u}-plasmon hybrid. However, the evidence for plasmon excitations in molecules seems rather weak, especially since they are supposed to be seen in electron-impact energy-loss experiments, and not in optical spectra [N19].

VI.B. Higher Aromatics

In benzene and the benzenoid derivatives (Section VI.A), the four lowest excited valence shell states are generated by promoting a pi electron from the highest filled degenerate MO ϕ_i to the lowest empty degenerate MO ϕ_{i+1}, i.e., from the configurations (ϕ_i, ϕ_{i+1}). The result is a low-lying transition in the quartz ultraviolet of very low oscillator strength and transverse polarization ($^1B_{2u}$), a band about 10 000 cm^{-1} higher with an oscillator strength about ten times larger and longitudinal polarization ($^1B_{1u}$), and finally a doubly degenerate state slightly higher still with an oscillator strength again ten times larger than that to the second state ($^1E_{1u}$). In studying the spectral relationships of catacondensed hydrocarbons, Platt and co-workers [K26, M3, P29, P30] found the same principle at work: In systems ranging from naphthalene to pentacene and triphenylene, the orbital structure is as if there were an upper, degenerate, filled MO set ϕ_i and an empty, degenerate set ϕ_{i+1}, leading to a first $\pi \to \pi^*$ excitation in the quartz ultraviolet of very low oscillator strength and longitudinal polarization (correlating with $^1B_{2u}$ in benzene), a stronger band of in-plane polarization opposite to the first and about ten times more intense (correlating with $^1B_{1u}$ in benzene), and finally one or two intense bands, depending upon symmetry, which correspond to the components to $^1E_{1u}$ in benzene. The initial fourfold excited-state degeneracy in benzene is lifted by configuration interaction, which is a two-electron effect, whereas in the less symmetric hydrocarbons in which there is no real degeneracy, there are instead orbital separations at the one-electron level; in the end, the effects of configuration interaction and of the orbital-energy separations are the same, i.e., a splitting into four states.

Platt gives this set of four bands in aromatic molecules the general labels ^1L_b, ^1L_a and ^1B_b, ^1B_a, and finds that they maintain their intensity ratios with frequencies which move in a regular way through a very long

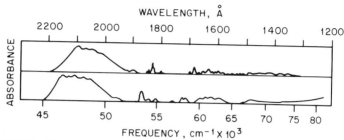

Fig. VI.B-1. Optical absorption spectra of naphthalene (upper) and 2-methyl naphthalene (lower) in the vapor phase [H6].

series of hydrocarbons. The ordering however, is not inviolate, since the ^1L_b and ^1L_a bands cross over between naphthalene and anthracene.† In aromatic systems with large numbers of pi MOs, the valence shell configurations (ϕ_{i-1}, ϕ_{i+1}) also contribute relatively low-lying valence shell states, called ^1C_a and ^1C_b by Platt.

The four singlet L and B bands of naphthalene and the higher aromatics in general fall in the region below 50 000 cm^{-1}, and having been extensively studied and reviewed, little more will be said about them here. The C bands are found in the 40 000–60 000-cm^{-1} region. Due to the lower ionization potentials in the higher aromatics, the lowest Rydberg excitations come at frequencies below 50 000 cm^{-1} and so are interweaved within the L, B, C patterns of $\pi \rightarrow \pi^*$ valence shell excitations. However, they can be recognized by their term values and by comparing the vapor- and condensed-phase spectra (Section II.C).

The vacuum-ultraviolet spectra of naphthalene and 2-methyl naphthalene were first described by Hammond et al. [H6] (Fig. VI.B-1). In naphthalene, they identified a Rydberg absorption beginning at 54 600 cm^{-1} and another at 56 180 cm^{-1} with a series limit of 65 300 cm^{-1}. The corresponding Rydbergs and ionization potential are about 800 cm^{-1} lower in 2-methyl naphthalene. They also identified continua centered at 47 600, 61 700, and 67 300 cm^{-1} (vert.) in naphthalene which they felt formed a series converging upon a second ionization potential at **75 000** cm^{-1}. Photoelectron spectroscopy has since confirmed the lowest ionization potential, placing it at 65 600 cm^{-1} (advert.) [B60, C22, E1]; however, since the next two come at 71 600 and 81 460 cm^{-1} (vert.), the con-

---

† In the limit of very large splittings between the one-electron levels, the ^1L and ^1B states become strongly mixed, with all bands then having about equal intensity. However, in the usual case, the two-electron configuration interaction effect is the predominant one, and the ^1B bands are the strongest [M47, P30].

tinua in the optical spectrum of naphthalene are not part of a Rydberg series, but instead are valence shell bands.

The valence shell transitions of naphthalene up to $\sim$54 000 cm^{-1}, the onset of the prominent Rydberg excitations, are covered in the recent work of George and Morris [G9]. We refer the interested reader to their paper and to those of Klevens [K27, M3] for discussions of these bands. The Rydberg spectrum of naphthalene has generated a great deal of interest recently. Angus et al. [A9] made a special study of these bands in naphthalene in the 50 000–70 000-cm^{-1} region and discovered five series converging upon 65 620 $\pm$ 40 cm^{-1} (Fig. VI.B-2). Beginning with $n = 4$, the quantum defects for these series are 0.94, 0.88, 0.82, 0.67, and 0.53, all $\pm$0.01. The five bands having $n = 4$ are labeled A–E, and as one goes from $n = 4$ to $n = 8$, the A–E interval decreases in a regular way from 2580 to 210 cm^{-1}. The term values of the $n = 4$ members, 11 700, 11 300, 10 990, 9900, and 9120 cm^{-1} (advert.) look much like those expected for the five members of the 3d manifold, in which case the quantum defects become -0.06, -0.12, -0.18, -0.33, and -0.47 for the A–E series. Scheps et al. [S16] have assigned the A, B, and C bands of these series as having $^1B_{1u}$, $^1B_{3u}$, and $^1B_{2u}$ upper states, respectively. Virtually the same data have been arranged by Kitagawa into three Rydberg series going to the 65 600 cm^{-1} ionization potential and beginning with $n = 4$, but with δ values of 1.08, 0.81, and 0.47, indicating ns and np series [K22]. The alternate assignment, in which n begins at three and the quantum defects instead are negative, can be defended, for the ab initio calculation of Buenker and Peyerimhoff [B66] assigns the uppermost filled MO as $1a_u$, so transitions will be electronically allowed from there to three of the five nd components but not to ns or np orbitals. Since Angus et al. assure us that the stronger bands are allowed (the earlier members have $f \simeq 0.01$ [K22]), the series must be nd and the term values undoubtedly identify the members at $\sim$55 000 cm^{-1} as $n = 3$. The Rydberg transitions to nd are quite vertical (as is the $1a_u$ photoelectron band) with only a few vibrational quanta and no progressions discernible. Some 430-cm^{-1} intervals were noted and assigned to the b_{3g} vibration.

In their pursuit of the $n = 4$, positive-δ series, Angus et al. searched for the $n = 3$ members of the five series and siezed upon the weak features at $\sim$45 000 cm^{-1} (vert.) to complete the series [A13]. These weak bands are not seen in the condensed-phase spectra, as is appropriate for Rydberg excitations, though they might be covered by the stronger $\pi \rightarrow \pi^*$ band at 47 400 cm^{-1} (vert.). We maintain that this assignment is incorrect. More precisely, two origins are found at 45 070 and 45 390 cm^{-1} (advert.) which are identified by Angus et al. as $n = 3$ of the B and C series. Note, however, that three origins are expected (as observed

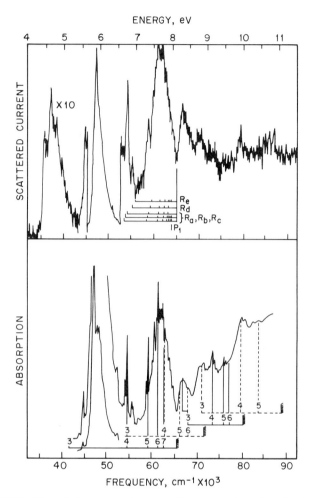

Fig. VI.B-2. The electron-impact energy-loss (upper) and optical absorption (lower) spectra of naphthalene vapor [H34, AD106]. The electron-impact spectrum was taken with electrons incident at 100 eV and scattered through zero angle.

for all other n of the series), and that from $n = 8$ to $n = 4$, the B–C interval increases regularly from 35 to 325 cm^{-1}, so that by extrapolation, one would expect a B–C interval of ∼500 cm^{-1} for their $n = 3$ components, but only 300 cm^{-1} is observed. Our guess is that these two bands with term values of 20 530 and 20 230 cm^{-1} are either the forbidden Rydberg excitations to 3s or 3p, or, as suggested by Kitagawa [K22], they are the hot bands of the adjacent valence shell excitation. In either event, there seems to be no good reason for associating them with the higher A–E bands.

The naphthalene Rydberg spectrum has been studied from a different point of view by Jortner and Morris [J19] and Scheps et al. [F7, S16], who find very obvious antiresonances between the sharp Rydberg transitions and the underlying continuum in the 58 000–64 000-cm^{-1} region (Fig. I.A-11). These are especially noticeable for the A, B, and C components of the $n = 5$ band (using the original numbering), the vibronic components of these bands increased by 1360 cm^{-1}, and the $n = 6$ members of the same three transitions. Further analysis of the problem is difficult and inconclusive because one knows neither how many continua are involved in the coupling, nor their symmetries or vibronic polarizations.

The Rydberg excitations of naphthalene are also involved in some interesting rare gas matrix spectra [A11, A12]. Angus and Morris worked down to 80 000 cm^{-1} in rare gas matrices (1:200) at 20°K and observed two very weak lines at 48 700 and 49 400 cm^{-1} (in Kr) (Fig. VI.B-3), which do not fit into the vibronic analysis of the superposed $\pi \to \pi^*$ band. In the same spectrum, additional weak but narrow features appear at 54 600 and 56 500 cm^{-1} (vert.) and according to these authors are related to Rydberg excitations. They postulate that the "lowest" Rydberg excitations at 45 100 cm^{-1} in the gas-phase spectrum are shifted upward to the 49 000-cm^{-1} region, and are still molecular Rydberg in character even

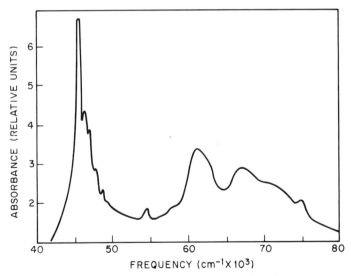

Fig. VI.B-3. The absorption spectrum of naphthalene doped into krypton (1:200, 20°K) [A12].

in the matrix. However, the higher Rydberg states which begin at ~55 000 cm^{-1} in the gas phase are no longer visible, and are replaced instead by Wannier excitons, the $n = 2$ and $n = 3$ members falling at 54 600 and 56 500 cm^{-1}, respectively. These bands are at their highest frequency in a krypton matrix, and are not seen in an n-heptane solution of naphthalene at room temperature [K26].

The vacuum-ultraviolet spectrum of naphthalene vapor has been extended by the studies of Koch and Otto [K33], who studied the electron-impact energy-loss spectrum with 30-keV incident energy and $\theta = 0°$, and Huebner et al. [H34], who used 100-eV electrons, also at $\theta = 0°$, but with a higher resolution (~300 cm^{-1}). The latter workers point out that the band at 66 900 cm^{-1} (vert.) energy loss has a Franck–Condon envelope resembling that of the third band in the photoelectron spectrum ($1b_{2g}$, 81 460 cm^{-1} vert.) [C22]. If this correlation is proper, this band could be assigned as the allowed $1b_{2g} \to 3p$ Rydberg excitation. Huebner et al. also report peaks at 70 980, 79 800, 86 300, and 96 800 cm^{-1} (vert.), but make no assignments. In more recent optical work on the deeper states of naphthalene, Rydberg series going to the higher ionization potentials (already determined by photoelectron spectroscopy [E4]) were sought, but the results are not very convincing. Koch et al., with data of much higher quality than is usually obtained by electron impact, searched for the allowed $2b_{3u} \to ns$ series going to the second ionization potential at 71 800 cm^{-1} (vert.), and devised a series with $\delta = 0.46$, which is more appropriate for an np series (forbidden from b_{3u}). In addition to the well-known optical bands, Koch and Otto observed two prominent losses centered at 97 000 and 135 400 cm^{-1} (vert.) in the electron-impact spectrum. These are unassigned as yet, but probably relate to the bands found at 91 900 and 130 000 cm^{-1} in benzene and at 90 700 and 131 400 cm^{-1} in anthracene [K33]. Interestingly, the 135 400-cm^{-1} band, a broad and intense feature, is also observed as such in the optical spectrum, showing that it is not a plasmon resonance, as suggested for this band in some of the higher acenes (Section I.A-3). Huebner et al. have summed the oscillator strengths of the naphthalene transitions and report 2.25 up to the first ionization potential and 8.6 up to 121 800 cm^{-1}.

Several new absorptions at high frequencies can be seen if the molecule is first excited to its lowest triplet or excited singlet state and the absorption then measured from there. Bonneau et al. [B33] have done this for naphthalene by first exciting to the $^1B_{3u}$ state (32 250 cm^{-1} vert.) and then observing the transition from there to another singlet state requiring an additional 23 250 cm^{-1}. This transient absorption is a strong one, and on this basis the transition is assigned as $^1B_{3u} \to {^1A_g}$, leading us then to expect the $^1A_g \to {^1A_g}$ band at 55 500 cm^{-1} (vert.) in the transition from the ground state. Hunziker [H38] instead put naphthalene in its lowest triplet state ($^3B_{2u}$, 21 250 cm^{-1}) and then observed the triplet–triplet absorption. From his results, one deduces that there is a $^1A_g \rightarrow {}^3A_g$ transition at 61 600 cm^{-1} (vert.) and a $^1A_g \rightarrow {}^3B_{1g}$ transition at 65 300 cm^{-1} (vert.). He also mentions a vibrationally structured band (triplet–triplet) beginning at 29 440 cm^{-1} (adiab.) which is present in the gas-phase spectrum, but not in solution. Thus there might well be a Rydberg triplet state 50 700 cm^{-1} above the ground state.

Compton et al. [C25] have determined the SF_6-scavenger spectrum (Section II.D) of 1-chloronaphthalene out to 88 000 cm^{-1} with modest resolution. Up to about 50 000 cm^{-1}, the spectrum looks just like that of naphthalene, and from this point upward, there are a large number of poorly resolved peaks. Of the two most prominent peaks, at 68 400 and 74 700 cm^{-1} (vert.), the first probably corresponds to the prominent peak at 70 700 cm^{-1} (vert.) in the SF_6-scavenger spectrum of naphthalene [C25], while the second is related to the D bands (3p $\rightarrow$ 4p) found in all chlorine-containing compounds at about 72 000 cm^{-1}. Pisanias et al. [P23] have obtained similar spectra for two of the azanaphthalenes, quinoline and isoquinoline. The spectra of these compounds are much broader than that of naphthalene, but the several $\pi \rightarrow \pi^*$ transitions up to $\sim$50 000 cm^{-1} can be discerned. Beyond that, quinoline has a broad, strong peak at 59 500 cm^{-1} (vert.), and the corresponding band in isoquinoline comes at 61 500 cm^{-1} (vert.).

The interpretation of the anthracene absorption spectrum in the vapor phase is somewhat confusing at present. Lyons and Morris [L39] first presented the spectrum from 50 000–66 000 cm^{-1} as revealing only a series of sharp, narrow bands between 50 000 and 53 000 cm^{-1}. These bands were analyzed by them as being the earlier members of a Rydberg series going to a limit of 54 930 cm^{-1} (vert.). Angus and Morris [A8] later reinvestigated this spectrum and observed many other Rydberg excitations out to 58 000 cm^{-1} which were unaccountably missing in the previous study. As in the naphthalene spectrum, they were able again to assemble five Rydberg series going to an ionization potential of 57 650 cm^{-1} (vert.). Several of the members show a splitting into a doublet or triplet which decreases as n increases. Though the 36 members of the various series were observed within ±50 cm^{-1} of the values calculated from the Rydberg formula, in fact, this limit is more than 2000 cm^{-1} below the first ionization potential determined by photoelectron spectroscopy (59 690 cm^{-1} vert.) [C22]. Thus the grouping of bands into a series as given by Angus and Morris must be incorrect, at least in part, as must the corresponding δ values. Kitagawa also studied the anthracene spectrum [K22] and was able to assemble three Rydberg series having 59 800 cm^{-1} (vert.)

VI. AROMATIC COMPOUNDS

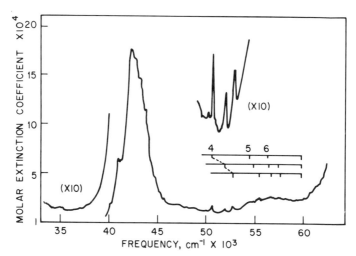

Fig. VI.B-4. The optical absorption spectrum of anthracene vapor, with three Rydberg series delineated going to the first ionization potential [K22].

as a limit, with $\delta = 0.98$, 0.50, and 0.37. The series begin at $n = 4$, assigned to the three sharp bands at 50 550, 51 950, and 52 740 cm^{-1} (vert.). These three bands by their quantum defects (but not their term values) would seem to be $\pi \rightarrow$ 4s, 4p, and 4d transitions, but since the highest pi orbital has $2b_{2g}$ symmetry, only the transition to 4p is allowed. An oscillator strength of ~0.01 in each of the bands suggests that they are allowed excitations, and so these bands may terminate at the three components of the 4p manifold instead. The corresponding transitions to 3p will have terms of ~20 000 cm^{-1} and so should appear near 40 000 cm^{-1}, at which frequency there is a prominent step (Fig. VI.B-4). There are several clear antiresonances in the 53 000–58 000-cm^{-1} region of the anthracene vapor spectrum as well as one or more underlying continua. In n-heptane solution, these features are missing, of course, but the transition to ^1B_a is seen at 54 000 cm^{-1} (vert.) [K26].

The crystal spectrum of anthracene has been studied in the vacuum ultraviolet by several groups by different methods yielding different results. Cook and Le Comber [C26] obtained the optical specular reflectance of the ab crystal face with unpolarized light down to 105 000 cm^{-1} and transformed the reflection spectrum into an absorption spectrum having two strong peaks at 50 000 and 79 800 cm^{-1} (vert.) and a weaker one at 60 500 cm^{-1} (vert.). Clark's reflectance work [C18, C19] and that of Koch and Otto [K35] cover the same spectral region, but were done with plane-polarized light and on several different crystal faces. Their finding is that the band at 53 000 cm^{-1} (vert.) is stronger in the b-axis spectrum of the ab face and is probably an exciton-split component of the very strong $^1A_g \to {}^1B_{1u}$ band at 40 000 cm^{-1} in the gas phase. Another band which is strong in the b-axis spectrum ($f = 0.3$) has its maximum at 63 000 cm^{-1} (65 000 cm^{-1} in the a-axis spectrum) and also must be $^1A_g \to {}^1B_{1u}$ with short-axis polarization. A weak band appearing in the a-axis spectrum at 57 300 cm^{-1} (vert.) has out-of-plane polarization and may be a forbidden $\pi \leftrightarrow \sigma$ excitation.† Going deeper into the vacuum ultraviolet using polarized synchrotron radiation, Koch et al. [K36] uncovered a strong band centered at $\sim 133\,000$ cm^{-1} having subsidiary maxima in the 105 000–200 000-cm^{-1} region. This peak is stronger in the a-axis spectrum and is thought to be an out-of-plane-polarized $\pi \to \sigma^*$ excitation. Kunstreich and Otto [K40] have performed electron energy-loss experiments on thin, oriented anthracene crystals and observed spectra at different scattering angles. In addition to the 50 000- and 160 000-cm^{-1} losses found in graphite and all other aromatic systems, anthracene also shows a peak at 96 800 cm^{-1} (vert.). However, the strong band found at 133 000 cm^{-1} in the reflection spectrum does not appear prominently in the energy-loss spectrum of the crystal, but possibly correlates with the strong band observed at 131 000 cm^{-1} in the vapor spectrum [K36]. All investigators to date assign the strong 160 000-cm^{-1} loss (Fig. I.A-12) to either a pi-plus-sigma plasmon or a pure sigma plasmon, but this seems doubtful since the corresponding bands in related molecules can be shown not to be plasmons.

Since the hexane-solution spectrum of phenanthrene to 52 000 cm^{-1} [U3] shows all of the features also reported in the vapor spectrum [K22], we cannot assign any of them to Rydberg excitations. This is somewhat disappointing since the sharp, vertical feature at 35 200 cm^{-1} (advert.) does look like a transition to 3s, but it must be valence shell since it also appears in the solution spectrum. Kitagawa [K22] has constructed a Rydberg series in phenanthrene vapor converging to 62 000 cm^{-1} (vert.) with the $n = 4$ member at 49 700 cm^{-1} (vert.). However, the threshold of the first photoelectron band of phenanthrene is 63 800 cm^{-1} [D13, E1] and so the good agreement obtained using the Rydberg formula with $\delta = 0.90$ is somewhat spurious. The electron-impact spectrum of phenanthrene vapor out to 200 000 cm^{-1} is presented by Koch and Otto [K33]. Kitagawa has also observed the absorption of phenothiazine vapor to 60 000 cm^{-1} and reports a Rydberg series converging to 61 510 cm^{-1} (advert.), having $n = 4$ at 50 080 cm^{-1} (advert.). As in all of the hetero-

---

† Actually, this could be a forbidden $\pi \to \pi^*$ band, say, made allowed by vibronic mixing with higher $\pi \to \sigma^*$ or $\sigma \to \pi^*$ excitations.

atomic polycyclics, the Rydberg excitations that originate with the uppermost filled pi MO are weak but sharp, and lie upon a valence-shell continuum.

The electron energy-loss spectra of several aromatic hydrocarbons present an interesting situation. Using 30-keV electrons at $\theta = 0°$, Koch and Otto [K33] report several energy-loss spectra of gas-phase aromatics having maximum scattering current at 48 000 cm^{-1} loss and then much lower currents at 80 000–200 000 cm^{-1}. Jäger [J3] has studied the same materials as 0.1-μ films with 35-keV electrons in transmission and finds all of the low-frequency losses of the gas-phase spectrum and in addition to this, a broad, extremely intense loss centered at $\sim$185 000 cm^{-1} (vert.) (Fig. I.A-13), which is totally absent in the free-molecule spectrum. Comparative data for anthracene, tetracene, chrysene, and pyrene show the extraneous band prominently. A similar 185 000-cm^{-1} loss is reported as well for 1,2-benzanthracene, 1,2-benzpyrene, and picene, but the gas-phase spectra have not been determined for comparison. According to Jäger, these high-frequency, high-intensity losses are collective excitations (plasmons, Section I.A-3), occurring in a volume of perhaps (100 Å)^3, and so are absent in the free-molecule spectra. The electron-impact spectra of solid films of coronene and hexabenzocoronene (described in Section I.A-3) show shoulders at 185 000 cm^{-1}.

Morris's spectrum of naphthacene from 20 000 to 54 000 cm^{-1} is very interesting with regard to the spectra discussed for naphthalene and anthracene [M53]. The uppermost filled MO in naphthacene is $2a_u$ as in naphthalene, and so Rydberg excitations to ns and np will be forbidden, but those to nd will be symmetry allowed. Its ionization potential is 56 540 cm^{-1} (advert.) [C22], and so ($2a_u$, 3s) and ($2a_u$, 3d) term values of typically 22 000 and 13 000 cm^{-1} would place these Rydberg bands at $\sim$35 000 and 43 000 cm^{-1}, respectively. Comparison of the vapor spectrum of naphthacene with that of its solution in n-hexane [K26, M53] reveals that of the many features present, two are very clear in the vapor spectrum but missing from the solution spectrum, which is good evidence that they have Rydberg upper states. They seem to be missing from the crystal spectrum as well [L40]. In corroboration of their Ryberg natures, the two bands in question appear in the vapor spectrum at 35 200 and 43 200 cm^{-1} (advert.). The highest filled MO in pentacene ($3b_{2g}$) has an ionization potential of 53 560 cm^{-1} (advert) [C22], and so the allowed Rydberg excitation to 3p is expected at 33 600 cm^{-1} (advert.) in the vapor spectrum.

The visible and ultraviolet absorption spectra of azulene vapor are nicely explained by the semiempirical pi-electron theories up to 48 000 cm^{-1}, but at this point, sharp, vertical excitations commence that are out-

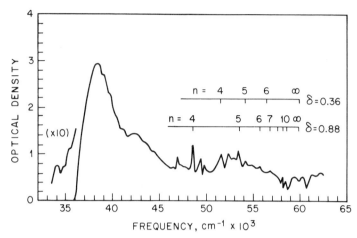

Fig. VI.B-5. The optical absorption spectrum of azulene vapor [C17, K21].

side the pi-electron manifold. Further, these sharp bands do not appear in the heptane-solution spectrum [M3, T4], though prominent in the gas phase. From these bands (Fig. VI.B-5), both Clark [C17] and Kitagawa et al. [K21] were able to assemble Rydberg series converging to 59 940 cm^{-1} (vert., $\delta = -0.09$) and 59 750 cm^{-1} (vert., $\delta = -0.12$), respectively. In both cases, n starts at three. The first ionization potential of azulene by photoelectron spectroscopy is 59 850 cm^{-1} (vert.) [E1], and according to an *ab initio* calculation, originates at the $2a_2$ π MO [B66]. The first member of the series has $f = 0.024$ and a term value of 12 490 cm^{-1} (vert.), indicating a $\pi 2a_2 \to 3d$ assignment. The corresponding $\pi 2a_2 \to 3p$ excitation will have a term value of $\sim$19 000 cm^{-1} and so should appear at about 40 000 cm^{-1}, at which frequency there is already a very intense $\pi \to \pi^*$ excitation. The nd Rydberg series members ride upon a weaker valence shell continuum centered at $\sim$53 700 cm^{-1} (vert.). The negative δ observed for the nd Rydberg series of azulene is somewhat unusual, but it should be pointed out that the five nd series of naphthalene all have negative quantum defects.

The absorption spectra of o-, m-, and p-terphenyl as solids at 77°K are quite indistinct. In p-terphenyl, a weak and a strong band are observed at 52 700 and 57 100 cm^{-1} (vert.), but in o- and m-terphenyl, there is general absorption in the same region but no maxima [M39]. In the gas phase, a strong peak is observed at 50 000 cm^{-1} (vert.) in p-terphenyl and several smaller blips, too, from which a Rydberg series was said to lead to an ionization potential of $\sim$61 300 cm^{-1} [K22].

Working with hexane solutions, Mullen and Orloff were able to follow the absorption spectrum of fluorenone down to 60 000 cm^{-1} [M56]. Due to the strong solvent effect, the spectrum is no doubt shorn of its Rydberg absorptions, and, as such, consists largely of a broad band of strong absorption ($\epsilon \simeq 25\,000$) in the 47 000–60 000-cm^{-1} region, with perhaps four or five strong peaks visible. The absorption is explained by them with a pi-electron calculation that places 14 allowed $\pi \to \pi^*$ ($^1A_1 \to {}^1A_1, {}^1B_2$) transitions in this frequency region. However, only five of them have a calculated oscillator strength greater than 0.25, and these presumably are the ones observed as distinct peaks.

CHAPTER VII

Inorganic Systems

This chapter is meant to include all inorganic systems not already incorporated into earlier chapters. Since almost all such systems have a heavy central atom surrounded by lighter atoms, one can logically separate the compounds into those in which the central atom is a nonmetal (Section VII.A) and those in which it is a metal (Section VII.B). Note, in regard to this, that though white phosphorus and silane might logically be expected to be found in Sections VII.A and VII.B, respectively, they are instead discussed in the appropriate sections dealing with compounds of phosphorus and silicon. All such inorganic compounds which instead could be placed in sections dealing with related compounds have been so treated, so that the number of inorganic systems discussed in this work is far larger than the following pages would indicate.

VII.A. Nonmetals

Of the three xenon fluorides XeF_2, XeF_4, and XeF_6, detailed vacuum-ultraviolet absorption data are available only for the first two, while photoelectron data are available for all three. Optical spectra of the difluoride and tetrafluoride are given in Fig. VII.A-1 [J17]. Beginning with the difluoride, two valence shell excitations appear, centered at 43 500

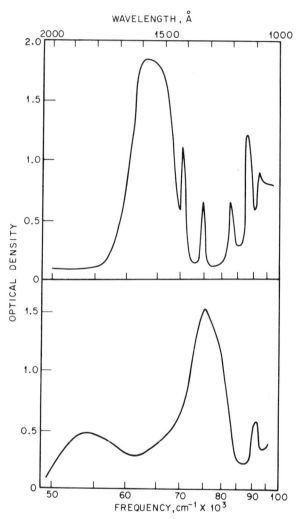

Fig. VII.A-1. Gas-phase absorption spectra of xenon difluoride (upper) and xenon tetrafluoride (lower) [J17].

cm^{-1} (vert.) and 63 300 cm^{-1} (vert.), followed by four or five bands of much smaller half-width which are no doubt Rydberg excitations.

Photoelectron spectroscopy [B46, B54, B57] and Gaussian orbital calculations [B14, B54] have been combined to show that the uppermost filled MOs in XeF$_2$ are $5\pi_u$ (59% Xe 5p, 31% F 2p) and $10\sigma_g$ (9% Xe 5s, 91% F 2p), while the lowest empty MO is $7\sigma_u$ (68% Xe 5p, 31% F 2p).

VII.A. NONMETALS

This leads to the expectation of a low-lying $5\pi_u \to 7\sigma_u$ ($^1\Sigma_g \to {}^1\Pi_g$) forbidden transition followed by a strongly allowed $10\sigma_g \to 7\sigma_u$ ($^1\Sigma_g \to {}^1\Sigma_u$) transition of the N → V variety. Indeed, these are the assignments first deduced by Jortner et al. [J17, P52, W31] for the 43 500- and 63 300-cm^{-1} bands of XeF$_2$. Detailed calculations by them showed that the weak band at 43 500 cm^{-1} ($f = 0.002$) is still too intense to be the $^1\Sigma_g \to {}^3\Sigma_u$ arising from the $10\sigma_g \to 7\sigma_u$ promotion, in spite of the heavy-atom enhanced spin–orbit coupling. Later, more exact calculations of the XeF$_2$ spectrum by Basch et al. confirm these assignments and also support the idea that a weak $3\pi_g \to 7\sigma_u$ excitation is buried in the stronger $10\sigma_g \to 7\sigma_u$ absorption [B14]. The N → V type excitation $10\sigma_g \to 7\sigma_u$ is calculated by Basch et al. to have an oscillator strength of 1.39, but only 0.45 is observed [P52].

Excitation of the uppermost $5\pi_u$ electron of XeF$_2$ into a Rydberg orbital leaves the core in a $^2\Pi$ state which is split into $^2\Pi_{3/2}$ and $^2\Pi_{1/2}$ components through the spin–orbit coupling on the xenon atom. This splitting in the XeF$_2^+$ ion is 3790 cm^{-1}. The structures of the four Rydberg excitations shown in Fig. VII.A-1 were reinvestigated by Brundle et al. [B54], who found the first two bands to show distinct vibronic structure with origins at 69 295 and 73 870 cm^{-1} (adiab.). Their difference, 4750 cm^{-1}, is rather larger than that at the ionization limit, but strongly suggests that this interval is due to spin–orbit coupling in the core [B54, W31]. The two other bands at 82 300 and 87 260 cm^{-1} (vert.) are split by 4960 cm^{-1}, and so probably form a second such spin-orbit-split pair. Comparison of the XeF$_2$ term values and those of the Xe atom (Table VII.A-I) convincingly shows that the four Rydberg excitations in question are the

TABLE VII.A-I

TERM ENERGIES IN XE AND XEF$_2$

Xe[a]		XeF$_2$	
Upper state	Term (cm^{-1})	Upper state	Term (cm^{-1})
5p(^{2}P$_{3/2}$)6s	30 400	$5\pi_u$($^2\Pi_{3/2}$)6s	30 865
5p(^{2}P$_{1/2}$)6s	31 433	$5\pi_u$($^2\Pi_{1/2}$)6s	30 080
5p(^{2}P$_{3/2}$)6p	19 322	—	—
5p(^{2}P$_{1/2}$)6p	19 317	—	—
5p(^{2}P$_{3/2}$)5d	16 628	$5\pi_u$($^2\Pi_{3/2}$)5d	17 860
5p(^{2}P$_{1/2}$)5d	16 567	$5\pi_u$($^2\Pi_{1/2}$)5d	16 600
5p(^{2}P$_{3/2}$)7s	12 551	—	—
5p(^{2}P$_{1/2}$)7s	12 590	—	—

[a] Taken from Reference [M50].

allowed excitations from $5\pi_u$ to 6s and 5d with $\Omega = \frac{3}{2}$ and $\Omega = \frac{1}{2}$ core configurations. Each of the transitions to 6s has an oscillator strength of 0.02, but values of 0.03 and 0.06 are found for the transitions to 5d [J17]. The transitions to 6p are $u \rightarrow u$ forbidden, but are expected at 80 840 and 84 650 cm^{-1} (vert.).

The spectrum of XeF$_4$ (Fig. VII.A-1) again only shows a small number of valence shell excitations, and only one possible Rydberg excitation, at 91 740 cm^{-1} (vert.) [J17, J18, M2]. The asymmetry of the weak absorption around 40 000 cm^{-1} suggests two overlapping bands, centered at 38 700 and 43 800 cm^{-1} (vert.) with respective oscillator strengths of about 0.003 and 0.009. This is followed by two stronger features centered at 54 400 cm^{-1} (vert., $f = 0.17$) and at 75 500 cm^{-1} (vert., $f = 0.8$). According to the *ab initio* calculation of Basch *et al.* [B14, B57], the lowest empty MO in XeF$_4$ is $8e_u$ (64% Xe 5p, 34% F 2p), while the three uppermost filled MOs are $10a_{1g}$ (16% Xe 5s, 84% F 2p), $5a_{2u}$ (48% Xe 5p, 52% F 2p), and $5b_{1g}$ (100% F 2p). All of these lower-lying excitations can be described within this manifold of four levels.

The analog of the $10\sigma_g \rightarrow 7\sigma_u$ (N $\rightarrow$ V) excitation of XeF$_2$ is the $10a_{1g} \rightarrow 8e_u$ ($^1A_{1g} \rightarrow {}^1E_u$) excitation of XeF$_4$, for which the oscillator strength is calculated to be 1.36. Since the equally intense $5b_{1g} \rightarrow 8e_u$ ($^1A_{1g} \rightarrow {}^1E_u$) promotion is calculated to be more than 15 000 cm^{-1} higher than the first, and all other bands are an order of magnitude weaker, it seems logical to assign these to the strong bands at 54 400 cm^{-1} and 75 500 cm^{-1}, respectively, possibly mixed by configuration interaction. Jortner *et al.* reverse these two assignments, placing $5b_{1g} \rightarrow 8e_u$ at lower frequency, but this is based upon a one-electron ordering which places $5b_{1g}$ above $10a_{1g}$, rather than below it as in the scheme of Basch *et al.* In a semiempirical calculation on XeF$_4$, Yeranos also concludes that the $10a_{1g} \rightarrow 8e_u$ transition comes at 54 400 cm^{-1} [Y8]. In XeF$_6$, it is predicted that the first allowed band ($8a_{1g} \rightarrow 8t_{1u}$) will be an intense one, with a frequency below 56 000 cm^{-1} [B14], possibly corresponding with one or the other of the broad bands observed at $\sim$46 000 and 55 600 cm^{-1} (vert.) [B18].

The two weak bands in XeF$_4$ in the 40 000-cm^{-1} region are discussed by Basch *et al.*, who suggest that this region might contain the triplet of the $10a_{1g} \rightarrow 8e_u$ excitation as well as the vibronically allowed $5a_{2u} \rightarrow 8e_u$ singlet excitation. Indeed, Pysh *et al.* [P52] have shown that the spin–orbit mixing for the lower states of XeF$_4$ is much more effective than in XeF$_2$, thus rationalizing a triplet assignment in the tetrafluoride.

The narrow feature at 91 740 cm^{-1} in XeF$_4$ is something of a problem. By its width, intensity, and frequency, we would be tempted to classify it as an allowed Rydberg excitation, and as such, would expect term values not too different from those given in Table VII.A-I for Xe and XeF_2. Since the first four vertical ionization potentials of XeF_4 are 105 300 ($10a_{1g}$), 107 900 ($5a_{2u}$), 116 100 ($5b_{1g}$), and 121 800 cm^{-1} ($1a_{2g}$) [B57], the term values of the 91 740-cm^{-1} band with respect to these are 13 600, 16 200, 24 400, and 30 100 cm^{-1}. Combining the term values of Table VII.A-I and the symmetry selection rules leads to the unique assignment $5a_{2u} \to 5d$, but leaves us wondering where the allowed $10a_{1g} \to 6p$ transition is to be found.

The SF_6 electron excitation spectrum (Section II.D) of XeF_4 has been recorded [B18], but it does not resemble the optical spectrum, and no explanation is evident for the differences. Similarly, the SF_6 excitation spectrum of XeF_6 has few features in common with its optical spectrum. This latter is described by Begun and Compton [B18] as consisting solely of broad, ill-defined features with peaks at 29 600, 39 400, 45 900, 55 600, and 70 800 cm^{-1} (vert.). The last of these has a term value of 30 200 cm^{-1} with respect to the first ionization potential ($8a_{1g}$) at 101 000 cm^{-1} (vert.), and according to Table VII.A-I, may be the forbidden $8a_{1g} \to 6s$ Rydberg excitation.

The electronic structure of the nitrate ion is essentially like that of the alkyl nitrates $RONO_2$ (Section V.B), with the exception that certain pairs of bands will be degenerate in the higher symmetry of the ion. According to the semiempirical calculation on the nitrate ion performed by McEwen [M19], two very weak $n_0 \to \pi_4^*$ transitions should appear at 39 500 cm^{-1} followed by the intense $\pi_2, \pi_3 \to \pi_4^*$ ($^1A_1' \to {}^1E'$) transition at 50 000 cm^{-1}. These figures are very close to what is observed; in the $NaNO_3$ crystal, the $n_0 \to \pi_4^*$ band appears weakly at 35 200 cm^{-1} (vert.) [F15], with the stronger $\pi_2, \pi_3 \to \pi_4^*$ bands coming at about 50 000 cm^{-1} (vert.). It is interesting that the corresponding $\pi \to \pi^*$ band in ethyl nitrate is found at 52 600 cm^{-1} (vert.). In accord with the $^1A_1' \to {}^1E'$ assignment, this band is completely in-plane-polarized in the nitrate ion [F15, U6, Y4]. That this band of the crystal is due only to the nitrate ion is verified by its appearance in the aqueous solution spectrum, and at very nearly the same frequency in all of the alkali metal nitrates (Fig. VII.A-2).

The spectra of the alkali metal nitrates recently published by Yamashita and Kato [Y3] are very interesting, for the $^1A_1' \to {}^1E'$ transition in $LiNO_3$ is split into two components separated by 6000 cm^{-1}, and in the heavier nitrates, the splitting decreases monotonically, amounting to only 1200 cm^{-1} in $CsNO_3$. Since the nitrate ions in both $LiNO_3$ and $NaNO_3$ occupy sites of trigonal symmetry, it would seem that the splitting is due either to Jahn–Teller interaction in the degenerate excited state, or to a Davydov splitting between the two ions in the unit cell.

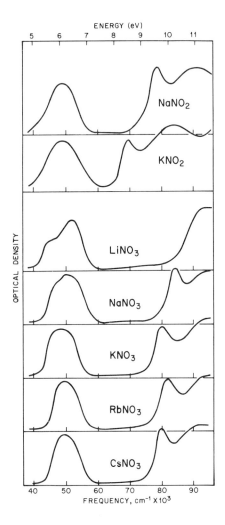

Fig. VII.A-2. Optical transmission spectra of several alkali metal nitrites and nitrates as evaporated thin films [Y3].

As pointed out by Yamashita [Y5], the latter possibility predicts that the transition to the lower component of the split is forbidden from the ground state, whereas the two components are observed to have about nearly equal intensity. The oscillator strength of the $^1A_1' \to {}^1E'$ transition in the nitrate ion is estimated to be 0.77 from the reflection spectrum, which is reduced to 0.49 by the local-field correction [Y5], while McEwen calculates 0.34, and the corresponding transition in ethyl nitrate has an oscillator strength of 0.23.

As seen in Fig. VII.A-2, the $^1A_1' \to {}^1E'$ transition in the nitrates is followed by a region of very high transparency extending to 80 000 cm^{-1}, where another strong transition begins. Since the frequency of this second band is dependent upon the cation and is very temperature sensitive, Yamashita and Kato assign it as a crystal transition rather than as being localized in the nitrate ion. It is largely polarized out of plane [Y5] and probably corresponds to a charge transfer from the nitrate ions to the alkali ions, which are stacked above and below the planes of the nitrate ions. From this point, the absorption remains intense out to 170 000 cm^{-1}. It is surprising to see how closely the nitrite ion spectrum resembles that of the nitrate ion. A close similarity was also noted between the spectra of organic nitrates and nitrites (Section V.B).

Since the carbonate and nitrate ions are isoelectronic and isostructural, one expects a certain similarity in their optical spectra, and this is observed. Studying a thin plate of calcite ($CaCO_3$) in transmission, Damany et al. [S14, U7] found a weak band at 54 000 cm^{-1} (vert.) having in-plane polarization. They assigned this as a forbidden $n_O \to \pi_4^*$ transition, analogous to that in the nitrate ion at 35 000 cm^{-1}. The same band appears at 56 000 cm^{-1} (vert.) in a solution of sodium carbonate in water, with a molar extinction coefficient of about 1000 [L23]. Turning to polarized reflection spectroscopy on calcite, magnesite ($MgCO_3$), and dolomite [$MgCa(CO_3)_2$], Uzan et al. [U4, U5, U8] found in each of these a broad, strong transition centered at 62 500 cm^{-1} (vert.) having in-plane polarization. This is no doubt the $\pi_2, \pi_3 \to \pi_4^*$ ($^1A_1' \to {}^1E'$) excitation, equivalent to that at 50 000 cm^{-1} (vert.) in the nitrate ion. Now the π_2, π_3 MOs are totally on the oxygen atoms in these ions, but a significant amount of $2p\pi$ from the central atom also enters the π_4^* wave function. Since the binding energy of a pi electron to the N^+ central atom of NO_3^- will be much larger than to the neutral C central atom of CO_3^{2-}, it is quite reasonable that both $\pi_2, \pi_3 \to \pi_4^*$ and $n_O \to \pi_4^*$ come at higher frequencies in the latter ion. In the 73 000–90 000-cm^{-1} region in the carbonates, there is considerable out-of-plane-polarized absorption [K38]; paralleling the situation in the alkali nitrates, it is probably due to anion → cation charge transfer. Inasmuch as the nitrate and carbonate ion spectra are so similar, one might hope to see the splitting of the $^1A_1' \to {}^1E'$ band in the carbonates, especially for the lightest cations. Such a splitting was not observed in $MgCO_3$, but possibly could be found in $BeCO_3$.

VII.B. Metals

The Rydberg spectra of the compounds considered in this section are readily identified using external perturbations (Sections II.B and II.C) and term values obtained with the help of photoelectron spectra. Once

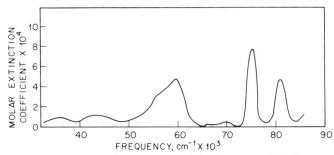

Fig. VII.B-1. Optical absorption spectrum of titanium tetrachloride vapor [B16].

this is done, the remaining bands can be classified as valence shell excitations, but their explicit assignment is difficult due to the close bunching of a large number of filled ligand levels and the presence of low-lying empty or partially filled nd levels. At this stage, about all one can do is point out the similarities between various spectra and suggest a variety of assignments for the more prominent bands. The situation will improve once the spectra of series of related compounds have been determined.

The spectrum of titanium tetrachloride vapor has been observed down to 85 000 cm^{-1} (Fig. VII.B-1) [B16, D16, D18]. Inasmuch as the various calculations on this molecule have not gone beyond the valence shell, all assignments so far neglect the possibility of Rydberg excitations. Semiempirical calculations on TiCl$_4$ yield the three uppermost filled MOs as $1t_1$, $3t_2$, and $1e$, each of which is located largely on the ligand atoms, with energies determined by both ligand–ligand and ligand–metal interactions [B16, G24, P2]. One can readily see the reality of this description by comparing the first three ionization potentials of titanium tetrachloride with those of carbon tetrachloride (Table VII.B-I). Evidently the first three ionization potentials in both molecules involve the chlorine

TABLE VII.B-I
Vertical Ionization Potentials (cm^{-1})
of Carbon Tetrachloride and
Titanium Tetrachloride[a]

MO	CCl$_4$	TiCl$_4$
$1t_1$	93 550	95 000
$3t_2$	⎧100 000⎫ ⎨101 600⎬ ⎩102 830⎭	103 100
$1e$	108 020	106 700

[a] From Reference [G24].

lone pair electrons. Now in all such chloro-containing molecules as carbon tetrachloride, there is a rich Rydberg spectrum, the most prominent feature of which is the D band (Section III.B-2). In a monochloro compound, the D band is an intense, vertical feature characteristic of the halogen atom, assigned as 3p → 4p in the atom, with a term value of very nearly 20 000 cm^{-1} (vert.). In the polyhalides, there will be several such D bands, each originating from a different halogen MO, but the term value seems to remain constant at 20 000 cm^{-1}. Applying this criterion to the spectrum of TiCl$_4$, we see immediately that the band at 75 400 cm^{-1} (vert.) is the D band originating at the 1t_1 MO (19 600 cm^{-1} vert. term value), while the band at 81 000 cm^{-1} (vert.) is the D band having 3t_2 as the originating MO.

The first B band in TiCl$_4$ (chlorine 3p → 4s) should have a term value of ∼26 000 cm^{-1}, again by comparison with the situation in CCl$_4$, and being $t_1 \to a_1$, will be electronically forbidden. These requirements are ably met by the weak band at 70 000 cm^{-1} (vert.; Fig. VII.B-1), which we assign as 1t_1 → 4s. The allowed 3t_2 → 4s band is expected at 78 000 cm^{-1} (vert.), where it is covered by the intense D band originating at 1t_1. The intense features at 48 500 and 52 500 cm^{-1} (vert.) in titanium tetrabromide [D18] are far too low to be Rydberg bands, which are expected at frequencies above 60 000 cm^{-1} in the vapor spectrum of this compound.

The analogy between the spectra of carbon and titanium tetrachlorides can be extended another step. At lower frequencies in the alkyl chlorides, a set of bands of moderate intensity appears which are known as A bands, having the valence shell assignment chlorine 3p → σ^*(C—Cl). The situation is somewhat different in TiCl$_4$, where the empty 3d orbitals are involved in the valence shell. In this case, the 3d manifold splits into a lower $\pi^*(e,$ Ti—Cl$)$ set and an upper $\sigma^*(t_2,$ Ti—Cl$)$ set. The transitions from the 3t_2 and 1t_1 filled MOs to $\pi^*(e)$ and $\sigma^*(t_2)$ have been identified by Becker et al. [B16, B17], Dijkgraaf and Rousseau [D16], Parameswaran and Ellis [P2], and DiSipio et al. [D18] with the bands observed at 36 000 and 43 000 cm^{-1} (vert.) in TiCl$_4$, and in a sense are analogous to the A bands of carbon tetrachloride.

The bands at 57 000–60 000 cm^{-1} in TiCl$_4$ have term values that are too large to allow Rydberg assignments; however, in CCl$_4$ itself, it was concluded that the intense $\sigma \to \sigma^*$ valence shell bands came at 72 500 and 78 500 cm^{-1} (vert.), and it seems likely that the bands at ∼60 000 cm^{-1} in TiCl$_4$ have a similar assignment. By its intensity, the upper state of the transition must be $^1T_2$, and so the two components at 58 000 and 60 000 cm^{-1} may be Jahn–Teller related. More specifically, the most intense valence shell transition (N → V) in such a d^0 tetrahedral molecule will be between MOs having the same AO components in each, with strong metal–ligand interaction, and differing in only one radial node. This transition is $9t_2\sigma(\text{Ti—Cl}) \to 10t_2\sigma^*(\text{Ti—Cl})$, and so is our choice for the 57 000–60 000 cm^{-1} band in TiCl$_4$, and for the corresponding bands in the oxo compounds discussed later.†

Spectra of the halide ions Cl$^-$, Br$^-$, and I$^-$ in aqueous solution are given by Scheibe (Fig. VII.B-2) [S11]. As expected, the absorption bands beyond 50 000 cm^{-1} show the spin–orbit splittings characteristic of the neutral atoms. Fox and Hayon [F11] have pushed these measurements to 62 000 cm^{-1} in certain solvents, and Rabinowitch [R4] explains these transitions as charge-transfer-to-solvent, without describing the terminating orbital further. With regard to these spectra, the interesting work of Bird and Day [B29] presents an anomalous situation. They found that the [N(n-C$_4$H$_9$)$_4$]$_2$ZnX$_4$ salts could be cast as transparent films and the spectra of the tetrahalozincate ions in them measured to beyond 60 000 cm^{-1} without interference from the cation (Fig. VII.B-2).‡ In these 3d^{10} systems, they found ultraviolet absorption frequencies which were suggestively like those of the corresponding halide ions in water. This led them to assign these bands as purely halogen atom transitions, which they took to be np $\to$ ($n + 1$)s. However, such a transition is a Rydberg excitation resembling the B bands of the alkyl halides, and does not survive in a condensed phase (Section II.C). If one accepts the thesis that there are no vacant valence shell MOs in the ZnX$_4^{2-}$ ions, then the valence shell upper states observed in these ions must be charge-transfer-to-solvent configurations, as in the case of the halide ions themselves. However, there seems to be a general feeling among spectroscopists in this area that the higher zinc orbitals such as 4s and 4p are not Rydberg orbitals, but part of the valence shell instead. If this is true, then there is no need to invoke the charge-transfer-to-solvent concept for the bands reported in Fig. VII.B-2. In the MnX$_4^{2-}$ ions, the same assignments may apply in the 50 000–60 000-cm^{-1} region, but complications may arise from the half-filled 3d shell.

Spectral studies of a few transition metal hexafluorides have been reported by McDiarmid (ReF$_6$, WF$_6$) [M14] and Tanner and Duncan (MoF$_6$, WF$_6$) [T7], none of which reached to 60 000 cm^{-1}. Besides the crystal field bands in the near infrared, the vapor spectrum of rhenium hexafluoride shows three bands near 50 000 cm^{-1} of much higher intensity.

† Dr. K. Sodoi makes the interesting suggestion that possibly the strong band in question is $3e(\pi) \to 4e(\pi^*)$, and that the $\sigma \to \sigma^*$ transition is considerably weaker, as seems to be the case in several organic pi-electron systems.

‡ The high transparency of the tetraalkyl ammonium ion is not unexpected, since the first intense ammonium ion band in NH$_4$Br comes at 73 000 cm^{-1} (vert.) [S50].

VII.B. METALS

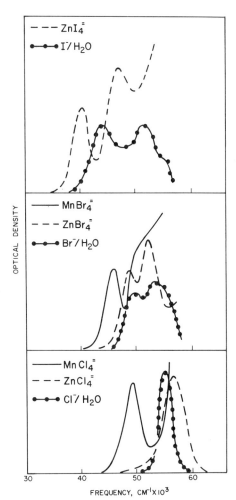

Fig. VII.B-2. Absorption spectra of the halide ions in aqueous solution [S11] and of the tetrahalometallates as thin films of the $[N(n\text{-}C_4H_9)_4]^+$ salts [B29].

These are a broad, structureless band centered at 47 700 cm^{-1} and two barely structured bands with electronic origins at 49 320 and 56 400 cm^{-1} (adiab.) sporting progressions of ~600–700 cm^{-1} (ν_1', the totally symmetric Re—F stretch). McDiarmid assigned both of the structured bands to a Rydberg series converging to 64 370 cm^{-1}, but the photoelectron spectroscopic value of the ionization potential is 89 920 cm^{-1} (adiab.) with a 730 cm^{-1} vibrational spacing visible [R19]. We see that the term value of the 49 320-cm^{-1} band (greater than 40 600 cm^{-1} adiab.) is perhaps too large for a Rydberg excitation, but the second at 56 400 cm^{-1} (adiab.) has a term value, 33 500 cm^{-1}, which is not unusual for a transition to

ns in a hexafluoride. (Compare with a term value of about 31 000 cm^{-1} (vert.) in sulfur hexafluoride, Section III.F.) However, in such a Rydberg excitation in ReF$_6$, the optical electron is undoubtedly the single 5d electron in the t_{2g} MO, which leads to a forbidden transition to ns, in contrast to the observation. Thus it looks as if the ReF$_6$ spectrum to 57 000 cm^{-1} is totally valence shell.

In MoF$_6$ and WF$_6$, there are no unpaired d electrons in the outermost shell and the lowest ionization would be from the fluorine lone pairs at about 120 000 cm^{-1}. Thus all transitions in these molecules below 85 000 cm^{-1} must be valence shell fluorine → metal charge transfer transitions. In MoF$_6$, one such strong transition ($f \simeq 0.1$) is centered at about 54 000 cm^{-1} (vert.) with 630 cm^{-1} vibrational intervals evident [T7], whereas in WF$_6$, the first strong peak seems to be just beyond 57 000 cm^{-1}.

Some extremely intense charge transfer bands having oscillator strengths approaching two near 50 000 cm^{-1} in the third-row transition metal hexahalides are discussed by Jorgensen [J15].

Electronic spectra of the isostructural and isoelectronic ions MnO$_4^-$ and CrO$_4^{2-}$ each shows a single strong band in the vacuum ultraviolet of very high intensity. In the permanganate ion, this band comes at 52 900 cm^{-1} (vert.) and has an oscillator strength of 0.6 [J7, M55], whereas in the chromate ion, it is found at 55 560 cm^{-1} (vert.) with an extinction coefficient of 28 000 in water solution [J5]. For each of these anions, the relative intensity of these strong bands in water is dramatically reduced in thin films of their salts, suggesting an excitonic coupling in the crystals leading to hypochromism. In the ReO$_4^-$ and TcO$_4^-$ ions in aqueous solution, the lower-frequency charge transfer bands are at 10 000–20 000 cm^{-1} higher frequencies than in MnO$_4^-$ and the strong band is not observed out to 55 000 cm^{-1}. According to Mullen et al. [M55], the 52 900-cm^{-1} band of MnO$_4^-$ is so strong that it must be assigned as an N → V excitation, i.e., $t_2(\pi + \sigma) \rightarrow t_2(3d)$. A similar assignment would apply to the 55 560-cm^{-1} band of the CrO$_4^{2-}$ ion.

The corresponding N → V transition in the sulfate ion must be at very high frequency, since the extinction coefficient is only up to 200 at 53 500 cm^{-1} [R4]. That of the perchlorate ion must also be at a very high frequency.

The two volatile oxides OsO$_4$ and RuO$_4$ both show beautifully detailed spectra in the quartz- and vacuum-ultraviolet regions (Fig. VII.B-3) [A3, F10]. Armed with the high-quality photoelectron spectra of these compounds [E9, F9], we can identify the low-lying Rydberg excitations in large part by their term values. The analysis given in Table VII.B-II is essentially that given by Foster et al. As was first mentioned with regard to the Rydberg spectrum of SF$_6$ (Section III.F), in such hetero-

VII.B. METALS

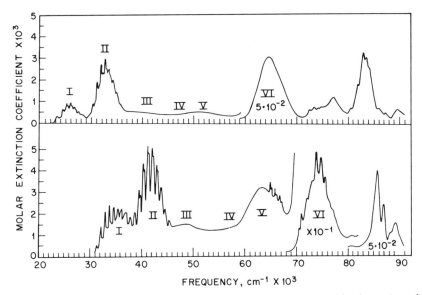

Fig. VII.B-3. Optical absorption spectra of ruthenium tetroxide (upper) and osmium tetroxide (lower) in the vapor phase [F10].

TABLE VII.B-II
RYDBERG EXCITATIONS IN OSMIUM AND RUTHENIUM TETROXIDES[a]

	Ionization potential	Absorption frequency	Term value	Upper orbital symmetry	Oscillator strength
OsO$_4$	99 400	65 000	34 400	3s	≪0.134[b]
		77 000	22 400	3p	—
	106 000	70 500	35 500	3s	—
		85 500	20 500	3p	0.356[c]
	108 900	73 900	35 000	3s	<0.1
		89 200	19 700	3p	—
	117 100	85 500	31 600	3s(?)	0.356[c]
RuO$_4$	97 500	75 700	21 800	3p	0.148[c]
	104 000	72 300	31 700	3s(?)	0.056
	{111 000 / 111 800}	75 700	{35 300 / 36 100}	3s	0.148[c]
		89 500	{21 500 / 22 300}	3p	—

[a] From References [A3, F10].
[b] The oscillator strength of band V plus the superposed Rydberg excitation is 0.134.
[c] This oscillator strength is probably that for two overlapped Rydberg excitations.

atomic systems, there is an ambiguity in describing the Rydberg orbitals, for one can use either the principal quantum numbers of the central atom or those of the surrounding ligand atoms. A suggestion with regard to this is given in Section I.A.-1, but for the moment, we will more simply just use the quantum numbers of the ligand atoms. Note then that whereas the transitions terminating at 3p have the usual term value of $\sim$20 000 cm^{-1} in these oxides, those terminating at 3s have the unusually high value of 35 000 cm^{-1}. This is probably due to the very high formal oxidation state of the central metal ions, which makes the oxygen atoms more positively charged than normal, and thereby increases the penetration energy on these atoms for ns orbitals. The only perturbation experiments to date show that bands I and II in OsO$_4$ (Fig. VII.B-3) are valence shell, as is also obvious from their term values [F10, L5].

According to the discussion in Section I.A-1, the oscillator strength of a Rydberg excitation will not exceed 0.08 per degree of spatial degeneracy regardless of the system, i.e., will not be greater than 0.24. Though the data are not very complete, and though there is reason to expect that some of the reported values refer to overlapping bands (Table VII.B-II), it does appear that the intensity rule is being obeyed in these tetroxides. Clearly, band VI is too intense to be a Rydberg excitation.

A semiempirical calculation of the electronic structures of OsO$_4$ and RuO$_4$ gives the expected results: the empty metal d orbitals split into a lower $2e(\pi)$ set and an upper $t_2(\sigma)$ set, while the three uppermost filled MOs ($1t_1$, $3t_2$, and $2a_1$) are purely oxygen lone pairs, with the metal–ligand bonding MOs ($2t_2$ and $1e$) below these in that order [F9]. As mentioned earlier, a bewildering array of allowed transitions can be constructed from this simple set of MOs, and no compelling valence shell assignments have appeared yet. Foster et al. identify the first five valence shell transitions with promotions from the first five filled MOs up to $2e$. This, however, does not take account of the fact that because of degeneracy, the five such promotions lead to *nine* distinct singlet excited states. One argument which can be used to sort the valence shell spectrum is that the truly intense features such as band VI in Fig. VII.B-3 (OsO$_4$, $f = 0.97$; RuO$_4$, $f = 0.43$ [A3]) must involve considerable metal–ligand delocalization in both orbitals, and so must originate at either $1e$ or $2t_2$. Since $1e \to 2e$ is forbidden in T_d, the most likely choices are $2t_2 \to 2e$ and $2t_2 \to 4t_2$, with the former being more likely. These intense bands in the tetroxides no doubt correspond to those at 50 000–55 000 cm^{-1} in the CrO$_4^{2-}$ and MnO$_4^-$ ions, and at 60 000 cm^{-1} in TiCl$_4$.

The only octahedral transition metal species studied spectroscopically in the vacuum ultraviolet are the nd^6 hexacarbonyls Cr(CO)$_6$, Mo(CO)$_6$, and W(CO)$_6$, shown in Fig. VII.B-4 [18]. In a sense, the electronic struc-

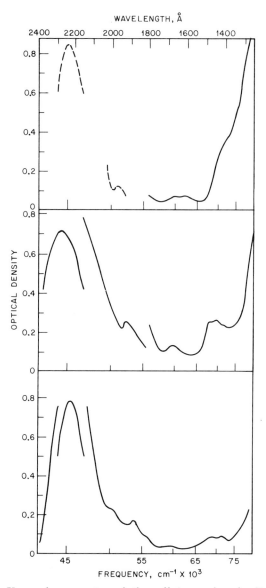

Fig. VII.B-4. Vapor-phase spectra of the nd^6 hexacarbonyls. Top: chromium hexacarbonyl (500 μ pressure in a 10-cm cell); dashed curves are qualitative and are given only to indicate the positions of maximum absorption. Middle: Molybdenum hexacarbonyl at 220 μ pressure (right) and 150 μ pressure (left), both in a 10-cm cell. Bottom: Tungsten hexacarbonyl at 150 μ pressure in a 10-cm cell [I8].

tures of these compounds are even more complicated than those of the oxides and halides considered earlier, for in the carbonyls, the empty ligand π^* MOs are available as terminating MOs in electronic transitions. This complexity is relieved by the fact that among the filled MOs, the predominantly metal t_{2g} MO has an ionization potential of about 67 000 cm^{-1} (vert.) while the occupied ligand MOs are at over 100 000 cm^{-1} [T21]. Thus almost all of the lower-lying transitions in the hexacarbonyls originate at the t_{2g} MO. Moreover, by symmetry, Rydberg excitations will be allowed only to the np orbitals and not ns and nd. With a typical term value of 20 000 cm^{-1}, the $t_{2g} \to$ 3p transitions will be buried beneath the very intense charge transfer band at $\sim$47 000 cm^{-1} in these compounds. No other Rydberg excitations have been identified in the hexacarbonyls.

The very intense valence shell excitations at 45 000–55 000 cm^{-1} in the hexacarbonyls must be of the N → V type, and have been assigned as $t_{2g}(\pi) \to t_{2u}(\pi^*)$ in accord with this [A1, B15, G22]. The weaker band on the high-frequency tail of the most intense feature is thought to be a forbidden charge transfer transition, $t_{2g}(\pi) \to t_{2g}(\pi^*)$.

There is very little information on metal-organic spectra in the vacuum ultraviolet, and even less in the way of convincing assignments. We mention several isolated examples, with the hope that this neglected area will receive more attention in the future.

Many of the metal alkyls show nicely structured bands in the 40 000–50 000-cm^{-1} region [T10, T11] and could be very interesting at higher frequencies.

The electron-impact energy-loss spectrum of ferrocene vapor has been recorded at high impact energies [K34, L31] and in addition to the usual visible-UV peaks, shows a sharp, intense feature at 52 000 cm^{-1} (vert.), followed by far broader peaks at 100 000, 129 000, and 170 000 cm^{-1} (vert.). The sharp band at 52 000 cm^{-1} is thought to be a $\pi \to \pi^*$ transition within the cyclopentadienyl rings, and the peaks at higher frequencies are not unlike those found in the hydrocarbons at similar frequencies.

Rather interesting optical spectra of thin films of phthalocyanines have been reported by Schechtman and Spicer (Fig. VII.B-5) [S10]. The visible bands (13 000–16 000 cm^{-1}) and the Soret band (30 000 cm^{-1}) are $\pi \to \pi^*$ excitations within the phthalocyanine moeity and have been intensively studied experimentally and theoretically. In each of the compounds studied, beyond the Soret band, there is a complex of four maxima (1–4) which come at very nearly the same frequencies in all (except for the mixture of chlorinated isomers). Obviously, these bands do not involve central metal orbitals, and Schechtman and Spicer tentatively sug-

VII.B. METALS

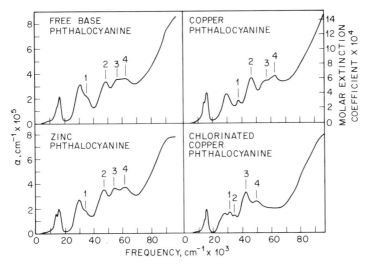

Fig. VII.B-5. Optical absorption spectra of thin films of various phthalocyanines [S10]. That labeled chlorinated copper phthalocyanine is of a mixture of variously chlorinated species.

gest that these four bands are localized within the benzene rings of the phthalocyanines and are closely related to the four ultraviolet transitions of benzene ($^1B_{2u}$, $^1B_{1u}$, $^1E_{1u}$ upper states, Section VI.A-1).

The intense visible colors of nickel dimethyl glyoxime and its isomorphous analogs were long thought to be due to charge transfer transitions between metal ions along the chains. More recent work has suggested that these out-of-plane-polarized bands were instead intramolecular charge transfer excitations. In the most recent study, Ohashi et al. [O2] also take this view, and working with single crystals in polarized light, found a second band with out-of-plane polarization in the neighborhood of 53 000 cm^{-1} (vert.). These workers claim that it is this band in the vacuum ultraviolet which is the $3d_z \rightarrow 4p_z$ charge transfer between adjacent metal ions.

Using X-ray spectroscopy, Sadovskii et al. [S2] have studied the $1s \rightarrow 4p$ metal-ion transition in several cobalt and copper complexes having salicylaldimine ligands. Only relative absorption energies are given, but in the cobalt complexes, a very broad band is observed, which appears at about 80 000 cm^{-1} lower frequency in the corresponding copper complexes. The shapes of these bands were used to infer distortions away from the ideal "tetrahedral" ligand configuration in the ground state.

CHAPTER VIII

Biochemical Systems

The biological molecules in this chapter are in general polyfunctional, and so it was easier to place them in their own category rather than decide to which other category they might best fit. Since these molecules are usually large and/or polymeric with strong hydrogen bonding between them, the spectroscopic studies have been carried out with the sample either as a thin polycrystalline film or in aqueous solution. By this circumstance, the Rydberg excitations are cleared from the spectrum, and only valence shell transitions remain.

With the possible exception of the polypeptides (Section V.A-2), the most popular biological molecules for spectroscopic and theoretical study are the nucleic acid bases guanine, adenine, cytosine, and uracil, and their derivatives. Following a lengthy spectroscopy study of these molecules and their derivatives, Clark and co-workers [C15, C16] concluded that with the exception of a low-lying n → π* band, the next four transitions in these molecules are π → π* excitations related to the $^1B_{2u}$, $^1B_{1u}$, and $^1E_{1u}$ pi-electron excited states of benzene (Section VI.A-1). Of course, in the lower symmetry of the bases, the $^1E_{1u}$ state is split into two transitions, both intense. In accord with this idea, there do seem to be two weak bands in these compounds, followed by two strong bands (Fig. VIII-1). The benzenelike assignments of Berthod et al. [B27] are given in Table VIII-I. Though the theoretical calculations do support pi-electron assignments in these bases [B27, T6] and in one case are specifically phrased

286

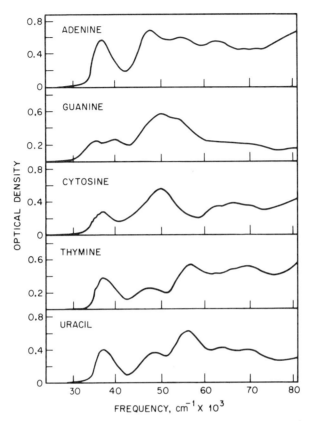

Fig. VIII-1. Optical absorption spectra of thin films of the nucleic acid bases at room temperature [Y1].

in terms of the benzene transitions [B27], it should also be noted that these calculations predict many more than four low-lying $\pi \to \pi^*$ transitions, and if all of the $\pi \to \pi^*$ transitions which are predicted could be observed, the spectra would not look so benzenoid.

Spectra of the nucleic acids as solid thin films to 80 000 cm^{-1} reveal several other strong transitions beyond those to the components of $^1E_{1u}$ (Fig. VIII-1) [T6, Y1]. Using the novel technique of electron energy-loss spectroscopy by transmission through thin films, Isaacson [I5, I6] confirmed the transitions found optically and showed that there is a rapid rise of absorption beginning at about 80 000 cm^{-1} and peaking at about 105 000 cm^{-1}, following by smaller subsidiary maxima. His measurement of the oscillator strength and how it accumulates throughout the spectrum suggests that the strong and plentiful $\sigma \to \sigma^*$ transitions commence at

TABLE VIII-I
SPECTRAL ASSIGNMENTS OF THE NUCLEIC ACID BASES
AS RELATED TO THE BENZENE SPECTRUM[a]

Molecule	$^1B_{2u}$	$^1B_{1u}$	$^1E_{1u}$	
Cytosine	36 300 (0.2)	41 900 (0.2)	49 200	(0.6)
			54 000	
9-Methyl guanine	36 300 (0.1)	39 500 (0.3)	48 400 ⎫	(1.1)
			53 200 ⎭	
9-Methyl hypoxanthine	35 500 (0.05)	40 300 (0.2)	50 000 ⎫	(0.9)
			54 000 ⎭	
Uracil	38 700 (0.2)	—	49 200	(0.3)
			54 800	
Adenine	38 700[b]	38 700[b]	48 400	(0.4)
			54 000	
Purine	37 900 (0.1)	41 900 (0.05)	50 000 ⎫	(0.6)
			53 200 ⎭	

[a] From Reference [B27]. Vertical values with frequencies in cm⁻¹; oscillator strengths in parentheses.
[b] Combined oscillator strength for the two bands is 0.3.

about 80 000 cm⁻¹ in these materials, and are possibly responsible for the maxima between 100 000 and 200 000 cm⁻¹. In this regard, it should be remembered that in both the amides and phenyl compounds (Sections V.A.-1 and VI.A-2), a strong absorption at high frequencies was observed in the solid where none was present in the free-molecule spectrum. These transitions are thought to be excitations into the conduction band rather than single-molecule excitations, and a similar situation may hold for the solid nucleotide bases.

Johnson and Rymer [J4] have used the same technique as Issacson to study the energy losses in films of calf thymus nucleic acid, using an impact energy of 150 keV and a spectral resolution of 1 eV. They found an intense, broad excitation at 194 000 cm⁻¹ (vert.) which they concluded was a volume plasmon (Section I.A-3) resulting from the excitation of a collective electronic oscillation. However, Issacson observed similar losses in films of the pure bases and argued against plasmon losses.

In a related study, Onari reports the spectra of the homopolynucleotide salts [O8, O11]. Correction of these spectra by subtraction of the absorption due to D-ribose-5-phosphate yields polymer spectra which are very close to those of the unpolymerized crystal spectra in Fig. VIII-1. Spectra of the polyribonucleotides poly(U + A), a double-stranded helical complex, and thymus DNA are reported to 66 700 cm⁻¹ [B39, P35] and show a complex pattern of strong bands in the $^1E_{1u}$ region suggesting exciton coupling.

Nelson and Johnson [N12] have studied the circular dichroism spectra of various sugars in aqueous solution out to 61 000 cm^{-1}. In D-glucose and D-xylose, positive CD bands were observed with maxima at 59 000 and 59 900 cm^{-1} (vert.), respectively, while in D-galactose, a negative band was observed at 56 000 cm^{-1} (vert.) and a second, stronger band peaks beyond 61 000 cm^{-1}. Since these substances exist as anomeric mixtures in solution, the interpretation of the CD results is especially difficult, and no certain assignments can be made as yet. However, it is clear that there is a broad valence shell transition in ethers at about 58 000 cm^{-1} (vert.) which can be assigned as $n_O \to \sigma^*$ (C—O—C) (Section III.E-3). Since the sugars are in their pyranose forms, they do contain the ether grouping and should display a valence shell excitation in the region in which the CD maxima are observed.

Considerable work has been done on biological molecules containing the isolated C=C double bond as their only chromophore. The optical absorption [T19] and ORD-CD spectra [F4, F5, S22, Y9, Y10, Y11] of several steroids and triterpenoids have been studied in paraffin solution, and the pinenes have had a detailed study in the gas phase, which allows deeper penetration (Section IV.A-2). The vacuum-ultraviolet spectra of solid films of several phthalocyanines appear in reference [S10] and are discussed further in Section VII.B.

Addendum

This addendum is meant to serve two purposes. First, it brings the literature coverage for the topics in both Volumes I and II up to January 1974, and allows the addition and discussion of many other, earlier papers which were inadvertently overlooked. Referral to references already listed in Volumes I and II or to pages in these volumes is preceded by the respective volume number. Second, some errors that were committed in Volume I and have been uncovered in the time since that volume went to the printer are hereby corrected. No such period of grace has been available for Volume II, however, and the reader is so forewarned.

AD.I.A-1. Rydberg States in Atoms and Molecules

The constancy of the Rydberg term value and its relative lack of dependence upon chemical bonding and molecular geometry means that Rydberg potential surfaces will follow the ionic surface, keeping a fixed energy below it. This is nicely shown in the calculations of the twisting potentials in ethylene [II-B69], where the calculated twisting curves for four different Rydberg states accurately parallel the twisting potential of the related ionic state. Note, however, that selective Rydberg/valence shell mixing, as in the (ϕ_i, 3s) states of water, can complicate this otherwise simple situation. Phenomenologically, the simple case of Rydberg potential surfaces resembling that of the ion will be reflected in a virtu-

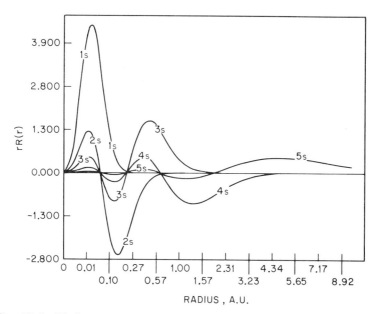

Fig. AD-1. Display of the rubidium atom ns functions $rR(nlr)$ versus the radial coordinate in atomic units [AD55].

ally nonbonding Rydberg orbital, and closely similar Rydberg and photoelectron band envelopes.

With respect to inner-orbital recapitulation for orbitals having real precursors in the core, C. Froese Fischer has calculated the Hartree–Fock radial wave functions for the He to Ra atoms, and these clearly illustrate the point [AD55]. For example, in Fig. AD-1, the contribution of the 1s–4s core AOs to the construction of the 5s AO of rubidium is most apparent. The phrase "excited precursor" is preferred by Mulliken to the earlier description "virtual precursor" for unoccupied precursor orbitals [AD138].

An interesting relationship between the generalized oscillator strength [Eq (IV.3)] of a singlet–singlet excitation and the corresponding singlet–triplet split has been derived [AD115], and may be of use in predicting the singlet–triplet Rydberg frequencies. There is probably a simple relation as well between the singlet–triplet split and the electronically allowed optical oscillator strengths of Rydberg excitations.

Fano discusses the X-ray spectroscopic evidence for outer- and inner-well states in certain classes of polyatomic molecules, and their relationship to the situations in atoms and metals [AD53]. It is not yet clear how these inner- and outer-well states of molecules differ from the more common valence shell and Rydberg states, respectively.

AD.I.C-1. Observation of Rydberg Trends in Molecular Spectra

A very good review of the different types of spectra encountered in the vacuum ultraviolet and practical points of view for their explanation is given by Sandorfy [AD164]. The discussion deals heavily in Rydberg excitations.

A general correlation of the lower nR Rydberg term values of many organic molecules has been presented [AD126]. This work correctly demonstrates the invariance of the $(\phi_i, 3p)$ term value by obtaining a straight line relationship of the expected slope by plotting the ionization potential versus the excitation frequency. Though a similar linearity was also claimed for the $(\phi_i, 3s)$ configurations, we feel that the line has been constructed using several transitions that do not terminate at 3s.

In an interesting series of papers, Lindholm explores the behavior of the Rydberg term values for diatomic molecules [AD122] in a way which closely parallels our work for the larger systems. In this, Lindholm demonstrates that the term value is independent of "the nature of the molecule" and independent of the originating MO, and further stresses the utility of photoelectron spectroscopy for interpreting molecular Rydberg spectra. In this sense, our work is but an extension of Lindholm's ideas in this area.

AD.II.C. Vacuum-Ultraviolet Spectra in Condensed Phases

The recent compilation of matrix spectra by Gedanken *et al.* [AD63] allows some generalizations to be drawn. The most pertinent data are listed in Table AD-I. In the equation for the matrix absorption frequency [Eq (II.2)], G is the Rydberg (Wannier) term value for $n = 1$. As seen in the table, its value is almost totally independent of the guest molecule, and depends only upon the host. This constancy of the term value in a matrix nicely parallels that in the case when the same chromophores in the gas phase are surrounded not with large numbers of matrix atoms, but with large numbers of alkyl groups (Fig. I.C-3) or fluorine atoms (Fig. I.C-4). In matrices, 10-50% of the term value is attributed to central cell corrections (nonorthogonality), whereas we phrase it instead as due to penetration (also related to nonorthogonality, p. I-15) when speaking of the limiting term value in gaseous molecules.

TABLE AD-I

Matrix Spectral Parameters for $n = 1$ Excitons[a]

	Ne	Ar	Kr	Xe	CF_4	N_2	D_2
Term value							
Xe	35 200	19 200	13 800	—	—	—	—
CH_3I	35 200	17 600	16 800	—	—	—	—
C_2H_4	35 200	16 800	15 200	7000	—	—	—
C_6H_6	35 600	19 200	16 800	8000	—	—	—
Gas–matrix shift							
CH_3I	8 950	6 480	5 095	—	12 880	—	—
C_2H_2	6 910	6 650	4 210	—	8 700	—	—
C_2H_4	6 330	5 720	4 160	2850	—	—	—
C_6H_6	6 800	5 640	2 920	1400	—	5520	—
Central cell correction							
Xe	18 700	8 100	4 400	500	8 800	—	6000
CH_3I	17 300	4 700	5 000	—	—	—	—
C_2H_2	14 000	10 700	5 000	—	—	—	—
C_2H_4	15 200	6 400	5 300	1600	—	—	—
C_6H_6	24 000	12 500	10 000	2800	—	—	—
Linewidth							
CH_3I	1 140	760	550	—	2 320	—	—
C_2H_2	1 590	760	345	—	1 250	—	—
C_2H_4	—	—	1 010	620	—	—	—
C_6H_6	2 760	1 240	1 070	570	—	3580	—
Matrix polarization							
Xe	−4 600	−10 300	−8 900	—	—	—	—
CH_3I	−5 000	− 4 800	−6 000	—	—	—	—
C_2H_4	−5 400	− 8 600	−8 100	—	—	—	—
C_6H_6	−5 300	− 3 500	−3 600	9000	—	—	—

[a] From Reference [AD63]. Values in cm^{-1}.

The gas-phase ionization potential is in part reduced in the matrix by the stabilization given to the ion by the polarization of the matrix. As such, it should increase from Ne to Xe, which is the general trend. The gas-to-matrix high-frequency shift as the 3s Rydberg orbital becomes the $n = 1$ intermediate exciton is largest for Ne and CF_4 and least for Xe as host matrix. The variation of the $n = 1$ linewidth varies in the same way as the frequency shift, a fact which remains to be explained.

In a review article, Baranovskii [AD10] discusses the importance of both valence shell and Rydberg excitations in the X-ray absorption spectra of molecular systems. However, his analysis seems to ignore the condensed-phase effect expected for the Rydberg excitations in solids of low electron mobility. X-ray absorption in solids is also discussed by Kunz

[AD112], who mentions that the spectra of gases and solids are very similar, except close to threshold, i.e., in the region of Rydberg absorption.

AD.II.E. Instrumentation

An excellent review of the field of vacuum-ultraviolet instrumentation, from the early discoveries of Schumann to the present-day fields of high-temperature plasma and rocket spectroscopies, is given by Tousey [AD179], while Hunter [AD85, AD86] has described various facets of instrumentation which have come on the scene in the last ten years. Several aspects of recent instrumentation and vacuum-ultraviolet technique are also discussed in the proceedings of the NATO Summer School on the subject "Chemical Spectroscopy and Photochemistry in the Vacuum Ultraviolet" [AD30].

AD.II.E-1. Light Sources

What must be called a medium-pressure lamp using a pulsed discharge in ~ 2 atm of Xe or Ar gas has been described [AD109]. Using a 10-μsec pulse, continua were generated extending to 63 000 cm^{-1} in Xe and to 83 000 cm^{-1} in Ar gas. The radiation is equivalent to that from a black body at 23 000°K. Thus, depending upon the mode of excitation, the rare gas lamps can be made to emit continua from pressures of 100 mm to 20 atm. The mechanics of the BRV continuum source (sliding spark with uranium anode) is discussed by Fox and Wheaton [AD59], who found the continuum is radiated from a fast-pinch discharge. A variety of this lamp involves a sliding spark over the insulator Plexiglass, a continuum again being produced [AD151].

Simonenko [AD169] has described a standard source for the vacuum ultraviolet using square pulse electrical excitation in flowing helium gas. Such a plasma is at 35 000°K and is in thermodynamic equilibrium. Also using a discharge in flowing helium gas, Sauvageau et al. [AD165] describe the practical aspects of operating the Hopfield continuum light source (100 000–167 000 cm^{-1}) and construction of its power supply, as well as the technique involved in making semiquantitative intensity measurements in a windowless cell.

A very detailed and practical guide to the characteristics of synchrotron radiation and its advantages and disadvantages is presented by Taylor in the NATO volume [AD178].

The laser spark has been investigated for emission in the vacuum ultraviolet and found to be potentially useful as a light source. When the output of a Nd-glass or ruby laser (1–10 J) is focused upon the surface of a high-Z metal, a 1-mm plasma ball is produced having a temperature above 100 000°K. Such a plasma can emit 3.5×10^{22} photons/sec cm sr Å at 83 000 cm^{-1} [AD21, AD22] and has its peak emission at 500 000 cm^{-1} [AD49]. When used with a light collection system, such a plasma could easily be used with photographic or photoelectric detection. Plasmas formed of metals of low Z give only line spectra rather than continua.

Finally, those spectroscopists using microwave-driven rare gas discharge lamps will want to read the paper of Stanley et al. [AD172], who report that the common sorts of cavities used to couple to the lamps (Evanson, Broida, axe head, etc.) leak microwave radiation at levels far above the present national safety standards (which themselves are probably too high).

AD.II.E-4. Frequency Shifters

Burton and Powell [AD26] report further on 1,1,4,4-tetraphenyl butadiene as an ultraviolet wavelength shifter. They find a film of 1 mg/cm^2 is stable over long periods of time with two to three times more sensitivity than sodium salicylate; its emission is well matched to S-11 and S-20 photocathodes. Another phosphor which has been used successfully in the vacuum ultraviolet is "liumogen" (2,2'-dihydroxy-1,1'-napthaldiazine) [AD110]. This material seems especially well suited for work involving matrices at low temperatures.

AD.II.E-5. Polarizers

The useful range of the biotite polarizer has been demonstrated by Matsui and Walker [AD129] to extend to 112 000 cm^{-1}, where the reflectivity at Brewster's angle is 30% with a 92% degree of polarization after two reflections. Such a biotite polarizer was used by them to measure the polarization of light at the exit slits of various commercial vacuum-ultraviolet monochromators.

Chandrasekharan and Damany [AD29] have determined the birefringence of sapphire, MgF$_2$, and quartz, allowing the construction of quarter-wave retardation plates which are achromatic at several frequencies. A somewhat different type of retardation plate is described by Metcalf and Baird [AD131], who mechanically stress an LiF crystal. Such a retardation plate can be used for converting linear to circularly polarized light in the vacuum ultraviolet.

AD.II.E-6. Optical Spectrometers

Grojean [AD68] has compared the performance of the commercial McPherson double-beam spectrometer (model RS-225) with that built by Korn and Braunstein [I-K21] and found them to be very similar. Details of the construction and operation of a double-beam spectrometer suitable for both transmission and reflection measurements down to 200 000 cm^{-1} are given by Dickinson and Ellis [AD41]. A BRV (uranium anode) source is used, and variable-angle specular reflectance is available in an ultrahigh-vacuum sample chamber. In spite of sophisticated electronic compensation and signal handling, the spectrometer requires 3 hr to scan 1000 Å. A brief review of the status of the reflection spectroscopy of liquids in the vacuum ultraviolet has been given by Birkhoff et al. [AD17].

AD.II.E-7. Filters and Windows

Pure LaF$_3$ has an absorption edge at $\sim$77 000 cm^{-1} at 100°K. When doped with Ce^{3+} ion at the 1% level, a window results at 56 800 cm^{-1} having a half-width of 5700 cm^{-1} and 40% maximum transmission [AD51]. The width and transmission of the window are dependent upon dopant level, temperature, and crystal thickness.

Interference filters for the vacuum ultraviolet have been described recently. Fairchild [AD52] reports that Al–MgF$_2$ multilayer filters have transmissions which fall from 23% at 52 500 cm^{-1} to 11% at 80 000 cm^{-1} with a half-width of $\sim$10 000 cm^{-1}. Baldini and Rigaldi [AD8] discuss theoretically the optimum parametrization for such a metal–dielectric multilayer filter and describe, for example, the construction of a filter having 70% transmission at 62 500 cm^{-1} with a half-width of only 1200 cm^{-1}. Of course, the transmission frequency of the filter can be adjusted between 50 000 and 80 000 cm^{-1} by the proper choice of parameters, and similar constructions can be used for polarizing light in the vacuum ultraviolet. Multilayer aluminum-magnesium fluoride–aluminum interference filters have also been constructed by Bates and Bradley [AD14], who typically attained 25% transmission in a band half-width of 6000 cm^{-1}.

Single crystals of BeO (1 mm) have been investigated optically in the vacuum ultraviolet [AD153] and found to have transmission to $\sim$77 500 cm^{-1}, with great apparent resistance to radiation damage. The optical transmission of BeO appears to be about like that of MgF$_2$; the transmission of crystals of BeF$_2$ should be better than either of these.

Samples of synthetic quartz of very high purity have been investigated at room temperature and found to have a vacuum-ultraviolet cutoff frequency dependent upon the concentrations of iron impurities. In the cleanest samples, an α (extinction coefficient per centimeter of length) of 1.5 was achieved at 68 500 cm^{-1} [AD9], with only a very small dependence upon the choice of crystal face illuminated [AD64]. Surprisingly, natural quartz transmits to the same frequency. Reflection studies on both natural and synthetic quartz crystals show peaks at 83 000 and 93 000 cm^{-1} regardless of purity [AD64]. On irradiation in the 50 000–67 000-cm^{-1} region with a Xe lamp, Suprasil W rapidly develops an absorption band centered at 38 500 cm^{-1}, while retaining most of its transparency at higher frequencies [AD113]. Possibly such irradiated quartz could be used as a filter to separate the mercury arc lines at 39 400 and 52 800 cm^{-1}.

Hunter and Malo [AD84] have measured the ultraviolet cutoffs of several common window materials in the temperature range 10–370°K (Fig. AD-2). In each of these materials (except LaF$_3$), the cutoff moves to higher frequency with decreasing temperature, the gain of transmission limit amounting to about 50 Å from room temperature to 10°K. They also report fluorescence from CaF$_2$, BaF$_2$, and LaF$_3$ upon vacuum-ultraviolet illumination.

The use of cellulose nitrate (celluloid) as a window material for spectroscopy in the Hopfield region has been explored by O'Bryan [AD143], who found that a 100-Å-thick film of the material was 35% transmitting at 100 000 cm^{-1}, with a near-monotonic increase to 79% transmitting at 333 000 cm^{-1}.

AD.II.E-8. Detectors

New information on detectors is sparse indeed. The construction of standard ultraviolet-sensitive detectors is presented in two recent papers [AD27, AD57], and another describes the conversion of an end-on photomultiplier tube to side-viewing geometry [AD148]. An inexpensive underwater camera has been adapted for use under vacuum as a detector coupled to a McPherson monochromator [AD32]. Hunter, in a recent review article, compares the performances of various types of vacuum-ultraviolet detectors [AD86].

AD.III.A-1. Methane

The trapped-electron-impact spectrum of methane shows several apparent excitations to triplet states since it is a threshold technique and integrates over all scattering angles. The spectrum of Dicoum *et al.*

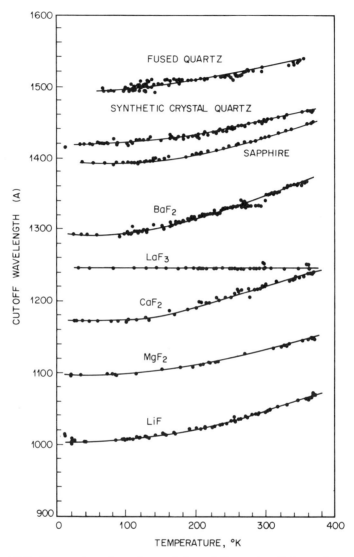

Fig. AD-2. Temperature dependence of the cutoff wavelength for various window materials in the vacuum ultraviolet [AD84].

[AD42] reveals the triplet at 71 000 cm^{-1} earlier found by Brongersma and Oosterhoff [I-B59] as well as another at $\sim$75 000 cm^{-1}. Further structure is seen in the 80 000–115 000-cm^{-1} region, much but not all of which correlates with features in the high-energy electron-impact spectrum (Fig. III.A-2). There are two obvious antiresonances in the trapped-electron spectrum between 105 000 and 113 000 cm^{-1}. In a paper regrettably overlooked, Narayan [AD141] points out the similarities in the $1t_2$ photoelectron band of methane and the bands near 80 000 cm^{-1} in the optical and electron impact spectra, and concludes that the latter are Jahn–Teller components of the $1t_2 \to 3s$ excitation. This parallels our explanation.

The near-Hartree–Fock calculations of the geometry of the Jahn–Teller-unstable ion CH_4^+ are unanimous in giving the lowest energy configuration D_{2d} geometry, followed by C_{2v}. However, in a calculation using an even larger and more flexible basis with configuration interaction, Meyer found that the most stable geometry was C_{2v}, followed by D_{2d} and C_{3v} [AD132]. This is an important point, since the geometry of the $^2T_2$ ion will also be that of the $(1t_2, ns)$ Rydberg states. A Slater orbital calculation on the band structure of solid methane in the tight binding approximation yields bands about 20 000 cm^{-1} wide centered approximately about the free-molecule $2a_1$ and $1t_2$ binding energies. The valence shell band-to-band gap is 197 000 cm^{-1} [AD155].

In a study of superexcited states in methane, Nishikawa and Watanabe [AD142] find the ionization efficiency is unity above 130 000 cm^{-1}, but from 105 000 cm^{-1} to this frequency there is appreciable excitation to superexcited states. These arise in the following way; due to strong Jahn–Teller distortion, the *vertical* transitions to Rydberg states close to the $1t_2$ ionization potential correspond to strong vibrational excitation in these excited states. Such highly excited vibronic Rydberg states may have energies above the *adiabatic* $1t_2$ ionization potential, and so are superexcited. Such states will be less prominent above the $2a_1$ ionization potential, for such excitations will be far more vertical than those originating at $1t_2$. Indeed, experiments by Ehrhardt and Linder [AD50] show that the cross section for superexcitation in methane drops to a very low level at $\sim 145\,000$ cm^{-1} and remains there to well beyond the $2a_1$ ionization potential.

Bagus et al. [AD5] have performed a theoretical study of the carbon K X-ray absorption spectrum of methane, reaching conclusions in agreement with those given by us in the text. They calculate the $1s_C \to 3s$ and $1s_C \to 3p$ absorption frequencies to within ± 1000 cm^{-1} of Chun's observed frequencies, and find the corresponding triplet states to lie lower by 1600 and 800 cm^{-1}, respectively, reflecting the very small exchange between $n = 1$ and $n = 3$ orbitals. The $1a_1 \to 3p$ oscillator strength is calculated to be 1.9×10^{-2}, whereas $1a_1 \to 3s$ is forbidden electronically, but is mixed with $1a_1 \to 3p$ via ν_3' and ν_4' vibrations. This vibronic mixing is estimated to yield an oscillator strength for $1a_1 \to 3s$ which is 10%

of that to $1a_1 \to 3p$ in CH_4, and which increases to 20% in CD_4. It is estimated from experiment that the $1a_1 \to 3p$ oscillator strength is approximately 0.6×10^{-2}. The most probable decay mode for the superexcited $1s^{-1} 2t_2^1$ state is autoionization to $1t_2^{-2} 2t_2^1$, with the electron in $2t_2$ playing the role of spectator. Deutsch and Kunz [AD40] have also studied the X-ray absorption spectrum of methane theoretically, using a single-center calculation. They assign the absorption edge as $1a_1 \to 3p$, but this transition is about 20 000 cm^{-1} higher than the threshold assumed by them.

AD.III.A-2. Ethane

Arguments still rage over the symmetry of the highest filled orbital in ethane. The nature of the highest filled MO in the alkanes, which are the originating orbitals for Rydberg excitations, was explored by Pauzat et al. [AD152] by exciton theory, which accounts for both reorganization energy and changes of correlation energy upon excitation. Their work predicts that the C—H bond ionization precedes that from the C—C bond in ethane, whereas the order is reversed on going from propane to octane. Murrell and Schmidt [AD139] also studied the same problem using photoelectron spectroscopy and *ab initio* calculations, and concluded that because the uppermost MOs in neopentane, isobutane, and propane involved C—C bond orbitals, the same situation must exist in ethane.

Narayan's [AD141] comparison of the optical spectra of ethane and ethane-d_6 with their photoelectron spectra led him to the conclusion that it was the $\pi 1e_g$ MOs that were the originating orbitals for these transitions. However, he then went one step further and assigned the structured band at 75 800 cm^{-1} in ethane as $1e_g \to 3s$, paralleling the assignment of the first intense band of methane, whereas we feel this band of ethane should be assigned as $1e_g \to 3p\sigma$.

AD.III.A-3. Propane and the Higher Acyclic Alkanes

In an interesting work, Narayan [AD141] stresses the interrelationship of the vacuum-ultraviolet, electron-impact, and photoelectron spectra of the smaller alkanes. Though we agree with his assignment of the first intense band of methane as terminating at 3s, his parallel assignment of the first strong band of propane is incorrect, we feel. Instead, an assignment terminating at 3p is preferred on the basis of term values. The *ab initio* calculation of the molecular orbital ordering in propane [AD139] agrees with that used in Table III.A-I, except that the positions of $2b_1$ and $1b_2$ are reversed. The same types of calculations for neopentane and isobutane predict a surprisingly large interaction between geminal methyl groups.

Using the coincidence technique together with electron-impact energy-loss scattering, Ehrhardt and Linder [AD50] have found that, unlike methane, in n-heptane there is considerable population of superexcited states far beyond the first ionization potential. From 88 000 to 240 000 cm^{-1}, the cross section for superexcitation in n-heptane is at least half that for direct ionization.

Electron transmission and back-scattering spectra of very thin films of linear, branched, and cyclic alkanes are given but not interpreted [AD77]. Optical reflectivity studies have been made on the liquids $C_{11}H_{24}$, $C_{14}H_{30}$, and $C_{17}H_{36}$ [AD150]; each of these liquids displays an absorption edge at 60 500 cm^{-1}, with a plateau between 72 000 and 80 000 cm^{-1}. For the two larger alkanes, the reflectivity data extend to 160 000 cm^{-1}.

The systematic dependence of the fluorescence properties of alkanes upon molecular size and geometry has been determined by Rothman et al. [AD161] in an investigation spanning over 100 compounds. Luminescences as described in Fig. III.A-14 were excited in neat liquids at 60 500 and 68 000 cm^{-1}. Briefly, the results are these: (i) In n-alkanes, the fluorescence quantum yield is larger for excitation at 60 500 than at 68 000 cm^{-1}, and the fluorescence frequency of butane (46 500 cm^{-1} vert.) is noticeably lower than that of the other alkanes (48 200 cm^{-1} vert.). No emission was observed from propane or ethane liquids. (ii) On branching, the fluorescence maximum shifts to 45 500 cm^{-1} and the quantum yield drops to $\sim$10% of its value in the corresponding alkane. (iii) No fluorescence was detected in any acyclic, geminal dibranched alkane. (iv) In vicinal dibranched compounds, the fluorescence maximum is at 41 300 cm^{-1} and the quantum yield rises to approximately that of the corresponding linear alkane. (v) The relative positions of branching points in a molecule are important in determining the fluorescence frequency. And (vi) The frequencies of maximum emission seem to correlate with the boiling points of the liquids. The temperature dependences of the decalin and dodecane luminescences excited by pulsed X-rays have been measured [AD75]; CCl$_4$ quenches these at a rate somewhat faster than the diffusional rate.

AD.III.A-4. Simple Rings and Polycyclic Alkanes

We concluded from the intense $\phi_i \to 3s$ origin in adamantane that ϕ_i, the uppermost filled MO, must have t_2 symmetry. A recent MINDO/1 calculation by Worley et al. [AD188] confirms this, and additionally, they report vibrational quanta of 890 cm^{-1} excited in the first photoelectron band, presumably a C—C stretching motion. These intervals no doubt correspond to the $\sim$840-cm^{-1} intervals found by Raymonda [I-R20] throughout the optical spectrum. Note, however, that Worley et al. claim that there are no resolvable Jahn–Teller splittings anywhere in the adamantane photoelectron spectrum. Similar intensity and term value arguments in cyclohexane suggested that the uppermost filled MO had g symmetry for molecules in the chair conformation. Calculations by Hoffmann et al. [AD79] describe the uppermost filled MO in cyclohexane as e_y.

Because sharply structured absorption spectra in the alkanes appear only for rigidly constrained rings (ethane is a two-membered ring!), it was hoped that cubane would show a spectrum resembling that of adamantane. Optically, the cubane spectrum commences with a broad shoulder centered at 52 000 cm^{-1} (vert., $\epsilon = 1000$) having a term value of 21 300 cm^{-1} with respect to the first ionization potential at 73 300 cm^{-1} [I-R7]. This term value agrees nicely with the $(\phi_i, 3s)$ limiting term value expected for a large alkane; transitions to 3p are also expected in this area. Following this, there is another peak at 60 000 cm^{-1} (vert., $\epsilon = 3500$), and finally a massive feature ($\epsilon = 10\,000$) with an apparent peak at 71 000 cm^{-1}. No vibronic structure is seen optically to 78 000 cm^{-1}. Perhaps this lack of structure is understandable, for the first six bands in the photoelectron spectrum are structureless. However, since the seventh band (origin at 110 000 cm^{-1}) shows a beautiful progression of five vibrational quanta (810 cm^{-1}), the corresponding $\phi_7 \rightarrow$ 3s optical Rydberg band will commence at 88 000 cm^{-1} and could show the same vibrational structure.

Hirayama and Lipsky [AD78] report some rather peculiar behavior in the fluorescences of solid cyclohexane and solid bicyclohexyl as they undergo phase transitions. In solid cyclohexane, there is a phase change at 186°K, and as the solid is cooled below this point, the fluorescence maximum abruptly shifts from 49 800 cm^{-1} (Fig. III.A-14) to 52 600 cm^{-1}, while decreasing its intensity to about 10% of its value in the warmer phase. These authors feel that because the fluorescence originates with a state having "partial Rydberg nature," the 2800-cm^{-1} shift of the fluorescence to higher frequencies at the phase transition temperature is a consequence of the increased density in the colder phase. No shift in emission characteristics was noted as adamantane was passed through its phase transition at 290°K. Somewhat different behavior obtains for bicyclohexyl: As the temperature of the solid is raised from 77°K, the intensity of the fluorescence at 46 500 cm^{-1} (vert.) first rises to a maximum at 233°K, falls slowly to 273°K, and then decreases very rapidly as the crystals melt at that temperature. In the liquid, the fluorescence intensity is recovered, but the luminescence maximum has shifted to 44 100 cm^{-1} (vert.).

AD.III.B-1. Alkyl Monochlorides, Bromides, and Iodides

The absorption spectra of methyl iodide in solid and liquid krypton solutions are unique, for they show three members of the first two Wannier series converging upon the bottoms of the conduction bands. Gedanken et al. [AD62] found the $n = 1$, 2, and 3 members of the 5p iodine excitation in both the solid and liquid solutions, leaving the CH$_3$I$^+$ core in the $^2E_{3/2}$ and $^2E_{1/2}$ spin–orbit configurations. In the solid (Fig. AD-3), the bands are about one-half to one-third the width of the bands in solution, so that vibrational structure (ν_2') is observed in the solid, but not in the liquid. There is a shift to higher frequency on going from the liquid to the solid, with spin–orbit splits of 4795 (gas), 4975 (liquid), and 4725 cm^{-1} (solid). The corresponding ionization potentials in the solid are 65 500 ($^2E_{3/2}$) and 70 200 cm^{-1} ($^2E_{1/2}$) and in the liquid are 64 400 ($^2E_{3/2}$) and 69 400 cm^{-1} ($^2E_{1/2}$).

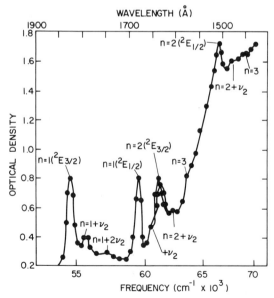

Fig. AD-3. Absorption spectrum of 1% methyl iodide in solid krypton at 35°K [AD62].

Gedanken et al. present a compilation of their work on methyl iodide in Ne, Ar, Kr, and CF_4 matrices [AD63]. The most remarkable feature of this work is that whereas the $E_{3/2}$–$E_{1/2}$ split in the Ar and Kr matrices is very close to that found for methyl iodide in the gas phase (4795 cm^{-1}), in a CF_4 matrix, this splitting increases to 6130 cm^{-1}, while in a neon matrix, the split has grown to 15 140 cm^{-1}. The 5p → 6s spin–orbit-split transitions of the Xe atom in a Ne matrix also show this large splitting enhanced by large nonorthogonality corrections. The lowest ionization potentials of methyl iodide in Ne, Ar, and Kr matrices are 76 600, 69 150, and 65 600 cm^{-1} (vert.), respectively.

Though the general feeling is that the electronic mobility in organic matrices is far too low to support Rydberg states [I-R12], the fact that the Rydberg excitations of methyl iodide are clearly visible in a methane matrix (Table III.B-III) is perhaps a little less surprising once it is realized that the electron mobility in liquid methane (300 cm^2/V sec) is almost as large as that for liquid argon (450 cm^2/V sec), whereas for the straight-chain alkanes, the mobility is approximately 0.1 cm^2/V sec [AD31]. By this criterion, neopentane (μ_L = 70 cm^2/V sec) may be a good matrix for Rydberg excitations, though its transmission range will not extend very far into the vacuum ultraviolet.

Many of the ionization potentials quoted in Section III.B-1 are unpublished values taken from our photoelectron spectra. Published values of these can now be found in the work of Kimura et al. [AD102] and Heilbronner and co-workers [AD24, AD72].

On rereading the text, I find that insufficient credit has been given the paper [I-M34]. In this pioneering work, Mulliken lays out very clearly the origins of the various bands discussed here, and the trends in their term values as the hydrogen halides are alkylated. Very little beyond the gathering of more confirmatory data can be added to the topic of alkyl halide spectra as discussed in this wide-ranging paper.

AD.III.B-2. Di-, Tri-, and Tetrachloro-, Bromo-, and Iodoalkanes

The spectra of the various chloromethanes have been photographed by Russell et al. [I-R30], and are presented in Fig. AD-4, for comparison with one another and with Fig. III.B-7. With respect to the orbital ordering in the chloromethanes, Hopfgarten and Manne [AD80] have proposed somewhat different assignments for some of the orbital ladders given in Fig. III.B-7. Using semiempirical calculations of the X-ray emission profiles (valence shell MOs → C K, Cl Kβ, and Cl L$_{II,III}$), and comparing these with the experimental curves, they conclude that in chlo-

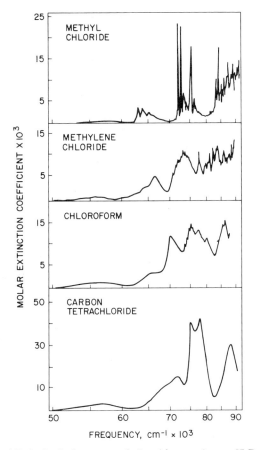

Fig. AD-4. Optical spectra of the chloromethanes [I-R30].

roform, the ordering is $3a_1$ followed by $1a_2$ rather than the reverse, and in methylene chloride, the first degenerate pair is $2b_1, 2b_2$ and the second is $3a_1, 1a_2$, rather than as given in the figure.

AD.III.B-3. Fluoroalkanes

The absorption and photoelectron spectra of ethyl fluoride have been recorded and the former interpreted in terms of Rydberg excitations [AD189]. Ionizations from the $\sigma(\text{C—C})$ a' and $\pi(\text{CH}_3)$ a'' MOs are nearly degenerate, according to the photoelectron spectrum, coming at 100 100 and 103 800 cm^{-1} (vert.), respectively. These correspond to the $3a_{1g}$ and $1e_g$ MOs of ethane, which are even closer in energy. Down to

86 000 cm^{-1} in the optical spectrum of ethyl fluoride, four broad, structureless bands are seen, and they can be assigned as originating at the a' and a'' MOs and terminating at 3s and 3p MOs. The $a' \to$ 3s excitation is quite weak, and appears only as a shoulder at $\sim$73 000 cm^{-1}, whereas $a'' \to$ 3s is very intense at 77 500 cm^{-1} (vert.). These bands have term values of $\sim$26 500 cm^{-1}, which is rather lower than expected, since the $(\phi_i, 3s)$ term values in propane have an average value of 27 250 cm^{-1} (Table III.A-I), and those for ethyl fluoride might be expected to be somewhat higher than these. The remaining two bands at 80 600 and 84 700 cm^{-1} (vert.) in ethyl fluoride originate at a' and a'' and terminate at 3p, giving term values of 19 500 and 19 100 cm^{-1}, respectively. These term values compare well with those of the fluoroalkanes listed in Table III.B-V, with due allowance being given for the higher penetration energy in the more highly fluorinated molecules.

The optical absorption spectra of the fluoromethanes have been extended to 165 000 cm^{-1} by Sauvageau and co-workers [AD165], using the Hopfield continuum as light source and a windowless absorption cell. Their spectra are in one-to-one correspondence with the bands already reported for this region using electron-impact spectroscopy [I-H18], with the exception, however, that the optical intensities are not proportional to the scattered current in the electron-impact spectra. For example, the last band observed optically in the 130 000–140 000-cm^{-1} region is the most intense in the spectrum, whereas it is generally very weak in the electron-impact spectrum.

Kaufman et al. [AD99] have challenged the generally accepted view that the lowest ionization potentials of tetrafluoromethane and carbon tetrachloride as measured by photoelectron spectroscopy are direct ionizations. They claim instead that these peaks represent excitations to neutral fragments which are subsequently ionized in some unspecified way. According to this, the regularity of the first few term values in these molecules would be spurious, since the photoelectron spectra are said not to give the proper ionization potentials.

Our attempts to interpret the fluorine K X-ray absorption spectra of the fluoromethanes in terms of Rydberg excitations having term values resembling those of the ultraviolet spectra (Table III.B-V) are not totally successful. In addition to this, LaVilla's data [AD116] are not tabulated, and the experimental term values can be read from his spectra only with difficulty. For methyl fluoride, two bands are observed with term values of 32 000 and 13 700 cm^{-1} (vert.), which would seem to identify them as terminating at 3s and 3d, respectively. However, by symmetry, these would be expected to be weak, whereas that to 3p would be intense (see the carbon K spectrum of methane, p. I-116). In fluoroform and methylene fluoride only a single line with a term value of 19 000 cm^{-1} is observed, this being $1s_F \rightarrow 3p$. Again in carbon tetrafluoride, only one line is observed, this with a term value of 12 000 cm^{-1}, implying $1s_F \rightarrow 3d$. All of the term values deduced from the X-ray spectra are of just the sizes previously found in the ultraviolet spectra; however, the capricious pattern of intensities is still puzzling. LaVilla assigned all of these bands to valence shell excitations.

The study of the optical and photoelectron spectra of the series of fluorochloromethanes by Doucet et al. [AD45] is extremely interesting, both in its own right and for the light it sheds on our attempts to understand the spectrum of CF$_3$I. The spectra of these CH$_x$F$_y$Cl$_z$ compounds follow the usual pattern of alkyl halide absorption, i.e., all of the transitions below the first ionization potential originate with the heavy-halide lone-pair electrons, beginning with a very weak set of valence shell A bands, followed by more intense B and C Rydberg excitations (unresolved) terminating at the lowest ns level, and then a D band corresponding to a Rydberg excitation terminating at the lowest np level. In those compounds having more than one chlorine atom, there is a slight Cl–Cl splitting which leads to overlapping sets of A, B, C, D manifolds, each originating at a separate lone-pair MO.

Starting with the A bands, we see a rather odd situation. On comparing CH$_3$I with CF$_3$I, and CH$_3$Br with CF$_3$Br, in both pairs of compounds the exchange of CH$_3$ by CF$_3$ shifts the A band by less than ± 1000 cm^{-1}, whereas in the chlorine series CH$_3$Cl, CH$_2$FCl, CHF$_2$Cl, and CF$_3$Cl, the A-band frequencies are 58 000, 62 500, 66 200, and 71 500 cm^{-1} (vert.). In the fluorinated chloromethanes containing two or three chlorine atoms, two A bands are seen in the 55 000–65 000-cm^{-1} region. This regular shift of 13 500 cm^{-1} for the A-band frequency in the fluorochlorides is followed rather nicely by the shift of the chlorine lone-pair ionization potential by 15 000 cm^{-1} in the same series. However, this is not to imply that the A bands are Rydberg excitations, for in the fluorobromides and fluoroiodides, the CH$_3$X–CF$_3$X series ionization potential shifts are 7000 and 12 000 cm^{-1}, respectively, while the A bands are totally unresponsive to the fluorination. Taking a simple one-electron view, it is as if the lone-pair ionization potentials and σ^*(C—X) MO energies shift in unison in the bromides and iodides, whereas in the chlorides, the ionization potentials shift and the σ^*(C—Cl) MO energies remain constant. There is no obvious reason why this should be so.

In methyl chloride and methylene chloride, the B, C-band term values are 27 300 and 26 000 cm^{-1}, respectively, and fluorination of these molecules, in general, would be expected to increase their term values due to the high penetration energy at fluorine. This is borne out, for in

TABLE AD-II
Term Values in the Alkyl Halides and Polyhalides[a]

Compound	Ionization potential	B,C-Band absorption frequency	B,C-Band term value	D-Band absorption frequency	D-Band term value
C_2H_5F	100 100	73 000	27 100	80 600	19 500
	103 800	77 500	26 300	84 700	19 100
H_2CFCl	94 700	65 310	29 300	72 780	21 900
				74 850	19 900
	99 500	71 120	28 400	—	—
HCF_2Cl	101 600	74 500	27 100	80 650	21 000
CF_3Cl	104 900	78 100	26 800	—	—
CF_3Br	96 800	70 500	26 300	76 900	19 900
$HCFCl_2$	100 800	70 000	30 800	80 000	20 800
	105 500	75 000	30 500	84 000	21 500
CF_2Cl_2	99 200 | 101 500	74 000	{25 200 | 27 500}	79 100	{20 100 | 22 400}
$CFCl_3$	96 000	—	—	—	—
	98 400	71 000	27 400	81 200	17 200
	104 900	75 500	29 400	—	—
	108 900	78 800	30 100	—	—

[a] Values in cm^{-1} (vert.).

CH_2FCl and $CHFCl_2$, the B, C term values are 29 300 and $\sim$30 700 cm^{-1}, respectively. However, on taking the next step in the series, a most unexpected thing happens. In CHF_2Cl and CF_2Cl_2, the B, C term values have *decreased* to 27 100 and $\sim$26 500 cm^{-1} (Table AD-II), and finally in CF_3Cl, the term value is only 26 800 cm^{-1}. In this fully fluorinated chloromethane, a B,C term value of approximately 32 000 cm^{-1} would otherwise be expected. The decrease of the ns term value on comparing CF_3Cl with CH_3Cl is reflected in the bromides as well, where the term value for CH_3Br is 29 000 cm^{-1}, but that for CF_3Br is only 26 300 cm^{-1}.

In Section III.B-1, we were confronted with the apparent anomaly of a B,C term value for CF_3I which was less than that for CH_3I, and in an effort to avoid facing this, tried to reassign the entire B, C, D manifold. It now appears that there is a precedent for such an anomalous situation. That something drastic has happened to the B, C bands in the difluoro and trifluoro compounds is evident as well in the B, C band shapes and intensities: In CH_3Cl and CH_2FCl, where the term values are "regular," the excitations are structured and have molar extinction coefficients of approximately 300 in each, whereas in CHF_2Cl, CF_3Cl and CF_3Br, where the B, C term values are anomalously low, the bands are structureless and have molar extinction coefficients of 20 000, 17 000, and 12 000, respectively. Further, in $CHFCl_2$, where the term value is normal, the extinction coefficient is again only 2000, but in CF_2Cl_2, where the term value is abnormal, the extinction coefficient is about 10 000. Thus it would appear that in the difluorides and trifluorides, there is a configuration mixing which simultaneously intensifies the B, C bands by a factor of from three to seven while lowering the term values 3000–6000 cm^{-1} below expectations. The state in question must lie below the B, C levels, be a valence shell excitation of appreciable oscillator strength, and must transfer virtually all of its intensity to B, C upon mixing. This would be a good point of departure for semiempirical calculations.

As seen in Table AD-II, the D bands (lone pair → np) in the fluorochloromethanes retain their regular term values of 20 000 ± 1000 cm^{-1}, as in almost all compounds of whatever composition.

AD.III.C-1. Boron Hydrides and Halides

The X-ray absorption spectrum of diborane has been published by Zimkina and Vinogradov [AD194], and with the boron 1s ionization potential obtained from the ESCA spectrum [I-A6], the term values can be estimated. The spectrum consists of two sharp peaks at 1 536 000 and 1 565 000 cm^{-1} (vert., 190.5 and 194.1 eV, respectively) and two far broader ones at 1 592 000 and 1 634 000 cm^{-1} (vert., 197.4 and 202.7 eV, respectively). The latter two are above the 1s binding energy of 1 585 000 cm^{-1} (196.5 eV). A splitting of only 80 cm^{-1} is predicted between the two boron 1s MOs [I-S37]. The first sharp band in the X-ray absorption spectrum has a term value of 48 400 cm^{-1}, which is approximately twice that expected for a transition to 3s in diborane, and so is almost certainly a valence shell excitation. The second sharp band has a term value of 19 300 cm^{-1}, and so could be the $1s_B$ → 3p Rydberg excitation allowed from the a_g combination of 1s AOs.

It is clear now that the discussion on trimethyl borane is somewhat in error. In the text, the two photoelectron bands at 86 100 and 91 100 cm^{-1} (vert.) were thought to be the e' and a_1' components of the B—C sigma-bond MOs. The *ab initio* calculation of these MOs for trimethyl borane [AD3] instead shows that these two features are the Jahn–Teller-split components of the e' ionization, and that the a_1' ionization comes at ∼105 000 cm^{-1}, where it is quasidegenerate with the $\pi(e')$ ionization potential. With this reassignment of the photoelectron spectrum, the optical spectrum then can be reassigned as follows. The two features at 53 000 and 60 4000 cm^{-1} in the optical spectra are the Jahn–Teller components of the forbidden $e' \to a_2''$ valence shell excitation, whereas the intense peak at 75 500 cm^{-1} (vert.) corresponds to the allowed $a_1' \to a_2''$ promotion. In this way, the photoelectron and optical splittings are satisfactorily related; however, the band at 60 400 cm^{-1} should be forbidden, but is rather intense. Such an intensity enhancement could result either from vibronic mixing with the allowed 75 500 cm^{-1} band, or overlapping with the allowed $e' \to 3s$ Rydberg band which is expected at $\sim$62 000 cm^{-1} (vert.).

Optical spectra of the important compounds boron trichloride and boron tribromide have recently been reported [AD156]. As seen from Fig. AD-5, the optical spectrum of boron trichloride does not resemble very closely the SF$_6$-scavenger spectrum reported earlier for it. Still, the explanation follows rather closely that given in the text. With an ionization potential of 96 500 cm^{-1} (vert., a_2') [I-P29], the intense absorption at 57 900 cm^{-1} (vert.) in boron trichloride has a term value of 38 530 cm^{-1}, which clearly marks it as an allowed valence shell excitation, probably $a_2' \to \sigma^* e'$. Following this, there is a weak shoulder at 68 000 cm^{-1} and an intense feature at 73 400 cm^{-1} (vert.), which have term values

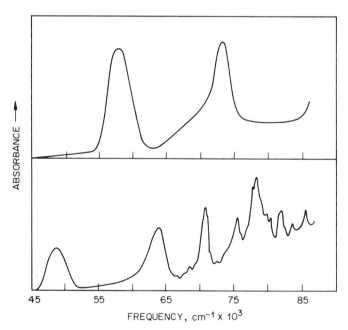

Fig. AD-5. Optical spectra of boron trichloride (upper) and boron tribromide (lower) in the gas phase [AD156].

of 28 500 and 23 100 cm^{-1}, respectively, with respect to the first ionization potential. These figures compare well with the B and D Rydberg band term values in chloroform (26 300 and 22 700 cm^{-1}, respectively), and so suggest $a_2' \to 4s$ and $a_2' \to 4p$ assignments, using quantum numbers appropriate to the peripheral atoms. The 73 400-cm^{-1} band also has a term value of 26 600 cm^{-1} with respect to the e' ionization potential at 100 000 cm^{-1} and so may also contain the allowed $e' \to 4s$ band. Maria et al. [AD125] have found a weak band at 48 200 cm^{-1} (vert.) in gaseous boron trichloride ($\epsilon = 27.3$) which they assign as having a triplet (n_{Cl}, π^*) valence shell upper state.

The optical and photoelectron spectra of boron tribromide are more complicated than those of boron trichloride, and many more overlapping assignments are possible. Potts et al. [I-P29] quote the first four ionization potentials of boron tribromide as 86 000 (a_2'), 91 500 (e'), 94 600 (e''), and 106 000 cm^{-1} (vert., $1a_2''$). The first three features of the boron tribromide optical spectrum parallel those in boron trichloride and have similar assignments: 48 650 cm^{-1} ($a_2' \to \sigma^*e'$), 59 000 cm^{-1} (27 000 cm^{-1} term value, forbidden $a_2' \to 5s$), and 63 630 cm^{-1} (22 400 cm^{-1} term value with respect to the e' ionization and therefore assignable as $e' \to 5s$). The sharp peak at 70 840 cm^{-1} (vert.) has the appropriate term value for $e' \to 5p$, while the bands at 75 720 cm^{-1} has a term value appropriate for $a_2' \to 4d$ (10 300 cm^{-1}), and that at 78 490 cm^{-1} appears to be the allowed $1a_2'' \to 5s$ band with a term value of 27 500 cm^{-1}. These assignments are in only partial agreement with those given by Planckaert et al. [AD156].

In boron trifluoride, Maria et al. [AD125] also have found a band at 63 600 cm^{-1} (vert.) having a molar extinction coefficient of 3.6. It may be due to an impurity since this substance is so reactive. The fluorine K absorption spectrum of boron trifluoride is presented by Zimkina and Vinogradov [AD194].

AD.III.C-2. Boron–Nitrogen Compounds

A photoelectron study of trimethylamine borane and its -d_3 derivative by Lloyd and Lynaugh [AD123] reveals that the first band originates from the $5e$ MO of the BH$_3$ group and shows a 5000-cm^{-1} Jahn–Teller split accompanied by progressions of the e deformation. This is in accord with the optical spectrum, except for the fact that the 5000-cm^{-1} splitting is not seen optically (Fig. III.C-6).

Bernstein and Reilly [AD15] also have analyzed the vibronic structure of the 50 000-cm^{-1} band of borazine, pointing out that the forbidden transition to $^1A_1'$ will gain intensity vibronically via a_2'' out-of-plane vibrations, but that absorption to $^1A_2'$ is not assisted by vibrations of this symmetry. In their analysis, they find five totally symmetric progressions, each attached to a false origin. The allowing modes responsible for the false origins are in three cases deduced to involve a_2'' vibrations, thus securing a $^1A_1' \rightarrow {}^1A_1'$ assignment for this band of borazine. In contrast, a CNDO calculation by Kuehnlenz and Jaffe [AD111] gives $^1A_2'$ as the lower excited state of borazine, as it is in benzene.

AD.III.D-1. Amines

In the text, the possibility was considered that the $a_2'' \rightarrow 3s$ Rydberg band in the optical spectrum of ammonia (Fig. AD-6), was resting upon a valence shell continuum, but this was discounted because the photoelectron spectrum of the a_2'' ionization also seemed to show the underlying continuum. However, Rabalais et al. [AD157] have since redetermined the a_2'' photoelectron band shape with higher resolution (Fig. AD-6), thereby resolving the vibrational structure almost to the baseline. Thus we see that there is something underlying the $a_2'' \rightarrow 3s$ optical band which is *not* related to the a_2'' ionization process, i.e., a valence shell band.

Herzberg and Longuet-Higgins [AD76] describe how the $(a_2'', 3s)$ Rydberg state of ammonia is weakly predissociated by mixing with a valence shell state which is repulsive with respect to dissociation into NH_2 and H. They also point out, however, that the mixing is small due to the small electronic overlap between the two states; otherwise, the Rydberg state would be completely dissociated, as in water.

The electron-impact spectrum of ammonia has also been reported by Lindholm [AD121], who also assigned it using intuitive term value arguments. In particular, he assigned the broad peak at 92 000 cm^{-1} (Fig. III.D-1) as $1e' \rightarrow 3sa_1$. Taking the broad $1e'$ ionization potential as 129 000 cm^{-1} (vert.) results in a term value of 37 000 cm^{-1} for this band, which agrees nicely with the value of 35 750 cm^{-1} measured for the $(a_2'', 3s)$ state. This independence of the $(\phi_i, 3s)$ term values upon the originating MO in ammonia parallels the situation in water (Section III.E-1). Lindholm has assigned higher features as $1e' \rightarrow 4s$ and $1e' \rightarrow 5s$, but they do not appear in our spectrum (Fig. III.D-1).

Very good figures of the higher bands in the optical spectrum of ammonia are given in the work of Watanabe and Sood [I-W19]. They, along with Hudson [I-H67], have carefully reviewed the intensity measurements in ammonia, and the complete partial oscillator strength spectrum of ammonia has been determined by coincidence electron-impact ionization measurements. After this is joined to the bound-state spectrum

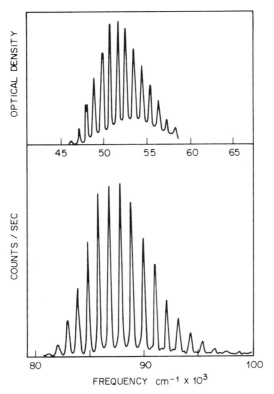

Fig. AD-6. Comparison of the Franck–Condon envelopes of the $a_2'' \rightarrow 3s$ optical transition of ammonia (upper) [I-T1], and the corresponding a_2'' band in the photoelectron spectrum (lower) [AD157].

and normalized to eight, the number of valence shell electrons in ammonia [AD183], it is found that about half of the oscillator strength appears in the ionization spectrum, and half in the bound spectrum.

The spectrum of pyrrolidine has recently been recorded photoelectrically [AD164], and is shown in Fig. AD-7. Assignment of this spectrum must await the determination of the photoelectron spectrum. Several sharp lines are reported in the 57 000–60 000-cm^{-1} region of N-methyl pyrrolidine [AD133]. Inasmuch as the ionization potential of such a tertiary amine is probably about 65 000 cm^{-1}, it is very puzzling as to what these bands could be.

Our interpretation of the DABCO spectrum has been shown faulty in part by the recent high-resolution study and analysis of this spectrum by Hamada et al. [AD69]. The weak bands near 38 600 cm^{-1} in DABCO

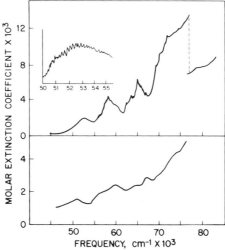

Fig. AD-7. Optical absorption spectra of tetrahydrofuran (upper) and pyrrolidine (lower) in the vapor phase [AD164].

have been shown by them to be hot bands rather than a separate transition as previously thought. The excitation to 3p has its origin at 39 782.8 cm^{-1} (advert.), but rests upon a continuum which may be the corresponding Rydberg excitation to 3s; this would give the excitations to 3s and 3p equal term values, which is acceptable for highly alkylated chromophores. In the $(a_1', 3p)$ upper state, there are four geometric minima, two from torsion of one C_3N fragment against the other about the threefold axis, and two from the out-of-plane umbrella motions of the two C_3N groups. The former has a barrier of ~ 30 cm^{-1} and the latter has a barrier of ~ 1600 cm^{-1}.

The humps in the rather ill-defined spectrum of hydrazine [I-S24] can be interpreted with the simple concept of term values computed from the optical and photoelectron spectra [AD159]. Ionization from the uppermost orbital of hydrazine (5a) requires 80 000 cm^{-1} (vert.) [AD147], and so the first optical band at 52 600 cm^{-1} (vert.) has a term value of 27 400 cm^{-1} and can reasonably be assigned as $5a \rightarrow 3s$. As such, it corresponds to the band at 46 140 cm^{-1} in ammonia and at 68 000 cm^{-1} in ethane. The 5a MO of hydrazine is but one component resulting from lone pair–lone pair splitting; the second component is the 4b MO, with an ionization potential of 85 800 cm^{-1} (vert.). A (4b, 3s) term value of 27 500 cm^{-1} would place the $4b \rightarrow 3s$ excitation frequency at 58 400 cm^{-1}, at which frequency a broad feature is found. However, this band also has a term value of 21 600 cm^{-1} with respect to the 5a MO, and so is probably a combination of $4b \rightarrow 3s$ and $5a \rightarrow 3p$ excitations. Similarly, the band at 66 300 cm^{-1} in the optical spectrum has term values appropriate for the 5a → 3d, 5a → 4s, and 4b → 3p excitations.

Little can be gleaned from the absorption spectrum of NF$_3$ in the 400-eV region [AD194]. With a nitrogen 1s ionization potential of 3 340 700 cm^{-1} (414.2 eV) [AD54], the first two absorption features at 2 228 700 (400.3 eV) and 3 278 700 cm^{-1} (vert., 406.5 eV) have term values which are too large to allow Rydberg assignments, and so one is driven to 1s$_N$ → σ* (N—F) valence shell assignments for these bands. Group theoretically, these would be $^1A_1$ → $^1A_1$ and $^1A_1$ → ^1E transitions in the C$_{3v}$ point group, and as expected from this, the band assigned as terminating at ^1E is observed to be considerably more intense than that to $^1A_1$. The final two bands at 3 316 600 (411.2 eV) and 3 331 100 cm^{-1} (vert., 413.0 eV) have term values of 24 200 and 9700 cm^{-1}, respectively; the first of these could be the 1s$_N$ → 3p Rydberg band, allowed by symmetry, but observed to be much weaker than the preceding valence shell excitations, as expected. The transition to 3s would be forbidden in the planar molecule, and so could be too weak to be seen.

AD.III.D-2. Compounds of Phosphorus, Arsenic, and Antimony

Ab initio calculations on the PH$_3$ and PF$_3$ molecules and their ground-state ions have been carried out with the aim of settling the outstanding problems of their geometries in excited states [AD1]. It was found that PH$_3^+$ is puckered in its ground state, but with a barrier so low that only a few vibrations are below it, in agreement with the conclusion of Maier and Turner [I-M2]. However, whereas the latter workers used the same arguments to show that PF$_3$ is exactly planar, the calculations instead predict a puckered structure with a very large barrier to inversion, so large that all of the Franck–Condon accessible levels are below the barrier. The small barrier in PH$_3$ and the larger barrier in PF$_3$ explain why the inversion frequency is halved in the former ion but not in the latter.

The ionization potential of methyl phosphine is 77 430 cm^{-1} (vert.) [AD36]; therefore the two bands at 49 700 and 53 500 cm^{-1} in the optical spectrum have term values of 27 730 and 23 930 cm^{-1}, respectively. Since the lowest *n*s term value of PH$_3$ (30 000 cm^{-1}) is very nearly equal to that of CH$_4$ (31 600 cm^{-1}), one expects the lowest *n*s term value of CH$_3$PH$_2$ to be much like that of CH$_3$CH$_3$ (29 500 cm^{-1}). On this basis, it is clear that it is the 49 700-cm^{-1} band which has the (n$_P$, 4s) upper state. Possibly, the 53 500-cm^{-1} band is the corresponding (n$_P$, σ*) valence shell conjugate excitation.

The photoelectron spectrum of PF$_2$Cl [AD37] places the chlorine lone-pair ionization potential at 92 600 cm^{-1} (vert.). This value yields a term value of 34 100 cm^{-1} for the strong feature at 58 500 cm^{-1} (vert., Fig. III.D-10), thus identifying it as the chlorine 3p → 4s Rydberg excitation. The two bands preceding it are lone pair → σ^* A bands involving chlorine and perhaps phosphorus lone pairs, while the band at 70 000 cm^{-1} (vert.) has the expected term value for the chlorine D band (3p → 4p).

Russell [AD162] has recorded the optical spectra of OPF$_3$, OPCl$_3$, and the corresponding mixed chlorofluorides. In OPF$_3$, there is observed only a single, broad band ($\epsilon \sim 2000$) at 71 000 cm^{-1}, with a term value of 38 000 cm^{-1} with respect to the oxygen lone-pair ionization potential at 109 000 cm^{-1} [I-B13], while in OPCl$_3$, the corresponding band comes at 62 500 cm^{-1} (vert.) and the ionization potential (oxygen and chlorine lone pair combined [I-H48]) comes at 97 700 cm^{-1} [I-B13]. The term values are far too large for Rydberg excitations, and so these must be lone pair → σ^* excitations, as in trimethylamine-N-oxide (Section III.E-3).

The reflection spectra of KH$_2$PO$_4$ and NH$_4$H$_2$PO$_4$ single crystals show identical patterns of peaks at 78 500, 88 400, and 107 000 cm^{-1} (vert.) [AD7]. These bands are apparently characteristic of the H$_2$PO$_4^-$ ion, and further, the NH$_4^+$ ion is seen to be quite transparent at high frequencies (see also the footnote on p. II-278). These bands are most likely the N → V$_n$ excitations among the π(P—O) and π^*(P—O) orbitals.

AD.III.E-1. Water

Experimental data on water continue to accumulate. Reflectivity data on the liquid are described in [AD101, AD149, AD160], while Trajmar et al. [AD180] give a detailed description of their electron-impact studies on the vapor, and high-resolution optical studies of H$_2$O and D$_2$O in the Hopfield region are also reported. [AD98]

Activity centering around the 36 000-cm^{-1} triplet band of water is reaching fever pitch without a satisfactory explanation in sight. Since the energies of the dissociation fragments OH ($^2\Pi$) + H (^{2}S) and O (^{3}P) + H$_2$ ($^1\Sigma_g^+$) are at about 40 000 cm^{-1}, Hosteny et al. [AD81] point out that the excitation must be to a bound state, yet their MC-SCF calculation on the (1b_1, 3s) ^{3}B$_1$ state of water does not give any bound nuclear configurations. Yeager et al. [AD190] similarly were unable to get any triplet Rydberg state of water below 55 500 cm^{-1} in a calculation which otherwise did a very nice job on the Rydberg spectrum. Their calculation gave a (1b_1, 3s) singlet–triplet split of 2400 cm^{-1}, thereby assigning the triplet observed at 58 100 cm^{-1} as originating from this configuration. However, the calculation by Yeager et al. deals only in Rydberg excitations, and as stated in the text, it is a possibility that the low triplet in water results from a large singlet–triplet split between valence shell configurations. Indeed, Truhlar has performed a variety of calculations in valence shell basis sets and finds the valence shell $(1b_1, \sigma^*)$ singlet–triplet split to be about 18 000 cm^{-1} [AD181]. The reported presence of this band in liquid water at 36 000 cm^{-1} [AD114] may also be used to argue that it is *not* a Rydberg triplet.

Recently, Sanche and Schulz [AD163] have succeeded in observing negative-ion states of water having the $(1b_1)^1(nR)^2$ configuration, where nR is a higher Rydberg orbital of the molecule. As shown in Fig. AD-8,

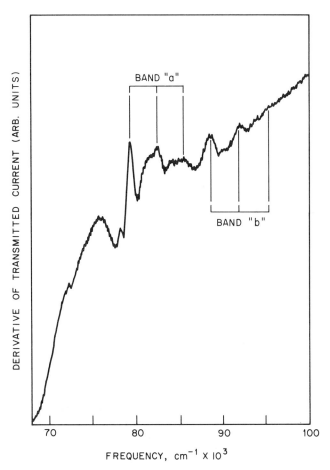

Fig. AD-8. Electron transmission spectrum of water vapor showing the temporary negative-ion bands [AD163].

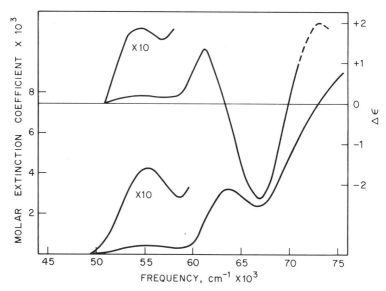

Fig. AD-9. The circular dichroism (upper) and optical absorption spectra (lower) of (+)-sec-2-butanol in the vapor phase [AD171].

the two bands show a 3200-cm^{-1} vibrational progression (ν_1') just as in the $^2B_1$ positive ion (Table III.E-I). Thus as regards force constants in the core, the two Rydberg electrons in the negative ion are as good as ionized. Bands **a** and **b** in water have their origins at 79 500 and 89 100 cm^{-1}. As anticipated from the sharp line shapes, these transitions are above the $(1b_1, 3s)$ level, and so do not involve the $(1b_1)^1 3s^2$ configuration. This configuration has its threshold at 44 500 cm^{-1}, but dissociates into H$^-$ (1s)^2 and OH ($^2\Pi$). It is said by Sanche and Schulz that the **a** band is converging upon the $(1b_1, 3pb_1)$ neutral-molecule level at 82 038 cm^{-1} and may have the $(1b_1)^1(3pb_1)^2$ configuration. Band **b** may be the corresponding $(1b_1)^1(4pb_1)^2$ excitation, but we cannot be too certain about this. Hydrogen sulfide also shows an **a** band in its electron transmission spectrum.†

AD.III.E-2. Alcohols

Circular dichroism spectral data on the basic chromophores of the vacuum ultraviolet are always welcome, especially when unencumbered by absorbing substituents. Thus the work of Snyder and Johnson [AD171] on sec-2-butanol in the gas phase (Fig. AD-9) is especially

† See Addendum for Section IV.A-1 for a further discussion of these negative-ion states.

important. Note that the pattern of absorption intensities and frequencies in this alcohol fits with those of other alcohols already analyzed in terms of Rydberg excitations. The advantages of circular dichroism spectroscopy is that bands have signs, so, for example, one sees that the single absorption band at 64 000 cm^{-1} in sec-2-butanol in fact consists of two oppositely rotating bands, centered at 62 000 and 64 500 cm^{-1} (vert.). These are preceded by a weak band at 55 300 cm^{-1} (vert.). Snyder and Johnson report rotatory strengths (e^2 Å^2 × 10^6) for these three bands of 1.7, 5.2, and —4.4, respectively. The signs of the rotations are reproduced by independent-systems calculations in which the bands are assigned as $n_O \rightarrow \sigma^*$(O—H), $n_O \rightarrow \sigma^*$(C—O), and $n_O \rightarrow$ 3s, and it is clearly pointed out that under all assumptions (within the model), the $n_O \rightarrow$ 3s transition will have a negative rotatory strength, whereas the first two bands of sec-2-butanol are observed to be positive.

Following our analyses of the other alcohols, we have determined the first ionization potential of sec-2-butanol (83 000 cm^{-1} vert.), and so find term values of 27 700, 21 000, and 18 500 cm^{-1} for the first three bands, all of which suggests that the band at 55 300 cm^{-1} is $n_O \rightarrow$ 3s and the other two bands at 62 000 and 64 500 cm^{-1} are $n_O \rightarrow$ 3p components, corresponding to the two $n_O \rightarrow$ 3p bands of methanol (Fig. III.E-4). This conflict concerning the rotatory sign of $n_O \rightarrow$ 3s means either that the entire concept of Rydberg assignments as presented in these volumes is in error, or that independent-systems calculations for Rydberg excitations are woefully inadequate.

The close similarity which is apparent in the optical spectra of CH$_3$OH and CH$_3$SH is studied further by Ogata et al. [AD144], who find that the lone pair in CH$_3$SH is somewhat more localized than is that in CH$_3$OH (as reflected in the widths of the lone-pair bands in the photoelectron spectra), while that for CH$_3$NH$_2$ is much more delocalized than in either of the others.

Analysis of the reflectivity of liquid glycerol up to 180 000 cm^{-1} yields two peaks at 68 500 and 104 000 cm^{-1} in the plot of ϵ_2, whereas in the energy-loss function, a third peak at 164 000 cm^{-1} (vert.) is also observed [AD74]. Up to 180 000 cm^{-1} the f-number integrates to 17. The authors conclude that the first two excitations are one-electron promotions, but that the third is a volume plasmon (Section I.A-3).

AD.III.E-3. Ethers

Gray et al. [AD66] have made a detailed vibronic analysis of the $n_O \rightarrow$ 3s Rydberg band of tetrahydropyran, which places the origin at

51 908 cm⁻¹ (advert.). Hot-band intervals corresponding to the ground-state excitation of 13 different fundamentals were uncovered, as well as the excitation of ten different fundamentals in the excited state. Since all of the vibrational progressions are short and highly vertical, there is no significant geometry change along any one coordinate in the upper state. In accord with the $n_O \to 3s$ assignment proposed by its term value, the rotational envelopes of these bands of tetrahydropyran are C type, giving a polarization perpendicular to the

plane.

The $n_O \to 3s$ band of tetrahydrofuran has also been analyzed by the Dundee group [AD39], with a surprising result. In the 48 700–54 000-cm⁻¹ region, they find an origin at 50 188 cm⁻¹, about which are clustered many sharp bands involving multiple excitation and deexcitation of the pseudorotation mode with frequencies ranging from 60 to 260 cm⁻¹. They conclude that the (n_O, 3s) upper state is planar, as is the corresponding state in trimethylene sulfide (Section III.F). Beginning at 51 400 cm⁻¹ the bands broaden noticeably (insert, Fig. AD-7) and a second electronic transition commences, with vibrational frequencies above 200 cm⁻¹. Because the presence of a second electronic state at this frequency with sharp structure is quite unexpected, there is good reason to investigate this system further.

A Rydberg series analysis of the tetrahydrofuran spectrum (Fig. AD-7) goes very smoothly, uncovering an ns series ($\delta = 0.94$), two np series ($\delta = 0.64$ and 0.52), and an nd series ($\delta = 0.08$) [AD44]. The assignments of the $n = 3$ members of these series are in agreement with the listing in Table III.E-II. Interestingly, the preliminary vibrational analysis given the 49 000–55 000-cm⁻¹ band by these workers does not involve a second electronic origin.

Kobayashi and Nagakura have discussed the mechanisms for lone pair–lone pair splitting in the dioxanes using CNDO calculations and photoelectron spectroscopy. They conclude that in both 1,3- and 1,4-dioxane, the uppermost filled MOs in the chair conformers have the equatorial orientation [AD105].

An excellent example of the application of the term-value concept to the interpretation of the higher states of polyatomic molecules is given by Tam and Brion [AD176]. In their study of the electron-impact energy-loss spectra of several alcohols and ethers, they successfully assigned dozens of bands as terminating at 3s, 3p, or 3d, using the ionization potentials derived from photoelectron spectroscopy and the known behavior of the term values upon alkylation. Their analyses of these spectra, though more complete, agree closely with that proposed in Section III.E. The electron-impact spectra are inherently of lower resolution than the optical spectra of Figs. III.E-5, III.E-6, and III.E-10, but extend to higher frequencies.

AD.III.F. Compounds of Sulfur, Selenium, and Tellurium

Parallel studies of the Rydberg [AD20] and photoelectron spectra [AD61] of ethylene sulfide supersede the older work. The very sharp break in the vibronic structure at 73 000 cm^{-1} (Fig. III.F-5) is confirmed as the first ionization potential. In the optical study, three Rydberg series were assembled: (i) an ns series having $\delta = 1.72$ with its first member ($n = 4$) at 52 010 cm^{-1} (advert.). This band was assigned as the first Rydberg band in the spectrum because it "lined up" with others in the larger cyclic sulfides when their ionization potentials were aligned. However, our view is that the ns members will not line up, only the np and nd members will, and that the $\delta = 1.72$ series in fact is $2b_1 \rightarrow n$s, with the transition to 4s coming at 47 000 cm^{-1} (advert.; Table III.F-I and Fig. III.F-6). When assigned in this way, the $(2b_1, 4s)$, $(2b_1, 4p)$, and $(2b_1, 3d)$ term values for ethylene sulfide are 26 000, 21 000, and 15 000 cm^{-1}, which agree with the trends shown by other sulfides (Fig. III.F-6) and compare well with those of ethylene oxide, 26 800, 21 600, and 13 000 cm^{-1} (Table III.E-II). (ii) A second series has $\delta = 1.34$ and places the $n = 4$ member at 57 490 cm^{-1} (advert.); we prefer to assign it to a 3d upper orbital, again on the basis of term values. (iii) The third series has $\delta = 0.05$ and starts with $n = 4$. It is probably nd, with the transition to 3d coming at $\sim$60 000 cm^{-1}. Vibrational intervals of 1050 cm^{-1} appear throughout the optical spectrum, and correspond to the methylene bending mode (1090 cm^{-1}) found in the first photoelectron band.

The optical spectra of the interesting series of molecules $(CH_3)_2S$, $(CH_3)_2Se$, and $(CH_3)_2Te$ (Fig. AD-10) parallel those of the corresponding hydrides. Scott et al. [AD167] find several Rydberg series in each, with lowest ionization potentials of 70 228, 67 753, and 63 933 cm^{-1} (advert.), respectively, in agreement with the values reported in the photoelectron spectra [AD35]. In $(CH_3)_2S$, the $b_1 \rightarrow 4s$ assignment is given to the band at 43 879 cm^{-1} (advert.) rather than to the band at 49 000 cm^{-1}, just as in the text. This latter band is probably a valence shell excitation $(b_1 \rightarrow \sigma^*)$ because its position is irregular with respect to the other Rydberg excitations. The lowest $(b_1, n$s$)$ term values in the series of dimethyl chalcogenides are 26 350, 25 250, and 23 930 cm^{-1}, respec-

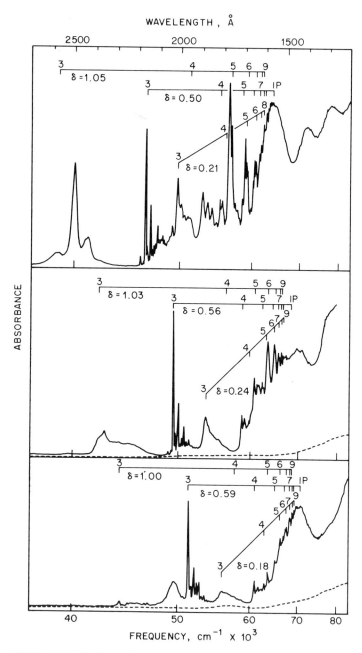

Fig. AD-10. Optical absorption spectra of dimethyl telluride (upper), selenide (middle), and sulfide (lower) in the vapor phase. The dashed lines show the spectrometer baseline [AD167].

tively, these being somewhat smaller than the value of $\sim$27 000 cm^{-1} for the corresponding hydrides, due to penetration into the alkyl groups. As expected from the situation in the group V and group VI hydrides (p. I-27), as the lowest n increases by unity in the dimethyl compounds at each step down the column, there is a corresponding increase of δ such that $n - \delta$ remains rather constant and so then do the term values. Excitations to the lowest np levels appear as sharp bands in the region near 50 000 cm^{-1}, with term values of 19 060, 18 510, and 17 700 cm^{-1} in the sulfur, selenium, and tellurium compounds. As seen from the figure, the excitations to nd resemble those to ns, i.e., excitation to the lowest member of the series is quite broad, while higher members are narrow. This strongly suggests that valence shell conjugate configurations are mixed into the lowest ns and nd members, but not into higher ones or into np series members.

Milazzo [AD133] describes the excitations to 4s, 4p, and 3d in tetrahydrothiophene taken under high resolution. Though the spectrum has been accurately measured, none of the vibrational intervals have been assigned. The vertical excitation values as measured by him are 43 952, 47 761, and 55 891 cm^{-1} for excitations to the first three Rydberg states. In tetrahydroselenophene, a similar absorption pattern is observed, with the transitions to 5s and 5p coming at 41 835 and 44 561 cm^{-1} (vert.) and two origins for transitions to 4d coming at 46 225 and 49 690 cm^{-1} (advert.).

The X-ray absorption spectra of SF$_6$ gas [AD193] originating with the sulfur 1s and 2p and fluorine 1s levels have been interpreted by Gianturco et al. [AD65] in terms of valence shell transitions which terminate in the manifold of π^*(S—F) and σ^*(S—F) virtual levels, a_{1g}, t_{1u}, t_{2g}, t_{1u}, and e_g, in that order. However, certain of these bands are below the corresponding ionization potentials, and may be Rydberg excitations instead.

AD.III.G. Silanes, Germanes, and Stannanes

A single-center calculation of the core excitation spectrum of silane by Deutsch and Kunz [AD40] is said by them to confirm the original assignments [I-H24, I-H27] of the absorption lines at 830 370 and 835 210 cm^{-1} (102.95 and 103.55 eV, respectively) as 3p → 4p. However, the corresponding term value is 33 400 cm^{-1}, which is far too large for an excitation terminating at 4p. This is also pointed out by Bagus et al. [AD5], who compared the silane spectrum with that of the silicon atom.

AD.IV.A-1. Ethylene

Theoretical work on the excited states of ethylene continues unabated, the prime question still being the spatial extent of the π^* orbital in the (π, π^*) singlet state. In Basch's latest work on the subject [AD13], he uses the MC SCF technique in a very flexible basis containing both diffuse orbitals and more compact polarization functions in addition to the usual double-zeta valence shell basis. He finds that as (σ, σ^*) mixing progresses, the (π, π^*) state contracts slowly, and suggests that a "complete" calculation would make the singlet (π, π^*) configuration valence shell. Thus, he tends to support the conclusion of Ryan and Whitten [II-R29] in this regard, as opposed to that of Bender et al. [II-B23]. In a general discussion of $\pi \rightarrow \pi^*$ V states in various small molecules, Mulliken [AD137] proposes that these formally ionic states will always incorporate large amounts of (σ, σ^*) configurations, thereby shrinking the π^* orbital considerably.

Ab initio calculations on ethylene in far smaller basis sets than that used by Basch are also of interest. Working in a minimal basis set, Tanaka also finds (σ, σ^*) configurations are very important when mixed with the singlet (π, π^*) configuration, making it valence shell in size and of the proper energy [AD177]. A much more extensive calculation by Fischer-Hjalmars and Kowalewski [AD56], again using a small basis and Rydberg AOs, illustrates several principles of large-molecule Rydberg spectra. Their calculations illustrate first the general decrease of both singlet–triplet splits and the oscillator strength with increasing n. Calling the in-plane and out-of-plane pi MOs π^* and π'^*, it was found that the singlet–triplet split for $(\pi, np\pi^*)$ configurations is larger than that for $(\pi, np\pi'^*)$, and similarly the splits for $(\pi, nd\pi^*)$ are larger than those for $(\pi, nd\pi'^*)$, as can be easily understood by considering the overlap density resulting in electron exchange. Note, however, that the quantum defects and singlet–triplet splits do not necessarily parallel one another, contrary to the suggestion in Section I.A-1. Thus, Fischer-Hjalmars and Kowalewski find quantum defects in the np manifold of 0.7 for $3p\pi'^*$, 0.6 for $3p\sigma$, and 0.5 for $3p\pi^*$, which does not follow the pattern of singlet–triplet splits. The quantum defects in the np series reflect the relative amounts of penetration possible into the core with various orientations of the Rydberg orbital, whereas the singlet–triplet splits are governed by the overlap of the Rydberg orbital with the half-filled MO in the core.

Brongersma [AD25] brings up an interesting point regarding the supposed triplet state at 53 200 cm^{-1} in ethylene uncovered by Nicolai [II-N13] by Li^+ ion bombardment. If this were a triplet excitation strongly intensified by electron exchange with the projectile, then spin conservation would result in the Li⁺ ion being excited to the (1s, 2s) state in the collision. However, since this excitation from $1s^2$ requires 60 eV, this clearly cannot be the mechanism, and the spin multiplicity of the band is again called into question.

Absorption cross-section and ionization cross-section data for ethylene are discussed critically in [I-H67].

Gedanken et al. [AD63] have presented the results of a comprehensive study of the ethylene spectrum in rare gas matrices. In Ne, Ar, Kr, and Xe matrices, they locate the $n = 1$ intermediate exciton together with a few quanta of the ν_2' vibrational excitation, and in solid Xe, the torsional motion $2\nu_4'$ is also observed. Transitions to the higher states having $n = 2$ and $n = 3$ are reported for ethylene in Ar and Kr matrices, again with accompanying vibrational quanta of ν_2'. In these matrices, the gas-phase ionization potential of ethylene (84 930 cm⁻¹) is reduced to 84 100 (Ne), 73 500 (Ar), 71 500 (Kr), and 65 500 cm⁻¹ (Xe).

Using the trapped electron technique, or what is equivalently called electron transmission spectroscopy, a very interesting state of the ethylene negative ion has been uncovered [AD38, AD163]. Beginning at 53 200 cm⁻¹ in ethylene gas, four quanta of ν_2', the C—C stretch (1330 cm⁻¹), are observed in a transition to a negative-ion state having a $^2B_{2u}$ core and two electrons in the $3sa_{1g}$ Rydberg orbital. This temporary negative-ion state (10^{-10}–10^{-15} sec) has a term value of 4100 cm⁻¹ with respect to ionization producing the neutral molecule in the triplet (π, 3s) state, and Sanche and Schulz [AD163] point out that this term value is universal for first-row negative ions (atoms, diatomic, and polyatomic molecules) in the $3s^2$ configuration. This is not at all surprising in view of our discussion of term values (Section I.C), which also suggest that higher values ($\sim$5000 cm⁻¹) could be found in highly fluorinated systems.

These negative-ion resonance frequencies can be estimated easily by placing the neutral molecule (ϕ_i, 3s) Rydberg state appropriately below the observed ϕ_i ionization potential, using the term value rules of Section I.C, and then placing the negative-ion $3s^2$ state another 4100 cm⁻¹ below that. Thus the $(1b_{2g}\sigma)^1 3s^2$ negative-ion state of ethylene should come at 103 200 − 28 000 − 4100 = 71 100 cm⁻¹. Dance and Walker [AD38] have observed another negative-ion resonance in ethylene at 62 000 cm⁻¹ (vert.) which is a poor match for the $(1b_{2g})^1(3s)^2$ state, and more likely is converging to $(1b_{2u}\pi, 3p)$, the negative ion being $(1b_{2u})^1 3s^1 3p^1$. The nature of these doubly occupied Rydberg orbitals in negative ions has been studied theoretically for NO⁻ [AD118], where Lefebvre-Brion found the negative-ion Rydberg orbitals to be much more diffuse than are those for the neutral molecule.

AD.IV.A-2. Alkyl Ethylenes

According to Sanche and Schulz [AD163], the negative-ion states $\pi^1(3s)^2$ will appear as resonances in the trapped-electron spectra of olefins at approximately 4000 cm^{-1} below the $\pi \to 3s$ neutral-molecule excitation. Using the data of Table IV.A-III, one predicts the resonances in propylene, butene-1, cis-butene-2, and trans-butene-2 to come at 49 000, 49 300, 44 300, and 45 400 cm^{-1} (vert.), respectively. Unfortunately, these resonances could not be found by Dance and Walker [AD38], though they did find the corresponding band in ethylene. However, comparison of the optical $\pi \to 3s$ threshold frequency in these alkyl ethylenes with the apparent threshold frequencies observed by electron impact led them to postulate the presence of an optically forbidden transition in propylene and butene-1 at about 50 000 cm^{-1}. Possibly, this is the transition to the $(\pi, 3s)$ triplet state.

Iverson et al. [AD88] compare the spectra of propylene and butene-1 with that of methyl vinyl silane, finding the (π, π^*) band in the latter to be lower by less than 1000 cm^{-1}. Each of these olefins also shows a valence shell band at $\sim$71 000 cm^{-1} which Iverson et al. attribute to absorption within the methyl groups in these compounds; however, it may correlate instead with the band at $\sim$78 000 cm^{-1} in ethylene itself (Fig. IV.A-1).

Basch presents more complete calculations on the spectrum of cyclopropene which confirm the earlier assignments [AD12]. Once again, a low-lying (σ, π^*) valence shell excitation is found to lie about 6000 cm^{-1} below the (π, π^*) singlet state, as observed. Interestingly, the π^* MO in the (π, π^*) singlet state is a diffuse orbital as it is in parallel calculations on ethylene, but the π^* MO in the (σ, π^*) configuration is totally valence shell. This result emphasizes the ionicity inherent in the (π, π^*) state, and the need for (σ, σ^*) configurations to counter this effect and pull in the π^* orbital. Basch also mentions that the $(\pi, 3s)$ and lowest (σ, π^*) configurations have the same symmetry, and are probably mixed, being in the same frequency range.

Credit must be given to Carr and Stücklen [AD28], who in 1939 identified the lowest Rydberg bands of 17 alkyl olefins as related to that at 57 340 cm^{-1} in ethylene.

AD.IV.A-3. Haloethylenes

The photoionization mass spectrum of vinyl chloride has been determined by monitoring the parent ion [AD158]. In the region between the ionization threshold (80 000 cm^{-1}) and 96 000 cm^{-1}, several peaks were observed corresponding to the excitation to autoionizing Rydberg states, and the peaks were placed in three series converging to 95 100 cm^{-1}, the ionization potential corresponding to the loss of a chlorine in-plane 3p electron. Inasmuch as the lowest members of the ns ($\delta = 0.82$), np ($\delta = 0.48$), and nd ($\delta = 0.17$) series are below the first ionization potential of vinyl chloride, they are not observed in the photoionization mass spectrum. Our suggestion in the text that the 86 500 cm^{-1} band in the optical spectrum of Sood and Watanabe converges upon the ionization potential at 109 400 cm^{-1} is probably wrong, because Reinke et al. [AD158] have assigned it as the $n = 4$ member of a five-member nd series going to the second ionization potential at 95 100 cm^{-1}.

AD.IV.B. Azo and Imine Compounds

Paraffin solution spectra of the related trans-azo compounds $(CH_3)_3C-N=N-C(CH_3)_3$, $(CH_3)_3C-N=N-Si(CH_3)_3$, and $(CH_3)_3Si-N=N-Si(CH_3)_3$ show the expected $n_+ \to \pi^*$ band in the visible region and another peak at 50 000–52 000 cm^{-1} (vert.) having a molar extinction coefficient of 1000–2000 in each [AD187]. This description of the second band closely resembles that of band I of the azoalkanes in the gas phase (Fig. IV.B-1), which we had earlier assigned as a Rydberg excitation to 3p. Since a Rydberg transition is not expected to survive in paraffin solution (Section II.C), we must postulate the existence of both Rydberg and valence shell excitations in the band I region of azoalkanes. The valence shell band at $\sim$50 000 cm^{-1} in these compounds is most likely the weakly allowed $n_- \to \pi^*$ excitation, as predicted by the semiempirical calculations of Haselbach and Schmelzer [AD71].

In difluorodiazirine, the Rydberg transitions from the n_- MO to 3s and 3p Rydberg orbitals should come about 26 000 and 20 000 cm^{-1}, respectively, below the n_- ionization potential at 95 700 cm^{-1} (vert.). These criteria are ably met by the strong bands observed at 70 000 and 75 700 cm^{-1} (vert.) [AD159].

An ab initio calculation of the spectrum of formaldazine ($H_2C=NH$) places the N $\to$ V excitation at 83 500 cm^{-1} (vert.) [AD124], whereas in alkylated imines, the observed frequency is 58 000 cm^{-1} (Fig. IV.B-3).

AD.IV.C. Aldehydes and Ketones

Using the equations-of-motion method, Yeager and McKoy [II-Y7] have investigated the formaldehyde spectrum with most interesting results. First of all, the calculation places the singlet $\pi \to \pi^*$ excitation at 80 000 cm^{-1} with an oscillator strength of only 0.1. The unexpectedly low oscillator strength is a result of the mixing of the (π, π^*) valence shell configuration with the $(2b_2, nb_2)$ and $(1b_1, nb_1)$ Rydberg configurations, as predicted by Mentall et al. [II-M22]. This admixture also expands the (π, π^*) wave function, but not to anything resembling a real Rydberg orbital. The splitting of the (n$_O$, 3p) manifold places the a_1 component lowest and the b_1 component highest.

Maria et al. [AD126] agree with us in placing the $\pi \to \pi^*$ excitation of phosgene at 65 000 cm^{-1} (vert.), but they then place the n$_O$ → 3s transition at 59 500 cm^{-1}, whereas we prefer 66 800 cm^{-1} on the basis of its term value (Table IV.C-I) and the similarity of the optical and photoelectron band envelopes.

The electron-impact spectrum of cyclopropenone [AD70] shows many resemblances to that of acetone (Fig. IV.C-5). In the cyclopropenone photoelectron spectrum, the three sharpest features are ionization from the oxygen lone pair ($4b_2$, 76 600 cm^{-1} vert.), from the olefinic pi bond ($2b_1$, 89 500 cm^{-1} vert.), and from the second lone pair on the oxygen atom ($7a_1$, 129 000 cm^{-1} advert.). These levels figure prominently in the electron-impact spectrum, where an ns series and the $n = 3$ member of the np series going to the $4b_2$ ionization potential are observed, along with the $n = 3$ and 4 members of the ($2b_1$, ns) series and ns and np series going to the $7a_1$ ionization potential. The vibronic envelopes in the electron energy-loss and photoelectron spectra of cyclopropenone clearly show that the oxygen lone pair ($4b_2$) is far more delocalized than in acetone. The (ϕ_i, 3s) term values of cyclopropenone, 26 000, 26 300, and 25 300 cm^{-1}, compare well with the values 27 150, 27 500, and 26 200 cm^{-1} deduced for acetone. In acetone, the (n$_O$, 3p) term value is 18 310 cm^{-1}, whereas in cyclopropenone, values of 20 700 and 21 300 cm^{-1} are observed.

According to Duncan et al. [AD46], the spectrum of methyl ethyl ketone is continuous in the region beginning at the extreme frequency of Fig. IV.C-7 and stretching to 128 000 cm^{-1}. Their analysis of the structured part of the spectrum uncovered the excitation of C—O stretching and C—C—C bending motions.

CNDO calculations on carbonyl cyanide [AD185] seem to give a reasonable interpretation to the spectrum: The band at 52 700 cm^{-1} is assigned as n$_O \to \pi^*$, where π^* is here an in-plane π orbital of the CN group, and several other n$_O \to \pi^*$ transitions contribute to the absorption in the 58 000–62 000-cm^{-1} region. Walsh [II-W3] also describes the spectrum of furfuraldehyde as showing several Rydberg excitations originating with the π orbitals of the furan ring, but none originating with the lone pair on oxygen. The attribution of Fig. IV.C-11 is incomplete; the correct reference is [AD145].

AD.IV.D. Acetylenes

Jungen has reanalyzed the gas-phase acetylene spectrum [AD97] and changed slightly the assignments of Price [II-P37]. The puzzling nR' series is assigned by him as terminating at $nd\pi$ with the $n = 3$ member at 74 747 cm^{-1} (advert.), rather than at 74 498 cm^{-1} as in the earlier assignment. This latter band is then assigned as the onset of the forbidden $n = 3$ member of the $\pi_u \to np\sigma$ series, made allowed by a single quantum of ν_5', the cis bending mode. This, of course, places the 3p transition in the expected spectral region, but still leaves the transition to 3d with an unusually large term value.

In our previous crystal work on acetylene (Fig. IV.D-5), it was found that the vibrational structure at frequencies below about 54 000 cm^{-1} (allowing for a matrix shift) was washed out in the crystal spectra, in contrast to those above this frequency, implying the presence of a second state with origin at about 54 000 cm^{-1}. This has been confirmed by Foo and Innes [AD58], who compared the ^{12}C$_2$H$_2$ and ^{13}C$_2$H$_2$ gas-phase spectra and found a line at 54 116 cm^{-1} in each, whereas all other pairs of lines were shifted by 10–30 cm^{-1}. It was also found that 54 116 cm^{-1} is the origin for a long progression in 720 cm^{-1}, the trans bending mode, ν_3'. Thus a new band system originates at 54 116 cm^{-1}; according to Foo and Innes, the bands in the region from 51 900 to 55 000 cm^{-1} can be placed as the higher vibronic members of the $\tilde{X} \to \tilde{A}$ valence shell excitation, but our experiment suggests considerable Rydberg character.

In a more comprehensive study of the matrix spectra of acetylene than previously reported, Gedanken, et al. [AD63] report on acetylene in Ne, Ar, Kr, and CF$_4$ matrices. The $n = 1$ intermediate exciton has an origin which varies from 70 030 cm^{-1} in Kr to 74 520 cm^{-1} in CF$_4$, and has two quanta of ν_2' appended. In Ar, Kr, and CF$_4$, the ν_2' vibration has the normal value of $\sim$1900 cm^{-1} (1849 cm^{-1} in the gas phase), whereas in the Ne matrix this vibration is depressed to 1400 cm^{-1}. In Kr, the $n = 1$ origin has a half-width of only 300 cm^{-1} due to unusually weak electron–phonon coupling. Following the $n = 1$ line, there is observed a cluster of bands which seem related to the B and C bands of Fig. IV.D-1, but which show a matrix shift of only 400 cm^{-1} in Kr, whereas 4000 cm^{-1} is more likely to be expected for Rydberg excitations. These are tentatively considered by Gedanken et al. as valence shell excitations with Rydberg admixture, as first claimed by Wilkinson [II-W28]. Transitions to

$n = 2$ are observed in the Ne matrix with more normal ν_2' intervals of 1710 and 1540 cm^{-1}.

Hudson gives a detailed and critical analysis of the absorption and ionization cross sections in acetylene gas [I-H67], and the intensities of the first two bands in the solution spectra of octyne-1 and octyne-2 have been measured by Platt et al. [II-P28].

The first two bands of the polyacetylenes H(C≡C)$_n$H have been measured in the gas phase [AD103] and resemble those of the bis-trimethylsilyl derivatives earlier described by Boch and Seidl [II-B38] (Section III.G). In the parent compounds, the $^1\Sigma_g^+ \rightarrow {}^1\Delta_u$ bands are found at 40 500, 33 300, and 29 000 cm^{-1} (adiab.) in the compounds having $n = 2$, 3, and 4, while the $^1\Sigma_g^+ \rightarrow {}^1\Sigma_u^+$ bands in the same compounds are found at 60 700, 54 600, and 48 300 cm^{-1} (adiab.), respectively. The lower-frequency bands are weak, but the N → V bands following these are extremely intense.

The absorption of diacetylene has been reinvestigated by Smith, who has given a much more complete analysis [AD170]. The new approach stems in large part from the comparison of the C$_4$H$_2$ and C$_4$D$_2$ spectra, where the electronic origins are shifted to the violet on deuteration, while excited ν_2' vibrations are shifted to the red. Using these criteria, Smith was able to assemble two Rydberg series going to the first ionization potential at 82 110 cm^{-1} (advert.), having $\delta = 0.50$ and 0.00. Each series consisted of the $n = 3$–7 members. Three members of a third series having $\delta = 0.82$ were also observed converging upon the second ionization potential at 102 000 cm^{-1} (advert.). Now the difficulty arises in assigning these series. Rydberg transitions converging upon the first ionization potential of diacetylene originate at the $1\pi_g$ MO, and so are allowed to np and nf upper orbitals only. Since the series start at $n = 3$, whereas nf series necessarily start at $n = 4$, Smith assigned them to npσ ($\delta = 0.50$) and npπ ($\delta = 0.00$) upper orbitals instead, while noting how unusual it is to have an np series with $\delta = 0.00$. Three other explanations immediately suggest themselves, each no more unbelievable than the claim that an np series can have $\delta = 0.00$: (i) The red-shift test proposed by Smith for finding vibronic bands really only works for the ν_2' vibrational components, so that one quantum of a nontotally symmetric vibration could give a false origin. According to this argument, the series having $\delta = 0.00$ is an nd series, made allowed vibronically. (ii) The $\delta = 0.00$ series is really an allowed nf series, and the supposed $n = 3$ member, which is off the Rydberg formula by 765 cm^{-1} while higher members are off the formula by only ±20 cm^{-1}, is not part of the series. This would allow the nf series to start at $n = 4$, as it should. (iii) The $\pi_g \rightarrow$ 3d forbidden transition is made allowed by rotational–electronic coupling

[II-H20]. This seems least likely of the three, for reasons of precedence. Inasmuch as the identical problem of Rydberg series identification also are found in benzene and acetylene, the solution to this problem could have general repercussions for our understanding of Rydberg spectra. The Rydberg series having $\delta = 0.82$ has ns upper orbitals, and, as expected, $1\pi_u \to n$s is electronically allowed.

A number of other bands in the diacetylene spectrum remain to be identified. For example, a weak band at 58 540 cm^{-1} (adiab.) is either a Rydberg excitation or a $\pi \to \pi^*$ valence shell excitation, according to Smith; this could be settled using the high-pressure effect (Section II.B). This band has a term value of 23 600 cm^{-1} with respect of the first ionization potential,† which is close to the $(1\pi_u, 3s)$ term value of 24 000 cm^{-1} observed for excitations to the second ionization potential.

The vacuum-ultraviolet spectrum of a closely related molecule, cyanoacetylene, has been described by Narayan [AD140] and Okabe and Dibeler [AD146]. The latter workers additionally determined the luminescence efficiency resulting from the formation of excited CN radicals upon ultraviolet illumination, and showed that it followed the absorption curve very closely. In cyanoacetylene, a broad feature centered at 69 000 cm^{-1} has a term value of 24 500 cm^{-1} with respect to the first ionization potential at 93 500 cm^{-1} (advert.) [AD6], and is probably the $2\pi e_1 \to 3s$ Rydberg band. Its term value matches closely that for $(\pi, 3s)$ configurations in diacetylene. As Narayan showed, the next band in cyanoacetylene is the first member of a five-member series going to the first ionization potential, and furthermore, it has a Franck–Condon envelope identical to that of the first photoelectron band. However, he assumed the first member of the series to have $n = 4$, in which case $\delta = 1.4$ for the series, as would be appropriate only for an ns series in a highly fluorinated molecule. On the other hand, the term value of the first member of this series, 16 100 cm^{-1} (advert.), is appropriate for 3p, and we feel that the series in question should be assigned as $2e_1 \to n$p, with the ns series starting instead at 69 000 cm^{-1}.

The photoionization yield of the HC$_2$CN$^+$ ion has also been measured [AD146], and between the ionization threshold and one quantum of the ν_2 vibration in the ion (2180 cm^{-1}), there are observed a great many autoionizing Rydberg states ($n = 8$–15) which were placed in two Rydberg series having $\delta = 0.95$ and 0.55. They are obviously ns and np series converging upon the vibrationally excited ion. As happens in many systems, it seems that the excitation to 3s in cyanoacetylene is broad due to mixing

---

† The term value is 25 600 cm^{-1} if one assumes that a 2000-cm^{-1} vibration is making it allowed, and that 58 540 cm^{-1} is really a false origin.

with its valence shell conjugate, whereas transitions to higher ns members are sharp. Two valence shell transitions of cyanoacetylene in the quartz ultraviolet [$^1\Sigma^+ \to {}^1A''$ (C_s) and $^1\Sigma^+ \to {}^1\Delta$ or $^1\Sigma^-$] have been analyzed in detail [AD91, AD92].

Theoretical work by Scott et al. [AD168] attempts to generalize the ordering and splitting of levels in linear molecules with electronic configurations such as $\pi_1^m \pi_2^n$, $\sigma^1 \pi_1^m \pi_2^1$, etc. Thus, for $\pi_1^3 \pi_2^1$, the theory predicts the ordering $^1\Sigma^-$, $^1\Delta$, and $^1\Sigma^+$; however, *ab initio* calculations often place $^1\Sigma$ and $^1\Delta$ as very close together, and their order possibly could be reversed if π_2 were a Rydberg orbital, or if even a minimal amount of configuration interaction were operative.

AD.IV.E. Nitriles

Okabe and Dibeler [AD146] have determined that the efficiency of CN luminescence produced by irradiating acetonitrile in the vacuum ultraviolet follows the absorption spectrum closely, with the Rydberg excitation at 77 370 cm^{-1} found by Cutler [II-C30] readily apparent. The $2e \to 3p\sigma$ origin displays quanta of ν_1' (CH stretch), ν_2' (CN stretch), and ν_3' (CH$_3$ deformation), and is preceded by a broad, somewhat structured feature beginning at 68 700 cm^{-1} and peaking at $\sim$73 500 cm^{-1}. Its term value of 25 100 cm^{-1} suggests that this is the allowed $2e \to 3s$ excitation. A very similar band is found at 69 000 cm^{-1} (vert.) in cyanoacetylene, with a term value of 24 500 cm^{-1} [AD146].

The plateau and intense peak in the 45 000–60 000-cm^{-1} region of acrylonitrile appear in the n-heptane solution spectrum [II-K30] and so are valence shell excitations. Though Halper et al. [II-H4] assign the shoulder as $n_N \to \pi^*$, a CNDO-type calculation by Liebovici [AD119] instead suggests a weakly allowed $\pi \to \pi^*$ assignment.

Tam and Brion [AD175] report the electron-impact spectrum of HCN, in which a broad, low band (71 700 cm^{-1} vert.) precedes the sharp Rydberg parade. This band corresponds to that in the alkyl nitriles at $\sim$60 000 cm^{-1}, and as mentioned in the text, is no doubt a valance shell transition. The dependence of the intensity of this band of HCN upon impact voltage and scattering angle led Tam and Brion to assign it as electronically forbidden, probably $\pi \to \pi^*$ ($^1\Sigma^+ \to {}^1\Sigma^-$).

AD.V.A-1. Amides

A semiempirical calculation on trifluoroacetamide [AD111] confirms that the $N \to V_1$ and $N \to V_2$ (Q) transitions are separated by 16 000 cm^{-1} (see Fig. V.A-15), with the oscillator strength of the second band only 66% that of the first. Unfortunately, the properties of the important $n_O \to \sigma^*$ excitations were not calculated.

The circular dichroism spectrum of the optically active amide 3-methyl pyrrolidine-2-one in hydroxylic solvents shows oppositely rotating $n_O \to \pi_3^*$ and $\pi_2 \to \pi_3^*$ transitions, but in hexane as solvent, more structure is observed due to hydrogen-bonding association [AD67]. Hydrogen bonding of amides in the gas phase was studied by Kaya and Nagakura [AD100], who found that the $N \to V_1$ band of ring dimers came at 6000 cm^{-1} higher frequency than that of the monomer, whereas in chain dimers, the shift was 3000 cm^{-1} in the opposite direction.

The aqueous and methanol solution spectra of alkyl hydrazides formed by substituting an amino proton of the amide group with a second amino group ($RCONHNH_2$) are reported to have intense (ϵ = 4000–12 000), structured bands centered at 53 000 cm^{-1} [AD127]. The strong bands are undoubtedly related to the $N \to V_1$ bands of the parent amides, but it is strange that the bands fall at identical frequencies for R = H and CH_3, whereas in formamide and acetamide, the corresponding $N \to V_1$ frequencies are separated by 4000 cm^{-1}.

The solution study of Turner on the cyclic imides [AD182] reveals that in succinimide and its alkyl derivatives, the $N \to V_1$ band comes at 52 300 cm^{-1} (vert.), with extinction coefficients (ϵ = 10 000–15 000) considerably larger than those of amides, but which decrease with alkylation, just as in the amides and acids.

AD.V.A-2. Polymeric Amides

Recent circular dichroism work on polypeptides has uncovered an anomalous situation. The theory of the exciton splitting in α-helical polypeptides predicts the presence of a strongly negative band at 55 600 cm^{-1} (vert.), whereas the experimental CD spectra of poly-L-alanine films [AD192] and solutions of poly-N-methyl glutamate [AD95], which are themselves nearly identical, show a strongly positive band at the frequency in question. Furthermore, the absorption spectra (Figs. V.A-8 and V.A-10) show absorption minima rather than maxima at 55 600 cm^{-1}. Since the calculations are otherwise quite successful, it was suggested that perhaps a weakly absorbing but strongly rotating band falls at 55 600 cm^{-1}, which is outside of the conventional n_O, π_2, π_3^* manifold, i.e., the $n_O \to \sigma^*$ band discussed in Section V.A-2.

The reflection spectrum of triglycine sulfate crystals down to 74 000 cm^{-1} is reported [AD184], and shows a broad band at $\sim$65 000 cm^{-1} (vert.). Possibly this is related to the band at 60 000 cm^{-1} in triglycyl glycine (Fig. V.A-13).

AD.V.A-3. Acids, Esters, and Acyl Halides

The optical spectra of trifluoroacetamide, trifluoroacetic acid, and trifluoroacetyl fluoride (Fig. V.A-15) have been calculated by the semi-empirical CNDO method with interesting results [AD111]. In each of these, two $\pi \to \pi^*$ excitations are predicted (N → V_1 and N → V_2) with a splitting which is 16 000 cm^{-1} in the amide, 8900 cm^{-1} in the acid, and only 4000 cm^{-1} in the acyl fluoride. Moreover, the N → V_2 frequency is predicted to be very constant at ~80 000 cm^{-1}, whereas the N → V_1 frequency increases rapidly through the series, as observed. This behavior is consistent with that deduced by the intramolecular charge transfer theory of Nagakura [II-N3, II-N4], in which the V_1 upper state is largely the charge transfer configuration while the V_2 state is largely the $\pi \to \pi^*$ excitation localized within the C=O group. Thus from this we expect that in the acid, the Q band is the broad feature at ~77 000 cm^{-1} supporting the R_2 feature, while in the acyl fluoride, Q and V_1 are nearly degenerate at 75 000–80 000 cm^{-1} (vert.) (Fig. V.A-15). The electronic spectrum of formic acid has been calculated in an *ab initio* way [AD154]; however, the authors have assigned several of the Rydberg bands to valence shell excitations on the basis of the frequency match.

Price and Evans [II-P41] report the members of a Rydberg series in formic acid, originating at the n_O MO and displaying progressions of 1450–1500 and 600–1000 cm^{-1}. Careful photoelectron spectroscopic work on formic acid and its deuterated derivatives now shows that the ionization band originating at n_O has a long progression of C=O stretching (~1470 cm^{-1} and no deuterium shift), while shorter progressions of 900–1000 cm^{-1} are shown to be in-plane O—H(D) bending [AD186].

Maria et al. [AD126] describe the spectrum of dimethyl carbonate as showing a band at 72 000 cm^{-1} (vert.) which they assign as $n_O \to 3p$, and an $n_O \to \pi^*$ band at about 66 000 cm^{-1} (vert., $\epsilon = 100$). As with ethylene carbonate (Section II.E-2), it would appear that dimethyl carbonate would make a good solvent for solution spectroscopy in the vacuum ultraviolet.

AD.V.B. Oxides of Nitrogen

The threshold electron-impact spectrum of nitromethane is reported by McAllister, who used ion cyclotron resonance for detecting zero-energy electrons [AD130]. With this technique, the $n_O \to \pi_3^*$ and $\pi_2 \to \pi_3^*$ bands were observed at their optical frequencies [II-N2], and additional peaks were recorded at 66 100, 75 000, and 84 600 cm^{-1} (vert.). The first of these has a term value of 25 200 cm^{-1} with respect to the $5a_1$ ionization potential, and so may be assigned as $5a_1 \to 3s$.

AD.V.C-1. Dienes

The problem concerning the admixture of diffuse orbitals into "valence shell" pi-electron excited states has been studied theoretically for butadiene [AD47]. As might be expected by comparison with ethylene, the lowest $^1A_g \to {}^1B_u$ excitation in butadiene, correlating with the intense $N \to V_1$ band observed at 47 800 cm^{-1} (vert.), is calculated to be Rydberglike (diffuse), and, as is also the case with ethylene, the external perturbation experiments on butadiene (Fig. V.C-2) suggest overwhelming valence shell character instead. This apparent contradiction will likely disappear once the *ab initio* calculations are performed with extensive configuration interaction. In the *ab initio* calculation, a valence shell $^1A_g \to {}^1A_g$ transition is again predicted to be the lowest in the pi-electron manifold (2000 cm^{-1} below $N \to V_1$). The calculation also assigns the feature at 58 700 cm^{-1} (vert.) [AD136] as a $\pi \to \pi^*$ Rydberglike band, having a ^1B_u upper state.

The most recent electron-impact spectrum of butadiene is somewhat confusing. Mosher *et al.* [AD135] find two triplet states at 26 000 cm^{-1} (^3B_u) and 39 500 cm^{-1} (^3A_g). Beyond the intense excitation to ^1B_u, they describe a band with vibronic components at 57 100, 58 700, and 60 100 cm^{-1} with Franck–Condon factors which in no way resemble the optical spectrum in that region (Fig. V.C-1), and assign it as most likely a second $^1A_g \to {}^1B_u$ $\pi \to \pi^*$ excitation, in agreement with the calculation quoted above. However, the sharp features in this region of the spectrum are already assigned by Price and Walsh [II-P43] as Rydberg, and this is verified by the spectrum of the solid (Fig. V.C-2). Note, though, that in this spectrum of solid butadiene there is an underlying valence shell band uncovered, and this may correspond to what is observed in the electron-impact spectrum. A new band at 88 000 cm^{-1} (vert.) in the electron-impact spectrum is also reported for butadiene [AD135].

AD.V.C-2. Heterocyclic Dienes

Spectra of the heterocyclic dienes and their reduced forms (tetrahydrofuran, etc.) are compared by Milazzo [AD134]. In the text, the possibility was briefly mentioned that the $1a_2$-above-$2b_1$ ordering of thiophene may be reversed in selenophene. This aspect has been investigated by

Schäfer et al. [AD166] by photoelectron spectroscopy, and they find that in selenophene, the orbitals in question are nearly degenerate, while in tellurophene, they are reversed and split by 4350 cm^{-1}.

AD.V.C-3. Higher Polyenes

The interesting question of the location of the first $^1A_g \to {}^1A_g$ excitation in polyenes has been investigated for 1,3,5-*trans*-hexatriene by the trapped electron method (Section II.D) and semiempirical calculations with full configuration interaction [AD104]. In addition to the bands found optically by Price and Walsh [II-P47], bands were also observed at 21 000 and 33 800 cm^{-1} (vert.), both of which are thought to have triplet upper states. There is no sign of the low-lying forbidden $\pi \to \pi^*$ excitation, but the reality of the questionable bands in the vicinity of 58 000 cm^{-1} in the optical spectrum is confirmed by the trapped electron spectrum, which shows two bands at 56 400 and 62 000 cm^{-1} (vert.).

AD.VI.A-1. Benzene

With regard to the problem of diffuse orbitals in large molecules, the calculation of Hay and Shavitt [AD73] on benzene is of some interest. Using diffuse π orbitals in the basis set and using only (π, π^*) configuration interaction, they find that the excited $^1B_{2u}$ and $^1B_{1u}$ states are strictly valence shell, but that $^1E_{1u}$ is diffuse, but not as diffuse as a true Rydberg state. The corresponding triplet states are all valence shell size. By analogy with ethylene, it is likely that the $^1E_{1u}$ state is a V state in the Mulliken sense [AD137] and, being highly ionic, places the π^* optical electron in a very large orbital, which would again shrink to valence shell size upon the application of extensive $(\pi, \pi^*)-(\sigma, \sigma^*)$ configuration interaction. Hay and Shavitt also described the missing $^1E_{2g}$ state as having considerable two-electron-excitation character and estimate that it comes 4000–8000 cm^{-1} beyond the transition to $^1E_{1u}$.† The triplet state observed 71 300 cm^{-1} above S_0 by $T_1 \to T_n$ absorption remains unidentified in this calculation.

Inagaki [AD87] has extended his transmission spectrum of liquid benzene to 87 000 cm^{-1} with an LiF cell, and finds essential agreement with the reflectance data of others. The transmission spectra show definite

† Being a two-electron excitation to a state of g parity, this transition in benzene is generically related to the forbidden $^1A_g \to {}^1A_g$ band of butadiene, presently of such great interest.

shoulders at 61 000 and 80 000 cm^{-1}, features also seen in the vapor absorption spectrum (Figs. VI.A-1 and VI.A-8).

The reality of the absorption feature between the $^1B_{2u}$ and $^1B_{1u}$ states of benzene proposed by Morris and Angus [II-M54] is confirmed by the work of Taleb et al. [AD174]. Working with solutions of benzene in perfluoro-n-hexane at 190°K as well as with Kr and Xe matrices, they uncovered a very clear, structured feature ($\epsilon \simeq 1000$) in the region 45 500–47 600 cm^{-1}, which they assigned as $^1A_{1g} \rightarrow {}^1E_{2g}$, made allowed by vibronic interaction. They also equate this band with one found in the same region in hexamethyl benzene, but this latter band is a Rydberg transition to 3s as judged by its term value and behavior in condensed phases (Section VI.A-2). On the other hand, the band in benzene is far too low to be considered as a Rydberg excitation to 3s (29 000 cm^{-1} term value). See Section AD.V.A-2 for further comment on this band.

The sharp feature at 46 565 cm^{-1} in the spectrum of crystalline benzene (Fig. VI.A-2) has received considerable attention in the last year. Brillante et al. [AD23] studied the polarized absorption on the (100) face of benzene and find the band in question to be completely polarized along the c axis, whereas the remainder of the spectrum is approximately twice as intense along the b axis. Pointing out that the origin and totally symmetric parts of the transition to $^1B_{1u}$ would be very nearly purely b-axis polarized, they suggest that the sharp line is probably the origin for a separate out-of-plane polarized excitation to $^1E_{2g}$. The presence of a "new" transition at $\sim$46 000–47 000 cm^{-1} in benzene thus seems to be gaining considerable support. The crystal spectrum of benzene has been confirmed by Bird and Callomon [AD16], who penetrated beyond the $^1E_{1u}$ absorption and report the polarization ratio as nearly constant from 47 000 to 59 000 cm^{-1}, implying that the bands in the 47 000–50 000-cm^{-1} region are coupled vibronically to the $^1E_{1u}$ transition.

Birks et al. [AD18] have studied the temperature dependence of the width of the 46 565-cm^{-1} band in crystalline benzene and several of its deutero derivatives. The width of 50 cm^{-1} obtained by extrapolation to 0°K implies a relaxation rate of 9.4×10^{12} sec^{-1} via two channels, one being relaxation to $^1B_{2u}$ and the other a channel leading to a nonplanar state which subsequently undergoes photochemistry.

Matrix spectra of benzene of higher quality than those in Fig. VI.A-6 have appeared recently [AD63] and with these the Wannier series have been extended to $n = 4$ in the various solid rare gases. These spectra also show the excitation of ν_2' and ν_{18}' vibrations, the second indicating a Jahn–Teller interaction. In matrices of Ne, Ar, Kr, and Xe, the first ionization potential of benzene assumes the values **74 000, 68 600, 65 900, and 62 500 cm^{-1}** (advert.), respectively.

The work of Yoshino et al. [II-Y12] on the vapor spectra of benzene and styrene down to 180 000 cm^{-1} has appeared in the formal literature [AD191]. Again, they point out that the two bands at 56 000 and 160 000 cm^{-1} in polystyrene are present in the vapor spectra of both benzene and styrene, and so are not due to collective excitations.

AD.VI.A-2. Alkyl Benzenes

Allen and Schnepp [AD2] add more fuel to the benzene fire with their report of the circular dichroism spectrum of 1-methyl indan. Though the absorption spectrum of this compound looks "normal" in the sense of having readily identifiable transitions to $^1B_{2u}$, $^1B_{1u}$, and $^1E_{1u}$, in addition to these, the circular dichroism spectrum shows unusual bands both preceding and following the excitation to $^1B_{1u}$. The first of these, at 46 000 cm^{-1}, is thought to be $\pi \rightarrow \sigma^*$, possibly a transition to 3s, whereas that at $\sim$50 500 cm^{-1} may be the transition to $^1E_{2g}$. Thus evidence from several directions on benzene and its derivatives suggests the presence of at least two and possibly three transitions in the $^1B_{1u}$ region. At this point, it seems most likely that the interlopers are the excitation to 3s and the $\pi \rightarrow \pi^*$ ($^1A_{1g} \rightarrow {}^1E_{2g}$) transition.

The complex spectra of paracyclophane and its multilayered relatives have been studied by Iwata et al. [AD90] to 59 000 cm^{-1} in paraffin solution. Considering the mixing of local excitations with a higher charge transfer configuration, they predict many more bands than are resolved. According to the calculation, the final states retain their free-molecule parentage largely, so that it can be said that in the double, triple, and quadruple layered compounds, the transitions to $^1E_{1u}$ fall at 53 000, 49 600, and 50 000 cm^{-1} (vert.), respectively.

AD.VI.A-3. Halobenzenes

In his study of synthetic polymer films, Onari reports the vacuum-ultraviolet spectra of various phenyl-chlorinated polystyrenes [II-O7]. These spectra resemble closely that of polystyrene itself and that of chlorobenzene, with the exception that the prominent Rydberg excitations normally seen for chlorine-containing compounds in the gas phase do not appear in the solid films (Section II.C). The polymer spectra hint strongly of a weak band on the high-frequency wing of the transition to $^1E_{1u}$, i.e., at $\sim$59 000 cm^{-1}.

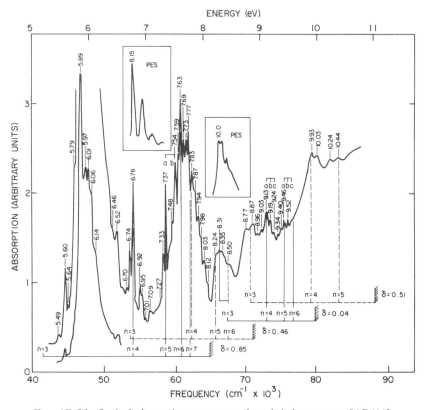

Fig. AD-11. Optical absorption spectrum of naphthalene vapor [AD106].

AD.VI.B. Higher Aromatics

The optical absorption spectra of the vapors of naphthalene [AD106]†
and anthracene [AD107] have been recorded using synchrotron radiation, by Koch et al. In naphthalene (Fig. AD-11), four Rydberg series
were enumerated going to the first four pi-orbital ionization potentials.
However, the series are highly irregular in their intensity distributions,
and some which should be forbidden as judged from the originating orbital symmetry and the quantum defect (such as the second, which seems
to be $b_{3u} \rightarrow 3p$) are quite intense. In anthracene vapor (Fig. AD-12),
Koch et al. assign the $b_{2g} \rightarrow 3pb_{1u}$ promotion to the step at 40 500 cm^{-1},
as suggested in the text. Note from the figure that this band is much
broader than the higher $b_{2g} \rightarrow n$p members, and so is probably mixed

† Credit for Fig. I.A-11 is more properly given to Koch et al. [AD106].

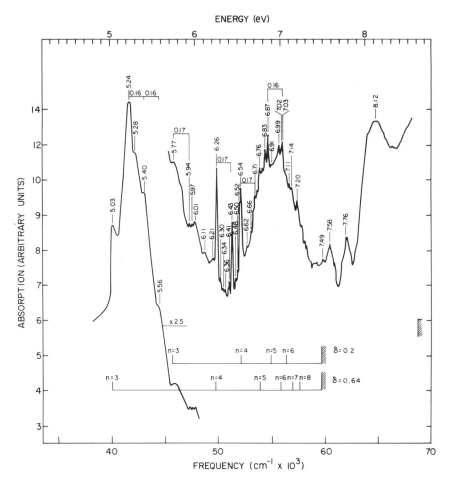

Fig. AD-12. Optical absorption spectrum of anthracene vapor [AD107].

somewhat with a valence shell excitation of the same symmetry. A second series in anthracene has $\delta = 0.2$ and is most likely $b_{2g} \to nd$. The same workers have studied the polarized reflection spectra of anthracene crystals [AD108], and from the observed polarization, they have identified $\pi \to \pi^*$ excitations at 27 800 ($^1B_{1u}$), 42 200 ($^1B_{2u}$), 46 700 ($^1B_{1u}$), 55 600 ($^1B_{1u}$), and 65 200 cm^{-1} (vert., $^1B_{1u}$).

The SF_6-scavenger spectrum of azulene vapor shows a very intense triplet at 19 000 cm^{-1} and a weak band at 38 000 cm^{-1}, but none of the Rydberg excitations easily seen optically at higher frequencies appear in the scavenger spectrum [AD82].

The vapor spectrum of biphenyl is of great interest insofar as it can be thought of as a dimer of benzene. In this molecule, a complex of two overlapping transitions comes at 39 000–45 000 cm^{-1}, followed by two more distinct transitions at 52 000 and 59 000 cm^{-1} (vert.) [II-C7, AD48]. In n-heptane solution, Klevens reports that these three regions have oscillator strengths of 0.47, 1.70, and $\sim$0.6, respectively [II-K27]. Edwards and Simpson have applied the independent-systems concept (Section III.A-3) to biphenyl and conclude that absorption in the 39 000–45 000-cm^{-1} region involves components of $^1B_{2u}$ and $^1B_{1u}$ absorption, whereas the two higher-frequency bands result from $^1E_{1u}$–$^1E_{1u}$ interactions between the phenyl rings. There are no obvious Rydberg transitions in the vapor spectra reported so far.

Electron energy-loss spectra of triphenylene and perylene in the vapor phase are dominated by intense losses centered at 120 000–130 000 cm^{-1}, with a less intense shoulder at $\sim$160 000 cm^{-1} [II-K34]. Exactly similar features are seen in the electron energy-loss spectra of solid films of pyrene, coronene, and hexabenzocoronene [I-G15], and in the optical spectrum of naphthalene vapor [AD106]. These intense pairs of features seem to be common aspects of saturated and unsaturated hydrocarbon spectra in both the gas and solid phases.

Catalogs of aromatic molecule spectra stretching from benzene to tetrabenzo-2,3,6,7,2′,3′,6′,7′-heptafulvalene can be found in the papers of Jones and Taylor [II-J12], Layton [AD117], and Klevens [II-K27]. Much of this is solution data and not all of it extends into the vacuum ultraviolet.

The spectrum of a film of polyvinyl carbazole [II-O7] shows a very rich spectrum from 28 000 to 63 000 cm^{-1}, with almost a dozen distinct transitions visible.

AD.VII.A. Nonmetallic Inorganic Systems

A most impressive spectral study of the deeper states of XeF$_2$ and XeF$_4$ has recently appeared [AD167]. Using synchrotron radiation, Comes *et al.* report the gas-phase and crystal spectra of these substances in the range 400 000–1 290 000 cm^{-1} (50–160 eV) as showing Rydberg and valence shell transitions originating at the core levels of the Xe and F atoms. The distinct features in XeF$_2$ begin at 495 100 and 510 500 cm^{-1} (61.38 and 63.29 eV) with $4d_{5/2} \to 7\sigma_u$ and $4d_{3/2} \to 7\sigma_u$ valence shell excitations. In the solid, these bands are shifted by about 3000 cm^{-1} to lower frequencies. In XeF$_4$, the virtual level analogous to $7\sigma_u$ in XeF$_2$ is the $8e_u$ MO; transitions from the $4d_{5/2}$ and $4d_{3/2}$ core levels to $8e_u$ are observed in the 500 000–530 000-cm^{-1} range (62–66 eV). In solid XeF$_4$, these bands show only a very small shift to lower frequencies. Obviously, there are no bands in the Xe atom spectrum that are analogous to these in XeF$_2$ and XeF$_4$.

The 4d ionization potentials are 568 600 and 584 800 cm^{-1} (70.5 and 72.5 eV) in XeF$_2$ and 588 800 and 604 900 cm^{-1} (73.0 and 75.0 eV) in XeF$_4$. Consequently, the 4d → nR Rydberg states fill the 540 000–585 000-cm^{-1} (67–73 eV) region of XeF$_2$ and the 565 000–605 000-cm^{-1} (70–75 eV) region of XeF$_4$, as in the Xe atom itself. These Rydberg bands in the gas phase are much narrower than the valence shell bands and do not appear in solid XeF$_4$, as expected. However, the corresponding bands of XeF$_2$ are observed in the solid, albeit at much higher frequencies and badly broadened.† The 4d → f-wave continua in both XeF$_2$ and XeF$_4$ peak at $\sim$766 000 cm^{-1} (95 eV) and a weak Rydberg excitation from 4p$_{3/2}$ is found in each at $\sim$1 170 000 cm^{-1} (145 eV). The Rydberg excitations in these molecules show splittings due both to ligand field effects (core splitting of both the originating and terminating orbitals) and to spin–orbit coupling.

In line with our contention that the Rydberg term values are independent of the originating MOs (Section I.C-1), it is of interest to compare the term values obtained in the 400 000–1 200 000 cm^{-1} (50–150 eV) regions of these compounds with those obtained in the 40 000–80 000 cm^{-1} region (5–10 eV). In Table VII.A-I, the (ϕ_i, 6s) term values for Xe and XeF$_2$ are seen to be 30 000–31 000 cm^{-1}. In XeF$_2$, the deeper configurations (4d$_{5/2}$, 6s), (4d$_{3/2}$, 6s), and (4p$_{3/2}$, 6s) have term values of 28 700, 28 200, and 28 900 cm^{-1}, respectively, in good agreement with those listed in the table. For XeF$_4$, transitions to 6s from the 4d levels are not observed ($g \to g$), but the (4p$_{3/2}$, 6s) configuration has a term value of 25 900 cm^{-1}. Thus there is a slight but unmistakable decrease of the (ϕ_i, 6s) term value as Xe is fluorinated. The corresponding Rydberg transitions from 4d to 6p are complicated by ligand field and spin–orbit effects in both XeF$_2$ and XeF$_4$, resulting in six components observed in each.

AD.VII.B. Metallic Inorganic Systems

In a very nice review of solution spectrophotometry in the vacuum-ultraviolet region, Fox [AD60] lists the absorption characteristics of aqueous solutions of several first-series transition metal ions, which we list in Table AD-III. Corresponding data for a few rare earth ions in water

† The fact that the Rydberg bands of XeF$_2$ can be seen in the solid suggests that the electron mobility in this phase may be quite large, as it is in solid Xe.

TABLE AD-III
ABSORPTION SPECTRA OF METAL ION HYDRATES IN WATER[a]

Ion	Absorption maximum (cm^{-1})	Molar extinction coefficient
Mn^{2+}	58 300	10^3
	63 300	$5–10 \times 10^4$
Fe^{2+}	58 140	10^3
	62 500	10^5
Co^{2+}	60 750	$1–3 \times 10^4$
Ni^{2+}	62 300	2.5×10^4
Cu^{2+}	48 500	$1.1–1.3 \times 10^4$
	59 000[b]	10^5
Eu^{3+}	53 200[c]	235
Ce^{3+}	50 000[c]	170
Pr^{3+}	53 000[c]	—
Tb^{3+}	55 000[c]	—
Yb^{3+}	~59 000[c]	—

[a] From Reference [AD60].
[b] A cluster of closely spaced bands.
[c] Taken from Reference [AD96].

have been reported by Jorgensen and Brinen [AD96]. The latter authors assign the bands in the 50 000–60 000-cm^{-1} region of these aquo ions to allowed 4f → 5d excitations in species of unknown hydration number and geometry. Since the bands in the corresponding regions of the transition metal ions are 10–1000 times more intense than those in the rare earths, it seems that they probably are not 3d → 4p excitations, but instead are metal ↔ water charge transfer.

In the text, there was demonstrated the close relationship between the spectra of CCl_4 and $TiCl_4$ in the vacuum ultraviolet. This relationship can now be extended to include VCl_4 and $SnCl_4$, two compounds recently studied by Iverson and Russell [AD89], along with $TiCl_4$. The interpretation of these optical spectra (Fig. AD-13) is aided considerably by the complementary photoelectron spectra taken from the work of Orchard and collaborators [II-G24, AD34]. The spectral work on $TiCl_4$ confirms the band frequencies but not the intensities given in Fig. VII.B-1, while showing that the band at 81 000 cm^{-1} in that figure is really a doublet with components at 80 580 and 82 780 cm^{-1} (vert.) and that another peak is to be found at 86 100 cm^{-1} (vert.). This latter band has a term value of 20 600 cm^{-1} with respect to the 1e ionization potential at 106 700 cm^{-1}, and so is the D band corresponding to excitation from 1e, i.e., 1e → 4p.

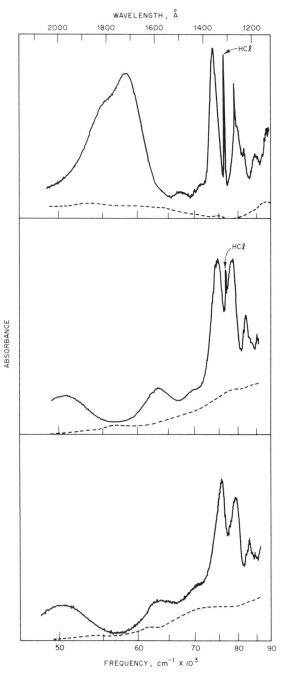

Fig. AD-13. Optical absorption spectra of titanium (upper), vanadium (middle), and tin (lower) tetrachloride vapors. The dashed lines represent the appropriate spectrometer baselines [AD89].

In VCl_4, the single electron occupying the vanadium $2e$ level has an ionization potential of 75 800 cm^{-1} (vert.), and excitations from there to 4s should be about 25 000 cm^{-1} below the ionization limit and electronically forbidden. This nicely describes the weak band seen at 50 800 cm^{-1} (vert.). Another weak band at 63 500 cm^{-1} has a term value (12 300 cm^{-1}) which suggests a $2e \rightarrow 4d$ assignment, but an assignment to a valence shell A band is also likely. A matrix spectrum of VCl_4 would settle this point. Intense D bands fall at 75 470 and 83 000 cm^{-1} (vert.) corresponding to $1t_1 \rightarrow 4p$ and $3t_2 \rightarrow 4p$ Rydberg excitations with term values of 19 400 and 20 800 cm^{-1}, respectively. These bands correspond directly to the D bands in $TiCl_4$ at 75 400 and 82 700 cm^{-1}, but it is interesting to note that the extinction coefficients in $TiCl_4$ are about a factor of ten larger. The weak shoulder at 69 900 cm^{-1} in VCl_4 is the B, C band originating at $1t_1$ and has a ($1t_1$, 4s) term value of 25 000 cm^{-1}, just as does the ($2e$, 4s) configuration. The band at 79 050 cm^{-1} in VCl_4 has a term value of 22 000 cm^{-1} with respect to an ionization potential reported at 101 000 cm^{-1} (vert.) and so is also a D band.

It is remarkable how similar the optical spectra of VCl_4 and $SnCl_4$ seem to be, considering that VCl_4 has an unpaired electron which should be active optically, and $SnCl_4$ does not. This resemblance may be more apparent than real, however, since the bands are about ten times more intense in $SnCl_4$, and the band at 50 800 cm^{-1} in VCl_4 is a Rydberg excitation while that at 50 300 cm^{-1} in $SnCl_4$ must be valence shell. As in VCl_4, the relatively weak shoulder at 69 900 cm^{-1} in $SnCl_4$ would seem to be a B, C pair, with a term value of 27 600 cm^{-1} with respect to the ionization potential at 97 500 cm^{-1}. Following this, the three intense bands in $SnCl_4$ at 75 600, 79 100, and 82 200 cm^{-1}, correspond to the three D bands at nearly the same frequencies in CCl_4, $TiCl_4$, and VCl_4. The corresponding term values are 21 900, 20 900, and 20 200 cm^{-1} in $SnCl_4$.

Solution and crystal spectra of the linear dicyanide complexes of Cu^+, Ag^+, and Au^+ ions reveal a number of bands from 35 000 to 55 000 cm^{-1} [AD128]. The more intense transitions in the region of 50 000 cm^{-1} (ϵ = 15 000–20 000) are assigned as $nd \rightarrow \pi^*$ (C≡N) metal-to-ligand charge transfer excitations, with spin–orbit coupling being of major importance in the gold complex.

The photoelectron spectra of OsO_4 and RuO_4 have also been determined by Diemann and Müller [AD43] with results in agreement with those of [II-F9]. However, they deduce an arrangement of MOs different from that given by Foster et al.

The general features of the X-ray spectra of octahedral and tetrahedral inorganic ions and molecules are presented by Baranovskii and Nakhmanson [AD11]. The $L_{II,III}$ and K emission spectra both involve transitions between the central atom core level and valence MOs, whereas for $L_{II,III}$ and K absorption spectra, they postulate that both valence shell and Rydberg levels are important. In absorbers of high symmetry, the intensities of the transitions between 1s(K) or $2p(L_{II,III})$ and the outer valence MOs are useful indicators of the symmetries of these outer orbitals. Thin films of the transition metal chlorides show nicely structured spectra in the chlorine K region (2820–2840 eV) [AD173]. Sugiura claims that the first feature in such spectra corresponds to the chlorine 1s → metal nd charge transfer excitation appropriately shifted by the Madelung potential. The spectra of $CuCl_2(3d^9)$ and $FeCl_3(3d^5)$ look remarkably alike, and the upper levels may correspond to Wannier excitons rather than to complex-ion levels.

AD.VIII. Biological Systems

A paper on the high-energy, electron-impact energy-loss spectra of guanine and cytosine by Johnson [AD93] supplements similar spectra on the other nucleic acids reported by Isaacson [II-15, II-16]. In the region below 80 000 cm^{-1}, the energy-loss spectra through thin films closely resemble the optical absorption spectra (Fig. VIII-1), followed by massive losses centered at ~120 000 cm^{-1} in both guanine and cytosine. These latter peaks are said to be one-electron excitations, modified by collective effects. The optical constants of crystalline cytosine have been measured by electron energy-loss spectroscopy, and compare well with values derived from optical reflectance data obtained using synchrotron radiation [AD94]. Hug and Tinoco [AD83] have further investigated the optical spectra of the nucleic acid bases in trimethyl phosphate solution, again using the benzene spectrum as a guide for the interpretation of the pyrimidine base spectra. Absorption and circular dichroism spectra of L-tryptophane and some of its derivatives in trifluoroethanol solution are reported by Auer [AD4].

Lewis and Johnson [AD120] have measured the circular dichroism of aqueous DNA solutions down to 60 000 cm^{-1}. Using both native and heat-denatured material from five different organisms, they found that the spectra differ from one another, reflecting different sequences and/or conformations, but nothing more specific could be said.

References

[A1] E. W. Abel, R. A. N. McLean, S. P. Tyfield, P. S. Braterman, A. P. Walker, and P. J. Hendra, Vibrational and electronic spectra and bonding in ionic transition metal hexacarbonyls, *J. Mol. Spectrosc.* **30,** 29 (1969).

[A2] G. J. Abruscato, R. G. Binder, and T. T. Tidwell, Steric crowding in organic chemistry. IV. Ultraviolet absorption spectra of crowded olefins, *J. Org. Chem.* **37,** 1787 (1972).

[A3] C. L. Adam, Ph.D. Thesis, Northeastern Univ., Boston, Massachusetts, 1969.

[A4] M. I. Al-Joboury and D. W. Turner, Molecular photoelectron spectroscopy. II. A summary of ionization potentials, *J. Chem. Soc.* 4434 (1964).

[A5] N. L. Allinger and M. A. Miller, Organic quantum chemistry. VII. Calculation of the near-ultraviolet spectra of polyolefins, *J. Amer. Chem. Soc.* **86,** 2811 (1964).

[A6] K. Allison and A. D. Walsh, *Chem. Inst. Can. Symp., Ottawa* (1957).

[A7] J.-M. André, M. Cl. André, G. Leroy, and J. Weiler, Theoretical study of isoelectronic systems: Diazomethane, ketene, and allene, *Int. J. Quantum Chem.* **3,** 1013 (1969).

[A8] J. G. Angus and G. C. Morris, Ionization potential of the anthracene molecule from Rydberg absorption bands, *J. Mol. Spectrosc.* **21,** 310 (1966).

[A9] J. G. Angus, B. J. Christ, and G. C. Morris, Absorption spectra in the vacuum ultraviolet and the ionization potentials of naphthalene and naphthalene-d_8 molecules, *Aust. J. Chem.* **21,** 2153 (1968).

[A10] J. G. Angus and G. C. Morris, Benzene–rare gas matrices: Evidence from electron photoemission spectra for Wannier impurity states, *Mol. Cryst. Liquid Cryst.* **11,** 309 (1970).

[A11] J. G. Angus and G. C. Morris, Wannier type impurity states in naphthalene-rare gas matrices and the energy of the quasi-free electron state, *Chem. Phys. Lett.* **5,** 480 (1970).

[A12] J. G. Angus and G. C. Morris, Naphthalene-rare gas solids: Absorption spectra from 30 000–80 000 cm⁻¹, *Mol. Cryst. Liquid Cryst.* **11,** 257 (1970).

[A13] J. G. Angus and G. C. Morris, The lowest free-molecule Rydberg transition of napthalene, *Aust. J. Chem.* **24,** 173 (1971).

[A14] L. Åsbrink, E. Lindholm, and O. Edqvist, Jahn–Teller effect in the vibrational structure of the photoelectron spectrum of benzene, *Chem. Phys. Lett.* **5,** 609 (1970).

[A15] L. Åsbrink, O. Edqvist, E. Lindholm, and L. E. Selin, The electronic structure of benzene, *Chem. Phys. Lett.* **5,** 192 (1970).

[A16] L. Åsbrink, C. Fridh, B. Ö. Jonsson, and E. Lindholm, Rydberg series in small molecules. XVI. Photoelectron, UV, mass and electron impact spectra of pyrimidine, *Int. J. Mass Spectrom. Ion Phys.* **8,** 215 (1972).

[A17] L. Åsbrink, C. Fridh, B. Ö. Jonsson, and E. Lindholm, Rydberg series in small molecules. XVII. Photoelectron, UV, mass and electron impact spectra of pyridazine, *Int. J. Mass Spectrom. Ion Phys.* **8,** 229 (1972).

[B1] R. D. Bach, Optical activity of *trans*-cyclo-octene, *J. Chem. Phys.* **52,** 6423 (1970).

[B2] A. D. Baker, C. Baker, C. R. Brundle, and D. W. Turner, The electronic structures of methane, ethane, ethylene, and formaldehyde studied by high-resolution molecular photoelectron spectroscopy, *Int. J. Mass Spectrom. Ion Phys.* **1,** 285 (1968).

[B3] A. D. Baker, D. P. May, and D. W. Turner, Molecular photoelectron spectroscopy. VII. The vertical ionization potentials of benzene and some of its monosubstituted and 1,4-disubstituted derivatives, *J. Chem. Soc. B* 22 (1968).

[B4] A. D. Baker, D. Phil. Thesis, Oxford Univ., 1968.

[B5] C. Baker and D. W. Turner, Photoelectron spectra of acetylene, diacetylene, and their deutero-derivatives, *Chem. Commun.* 797 (1967).

[B6] C. Baker and D. W. Turner, The photoelectron spectrum and ionization potentials of carbon suboxide, *Chem. Commun.* 400 (1968).

[B7] C. Baker and D. W. Turner, Photoelectron spectra of allene and ketene; Jahn–Teller distortion in the ionization of allene, *J. Chem. Soc. D (Chem. Commun.)* 480 (1969).

[B8] C. Baker, D. Phil. Thesis, Oxford Univ., 1969.

[B9] D. G. Barnes and W. Rhodes, Generalized susceptibility theory. II. Optical absorption properties of helical polypeptides, *J. Chem. Phys.* **48,** 817 (1968).

[B10] E. E. Barnes and W. T. Simpson, Correlations among electronic transitions for carbonyl and for carboxyl in the vacuum ultraviolet, *J. Chem. Phys.* **39,** 670 (1963).

[B11] H. Basch, M. B. Robin, and N. A. Kuebler, Electronic states of the amide group, *J. Chem. Phys.* **47,** 1201 (1967).

[B12] H. Basch, M. B. Robin, and N. A. Kuebler, Electronic spectra of isoelectronic amides, acids, and acyl fluorides, *J. Chem. Phys.* **49,** 5007 (1968); **50,** 5048 (1969).

[B13] H. Basch and V. McKoy, The interpretation of open-shell SCF calculations on the T and V states of ethylene, *J. Chem. Phys.* **53,** 1628 (1970).

[B14] H. Basch, J. W. Moskowitz, C. Hollister, and D. Hankin, A self-consistent field study of the series XeF_n, n = 2,4,6, *J. Chem. Phys.* **55**, 1922 (1971).

[B15] N. A. Beach and H. B. Gray, Electronic structures of metal hexacarbonyls, *J. Amer. Chem. Soc.* **90**, 5713 (1968).

[B16] C. A. L. Becker, C. J. Ballhausen, and I. Trabjerg, Investigation of the electronic structure of $TiCl_4$, *Theoret. Chim. Acta* **13**, 355 (1969).

[B17] C. A. L. Becker and J. P. Dahl, A CNDO-MO calculation of $TiCl_4$, *Theoret. Chim. Acta* **14**, 26 (1969).

[B18] G. M. Begun and R. N. Compton, Threshold electron-impact excitation and negative-ion formation in XeF_6 and XeF_4, *J. Chem. Phys.* **51**, 2367 (1969).

[B19] G. Bélanger and C. Sandorfy, The far-ultraviolet spectra of fluoro-ethylenes, *Chem. Phys. Lett.* **3**, 661 (1969).

[B20] G. Bélanger and C. Sandorfy, Far-ultraviolet spectra of fluoroethylenes, *J. Chem. Phys.* **55**, 2055 (1971).

[B21] S. Bell, T. S. Varadarajan, A. D. Walsh, P. A. Warsop, J. Lee, and L. Sutcliffe, Spectrum of carbon suboxide in the ultraviolet, *J. Mol. Spectrosc.* **21**, 42 (1966).

[B22] S. Bell, G. J. Cartwright, G. B. Fish, D. O. O'Hare, R. K. Ritchie, A. D. Walsh, and P. A. Warsop, The electronic spectrum of cyanogen, *J. Mol. Spectrosc.* **30**, 162 (1969).

[B23] C. F. Bender, T. H. Dunning, Jr., H. F. Schaefer, III, W. A. Goddard, III, and W. J. Hunt, Multiconfiguration wave functions for the lowest (π, π^*) excited states of ethylene, *Chem. Phys. Lett.* **15**, 171 (1972).

[B24] J. L. Bensing and E. S. Pysh, Higher energy excitons in polypeptide films, *Chem. Phys. Lett.* **4**, 120 (1969).

[B25] J. L. Bensing and E. S. Pysh, Polarized vacuum ultraviolet absorption of poly-L-alanine, *Macromolecules* **4**, 659 (1971).

[B26] R. S. Berry, Analog of n → π^* transitions in mono-olefins, *J. Chem. Phys.* **38**, 1934 (1963).

[B27] H. Berthod, C. Giessner-Prettre, and A. Pullman, Theoretical study of the electronic properties of biological purines and pyrimidines. II. The effect of configuration mixing, *Int. J. Quantum Chem.* **1**, 123 (1967).

[B28] D. Betteridge and A. D. Baker, Analytical potential of photoelectron spectroscopy, *Anal. Chem.* **42**, 43A (1970).

[B29] B. D. Bird and P. Day, Far ultraviolet spectra of metal halide complexes, *Chem. Commun.* 741 (1967).

[B30] J. B. Birks, Assignment of two electronic states of benzene, *Chem. Phys. Lett.* **3**, 567 (1969).

[B31] P. Bischof, J. A. Hashmall, E. Heilbronner, and V. Hornung, Photoelectron spectroscopic determination of the resonance effect between nonconjugated double bonds, *Helv. Chim. Acta* **52**, 1745 (1969).

[B32] H. Bock and H. Seidl, "d-orbital effects" in silicon-substituted π-electron systems. XI. Syntheses and properties of the isomeric bis-(trimethylsilyl)-1,3-butadienes, *J. Amer. Chem. Soc.* **90**, 5694 (1968).

[B33] R. Bonneau, J. Faure, and J. Joussot-Dubien, Singlet-singlet absorption and intersystem crossing from the $^1B_{3u}^-$ state of naphthalene, *Chem. Phys. Lett.* **2**, 65 (1968).

[B34] R. Bonneau, J. Joussot-Dubien, and R. Bensasson, Giant pulse laser photolysis of benzene and mesitylene, *Chem. Phys. Lett.* **3**, 353 (1969).

[B35] R. A. Boschi and D. R. Salahub, The far ultraviolet spectra of some branched-chain iodoalkanes, iodocycloalkanes, fluoroiodoalkanes, and iodoalkenes, *Mol. Phys.* **24**, 735 (1972).

[B36] R. A. Boschi and D. R. Salahub, The high resolution photoelectron spectra of some iodoalkanes, iodoalkenes, and fluoroiodohydrocarbons, *Can. J. Chem.* **52**, 1217 (1974).

[B37] R. Botter, V. H. Dibeler, J. A. Walker, and H. M. Rosenstock, Experimental and theoretical studies of photoionization-efficiency curves for C_2H_2 and C_2D_2, *J. Chem. Phys.* **44**, 1271 (1966).

[B38] C. R. Bowman and W. D. Miller, Excitation of methane, ethane, ethylene, propylene, acetylene, propyne, and 1-butyne by low-energy electron beams, *J. Chem. Phys.* **42**, 681 (1965).

[B39] J. Brahms, J. Pilet, H. Damany, and V. Chandrasekharan, Application of a new modulation method for linear dichroism studies of oriented biopolymers in the vacuum ultraviolet, *Proc. Nat. Acad. Sci. U.S.* **60**, 1130 (1968).

[B40] R. Bralsford, P. V. Harris, and W. C. Price, Effect of fluorine on the electronic spectra and ionization potentials of molecules, *Proc. Roy. Soc. London,* **258A**, 459 (1960).

[B41] G. R. Branton, D. C. Frost, T. Makita, C. A. McDowell, and I. A. Stenhouse, Photoelectron spectra of ethylene and ethylene-d_4, *J. Chem. Phys.* **52**, 802 (1970).

[B42] E. A. Braude, The conformations of conjugated *cyclo*-alkadienes, *Chem. Ind. (London)* 1557 (1954).

[B43] E. A. Braude, The labile stereochemistry of conjugated systems, *Experientia* **11**, 457 (1955).

[B44] C. L. Braun, S. Kato, and S. Lipsky, Internal conversion from upper electronic states to the first excited singlet state of benzene, toluene, *p*-xylene, and mesitylene, *J. Chem. Phys.* **39**, 1645 (1963).

[B45] W. Braun, A. M. Bass, and M. Pilling, Flash photolysis of ketene and diazomethane: The production and reaction kinetics of triplet and singlet methylene, *J. Chem. Phys.* **52**, 5131 (1970).

[B46] B. Brehm, M. Menzinger, and C. Zorn, The photoelectron spectrum of XeF_2, *Can. J. Chem.* **48**, 3193 (1970).

[B47] J. S. Brinen, R. C. Hirt, and R. G. Schmitt, Vapor absorption spectra of *s*-triazene and the methyltriazines in the 1650–2000 Å region, *Spectrochim. Acta* **18**, 863 (1962).

[B48] M. Brith, R. Lubart, and I. T. Steinberger, Reflection and absorption spectra of the higher $\pi \rightarrow \pi^*$ transitions of solid benzene, *J. Chem. Phys.* **54**, 5104 (1971).

[B49] H. H. Brongersma, J. A. v.d. Hart, and L. J. Oosterhoff, Interaction of low energy electrons with molecules, *in* "Fast Reactions and Primary Processes in Chemical Kinetics" (S. Claessen, ed.), p. 211. Wiley (Interscience), New York, 1967.

[B50] C. R. Brundle and D. W. Turner, The carbonyl π-ionization potential of formaldehyde, *Chem. Commun.* 314 (1967).

[B51] C. R. Brundle and D. W. Turner, Studies on the photoionization of the linear triatomic molecules: N_2O, COS, CS_2 and CO_2 using high-resolution photoelectron spectroscopy, *Int. J. Mass Spectrom. Ion Phys.* **2**, 195 (1969).

[B52] C. R. Brundle, D. W. Turner, M. B. Robin, and H. Basch, Photoelectron spectroscopy of simple amides and carboxylic acids, *Chem. Phys. Lett.* **3**, 292 (1969).

[B53] C. R. Brundle and M. B. Robin, Nonplanarity in hexafluorobutadiene as revealed by photoelectron and optical spectroscopy, *J. Amer. Chem. Soc.* **92**, 5550 (1970).

[B54] C. R. Brundle, M. B. Robin, and G. R. Jones, High-resolution HeI and HeII photoelectron spectra of xenon difluoride, *J. Chem Phys.* **52**, 3383 (1970).

[B55] C. R. Brundle, M. B. Robin, H. Basch, M. Pinsky, and A. Bond, Experimental and theoretical comparison of the electronic structures of ethylene and diborane, *J. Amer. Chem. Soc.* **92**, 3863 (1970).

[B56] C. R. Brundle, D. Neumann, W. C. Price, D. Evans, A. W. Potts, and D. G. Streets, Electronic structure of NO_2 studied by photoelectron and vacuum-UV spectroscopy and Gaussian orbital calculations, *J. Chem. Phys.* **53**, 705 (1970).

[B57] C. R. Brundle, G. R. Jones, and H. Basch, HeI and HeII photoelectron spectra and the electronic structures of XeF_2, XeF_4, and XeF_6, *J. Chem. Phys.* **55**, 1098 (1971).

[B58] C. R. Brundle and D. B. Brown, The vibrational structure in the photoelectron spectra of ethylene and ethylene-d_4, and its relationship to the vibrational spectrum of Zeise's salt $K[PtCl_3(C_2H_4)] \cdot H_2O$, *Spectrochim. Acta* **27A**, 2491 (1971).

[B59] C. R. Brundle, M. B. Robin, N. A. Kuebler, and H. Basch, The perfluoro-effect in photoelectron spectroscopy. I. Nonaromatic molecules, *J. Amer. Chem. Soc.* **94**, 1451 (1972).

[B60] C. R. Brundle, M. B. Robin, N. A. Kuebler, The perfluoro-effect in photoelectron spectroscopy. II. Aromatic molecules, *J. Amer. Chem. Soc.* **94**, 1466 (1972).

[B61] C. R. Brundle and M. B. Robin, unpublished results.

[B62] W. L. Buck, B. R. Thomas, and A. Weinreb, Optical properties of polystyrene films in the far ultraviolet, *J. Chem. Phys.* **48**, 549 (1968).

[B63] R. J. Buenker and J. L. Whitten, *Ab initio* SCF MO and CI studies of the electronic states of butadiene, *J. Chem. Phys.* **49**, 5381 (1968).

[B64] R. J. Buenker, Theoretical study of the rotational barriers of allene, ethylene, and related systems, *J. Chem. Phys.* **48**, 1368 (1968).

[B65] R. J. Buenker and S. D. Peyerimhoff, *Ab initio* study on the stability and geometry of cyclobutadiene, *J. Chem. Phys.* **48**, 354 (1968).

[B66] R. J. Buenker and S. D. Peyerimhoff, *Ab initio* SCF calculations for azulene and naphthalene, *Chem. Phys. Lett.* **3**, 37 (1969).

[B67] R. J. Buenker and S. D. Peyerimhoff, Combined SCF and CI method for the calculation of electronically excited states of molecules; Potential curves for the low-lying states of formaldehyde, *J. Chem. Phys.* **53**, 1368 (1970).

[B68] R. J. Buenker, S. D. Peyerimhoff, and W. E. Kammer, Combined SCF and CI calculations for the low-lying Rydberg and valence excited states of ethylene, *J. Chem. Phys.* **55**, 814 (1971).

[B69] R. J. Buenker, S. D. Peyerimhoff, and H. L. Hsu, A new interpretation for the structure of the V-N bands of ethylene, *Chem. Phys. Lett.* **11**, 65 (1971).

[B70] S. M. Bunch, G. R. Cook, M. Ogawa, and A. W. Ehler, Absorption coefficients of C_6H_6 and H_2 in the vacuum ultraviolet, *J. Chem. Phys.* **28**, 740 (1958).

[B71] C. S. Burton and H. E. Hunziker, Gas phase absorption spectra and decay of triplet benzene, benzene-d_6, and toluene, *Chem. Phys. Lett.* **6**, 352 (1970).

[B72] J. P. Byrne and I. G. Ross, Electronic relaxation as a cause of diffuseness in electronic spectra, *Aust. J. Chem.* **24**, 1107 (1971).

[C1]   J. H. Callomon, T. M. Dunn, and I. M. Mills, Rotational analysis of the 2600 Å absorption system of benzene, *Phil. Trans. Roy. Soc. London* **259A**, 499 (1966).

[C2]   E. P. Carr and M. K. Walker, The ultraviolet absorption spectra of simple hydrocarbons. I. n-heptene-3 and tetramethylethylene, *J. Chem. Phys.* **4**, 751 (1936).

[C3]   E. P. Carr and H. Stücklen, The ultraviolet absorption spectra of simple hydrocarbons. III. In vapor phase in the Schumann region, *J. Chem. Phys.* **4**, 760 (1936).

[C4]   E. P. Carr and H. Stücklen, The ultraviolet absorption spectra of the isomers of butene-2 and pentene-2, *J. Amer. Chem. Soc.* **59**, 2138 (1937).

[C5]   E. P. Carr and H. Stücklen, The ultraviolet absorption of simple hydrocarbons. IV. Unsaturated cyclic hydrocarbons in the Schumann region, *J. Chem. Phys.* **6**, 55 (1938).

[C6]   E. P. Carr and H. Stücklen, Extreme ultraviolet absorption spectra of simple hydrocarbons, in *Proc. Summer Conf. Spectrosc., 7th* p. 128. Wiley, New York, 1940.

[C7]   E. P. Carr, L. W. Pickett, and H. Stücklen, The absorption spectra of a series of dienes, *Rev. Mod. Phys.* **14**, 260 (1942).

[C8]   J. G. Carter, T. M. Jelinek, R. N. Hamm, and R. D. Birkhoff, Optical properties of polystyrene in the vacuum ultraviolet, *J. Chem. Phys.* **44**, 2266 (1966).

[C9]   B. Cetinkaya, G. H. King, S. S. Krishnamurthy, M. F. Lappert, and J. B. Pedley, Photoelectron spectra of electron-rich olefins and an isostructural boron compound; Olefins of exceptionally low first ionization potential, *Chem. Commun.* 1370 (1971).

[C10] D. Chadwick, A. B. Cornford, D. C. Frost, F. G. Herring, A. Katrib, C. A. McDowell, and R. A. N. McLean, Photoelectron spectra of some dihalo compounds, *in* "Electron Spectroscopy" (D. A. Shirley, ed.), p. 453. North-Holland Publ., Amsterdam, 1972.

[C11] C. H. Chang, A. L. Andreassen, and S. H. Bauer, The molecular structure of perfluorobutyne-2 and perfluorobutadiene-1,3 as studied by gas phase electron diffraction, *J. Org. Chem.* **36**, 920 (1971).

[C12] G. W. Chantry and D. H. Whiffen, Electronic absorption spectra of CO_2^- trapped in γ-irradiated crystalline sodium formate, *Mol. Phys.* **5**, 189 (1962).

[C13] G. C. Chaturvedi and C. N. R. Rao, Normal vibrations of the electronically excited states of HCN and C_2H_2, *Spectrochim. Acta* **27A**, 2097 (1971).

[C14] W. A. Chupka, J. Berkowitz, and K. M. A. Refaey, Photoionization of ethylene with mass analysis, *J. Chem. Phys.* **50**, 1938 (1969).

[C15] L. B. Clark, G. G. Peschel, and I. Tinoco, Jr., Vapor spectra and heats of vaporization of some purine and pyrimidine bases, *J. Phys. Chem.* **69**, 3615 (1965).

[C16] L. B. Clark and I. Tinoco, Jr., Correlations in the ultraviolet spectra of the purine and pyrimidine bases, *J. Amer. Chem. Soc.* **87**, 11 (1965).

[C17] L. B. Clark, Ionization potential of azulene, *J. Chem. Phys.* **43**, 2566 (1965).

[C18] L. B. Clark and M. R. Philpott, Anisotropy of the singlet transitions of crystalline anthracene, *J. Chem. Phys.* **53**, 3790 (1970).

[C19] L. B. Clark, Vacuum ultraviolet spectra of crystalline anthracene, *J. Chem. Phys.* **53**, 4092 (1970).

[C20] L. B. Clark, private communication, 1971.

[C21] P. A. Clark, Electronic transitions in methyl-substituted ethylenes, *J. Chem. Phys.* **48**, 4795 (1968).

[C22] P. A. Clark, F. Brogli, and E. Heilbronner, The π-orbital energies of the acenes, *Helv. Chim. Acta* **55**, 1415 (1972).

[C23] W. D. Closson and H. B. Gray, The electronic structures and spectra of the azide ion and alkyl azides, *J. Amer. Chem. Soc.* **85**, 290 (1963).

[C24] J. E. Collin and J. Delwiche, Ionization of acetylene and its electronic energy levels, *Can. J. Chem.* **45**, 1883 (1967).

[C25] R. N. Compton and R. H. Huebner, Collisions of low-energy electrons with molecules: Threshold excitation and negative ion formation, *in* "Advances in Radiation Chemistry" (M. Burton and J. L. Magee, eds.), Vol. II, p. 281. Wiley (Interscience), New York, 1970.

[C26] B. E. Cook and P. G. Le Comber, The optical properties of anthracene crystals in the vacuum ultraviolet, *J. Phys. Chem. Solids* **32**, 1321 (1971).

[C27] C. A. Coulson and Z. Luz, Bond length changes on ionization, *Trans. Faraday Soc.* **64**, 2884 (1968).

[C28] D. O. Cowan, R. Gleiter, J. A. Hashmall, E. Heilbronner, and V. Hornung, Interaction between the orbitals of lone-pair electrons in dicarbonyl compounds, *Angew. Chem. Int. Ed.* **10**, 401 (1971).

[C29] M. C. Crocker and A. Herzenberg, Electronic transitions with large generalized oscillator strengths in ethylene, *Mol. Phys.* **22**, 483 (1971).

[C30] J. A. Cutler, Absorption of the alkyl cyanides in the vacuum ultraviolet, *J. Chem. Phys.* **16**, 136 (1948).

[D1] D. F. Dance and I. C. Walker, Threshold electron impact excitation of acetylene, *Chem. Phys. Lett.* **18**, 601 (1973).

[D2] D. A. Demeo, Vacuum Ultraviolet and Photoionization Studies on Olefins, Ph.D. Thesis, Univ. of California at Los Angeles, 1969.

[D3] D. A. Demeo and M. A. El-Sayed, Ionization potential and structure of olefins, *J. Chem. Phys.* **52**, 2622 (1970).

[D4] P. J. Derrick, L. Åsbrink, O. Edqvist, B.-Ö. Jonsson, and E. Lindholm, Rydberg series in small molecules. X. Photoelectron spectroscopy and electronic structure of furan, *Int. J. Mass Spectrom. Ion Phys.* **6**, 161 (1971).

[D5] P. J. Derrick, L. Åsbrink, O. Edqvist, B.-Ö. Jonsson, and E. Lindholm, Rydberg series in small molecules. XI. Photoelectron spectroscopy and electronic structure of thiophene, *Int. J. Mass Spectrom. Ion Phys.* **6**, 177 (1971).

[D6] P. J. Derrick, L. Åsbrink, O. Edqvist, B.-Ö. Jonsson, and E. Lindholm, Rydberg series in small molecules. XII. Photoelectron spectroscopy and electronic structure of pyrrole, *Int. J. Mass Spectrom. Ion Phys.* **6,** 191 (1971).

[D7] P. J. Derrick, L. Åsbrink, O. Edqvist, B.-Ö. Jonsson, and E. Lindholm, Rydberg series in small molecules. XIII. Photoelectron spectroscopy and electronic structure of cyclopentadiene, *Int. J. Mass Spectrom. Ion Phys.* **6,** 203 (1971).

[D8] R. D. Deslattes and R. E. LaVilla, Molecular emission spectra in the soft X-ray region, *Appl. Opt.* **6,** 39 (1967).

[D9] M. J. S. Dewar and S. D. Worley, Ionization potential of *cis*-1,3 butadiene, *J. Chem. Phys.* **49,** 2454 (1968).

[D10] M. J. S. Dewar, M. Shanshal, and S. D. Worley, Calculated and observed ionization potentials of nitroalkanes and of nitrous and nitric acids and esters. Extension of the MINDO method to nitrogen-oxygen compounds, *J. Amer. Chem. Soc.* **91,** 3590 (1969).

[D11] M. J. S. Dewar and S. D. Worley, Photoelectron spectra of molecules. I. Ionization potentials of some organic molecules and their interpretation, *J. Chem. Phys.* **50,** 654 (1969).

[D12] M. J. S. Dewar, "The Molecular Orbital Theory of Organic Chemistry." McGraw-Hill, New York, 1969.

[D13] M. J. S. Dewar, E. Haselbach, and S. D. Worley, Calculated and observed ionization potentials of unsaturated polycyclic hydrocarbons; Calculated heats of formation by several semiempirical SCF MO methods, *Proc. Roy. Soc. London* **315A,** 431 (1970).

[D14] M. J. S. Dewar and J. S. Wasson, Long-range couplings between lone pair electrons and double bonds, *J. Amer. Chem. Soc.* **92,** 3506 (1970).

[D15] V. H. Dibeler and J. A. Walker, Photoionization of acetylene near threshold, *Int. J. Mass Spectrom. Ion Phys.* **11,** 49 (1973).

[D16] C. Dijkgraaf and J. P. G. Rousseau, Electronic transitions in $TiCl_4$, $TiCl_3OC_6H_5$, $TiCl_2(OC_6H_5)_2$, $TiCl(OC_6H_5)_3$, and $Ti(OC_6H_5)_4$, *Spectrochim. Acta* **25A,** 1831 (1969).

[D17] G. Di Lonardo, G. Galloni, A. Trombetti, and C. Zauli, Electronic spectrum of thiophene and some deuterated thiophenes, *J. Chem. Soc. Faraday Trans.* **II,** 2009 (1972).

[D18] L. Di Sipio, G. de Michelis, E. Tondello, and L. Oleari, Electronic spectra of $TiCl_4$, $TiBr_4$, and TiI_4, *Gazz. Chim. Ital.* **96,** 1785 (1966).

[D19] J. P. Doering and A. J. Williams, III, Low-energy, large-angle electron-impact spectra: Helium, nitrogen, ethylene and benzene, *J. Chem. Phys.* **47,** 4180 (1967).

[D20] J. P. Doering, Low-energy electron-impact study of the first, second, and third triplet states of benzene, *J. Chem. Phys.* **51,** 2866 (1969).

[D21] J. P. Doering and J. H. Moore, Jr., Observation of a singlet-triplet transition in gas phase pyridine by ion and electron impact, *J. Chem. Phys.* **56,** 2176 (1972).

[D22] W. E. Donath, Pariser-Parr calculations and electro-optical effects in benzene, *J. Chem. Phys.* **40,** 77 (1964).

[D23] B. Dumbacher, *Ab initio* SCF and CI calculations on the barrier to internal rotation of 1,3-butadiene, *Theoret. Chim. Acta* **23,** 346 (1972).

[D24]   A. B. F. Duncan, Far ultraviolet absorption spectra of some aliphatic ketones, *J. Chem. Phys.* **8,** 444 (1940).
[D25]   A. B. F. Duncan and R. F. Whitlock, Vacuum ultraviolet absorption spectrum of carbonyl cyanide, *Spectrochim. Acta* **27A,** 2539 (1971).
[D26]   T. M. Dunn and C. K. Ingold, Symmetry of the second excited singlet state of benzene, *Nature (London)* **176,** 65 (1955).
[D27]   T. M. Dunn, The spectrum and structure of benzene, *in* "Studies on Chemical Structure and Reactivity" (J. H. Ridd, ed.). Methuen, London, 1966.
[D28]   T. H. Dunning, Jr., W. J. Hunt, and W. A. Goddard, III, The theoretical description of the (π, π^*) excited states of ethylene, *Chem. Phys. Lett.* **4,** 147 (1969).

[E1]   J. H. D. Eland and C. J. Danby, Inner ionization potentials of aromatic compounds, *Z. Naturforsch.* **23A,** 355 (1968).
[E2]   J. H. D. Eland, Photoelectron spectra of conjugated hydrocarbons and heteromolecules, *Int. J. Mass Spectrom. Ion Phys.* **2,** 471 (1969).
[E3]   J. H. D. Eland, The photoelectron spectra of isocyanic acid and related compounds, *Phil. Trans. Roy. Soc. London* **268A,** 87 (1970).
[E4]   J. H. D. Eland, Photoelectron spectra and ionization potentials of aromatic hydrocarbons, *Int. J. Mass Spectrom. Ion Phys.* **9,** 214 (1972).
[E5]   M. F. A. El-Sayed, M. Kasha, and Y. Tanaka, Ionization potential of benzene, hexadeuterobenzene, and pyridine from their observed Rydberg series in the region 600–2000 Å, *J. Chem. Phys.* **34,** 334 (1961).
[E6]   M. A. El-Sayed, Effect of reducing the symmetry on the spectra of benzene in the 1500–2000 Å region: Spectra of pyridine, pyrimidine, and pyrazine, *J. Chem. Phys.* **36,** 552 (1962).
[E7]   V. R. Ells, Absorption spectrum of biacetyl between 1500 and 2000 Å, *J. Amer. Chem. Soc.* **60,** 1864 (1938).
[E8]   D. F. Evans, Effect of nitrogen under pressure on the Rydberg spectra of polyatomic molecules; The nature of the long-wavelength olefin bands, *Proc. Chem. Soc. London* 378 (1963).
[E9]   S. Evans, A. Hammett, and A. F. Orchard, Photoelectron spectra of osmium and ruthenium tetroxides (unpublished).

[F1]   L. M. Falicov, R. A. Harris, and P. B. Visscher, Pi-electron theory of acetylene, *J. Chem. Phys.* **52,** 3675 (1970).
[F2]   U. Fano, Quantum defect theory of l-uncoupling in H_2 as an example of channel-interaction treatment, *Phys. Rev.* **2A,** 353 (1970).
[F3]   S. Feinleib and F. A. Bovey, Vapour-phase vacuum-ultraviolet circular-dichroism spectrum of (+)-3-methylcyclopentanone, *Chem. Commun.* 978 (1968).
[F4]   M. Fétizon and I. Hanna, The circular dichroism of methylene-steroids, *Chem. Commun.* 462 (1970).
[F5]   M. Fétizon, I. Hanna, A. I. Scott, A. D. Wrixon, and T. K. Devon, Olefin stereochemistry: Analysis of methylene steroids and related compounds by the octant rule, *Chem. Commun.* 545 (1971).
[F6]   G. Fleming, M. M. Anderson, A. J. Harrison, and L. W. Pickett, Effect of ring size on the far ultraviolet absorption and photolysis of cyclic ethers, *J. Chem. Phys.* **30,** 351 (1959).
[F7]   D. Florida, R. Scheps, and S. A. Rice, Antiresonances in the Rydberg

spectrum of naphthalene. A new analysis, *Chem. Phys. Lett.* **15**, 490 (1972).
[F8] S. N. Foner, and R. L. Hudson, Diimide—identification and study by mass spectrometry, *J. Chem. Phys.* **28**, 719 (1958).
[F9] S. Foster, S. Felps, L. C. Cusachs, and S. P. McGlynn, Photoelectron spectra of osmium and ruthenium tetroxides, *J. Amer. Chem. Soc.* **95**, 5521 (1973).
[F10] S. Foster, S. Felps, L. W. Johnson, and S. P. McGlynn, Electronic spectra of ruthenium and osmium tetroxide, *J. Amer. Chem. Soc.* **95**, 6578 (1973).
[F11] M. F. Fox and E. Hayon, New CTTS absorption bands of iodide in the far-ultraviolet region, *Chem. Phys. Lett.* **14**, 442 (1972).
[F12] C. Fridh, L. Åsbrink, B. Ö. Jonsson, and E. Lindholm, Rydberg series in small molecules. XIV. Photoelectron, UV, mass, and electron impact spectra of s-triazine, *Int. J. Mass Spectrom. Ion Phys.* **8**, 85 (1972).
[F13] C. Fridh, L. Åsbrink, B. Ö. Jonsson, and E. Lindholm, Rydberg series in small molecules. XV. Photoelectron, UV, mass, and electron impact spectra of pyrazine, *Int. J. Mass Spectrom. Ion Physics* **8**, 101 (1972).
[F14] C. Fridh, L. Åsbrink, B. Ö. Jonsson, and E. Lindholm, Rydberg series in small molecules. XVIII. Photoelectron, UV, mass, and electron impact spectra of s-tetrazine, *Int. J. Mass Spectrom. Ion Physics* **9**, 485 (1972).
[F15] J. A. Friend and L. E. Lyons, The electronic spectrum of the nitrate ion and related molecules, *J. Chem. Soc.* 1572 (1959).

[G1] J. T. Gary and L. W. Pickett, The far ultraviolet absorption spectra of the isomeric butenes, *J. Chem. Phys.* **22**, 599 (1954).
[G2] J. T. Gary and L. W. Pickett, The far ultraviolet absorption spectra of selected isomeric hexenes, *J. Chem. Phys.* **22**, 1266 (1954).
[G3] A. Gedanken, B. Raz, and J. Jortner, Vacuum ultraviolet spectroscopy of deep lying impurity states in solid CF_4, *Chem. Phys. Lett.* **14**, 172 (1972).
[G4] J. Geiger and K. Wittmaack, High resolution electron scattering spectrometry of the electronic and vibrational spectra of ethylene, *Z. Naturforsch.* **20A**, 628 (1965).
[G5] U. Gelius, C. J. Allan, D. A. Allison, H. Siegbahn, and K. Siegbahn, The electronic structure of carbon suboxide from ESCA and *ab initio* calculations, *Chem. Phys. Lett.* **11**, 224 (1971).
[G6] U. Gelius, C. J. Allan, G. Johansson, H. Siegbahn, D. A. Allison and K. Siegbahn, The ESCA spectra of benzene and the isoelectronic series, thiophene, pyrrole and furan, *Phys. Scripta* **3**, 237 (1971).
[G7] U. Gelius, B. Roos, and P. Siegbahn, MO-SCF-LCAO studies of sulfur compounds. III. Thiophene, *Theoret. Chim. Acta* **27**, 171 (1972).
[G8] E. P. Gentieu and J. E. Mentall, Formaldehyde absorption coefficients in the vacuum ultraviolet (650 to 1850 angstroms), *Science* **169**, 681 (1970).
[G9] G. A. George and G. C. Morris, Intensity of absorption of naphthalene from 30 000 cm^{-1} to 53 000 cm^{-1}, *J. Mol. Spectrosc.* **26**, 67 (1968).
[G10] E. Gilberg, The K X-ray emission spectrum of chlorine in free molecules, *Z. Phys.* **236**, 21 (1970).
[G11] R. Gilbert and C. Sandorfy, The vacuum ultraviolet spectrum of fluorobenzene, *Chem. Phys. Lett.* **9**, 121 (1971).

[G12] R. Gilbert, P. Sauvageau, and C. Sandorfy, Far-UV and photoelectron spectra of 1,3,5-trifluorobenzene, *Chem. Phys. Lett.* **17,** 465 (1972).
[G13] R. Gilbert, P. Sauvageau, and C. Sandorfy, Valence shell transitions in the vacuum ultraviolet spectra of fluorobenzenes, *Can. J. Chem.* **50,** 543 (1972).
[G14] R. Gleiter, E. Heilbronner, and V. Hornung, Photoelectron spectra of azabenzenes and azanaphthalenes: I. Pyridine, diazines, *s*-triazine, and *s*-tetrazine, *Helv. Chim. Acta* **55,** 255 (1972).
[G15] T. S. Godfrey and G. Porter, Absorption spectrum of triplet benzene, *Trans. Faraday Soc.* **62,** 7 (1966).
[G16] C. F. Goodeve and S. Katz, Absorption spectrum of nitrosyl chloride, *Proc. Roy. Soc. London,* **A172,** 432 (1939).
[G17] K. Goto, Absorption coefficients of *trans*-dichloroethylene in the vacuum ultraviolet, *Sci. Light* **9,** 104 (1960).
[G18] K. Goto, Absorption coefficient of benzene in the vacuum ultraviolet, *Sci. Light* **11,** 116 (1962).
[G19] K. Goto, Absorption coefficient of tetrachloroethylene in the vacuum ultraviolet, *Sci. Light* **11,** 119 (1962).
[G20] J. Granier, N. Damany-Astoin, and M. Cordier, Absorption spectrum of benzene in the ultraviolet region, *C. R. Acad. Sci. Paris* **251,** 2672 (1960).
[G21] W. B. Gratzer, G. M. Holzwarth, and P. Doty, Polarization of the ultraviolet absorption bands in α-helical polypeptides, *Proc. Nat. Acad. Sci. U.S.* **47,** 1785 (1961).
[G22] H. B. Gray and N. A. Beach, The electronic structures of octahedral metal complexes. I. Metal hexacarbonyls and hexacyanides, *J. Amer. Chem. Soc.* **85,** 2922 (1963).
[G23] E. W. Greene, Jr., J. Barnard, and A. B. F. Duncan, Rydberg terms of acetylene, *J. Chem. Phys.* **54,** 71 (1971).
[G24] J. C. Green, M. L. H. Green, P. J. Joachim, A. F. Orchard, and D. W. Turner, A study of the bonding in the group IV tetrahalides by photoelectron spectroscopy, *Phil. Trans. Roy. Soc. London* **268A,** 111 (1970).
[G25] N. J. Greenfield and G. D. Fasman, Optical activity of simple cyclic amides in solution, *Biopolymers* **7,** 595 (1969).

[H1] T.-K. Ha and W. Hug, The interaction of nonbonding orbitals in dicarbonyls: *Ab initio* results on glyoxal, *Helv. Chim. Acta* **54,** 2278 (1971).
[H2] T.-K. Ha, *Ab initio* calculation of *cis-trans* isomerization in glyoxal, *J. Mol. Struct.* **12,** 171 (1972).
[H3] M. Hackmeyer and J. L. Whitten, Configuration interaction studies of ground and excited states of polyatomic molecules. II. The electronic states and spectrum of pyrazine, *J. Chem. Phys.* **54,** 3739 (1971).
[H4] J. Halper, W. D. Closson, and H. B. Gray, Electronic structures and spectra of cyanoethylene derivatives, *Theoret. Chim. Acta* **4,** 174 (1966).
[H5] J. S. Ham and J. R. Platt, Far U.V. spectra of peptides, *J. Chem. Phys.* **20,** 335 (1952).
[H6] V. J. Hammond, W. C. Price, J. P. Teegan, and A. D. Walsh, The absorption spectra of some substituted benzenes and naphthalenes in the vacuum ultraviolet, *Discuss. Faraday Soc.* **9,** 53 (1950).

[H7] V. J. Hammond and W. C. Price, Oscillator strengths of the vacuum ultraviolet absorption bands of benzene and ethylene, *Trans. Faraday Soc.* **51**, 605 (1955).

[H8] A. E. Hansen, Correlation effects in the calculation of ordinary and rotatory intensities, *Mol. Phys.* **13**, 425 (1967).

[H9] A. J. Harrison, C. L. Gaddis, and E. M. Coffin, Quantitative determination of extinction coefficients in the vacuum ultraviolet: Divinyl ether, *J. Chem. Phys.* **18**, 221 (1950).

[H10] E. Haselbach, J. A. Hashmall, E. Heilbronner, and V. Hornung, The interaction between the lone pairs in azomethane, *Angew. Chem. Int. Ed.* **8**, 878 (1969).

[H11] E. Haselbach, Jahn-Teller distortions in the radical cations of cyclopropane and allene, *Chem. Phys. Lett.* **7**, 428 (1970).

[H12] K. W. Hausser, R. Kuhn, A. Smakula, and M. Hoffer, Light absorption and double bonds. II. Polyene aldehydes and polyene carboxylic acids, *Z. Phys. Chem.* **29B**, 371 (1935).

[H13] C. H. Heathcock and S. R. Poulter, The effect of cyclopropyl substitution on the $\pi \rightarrow \pi^*$ transition of simple olefins, *J. Amer. Chem. Soc.* **90**, 3766 (1968).

[H14] V. Henri and L. W. Pickett, The ultraviolet absorption spectrum of 1,3-cyclohexadiene, *J. Chem. Phys.* **7**, 439 (1939).

[H15] R. B. Hermann, The ultraviolet spectrum of bicycloheptadiene, *J. Org. Chem.* **27**, 441 (1962).

[H16] G. Herzberg and G. Scheibe, Concerning the gas phase absorption spectra of the methyl halides and a few other methyl compounds in the ultraviolet and Schumann region, *Z. Phys. Chem.* **B7**, 390 (1930).

[H17] G. Herzberg, "Molecular Spectra and Molecular Structure," Vol. II, Infrared and Raman Spectra of Polyatomic Molecules. Van Nostrand Reinhold, Princeton, New Jersey, 1945.

[H18] G. Herzberg, The spectra and structures of free methyl and free methylene, *Proc. Roy. Soc. London* **262A**, 291 (1961).

[H19] G. Herzberg, Determination of the structures of simple polyatomic molecules and radicals in electronically excited states, *Discuss. Faraday Soc.* **35**, 7 (1963).

[H20] G. Herzberg, "Molecular Spectra and Molecular Structure," Vol. III, Electronic Spectra and Electronic Structure of Polyatomic Molecules. Van Nostrand Reinhold, Princeton, New Jersey, 1966.

[H21] A. Herzenberg, D. Sherrington, and M. Süveges, Correlations of electrons in small molecules, *Proc. Phys. Soc. London* **84**, 465 (1964).

[H22] M. Hirota, T. Hagiwara, and H. Satonaka, Infrared carbonyl absorptions and conformations of methyl-substituted acetones, *Bull. Chem. Soc. Japan* **40**, 2439 (1967).

[H23] R. Hoffmann, E. Heilbronner, and R. Gleiter, Interaction of nonconjugated double bonds, *J. Amer. Chem. Soc.* **92**, 706 (1970).

[H24] R. S. Holdsworth and A. B. F. Duncan, Intensities of electronic transitions in aliphatic ketones in the vacuum ultraviolet, *Chem. Rev.* **41**, 311 (1947).

[H25] J. M. Hollas and T. A. Sutherley, The geometry of the ground state of $C_2H_2^+$ from photoelectron spectroscopy compared with that of C_2H_2 in some Rydberg states, *Mol. Phys.* **21**, 183 (1971).

[H26] G. Holzwarth and P. Doty, The ultraviolet circular dichroism of polypeptides, *J. Amer. Chem. Soc.* **87**, 218 (1965).

[H27] B. Honig, J. Jortner, and A. Szöke, Theoretical studies of two-photon absorption processes. I. Molecular benzene, *J. Chem. Phys.* **46**, 2714 (1967).

[H28] M. Hori, K. Kimura, and H. Tsubomura, The electronic spectrum and the chemiluminescence of tetrakisdimethylamino ethylene (TDAE), *Spectrochim. Acta* **24A**, 1397 (1968).

[H29] G. Horváth and A. I. Kiss, The electronic spectra of five-membered heterocyclic compounds, *Spectrochim. Acta* **23A**, 921 (1967).

[H30] J. F. Horwood and J. R. Williams, Vapour phase carbonyl absorption in the far ultraviolet, *Spectrochim. Acta* **19**, 1351 (1963).

[H31] H. Hosoya, J. Tanaka, and S. Nagakura, Ultraviolet absorption spectra of monomer and dimer of benzoic acid, *J. Mol. Spectrosc.* **8**, 257 (1962).

[H32] H. Hosoya and S. Nagakura, The electronic structure and spectrum of tropone, *Theoret. Chim. Acta* **8**, 319 (1967).

[H33] M. J. Hubin-Fraskin and J. E. Collin, Electron impact excitation by the SF_6 scavenger technique. II. Benzene and ethylene, *Int. J. Mass Spectrom. Ion Phys.* **5**, 163 (1970).

[H34] R. H. Huebner, S. R. Mielczarek, and C. E. Kuyatt, Electron energy-loss spectroscopy of naphthalene vapor, *Chem. Phys. Lett.* **16**, 464 (1972).

[H35] R. H. Huebner, R. J. Cellota, S. R. Mielczarek, and C. E. Kuyatt, Electron energy-loss spectroscopy of acetone vapor, *J. Chem. Phys.* **59**, 5434 (1973).

[H36] C. M. Humphries, A. D. Walsh, and P. A. Warsop, Ultraviolet absorption spectrum of tetrachloroethylene, *Trans. Faraday Soc.* **63**, 513 (1967).

[H37] H. D. Hunt and W. T. Simpson, Spectra of simple amides in the vacuum ultraviolet, *J. Amer. Chem. Soc.* **75**, 4540 (1953).

[H38] H. E. Hunziker, Gas-phase absorption spectrum of triplet naphthalene in the 220–300 nm and 410–620 nm wavelength regions, *J. Chem. Phys.* **56**, 400 (1972).

[H39] D. A. Hutchinson, Structure of the butadiene positive ion, *Trans. Faraday Soc.* **59**, 1695 (1963).

[I1] T. Inagaki, Absorption spectra of pure liquid benzene in the ultraviolet region, *J. Chem. Phys.* **57**, 2526 (1972).

[I2] C. K. Ingold and G. W. King, Excited states of acetylene. Part I. Possibilities of interaction between σ-bond hybridization and π-electron excitation with resulting changes of shape during transitions, *J. Chem. Soc.* 2702 (1953).

[I3] K. K. Innes, Analysis of the near ultraviolet absorption spectrum of acetylene, *J. Chem. Phys.* **22**, 863 (1954).

[I4] K. K. Innes, J. P. Byrne, and I. G. Ross, Electronic states of azabenzenes: A critical review, *J. Mol. Spectrosc.* **22**, 125 (1967).

[I5] M. Isaacson, Interaction of 25 KeV electrons with the nucleic acid bases, adenine, thymine, and uracil. I. Outer shell excitation, *J. Chem. Phys.* **56**, 1803 (1972).

[I6] M. Isaacson, Interaction of 25 KeV electrons with the nucleic acid bases, adenine, thymine, and uracil. II. Inner shell excitation and inelastic scattering cross-sections, *J. Chem. Phys.* **56**, 1813 (1972).

[I7] H. Ito, Y. Nogata, S. Matsuzaki, and A. Kuboyama, Vacuum-ultraviolet absorption spectra of aliphatic ketones, *Bull. Chem. Soc. Japan* **42**, 2453 (1969).

[I8] A. Iverson and B. R. Russell, The vacuum ultraviolet spectra of three transition metal hexacarbonyls, *Chem. Phys. Lett.* **6**, 307 (1970).

[I9] A. A. Iverson and B. R. Russell, A medium resolution study of allene in the vacuum ultraviolet. I. Spectra and a preliminary ionization potential, *Spectrochim. Acta* **28A**, 447 (1972).

[J1] L. E. Jacobs and J. R. Platt, Does ultraviolet absorption intensity increase in solution? *J. Chem. Phys.* **16**, 1137 (1948).

[J2] J. H. Jaffe, H. Jaffe, and K. Rosenheck, New method of measuring linear dichroism in the ultraviolet; Application to helical polymers, *Rev. Sci. Instrum.* **38**, 935 (1967).

[J3] J. Jäger, Energy-loss of monochromatic electrons in condensed aromatic hydrocarbons, *Ann. Phys.* **22**, 147 (1969).

[J4] C. D. Johnson and T. B. Rymer, Existence of collective-excitation energy losses from an electron beam passing through biological materials, *Nature (London)* **213**, 1045 (1967).

[J5] L. W. Johnson and S. P. McGlynn, The electronic absorption spectrum of the chromate ion, *Chem. Phys. Lett.* **7**, 618 (1970).

[J6] L. W. Johnson, H. J. Maria, and S. P. McGlynn, Luminescence of the carboxyl group, *J. Chem. Phys.* **54**, 3823 (1971).

[J7] L. W. Johnson and S. P. McGlynn, Electronic absorption spectrum of $LiMnO_4 \cdot 3H_2O$ in $LiClO_4 \cdot 3H_2O$, *J. Chem. Phys.* **55**, 2985 (1971).

[J8] W. C. Johnson, Jr. and W. T. Simpson, Assignment of the second singlet in carbonyl, *J. Chem. Phys.* **48**, 2168 (1968).

[J9] W. C. Johnson, Jr., A circular dichroism spectrometer for the vacuum ultraviolet, *Rev. Sci. Instrum.* **42**, 1283 (1971).

[J10] W. C. Johnson, Jr., and I. Tinoco, Jr., Circular dichroism of polypeptide solutions in the vacuum ultraviolet, *J. Amer. Chem. Soc.* **94**, 4389 (1972).

[J11] N. Jonathan, K. Ross, and V. Tomlinson, The photoelectron spectra of dichloroethylenes, *Int. J. Mass Spectrom. Ion Phys.* **4**, 51 (1970).

[J12] L. C. Jones, Jr. and L. W. Taylor, Far ultraviolet absorption spectra of unsaturated and aromatic hydrocarbons, *Anal. Chem.* **27**, 228 (1955).

[J13] B.-Ö. Jonsson, E. Lindholm, and A. Skerbele, The electronic structure of pyridine, *Int. J. Mass Spectrom. Ion Phys.* **3**, 385 (1969).

[J14] B.-Ö. Jonsson and E. Lindholm, The electronic structure of benzene, *Ark. Fys.* **39**, 65 (1969).

[J15] C. K. Jorgensen and J. S. Brinen, Far ultraviolet absorption bands of osmium (IV), iridium (IV), and platinum (IV) hexahalides, *Mol. Phys.* **5**, 535 (1962).

[J16] J. Jortner, private communication, 1972.

[J17] J. Jortner, E. G. Wilson, and S. A. Rice, Theoretical and experimental studies of the electronic structure of the xenon fluorides, *in* "Noble—Gas Compounds" (H. H. Hyman, ed.), p. 358. Univ. Chicago Press, Chicago, Illinois, 1963.

[J18] J. Jortner, E. G. Wilson, and S. A. Rice, A far-ultraviolet spectroscopic study of xenon tetrafluoride, *J. Amer. Chem. Soc.* **85**, 815 (1963).

[J19] J. Jortner and G. C. Morris, Interference effects in the optical spectrum of large molecules, *J. Chem. Phys.* **51**, 3689 (1969).

[J20] C. Jungen, Rydberg series in the NO spectrum: An interpretation of quantum defects and intensities in the s and d series, *J. Chem. Phys.* **53**, 4168 (1970).

[K1] H. Kaiser, Vacuum UV Investigation of Molecules of Intermediate Size, Dissertation, Ludwig-Maximilians Univ., Munchen, 1970.

[K2] U. Kaldor and I. Shavitt, LCAO-SCF computations for ethylene, *J. Chem. Phys.* **48**, 191 (1968).

[K3] W. E. Kammer, Ab initio SCF and CI calculations of linear and bent acetylene, *Chem. Phys. Lett.* **6**, 529 (1970).

[K4] J. Karwowski, The electronic spectrum of benzene, *Chem. Phys. Lett.* **18**, 47 (1973).

[K5] B. Katz, M. Brith, A. Ron, B. Sharf, and J. Jortner, An experimental study of the higher excited states of benzene in a rare gas matrix, *Chem. Phys. Lett.* **2**, 189 (1968).

[K6] B. Katz and J. Jortner, Observation of molecular Rydberg states in rare gas solids, *Chem. Phys. Lett.* **2**, 437 (1968).

[K7] B. Katz, M. Brith, B. Sharf, and J. Jortner, Rydberg states of benzene in rare-gas matrices, *J. Chem. Phys.* **50**, 5195 (1969).

[K8] B. Katz, M. Brith, B. Sharf, and J. Jortner, Experimental study of the higher $\pi \rightarrow \pi^*$ transitions of benzene in low-temperature matrices, *J. Chem. Phys.* **52**, 88 (1970).

[K9] B. Katz, M. Brith, B. Sharf, and J. Jortner, Electronic spectra of some methyl substituted benzenes in a rare-gas solid, *J. Chem. Phys.* **54**, 3924 (1971).

[K10] K. Kaya, K. Kuwata, and S. Nagakura, The ultraviolet absorption spectra of nitramide and ethyl nitrate, *Bull. Chem. Soc. Japan* **37**, 1055 (1964).

[K11] K. Kaya and S. Nagakura, Vacuum ultraviolet absorption spectra of simple amides, *Theoret. Chim. Acta* **7**, 117 (1967).

[K12] K. Kaya and S. Nagakura, The electronic absorption spectra of the 2,5-diketopiperazine single crystal and evaporated film, *J. Mol. Spectrosc.* **44**, 279 (1972).

[K13] H. H. Kim and J. L. Roebber, Vacuum ultraviolet absorption spectrum of carbon suboxide, *J. Chem. Phys.* **44**, 1709 (1966).

[K14] H. S. Kimmel and W. H. Snyder, Vibrational spectrum of 2,3-dimethyl-2-butene, *J. Mol. Struct.* **4**, 473 (1969).

[K15] K. Kimura, H. Tsubomura, and S. Nagakura, The vacuum ultraviolet absorption spectra of aniline and some of its N-derivatives, *Bull. Chem. Soc. Japan* **37**, 1336 (1964).

[K16] K. Kimura and S. Nagakura, Vacuum ultraviolet absorption spectra of various mono-substituted benzenes, *Mol. Phys.* **9**, 117 (1965).

[K17] K. Kimura and S. Nagakura, Vacuum ultraviolet spectra of styrene, benzaldehyde, acetophenone, and benzonitrile, *Theoret. Chim. Acta* **3**, 164 (1965).

[K18] K. Kimura and S. Nagakura, Vacuum-ultraviolet absorption spectrum of α, α, α-trichlorotoluene, *J. Chem. Phys.* **47**, 2916 (1967).

[K19] G. H. Kirby and K. Miller, CNDO geometry of ethylene in the first singlet excited state, *Chem. Phys. Lett.* **3**, 643 (1969).

[K20] D. N. Kirk, W. Klyne, W. P. Mose, and E. Otto, Circular dichroism of ketones at 185–195 mμ, *Chem. Commun.* **35** (1972).

[K21] T. Kitagawa, Y. Harada, H. Inokuchi, and K. Kodera, Absorption spectrum of vapor phase azulene in the vacuum ultraviolet region, *J. Mol. Spectrosc.* **19**, 1 (1966).

[K22] T. Kitagawa, Absorption spectra and photoionization of polycyclic aromatics in the vacuum ultraviolet region, *J. Mol. Spectrosc.* **26**, 1 (1968).

[K23] M. Klasson and R. Manne, Molecular orbital interpretation of X-ray emission and photoelectron spectra. III. Chloroethylenes, *in* "Electron Spectroscopy" (D. Shirley, ed.), p. 471. North-Holland Publ., Amsterdam, 1972.

[K24] H. B. Klevens, J. R. Platt, and L. E. Jacobs, Ultraviolet spectra of fluorinated benzenes, *J. Amer. Chem. Soc.* **70**, 3526 (1948).

[K25] H. B. Klevens and J. R. Platt, Geometry and spectra of substituted anilines, *J. Amer. Chem. Soc.* **71**, 1714 (1949).

[K26] H. B. Klevens and J. R. Platt, Spectral resemblances of cata-condensed hydrocarbons, *J. Chem. Phys.* **17**, 470 (1949).

[K27] H. B. Klevens, Spectral resemblances between azulenes and their corresponding six-carbon ring isomers, *J. Chem. Phys.* **18**, 1063 (1950).

[K28] H. B. Klevens and L. J. Zimring, Absorption spectra and directing power of substituents in the fluorobenzenes, *J. Chim. Phys.* **49**, 377 (1952).

[K29] H. B. Klevens, Extreme ultraviolet absorption spectra of various polymers, *J. Polym. Sci.* **10**, 97 (1953).

[K30] H. B. Klevens and J. R. Platt, Survey of vacuum ultraviolet spectra of organic compounds in solution, *in* "Systematics of the Electronic Spectra of Conjugated Molecules: A Source Book" (J. R. Platt, ed.), p. 145. Wiley, New York, 1964.

[K31] F. W. E. Knoop, J. Kistemaker, and L. J. Oosterhoff, Low energy electron impact excitation spectra of 1,3,5-cyclo-heptatriene and 1,3,5,7-cyclo-octatetraene, *Chem. Phys. Lett.* **3**, 73 (1969).

[K32] F. W. E. Knoop, Excited States of Atoms and Molecules, Ph.D. Thesis, Univ. of Leiden, 1972.

[K33] E. E. Koch and A. Otto, Characteristic energy losses of 30 KeV electrons in vapours of aromatic hydrocarbons, *Opt. Commun.* **1**, 47 (1969).

[K34] E. E. Koch and A. Otto, Inelastic scattering of 30 KeV electrons in vapours of triphenylene, perylene and ferrocene, *Chem. Phys. Lett.* **6**, 15 (1970).

[K35] E. E. Koch and A. Otto, Optical anisotropy of anthracene single crystals for excitation energies from 4.5–11.5 eV, *Int. Symp. Vacuum Ultraviolet Spectrosc.*, Tokyo (1971).

[K36] E. E. Koch, S. Kunstreich, and A. Otto, Measurement of electron energy losses and VUV reflectivity of anthracene single crystals, *Opt. Commun.* **2**, 365 (1971).

[K37] E. E. Koch and A. Otto, Optical absorption of benzene vapor for photon energies from 6 eV to 35 eV, *Chem. Phys. Lett.* **12**, 476 (1972).

[K38] S. Kondo, H. Yamashita and K. Nakamura, Optical properties of calcite in the vacuum ultraviolet, *J. Phys. Soc. Japan* **34**, 711 (1973).

[K39] M. Krauss and S. R. Mielczarek, Minima in generalized oscillator strengths: C_2H_4, *J. Chem. Phys.* **51**, 5241 (1969).

[K40] S. Kunstreich and A. Otto, Anisotropy of electron energy loss spectra in anthracene single crystals, *Opt. Commun.* **1**, 45 (1969).

[K41] A. Kuppermann and L. M. Raff, Electron-impact spectroscopy, *Discuss. Faraday Soc.* **35**, 30 (1963).

[K42] A. Kuppermann and L. M. Raff, Differences between low-energy electron-impact spectra at 0° and at large scattering angle, *J. Chem. Phys.* **39**, 1607 (1963).

[L1] J. S. Lake and A. J. Harrison, Absorption of acyclic oxygen compounds in the vacuum ultraviolet. III. Acetone and acetaldehyde, *J. Chem. Phys.* **30**, 361 (1959).

[L2] R. F. Lake and H. Thompson, Photoelectron spectra of halogenated ethylenes, *Proc. Roy. Soc. London* **A315**, 323 (1970).

[L3] R. F. Lake and H. Thompson, The photoelectron spectra of some molecules containing the C≡N group, *Proc. Roy. Soc. London* **A317**, 187 (1970).

[L4] J. E. Lancaster, R. F. Stamm, and N. B. Colthup, The vibrational spectra of s-triazine and s-triazine-d_3, *Spectrochim. Acta* **17**, 155 (1961).

[L5] A. Langseth and B. Qviller, The ultraviolet absorption spectrum of osmium tetroxide, *Z. Phys. Chem.* **B27**, 79 (1934).

[L6] S. R. La Paglia and A. B. F. Duncan, Vacuum ultraviolet absorption spectrum of phosgene, *J. Chem. Phys.* **34**, 125 (1961).

[L7] S. R. La Paglia, Rydberg levels of polyatomic molecules: Oxygen-containing molecules, *J. Mol. Spectrosc.* **10**, 240 (1963).

[L8] G. C. Lardy, Ultraviolet absorption spectrum of biacetyl, *C. R. Acad. Sci. Paris* **176**, 1548 (1923).

[L9] E. N. Lassettre, A. Skerbele, M. A. Dillon, and K. J. Ross, High-resolution study of electron-impact spectra at kinetic energies between 33 and 100 eV and scattering angles to 16°, *J. Chem. Phys.* **48**, 5066 (1968).

[L10] V. W. Laurie, private communication, 1972.

[L11] R. E. LaVilla and H. Mendlowitz, Optical constants of thin films from the characteristic electron energy losses, *J. Phys. Radium* **25**, 114 (1964).

[L12] R. E. LaVilla and R. D. Deslattes, Chlorine Kβ X-ray emission spectra from several chlorinated hydrocarbon and fluorocarbon molecular gases, *J. Chem. Phys.* **45**, 3446 (1966).

[L13] R. E. LaVilla and R. D. Deslattes, Single and multiple vacancy effects in molecular X-ray spectra, *J. Phys.* **32**, C4–160 (1971).

[L14] M. Lawson and A. B. F. Duncan, Spectrum of deutero-acetone in the vacuum ultraviolet. A comparison with the spectrum of acetone, *J. Chem. Phys.* **12**, 329 (1944).

[L15] A. T.-H. Lee, Electronic Absorption Spectra of Olefins in the Ultraviolet and Vacuum Ultraviolet Spectral Regions, Ph.D. Thesis, Univ. of California at Los Angeles, 1968.

[L16] M. Legrand and R. Viennet, Circular dichroism and unsaturated bonds in cyclic systems, *C. R. Acad. Sci. Paris* **262C**, 1290 (1966).

[L17] C. Leibovici and J.-F. Labarre, Electronic spectra and structure of nitro compounds. II. Nitro-ethylene, *J. Chim. Phys.* **67**, 1664 (1970).

[L18] C. Leibovici, Valence shell calculations of the structure and spectrum of vinyl cyanide, *J. Mol. Struct.* **9**, 177 (1971).

[L19] M. Lenzi and H. Okabe, Photodissociation of NOCl and NO_2 in the vacuum ultraviolet, *Ber. Bunsenges.* **72**, 168 (1968).

[L20] C. C. Levin and R. Hoffmann, Role of torsion in the chirality of twisted olefins, *J. Amer. Chem. Soc.* **94**, 3446 (1972).

[L21] D. Lewis, P. B. Merkel, and W. H. Hamill, Low-energy electron-reflection spectrometry for thin films of aromatic and aliphatic molecules at 77°K, *J. Chem. Phys.* **53**, 2750 (1970).

[L22] H. Ley and B. Arends, The absorption of the carbonyl chromophore in the short wavelength ultraviolet, *Z. Phys. Chem.* **B12**, 132 (1931).

[L23] H. Ley and B. Arends, Absorption measurements in the short wavelength ultraviolet. II. Carboxylic acids, amines, and amino acids, *Z. Phys. Chem.* **B17**, 177 (1932).

[L24] A. D. Liehr, Butadiene Rydberg spectrum, *J. Chem. Phys.* **25**, 781 (1956).

[L25] A. D. Liehr, and W. Moffitt, Rydberg spectrum of benzene, *J. Chem. Phys.* **25**, 1074 (1956).

[L26] A. D. Liehr, Theory of Rydberg series in molecular spectra, *Z. Naturforsch.* **11A**, 752 (1956).

[L27] A. D. Liehr, Topological aspects of the conformational stability problem. II. Nondegenerate electronic states, *J. Phys. Chem.* **67**, 471 (1963).

[L28] A. D. Liehr, Topological aspects of the conformational stability problem. I. Degenerate electronic states, *J. Phys. Chem.* **67**, 389 (1963).

[L29] E. Lindholm and B.-Ö. Jonsson, Electronic structure of benzene, *Chem. Phys. Lett.* **1**, 501 (1967).

[L30] B. J. Litman and J. A. Schellman, The n-π* Cotton effect of the peptide linkage, *J. Phys. Chem.* **69**, 978 (1965).

[L31] W. A. Little, private communication, 1973.

[L32] B. B. Loeffler, E. Eberlin, and L. W. Pickett, Far ultraviolet absorption spectra of small ring hydrocarbons, *J. Chem. Phys.* **28**, 345 (1958).

[L33] K. R. Loos, U. P. Wild, and H. H. Günthard, Ultraviolet spectra and electronic structure of nitroethylene, *Spectrochim. Acta* **25A**, 275 (1969).

[L34] R. C. Lord and P. Venkateswarlu, The rotation–vibration spectra of allene and allene-d_4, *J. Chem. Phys.* **20**, 1237 (1952).

[L35] A. J. Lorquet and J. C. Lorquet, Electronic structure of ionized molecules. VII. Ethylene, *J. Chem. Phys.* **49**, 4955 (1968).

[L36] G. Loustauneau and G. Nouchi, Singlet–singlet and triplet–triplet absorption spectra of certain diazines in crystalline solution at 77°K, *C. R. Acad. Sci. Paris* **261**, 4693 (1965).

[L37] A. Lubezky and R. Kopelman, Electronic spectrum of ethylene single crystal; Search for the olefin mystery band, *J. Chem. Phys.* **45**, 2526 (1966).

[L38] G. Lucazeau and C. Sandorfy, On the far-ultraviolet spectra of some simple aldehydes, *J. Mol. Spectrosc.* **35**, 214 (1970).

[L39] L. E. Lyons and G. C. Morris, The absorption spectrum of anthracene vapor from 36 000 to 66 000 cm^{-1}, *J. Mol. Spectrosc.* **4**, 480 (1960).

[L40] L. E. Lyons and G. C. Morris, The intensity of ultraviolet light absorption by monocrystals. V. Absorption by naphthacene at 295 and 78°K of plane-polarized light of wavelengths 1850–3200 Å, *J. Chem. Soc.* 2764 (1965).

[L41] A. E. Lyuts, A. E. Cherkashin, and Yu. A. Kushnikov, Vacuum ultraviolet spectra of aliphatic carbonyl compounds, *Izv. Akad. Nauk Kazakhskoi SSR Ser. Khim.* **18**, 55 (1968).

[M1] H. E. Mahncke and W. A. Noyes, Jr., Ultraviolet absorption spectra of cis- and trans-dichloroethylenes, *J. Chem. Phys.* **3**, 536 (1935).

[M2] J. G. Malm, H. Selig, J. Jortner, and S. A. Rice, The chemistry of xenon, *Chem. Rev.* **65**, 199 (1965).

[M3] D. E. Mann, J. R. Platt, and H. B. Klevens, Spectral resemblances in azulene and naphthalene, *J. Chem. Phys.* **17**, 481 (1949).

[M4] H. J. Maria, D. Larson, M. E. McCarville, and S. P. McGlynn, Electronic spectroscopy of isoelectronic molecules. I. Nonlinear triatomic groupings containing eighteen valence electrons, with comments on the T_1 state of the peptide linkage, *Accounts Chem. Res.* **3**, 368 (1970).

[M5] H. J. Maria and S. P. McGlynn, Electronic states of oxalic acid and dimethyl oxalate. Absorption studies, *J. Mol. Spectrosc.* **42**, 177 (1972).

[M6] M. G. Mason and O. Schnepp, The absorption and circular dichroism spectra of ethylenic chromophores—trans-cyclooctene, α- and β-pinene, *J. Chem. Phys.* **59**, 1092 (1973).

[M7] S. F. Mason, The magnetic dipole character and the rotatory power of the 3000 Å carbonyl absorption, *Mol. Phys.* **5**, 343 (1962).

[M8] S. F. Mason and G. W. Vane, The nature of the long-wavelength absorption bands of simple unsaturated chromophores, *Chem. Commun.* 540 (1965).

[M9] O. Matsuoka and Y. I'Haya, The electronic structure and spectrum of nitrobenzene, *Mol. Phys.* **8**, 455 (1964).

[M10] R. McDiarmid and E. Charney, Far-ultraviolet spectrum of ethylene and ethylene-d_4, *J. Chem. Phys.* **47**, 1517 (1967).

[M11] R. McDiarmid, Vibrational intensity progression in the V ← N transition of ethylene, *J. Chem. Phys.* **50**, 1794 (1969).

[M12] R. McDiarmid, Rydberg progressions in cis- and trans-butene, *J. Chem. Phys.* **50**, 2328 (1969).

[M13] R. McDiarmid, On the 1795-Å transition of 2-methylpropene (isobutene), *J. Chem. Phys.* **55**, 2426 (1971).

[M14] R. McDiarmid, Higher electronic states of ReF$_6$, *J. Mol. Spectrosc.* **39**, 332 (1971).

[M15] R. McDiarmid, Origin of the V ← N transition of liquid ethylene, *J. Chem. Phys.* **55**, 4669 (1971).

[M16] J. R. McDonald, J. W. Rabalais, and S. P. McGlynn, Electronic spectra of the azide ion, hydrazoic acid, and azido molecules, *J. Chem. Phys.* **52**, 1332 (1970).

[M17] J. R. McDonald, V. M. Scherr, and S. P. McGlynn, Lower-energy electronic states of HNCS, NCS$^-$ and thiocyanate salts, *J. Chem. Phys.* **51**, 1723 (1969).

[M18] K. L. McEwen, Electronic structures and spectra of nitromethane and nitrogen dioxide, *J. Chem. Phys.* **32**, 1801 (1960).

[M19] K. L. McEwen, Electronic structure and spectra of some nitrogen–oxygen compounds, *J. Chem. Phys.* **34**, 547 (1961).

[M20] H. L. McMurry, The long wavelength spectra of aldehydes and ketones. I. Saturated aldehydes and ketones, *J. Chem. Phys.* **9**, 231 (1941).

[M21] R. McWeeny and T. E. Peacock, The electronic structure and spectra of some nitrogen heterobenzenes, *Proc. Phys. Soc. London* **A70**, 41 (1957).

[M22] J. E. Mentall, E. P. Gentieu, M. Krauss, and D. Neumann, Photoionization and absorption spectrum of formaldehyde in the vacuum ultraviolet, *J. Chem. Phys.* **55**, 5471 (1971).

[M23] A. J. Merer, The vacuum ultraviolet absorption spectrum of diazomethane, *Can. J. Phys.* **42**, 1242 (1964).

[M24] A. J. Merer and L. Schoonveld, Nonplanarity of the first Rydberg state of ethylene, *J. Chem. Phys.* **48**, 522 (1968).

[M25] A. J. Merer and R. S. Mulliken, Vibrational structure of the π^*-π electronic transition of ethylene, *J. Chem. Phys.* **50**, 1026 (1969).

[M26] A. J. Merer and R. S. Mulliken, Ultraviolet spectra and excited states of ethylene and its alkyl derivatives, *Chem. Rev.* **69**, 639 (1969).

[M27] A. J. Merer and L. Schoonveld, Electronic spectrum of ethylene. I. The 1744 Å Rydberg transition, *Can. J. Phys.* **47**, 1731 (1969).

[M28] A. J. Merer, R. S. Mulliken, and J. K. G. Watson, cited in Merer and Mulliken [M26].

[M29] P. H. Metzger and G. R. Cook, On the continuous absorption, photoionization and fluorescence of H_2O, NH_3, CH_4, C_2H_2, C_2H_4, and C_2H_6 in the 600 to 1000-Å region, *J. Chem. Phys.* **41**, 642 (1964).

[M30] A. Y. Meyer, B. Muel, and M. Kasha, The charge transfer band in the ultraviolet spectrum of cyclopropyl ketones, *J. Chem. Soc. D (Chem. Commun.)* 401 (1972).

[M31] G. Milazzo, On the ultraviolet absorption spectra of pyrrole and N-deuteropyrrole in the gas phase, *Spectrochim. Acta* **2**, 245 (1942).

[M32] G. Milazzo, Absorption spectrum of N-methyl pyrrole, *Gazz. Chim. Ital.* **74**, 152 (1944).

[M33] G. Milazzo, Absorption spectra of heterocyclic molecules. III. The ultraviolet absorption spectrum of thiophene, *Gazz. Chim. Ital.* **78**, 835 (1948).

[M34] G. Milazzo, Spectral relationships of a few pentatomic heterocyclic substances in the Schüler discharge tube, *Experientia* **8**, 259 (1952).

[M35] G. Milazzo, Emission spectrum of undecomposed N-methylpyrrole, *J. Chem. Phys.* **21**, 163 (1953).

[M36] G. Milazzo and E. Miescher, Absorption spectra of heterocyclic compounds. V. Absorption spectra of selenophene and N-methyl pyrrole in the Schumann region, *Gazz. Chim. Ital.* **83**, 782 (1953).

[M37] G. Milazzo, The emission spectrum of N-methyl pyrrole, *Rend. Dell' Istit. Sup. Sanita* **17**, 868 (1954).

[M38] G. Milazzo and E. Miescher, Absorption spectra of heterocyclic molecules. V. Absorption spectra of selenophene and N-methyl pyrrole in the Schumann region, *Rend. Dell' Istit. Sup. Sanita* **19**, 80 (1955).

[M39] G. Milazzo and P. de Gasperis, Spectra of *para, meta* and *ortho*-terphenyl in the ultraviolet region, *J. Chim. Phys.* **65**, 1171 (1968).

[M40] G. Milazzo, Molecular spectra of selenophene, *Gazz. Chim. Ital.* **98**, 1511 (1968).

[M41] K. J. Miller, Dependence of generalized oscillator strengths of ethylene on momentum transfer, *J. Chem. Phys.* **51**, 5235 (1969).

[M42] D. E. Milligan, M. E. Jacox, A. M. Bass, J. J. Comeford, and D. E. Mann, Matrix-isolation study of the reaction of F atoms with CO. Infrared and ultraviolet spectra of the free radical FCO, *J. Chem. Phys.* **42**, 3187 (1965).

[M43] E. Miron, B. Raz, and J. Jortner, A comment on the V excited state of the ethylene molecule, *Chem. Phys. Lett.* **6**, 563 (1970).

[M44] E. Miron, B. Raz, and J. Jortner, Intravalence and extravalence excitations of the ethylene molecule in liquid and solid rare gas matrices, *J. Chem. Phys.* **56**, 5265 (1972).

[M45] G. Moe and A. B. F. Duncan, Intensities of electronic transitions of acetylene in the vacuum ultraviolet, *J. Amer. Chem. Soc.* **74**, 3136 (1952).

[M46] W. Moffitt, Optical rotatory dispersion of simple polypeptides. II, *Proc. Nat. Acad. Sci. U.S.* **42**, 736 (1956).

[M47] F. Momicchioli and A. Rastelli, On the benzo-derivatives of five-membered heterocycles. Theoretical treatment and UV spectra, *J. Mol. Spectrosc.* **22**, 310 (1967).

[M48] R. K. Momii and D. W. Urry, Absorption spectra of polypeptide films in the vacuum ultraviolet, *Macromolecules* **1**, 372 (1968).

[M49] P. R. Monson and W. M. McClain, Polarization dependence of the two-photon absorption of tumbling molecules with application to liquid 1-chloronaphthalene and benzene, *J. Chem. Phys.* **53**, 29 (1970).

[M50] C. E. Moore, Atomic Energy Levels, Circ. Nat. Bur. of Std. 467, Washington, D.C., 1958.

[M51] J. H. Moore, Jr., An investigation of the low-energy singlet-triplet and singlet-singlet transitions in ethylene derivatives by ion impact, *J. Phys. Chem.* **76**, 1130 (1972).

[M52] K. Morokuma and H. Konishi, Diffuse orbitals in lower states of the oxygen molecule, *J. Chem. Phys.* **55**, 402 (1971).

[M53] G. C. Morris, The intensity of absorption of naphthacene vapor from 20 000 to 54 000 cm^{-1}, *J. Mol. Spectrosc.* **18**, 42 (1965).

[M54] G. C. Morris and J. G. Angus, Probable location of a $^1E_{2g}$ state of the benzene molecule, *J. Mol. Spectrosc.* **45**, 271 (1973).

[M55] P. Mullen, K. Schwochau, and C. K. Jorgensen, Vacuo ultraviolet spectra of permanganate, pertechnetate and perrhenate, *Chem. Phys. Lett.* **3**, 49 (1969).

[M56] P. A. Mullen and M. K. Orloff, The electronic spectrum of fluorenone between 1680 and 2350 Å, *J. Mol. Spectrosc.* **30**, 140 (1969).

[M57] P. A. Mullen and M. K. Orloff, Ultraviolet absorption spectrum of pyrrole vapor including the observation of low-energy transitions in the far ultraviolet, *J. Chem. Phys.* **51**, 2276 (1969).

[M58] P. A. Mullen and M. K. Orloff, The electronic spectrum of acrylonitrile, *Theoret. Chim. Acta* **23**, 278 (1971).

[M59] P. A. Mullen and M. K. Orloff, Ultraviolet absorption spectrum of pentaerythritol tetranitrate, *J. Phys. Chem.* **77**, 910 (1973).

[M60] R. S. Mulliken, Structure and ultraviolet spectra of ethylene, butadiene, and their alkyl derivatives, *Rev. Mod. Phys.* **14**, 265 (1942).

[M61] R. S. Mulliken, The lower excited states of some simple molecules, *Can. J. Chem.* **36**, 10 (1958).

[M62] R. S. Mulliken, Conjugation and hyperconjugation: A survey with emphasis on isovalent hyperconjugation, *Tetrahedron* **5**, 253 (1959).

[M63] R. S. Mulliken, Rydberg and valence-shell character as functions of internuclear distance in some excited states of CH, NH, H$_2$, and N$_2$, *Chem. Phys. Lett.* **14**, 141 (1972).

REFERENCES

[M64] J. N. Murrell, "The Theory of the Electronic Spectra of Organic Molecules." Methuen, London, 1963.

[N1] W. T. Naff, R. N. Compton, and C. D. Cooper, Electron attachment and excitation processes in selected carbonyl compounds, *J. Chem. Phys.* **57**, 1303 (1973).

[N2] S. Nagakura, Ultraviolet absorption spectra and π-electron structures of nitromethane and the nitromethyl anion, *Mol. Phys.* **3**, 152 (1960).

[N3] S. Nagakura, Intramolecular charge-transfer absorption spectra of formamide and acrolein, *Mol. Phys.* **3**, 105 (1960).

[N4] S. Nagakura, Electronic spectra and electron-transfer interaction between electron donor and acceptor, *in* "International Union of Pure and Applied Chemistry, Special Lectures," p. 79. Butterworths, London and Washington, D.C., 1963.

[N5] S. Nagakura, K. Kaya, and H. Tsubomura, Vacuum ultraviolet absorption spectra and electronic structures of formic acid, acetic acid and ethyl acetate, *J. Mol. Spectrosc.* **13**, 1 (1964).

[N6] S. Nagakura, M. Kojima, and Y. Maruyama, Electronic spectra and electronic structures of nitrobenzene and nitromesitylene, *J. Mol. Spectrosc.* **13**, 174 (1964).

[N7] Y. Nakato, M. Ozaki, A. Egawa, and H. Tsubomura, Organic amino compounds with very low ionization potentials, *Chem. Phys. Lett.* **9**, 615 (1971).

[N8] Y. Nakato, M. Ozaki, and H. Tsubomura, Ionization energies and Rydberg states of tetraaminoethylenes, *Bull. Chem. Soc. Japan* **45**, 1299 (1972).

[N9] Y. Nakato, M. Ozaki, and H. Tsubomura, Photoionization and Rydberg states of tetraaminoethylenes in organic solutions, *J. Phys. Chem.* **76**, 2105 (1972).

[N10] T. Nakayama and K. Watanabe, Absorption and photoionization coefficients of acetylene, propyne, and 1-butyne, *J. Chem. Phys.* **40**, 558 (1964).

[N11] R. C. Nelson and W. T. Simpson, Low resolution electronic spectrum of hexamethyl benzene with application to benzene, *J. Chem. Phys.* **23**, 1146 (1955).

[N12] R. G. Nelson and W. C. Johnson, Jr., Optical properties of sugars. I. Circular dichroism of monomers at equilibrium, *J. Amer. Chem. Soc.* **94**, 3343 (1972).

[N13] R. Nicolai, Energy loss of monochromatic 30 KeV lithium ions in argon and ethylene, *Z. Phys.* **228**, 16 (1969).

[N14] E. B. Nielsen and J. A. Schellman, Absorption spectra of simple amides and peptides, *J. Phys. Chem.* **71**, 2297 (1967).

[N15] S. Nishida, I. Moritani, E. Tsuda, and T. Teraji, Di-, tri-, and tetra-cyclopropylethylenes, *Chem. Commun.* 781 (1969).

[N16] G. Nordheim, H. Sponer, and E. Teller, Note on the ultraviolet absorption systems of benzene vapor, *J. Chem. Phys.* **8**, 455 (1940).

[N17] R. G. W. Norrish, H. G. Crone, and O. Saltmarsh, Primary photochemical processes. III. The absorption spectrum and photochemical decomposition of ketene, *J. Chem. Soc.* 1533 (1933).

[N18] E. D. Nostrand and A. B. F. Duncan, Effect of pressure on intensity of some electronic transitions in SF_6, C_2H_2 and C_2D_2 vapors in the vacuum ultraviolet region, *J. Amer. Chem. Soc.* **76**, 3377 (1954).

[N19] P. Nozières and D. Pines, Electron interaction in solids. Characteristic energy loss spectrum, *Phys. Rev.* **113**, 1254 (1959).

[O1] J. F. Ogilvie, Structural deductions from the vibronic spectra of ethene and ethene-d_4, *J. Chem. Phys.* **49**, 474 (1968).
[O2] Y. Ohashi, I. Hanazaki, and S. Nagakura, Spectroscopic study of the interaction between the central metal ions in the crystals of bis-(dimethylglyoximato) nickel (II) and related complexes, *Inorg. Chem.* **9**, 2551 (1970).
[O3] H. Okabe, Photodissociation of HN_3 in the vacuum-ultraviolet. Production and reactivity of electronically excited NH, *J. Chem. Phys.* **49**, 2726 (1968).
[O4] H. Okabe and A. Mele, Photodissociation of NCN_3 in the vacuum ultraviolet. Production of CN $B\ ^2\Sigma$ and NCN $A\ ^3\Pi$, *J. Chem. Phys.* **51**, 2100 (1969).
[O5] H. Okabe, Photodissociation of HNCO in the vacuum ultraviolet; Production of NCO $A\ ^2\Sigma$ and NH ($A^3\Pi$, ΠC^1), *J. Chem. Phys.* **53**, 3507 (1970).
[O6] R. M. O'Malley and K. R. Jennings, Detection of inelastic scattering of electrons by simple polyatomic molecules using the SF_6^- ion, *Int. J. Mass Spectrom. Ion Phys.* **2**, App. 1 (1969).
[O7] S. Onari, Vacuum ultraviolet absorption spectra of synthesized polymer films, *J. Phys. Soc. Japan* **26**, 500 (1969).
[O8] S. Onari, Vacuum ultraviolet absorption spectra of homo-polynucleotides, *J. Phys. Soc. Japan* **26**, 214 (1969).
[O9] S. Onari, Vacuum ultraviolet absorption spectra of homo-polypeptides, *J. Phys. Soc. Japan* **27**, 269 (1969).
[O10] S. Onari, Vacuum ultraviolet absorption spectra of polyamides, *Jap. J. Appl. Phys.* **9**, 227 (1970).
[O11] S. Onari and K. Watanabe, Ultraviolet absorption spectra of polyribonucleotides, *J. Phys. Soc. Japan* **28**, 1051 (1970).
[O12] S. Onari, Vacuum ultraviolet absorption spectrum of poly-L-leucine and its dependence on conformation, *J. Phys. Soc. Japan* **29**, 528 (1970).
[O13] S. Onari, Vacuum ultraviolet absorption spectra of polypeptides and their dependence on conformation, *J. Phys. Soc. Japan* **30**, 811 (1971).
[O14] M. K. Orloff, P. A. Mullen, and F. C. Rauch, Molecular orbital study of the electronic structure and spectrum of hexahydro-1,3,5-trinitro-s-triazine, *J. Phys. Chem.* **74**, 2189 (1970).
[O15] A. Otto and M. J. Lynch, Characteristic electron energy losses of solid benzene and ice, *Aust. J. Phys.* **23**, 609 (1970).

[P1] E. Pantos and T. D. S. Hamilton, Assignment of the vibrational bands of the $^1B_{1u} \leftarrow {}^1A_{1g}$ transition in solid benzene films, *Chem. Phys. Lett.* **17**, 588 (1972).
[P2] T. Parameswaran and D. E. Ellis, Photoelectron and optical spectra of $TiCl_4$ and VCl_4, *J. Chem. Phys.* **58**, 2088 (1973).
[P3] R. Pariser, Theory of the electronic spectra and structure of the polyacenes and of alternant hydrocarbons, *J. Chem. Phys.* **24**, 250 (1956).
[P4] L. J. Parkhurst and B. G. Anex, Polarization of the lowest-energy allowed transition of β-ionylidene crotonic acid and the electronic structure of the polyenes, *J. Chem. Phys.* **45**, 862 (1966).

[P5] J. E. Parkin and K. K. Innes, The vacuum ultraviolet spectra of pyrazine, pyrimidine, and pyridazine vapors. I. Spectra between 1550 Å and 2000 Å, *J. Mol. Spectrosc.* **15**, 407 (1965).

[P6] R. G. Parr, D. P. Craig and I. G. Ross, Molecular orbital calculations of the lower excited electronic levels of benzene, configuration interaction included, *J. Chem. Phys.* **18**, 1561 (1950).

[P7] R. H. Partridge, Vacuum-ultraviolet absorption spectrum of polystyrene, *J. Chem. Phys.* **47**, 4223 (1967).

[P8] J. C. Person, Isotope effect in the photoionization efficiency for benzene, *J. Chem. Phys.* **43**, 2553 (1965).

[P9] J. C. Person and P. P. Nicole, Isotope effects in the photoionization yields and the absorption cross-sections for ethylene and n-butane, *J. Chem. Phys.* **49**, 5421 (1968).

[P10] J. C. Person and P. P. Nicole, Isotope effects in the photoionization yields and the absorption cross sections for acetylene, propyne, and propene, *J. Chem. Phys.* **53**, 1767 (1970).

[P11] D. L. Peterson and W. T. Simpson, Polarized electronic absorption spectrum of amides with assignments of transitions, *J. Amer. Chem. Soc.* **79**, 2375 (1957).

[P12] S. D. Peyerimhoff, Relationships between AB_2 and H_nAB_2 molecular spectra and geometry: Accurate SCF MO and CI calculations for various states of $HCOO^-$, *J. Chem. Phys.* **47**, 349 (1967).

[P13] S. D. Peyerimhoff and R. J. Buenker, Theoretical study of cyclopropene and its C_3H_4 isomers, *Theoret. Chim. Acta* **14**, 305 (1969).

[P14] S. D. Peyerimhoff, R. J. Buenker, W. E. Kammer, and H. Hsu, Calculation of the electronic spectrum of formaldehyde, *Chem. Phys. Lett.* **8**, 129 (1971).

[P15] S. D. Peyerimhoff and R. J. Buenker, Vibrational analysis of the electronic spectrum of ethylene based on *ab initio* SCF-CI calculations, *Theoret. Chim. Acta* **27**, 243 (1972).

[P16] L. W. Pickett, Vibrational analysis of the absorption spectrum of furan in the Schumann region, *J. Chem. Phys.* **8**, 293 (1940).

[P17] L. W. Pickett, E. Paddock, and E. Sackter, The ultraviolet absorption spectrum of 1,3-cyclopentadiene, *J. Amer. Chem. Soc.* **63**, 1073 (1941).

[P18] L. W. Pickett and E. Sheffield, The ultraviolet absorption spectra of dioxadiene and dioxene, *J. Amer. Chem. Soc.* **68**, 216 (1946).

[P19] L. W. Pickett, N. J. Hoeflich, and T.-C. Liu, The vacuum ultraviolet absorption spectra of cyclic compounds. II. Tetrahydrofuran, tetrahydropyran, 1,4-dioxane, and furan, *J. Amer. Chem. Soc.* **73**, 4865 (1951).

[P20] L. W. Pickett, M. Muntz, and E. M. McPherson, Vacuum ultraviolet absorption spectra of cyclic compounds. I. Cyclohexane, cyclohexene, cyclopentane, cyclopentene and benzene, *J. Amer. Chem. Soc.* **73**, 4862 (1951).

[P21] L. W. Pickett, M. E. Corning, G. M. Wieder, D. A. Semenow, and J. M. Buckley, The vacuum ultraviolet spectra of cyclic compounds. III. Amines, *J. Amer. Chem. Soc.* **75**, 1618 (1953).

[P22] U. Pincelli, B. Cadioli and D. J. David, A theoretical study of the electronic structure and conformation of glyoxal, *J. Mol. Struct.* **9**, 173 (1971).

[P23] M. N. Pisanias, L. G. Christophorou, and J. G. Carter, Compound negative ion resonances and threshold-electron excitation spectra of quinoline and isoquinoline, *Chem. Phys. Lett.* **13,** 433 (1972).

[P24] M. N. Pisanias, L. G. Christophorou, J. G. Carter, and D. L. McCorkle, Compound-negative-ion resonance states and threshold electron excitation spectra of N-heterocyclic molecules: Pyridine, pyridazine, pyrimidine, pyrazine, and sym-triazine, *J. Chem. Phys.* **58,** 2110 (1973).

[P25] J. R. Platt, I. Rusoff, and H. B. Klevens, Absorption spectra of some organic solutions in the vacuum ultraviolet, *J. Chem. Phys.* **11,** 535 (1943).

[P26] J. R. Platt, and H. B. Klevens, Absolute absorption intensities of alkylbenzenes in the 2250–1700 Å region, *Chem. Rev.* **41,** 301 (1947).

[P27] J. R. Platt and H. B. Klevens, Further alkylbenzene spectra to 1750 Å, *J. Chem. Phys.* **16,** 832 (1948).

[P28] J. R. Platt, H. B. Klevens, and W. C. Price, Absorption intensities of ethylenes and acetylenes in the vacuum ultraviolet, *J. Chem. Phys.* **17,** 466 (1949).

[P29] J. R. Platt, Classification of spectra of cata-condensed hydrocarbons, *J. Chem. Phys.* **17,** 484 (1949).

[P30] J. R. Platt, Molecular orbital predictions of organic spectra, *J. Chem. Phys.* **18,** 1168 (1950).

[P31] J. A. Pople and J. W. Sidman, Intensity of the symmetry-forbidden electronic absorption band of formaldehyde, *J. Chem. Phys.* **27,** 1270 (1957).

[P32] W. J. Potts, Jr., Low-temperature absorption spectra of selected olefins in the farther ultraviolet region, *J. Chem. Phys.* **23,** 65 (1955).

[P33] W. J. Potts, Jr., Low-temperature absorption spectra of benzene, toluene and para-xylene in the farther ultraviolet region, *J. Chem. Phys.* **23,** 73 (1955).

[P34] M.-Th. Praet and J. Delwiche, Ionization energies of some cyclic molecules, *Chem. Phys. Lett.* **5,** 546 (1970).

[P35] J. W. Preiss and R. Setlow, Spectra of some amino acids, peptides, nucleic acids, and proteins in the vacuum ultraviolet, *J. Chem. Phys.* **25,** 138 (1956).

[P36] W. C. Price, The far ultraviolet absorption spectra of formaldehyde and the alkyl derivatives of H_2O and H_2S, *J. Chem. Phys.* **3,** 256 (1935).

[P37] W. C. Price, The absorption spectra of acetylene, ethylene, and ethane in the far ultraviolet, *Phys. Rev.* **47,** 444 (1935).

[P38] W. C. Price and R. W. Wood, The far ultraviolet absorption spectra and ionization potentials of C_6H_6 and C_6D_6, *J. Chem. Phys.* **3,** 439 (1935).

[P39] W. C. Price, The far ultraviolet absorption spectra and ionization potentials of the alkyl halides. I, *J. Chem. Phys.* **4,** 539 (1936).

[P40] W. C. Price, The far ultraviolet absorption spectra and ionization potentials of the alkyl halides. II, *J. Chem. Phys.* **4,** 547 (1936).

[P41] W. C. Price and W. M. Evans, The absorption spectrum of formic acid in the vacuum ultraviolet, *Proc. Roy. Soc. London* **162A,** 110 (1937).

[P42] W. C. Price and W. T. Tutte, The absorption spectra of ethylene, deuteroethylene and some alkyl-substituted ethylenes in the vacuum ultraviolet, *Proc. Roy. Soc. London* **174A,** 207 (1940).

[P43] W. C. Price and A. D. Walsh, The absorption spectra of conjugated dienes in the vacuum ultraviolet, *Proc. Roy. Soc. London* **174A**, 220 (1940).

[P44] W. C. Price and A. D. Walsh, The absorption spectra of the cyclic dienes in the vacuum ultraviolet, *Proc. Roy. Soc. London* **179A**, 201 (1941).

[P45] W. C. Price and D. M. Simpson, Absorption spectra of nitrogen dioxide, ozone, and nitrosyl chloride in the vacuum ultraviolet, *Trans. Faraday Soc.* **37**, 106 (1941).

[P46] W. C. Price and A. D. Walsh, The absorption spectra of triple bond molecules in the vacuum ultraviolet, *Trans. Faraday Soc.* **41**, 381 (1945).

[P47] W. C. Price and A. D. Walsh, The absorption spectra of hexatriene and divinyl acetylene in the vacuum ultraviolet, *Proc. Roy. Soc. London* **185A**, 182 (1945).

[P48] W. C. Price and A. D. Walsh, The absorption spectra of benzene derivatives in the vacuum ultraviolet. I, *Proc. Roy. Soc. London* **A191**, 22 (1947).

[P49] W. C. Price, J. P. Teegan, and A. D. Walsh, The absorption spectrum of ketene in the far ultraviolet, *J. Chem. Soc.* 920 (1951).

[P50] W. C. Price, R. Bralsford, P. V. Harris, and R. G. Ridley, Ultraviolet spectra and ionization potentials of hydrocarbon molecules, *Spectrochim. Acta* **14**, 45 (1959).

[P51] H. Prugger and F. Dörr, Absorption spectra of organic molecules with nonbonding electrons in the Schumann region: Exocyclic ketones, *Z. Elektrochem.* **64**, 425 (1960).

[P52] E. S. Pysh, J. Jortner, and S. A. Rice, Forbidden electronic transitions in XeF_2 and XeF_4, *J. Chem. Phys.* **40**, 2018 (1964).

[P53] E. S. Pysh, S. A. Rice, and J. Jortner, Molecular Rydberg transitions in rare-gas matrices—Evidence for interaction between impurity states and crystal states, *J. Chem. Phys.* **43**, 2997 (1965).

[Q1] F. Quadrifoglio and D. W. Urry, Ultraviolet rotatory properties of polypeptides in solution. I. Helical poly-L-alanine, *J. Amer. Chem. Soc.* **90**, 2755 (1968).

[Q2] F. Quadrifoglio and D. W. Urry, Ultraviolet rotatory properties of polypeptides in solution. II. Poly-L-serine, *J. Amer. Chem. Soc.* **90**, 2760 (1968).

[R1] J. W. Rabalais, J. R. McDonald, and S. P. McGlynn, Electronic states of HNCO, cyanate salts, and organic isocyanates. II. Absorption studies, *J. Chem. Phys.* **51**, 5103 (1969).

[R2] J. W. Rabalais, J. M. McDonald, V. Scherr, and S. P. McGlynn, Electronic spectroscopy of isoelectronic molecules. II. Linear triatomic groupings containing sixteen valence electrons, *Chem. Rev.* **71**, 73 (1971).

[R3] J. W. Rabalais, A photoelectron spectroscopic investigation of the electronic structure of nitromethane and nitrobenzene, *J. Chem. Phys.* **57**, 960 (1972).

[R4] E. Rabinowitch, Electron transfer spectra and their photochemical effects, *Rev. Mod. Phys.* **14**, 112 (1942).

[R5] C. Reid, Observation of discrete bands in the near ultraviolet absorption spectrum of liquid ethylene, *J. Chem. Phys.* **18**, 1299 (1950).

[R6] J. F. Rendina and R. E. Grojean, Electron impact spectrometer for gas analysis, *Appl. Spectrosc.* **25,** 24 (1971).

[R7] M. A. Robb and I. G. Csizmadia, Non-empirical LCAO-MO-SCF-CI calculations on organic molecules with Gaussian type functions. IV. Preliminary investigations on formamide, *Theoret. Chim. Acta* **10,** 269 (1968).

[R8] M. B. Robin and W. T. Simpson, Assignment of electronic transitions in azo dye prototypes, *J. Chem. Phys.* **36,** 580 (1962).

[R9] M. B. Robin, R. R. Hart, and N. A. Kuebler, Assignment of the ultraviolet mystery band of olefins, *J. Chem. Phys.* **44,** 1803 (1966).

[R10] M. B. Robin and N. A. Kuebler, Low-lying pi-sigma transition in the spectrum of norbornadiene, *J. Chem. Phys.* **44,** 2664 (1966).

[R11] M. B. Robin, R. R. Hart, and N. A. Kuebler, Electronic states of the azoalkanes, *J. Amer. Chem. Soc.* **89,** 1564 (1967).

[R12] M. B. Robin, H. Basch, N. A. Kuebler, B. E. Kaplan, and J. Meinwald, Assignments in the ultraviolet spectra of olefins, *J. Chem. Phys.* **48,** 5037 (1968).

[R13] M. B. Robin, H. Basch, N. A. Kuebler, K. B. Wiberg, and G. B. Ellison, Optical spectra of small rings. II. The unsaturated three-membered rings, *J. Chem. Phys.* **51,** 45 (1969).

[R14] M. B. Robin, F. A. Bovey, and H. Basch, Molecular and electronic structure of the amide group, *in* "The Chemistry of Amides" (J. Zabicky, ed.), p. 1. Wiley (Interscience), New York, 1970.

[R15] M. B. Robin and N. A. Kuebler, Pressure effects on vacuum ultraviolet spectra, *J. Mol. Spectrosc.* **33,** 274 (1970).

[R16] M. B. Robin, N. A. Kuebler, and C. R. Brundle, The perfluoro and He(II) intensity effects for identifying photoelectron transitions, *in* "Electron Spectroscopy" (D. A. Shirley, ed.), p. 351. North-Holland Publ., Amsterdam, 1971.

[R17] M. B. Robin, C. R. Brundle, N. A. Kuebler, G. B. Ellison, and K. B. Wiberg, Photoelectron spectra of small rings. IV. The unsaturated three-membered rings, *J. Chem. Phys.* **57,** 1758 (1972).

[R18] M. B. Robin, G. N. Taylor, N. A. Kuebler, and R. D. Bach, Planarity of the carbon skeleton in various alkylated olefins, *J. Org. Chem.* **38,** 1049 (1973).

[R19] M. B. Robin and N. A Kuebler, unpublished results.

[R20] J. L. Roebber, J. C. Larrabee and R. E. Huffman, Vacuum-ultraviolet absorption spectrum of carbon suboxide, *J. Chem. Phys.* **46,** 4594 (1967).

[R21] J. L. Roebber, Vacuum-ultraviolet absorption spectrum of carbon suboxide. II. The effect of low temperature and the effect of high pressure of an inert gas on the 1780-Å band, *J. Chem. Phys.* **54,** 4001 (1971).

[R22] J. Romand and B. Vodar, Absorption spectra of benzene in the vapor and condensed phase in the ultraviolet region, *C. R. Acad. Sci. Paris* **233,** 930 (1951).

[R23] E. J. Rosa and W. T. Simpson, Interlocking amide resonance in the DNA bases, *in* "Physical Processes in Radiation Biology" (L. Augenstein, R. Mason, and B. Rosenberg, eds.), p. 43. Academic Press, New York, 1964.

[R24] J. Rose, T. Shibuya, and V. McKoy, Application of the equations-of-motion method to the excited states of N_2, CO, and C_2H_4, *J. Chem. Phys.* **58,** 74 (1973).

[R25] J. B. Rose, T. Shibuya, and V. McKoy, Electronic excitations of benzene from the equations of motion method, *J. Chem. Phys.* **60**, 2700 (1974).

[R26] K. Rosenheck, H. Miller, and A. Zakaria, Polarized ultraviolet absorption spectra and electronic transitions of poly-L-proline, *Biopolymers* **7**, 614 (1969).

[R27] K. J. Ross and E. N. Lassettre, Intensity distribution in the energy-loss spectrum in ethylene, *J. Chem. Phys.* **44**, 4633 (1966).

[R28] I. I. Rusoff, J. R. Platt, H. B. Klevens, and G. O. Burr, Extreme ultraviolet absorption spectra of the fatty acids, *J. Amer. Chem. Soc.* **67**, 673 (1945).

[R29] J. A. Ryan and J. L. Whitten, A valence state description of the ethylene V state by configuration interaction theory, *Chem. Phys. Lett.* **15**, 119 (1972).

[S1] A. J. Sadlej, Vibronic absorption spectrum of the ethylene dimer, *Int. J. Quantum Chem.* **3**, 569 (1969).

[S2] A. P. Sadovskii, V. A. Kogan, and I. N. Lobanov, X-ray spectral study of Co(II) and Cu(II) inner complexes with salicylalarylimines, *J. Struct. Chem.* **11**, 630 (1970).

[S3] D. R. Salahub, Semi-empirical MO-CI calculations on excited states. IV. Unsaturated molecules, *Theoret. Chim. Acta* **22**, 330 (1971).

[S4] D. R. Salahub and R. A. Boschi, The far ultraviolet spectrum of iodoacetylene, *Chem. Phys. Lett.* **16**, 320 (1972).

[S5] P. Salvadori, L. Lardicci, M. Menicagli, and C. Bertucci, Cotton effects of the benzene chromophore in the 300–185 nm spectral region, *J. Amer. Chem. Soc.* **94**, 8598 (1972).

[S6] J. A. R. Samson, F. F. Marmo, and K. Watanabe, Absorption and photoionization coefficients of propylene and butene-1 in the vacuum ultraviolet, *J. Chem. Phys.* **36**, 783 (1962).

[S7] J. A. R. Samson, The electronic states of $C_6H_6^+$, *Chem. Phys. Lett.* **4**, 257 (1969).

[S8] C. Sandorfy, General and theoretical aspects of the imine group, in "The Chemistry of the Carbon–Nitrogen Double Bond" (S. Patai, ed.), p. 1. Wiley (Interscience), New York, 1971.

[S9] L. J. Schaad, L. A. Burnelle, and K. P. Dressler, The excited states of allene, *Theoret. Chim. Acta* **15**, 91 (1969).

[S10] B. H. Schechtman and W. E. Spicer, Near infrared to vacuum ultraviolet absorption spectra and the optical constants of phthalocyanine and porphyrin films, *J. Mol. Spectrosc.* **33**, 28 (1970).

[S11] G. Scheibe, The absorption of halogen ions in aqueous solution in the near-Schumann region (electron affinity spectra), *Z. Phys. Chem.* **B5**, 355 (1929).

[S12] G. Scheibe, F. Povenz, and C. F. Linström, Concerning the light absorption of a few hydrocarbon compounds in the Schumann ultraviolet, *Z. Phys. Chem.* **B20**, 283 (1933).

[S13] G. Scheibe and H. Grieneisen, Concerning the light absorption of certain hydrocarbons in the Schumann ultraviolet, *Z. Phys. Chem.* **B25**, 52 (1934).

[S14] J. Schellman, V. Chandrasekharan, H. Damany, and J. Romand, Dichroic absorption of calcite in the Schumann region, *C. R. Acad. Sci. Paris* **260**, 117 (1965).

[S15] J. A. Schellman and E. B. Nielsen, The spectra of amides. Theoretical considerations, *J. Phys. Chem.* **71,** 3914 (1967).

[S16] R. Scheps, D. Florida, and S. A. Rice, Interference effects in the Rydberg spectra of naphthalene and benzene, *J. Chem. Phys.* **56,** 295 (1972).

[S17] R. Scheps, D. Florida, and S. A. Rice, Comments on the Rydberg spectrum of pyrazine, *J. Mol. Spectrosc.* **44,** 1 (1972).

[S18] O. Schnepp, E. F. Pearson, and E. Sharman, Circular dichroism of (+)-3-methylcyclopentanone, *Chem. Commun.* 545 (1970).

[S19] O. Schnepp, E. F. Pearson, and E. Sharman, Circular dichroism of *trans*-cyclo-octene, *J. Chem. Phys.* **52,** 6424 (1970).

[S20] J. M. Schulman, J. W. Moskowitz, and C. Hollister, Ethylene molecule in a Gaussian basis. II. Contracted bases, *J. Chem. Phys.* **46,** 2759 (1967).

[S21] K. Schulten and M. Karplus, On the origin of a low lying forbidden transition in polyenes and related molecules, *Chem. Phys. Lett.* **14,** 305 (1972).

[S22] A. I. Scott and A. D. Wrixon, A symmetry rule for chiral olefins, *Chem. Commun.* 1182 (1969).

[S23] A. I. Scott and A. D. Wrixon, On the optical activity of *trans*-cyclo-octene, *Chem. Commun.* 43 (1970).

[S24] J. D. Scott and B. R. Russell, The vacuum ultraviolet absorption spectrum of 1,1-dichloro-2,2-difluoroethylene, *Chem. Phys. Lett.* **9,** 375 (1971).

[S25] J. D. Scott and B. R. Russell, Vacuum ultraviolet spectral studies of several chlorofluoroethylenes, *J. Amer. Chem. Soc.* **94,** 2634 (1972).

[S26] J. D. Scott and B. R. Russell, Vacuum ultraviolet absorption spectra of methyl-substituted allenes, *J. Amer. Chem. Soc.* **95,** 1429 (1973).

[S27] C. J. Seliskar and S. P. McGlynn, Electronic spectroscopy of maleimide and its isoelectronic molecules. I. Maleimide and *N*-alkylmaleimides, *J. Chem. Phys.* **55,** 4337 (1971).

[S28] C. J. Seliskar and S. P. McGlynn, Electronic spectroscopy of maleimide and its isoelectronic molecules. II. Maleic anhydride and 4-cyclopentene-3,5-dione, *J. Chem. Phys.* **56,** 275 (1972).

[S29] C. J. Seliskar and S. P. McGlynn, Electronic spectroscopy of maleimide and its isoelectronic molecules. III. Quantum chemical calculations, *J. Chem. Phys.* **56,** 1417 (1972).

[S30] D. Semenow, A. J. Harrison, and E. P. Carr, Absorption intensities of the isomeric pentenes in the vacuum ultraviolet, *J. Chem. Phys.* **22,** 638 (1954).

[S31] Yu. L. Sergeev, M. E. Akopyan, F. I. Vilesov, and V. I. Kleimenov, Photoionization processes in phenyl halides, *Opt. Spectroc.* **29,** 63 (1970).

[S32] S. Shih, R. J. Buenker, and S. D. Peyerimhoff, Nonempirical calculations on the electronic spectrum of butadiene, *Chem. Phys. Lett.* **16,** 244 (1972).

[S33] T. Shimanouchi, Y. Abe, and M. Mikami, Skeletal deformation vibrations and rotational isomerism of some ketones and olefins, *Spectrochim. Acta* **24A,** 1037 (1968).

[S34] J. W. Sidman, Electronic spectra and structure of the carbonyl group, *J. Chem. Phys.* **27,** 429 (1957).

[S35] J. W. Sidman, Electronic transitions due to nonbonding electrons in carbonyl, aza-aromatic, and other compounds, *Chem. Rev.* **58,** 689 (1958).

[S36] P. Siegbahn, Ab initio calculations on furan with a new computer program, *Chem. Phys. Lett.* **8**, 245 (1971).

[S37] C. Sieiro and J. I. Fernández-Alonso, Theoretical interpretation of the electronic spectra of nitrobenzene and benzoic acid, *Chem. Phys. Lett.* **18**, 557 (1973).

[S38] A. Skerbele and E. N. Lassettre, Electron-impact spectra, *J. Chem. Phys.* **42**, 395 (1965).

[S39] A. L. Sklar, Theory of color of organic compounds, *J. Chem. Phys.* **5**, 669 (1937).

[S40] D. R. Smith and J. W. Raymonda, Rydberg states in fluorinated benzenes: Hexa-, penta-, and monofluorobenzene, *Chem. Phys. Lett.* **12**, 269 (1971).

[S41] L. C. Snyder and H. Basch, "Molecular Wave Functions and Properties: Tabulated from SCF Calculations in a Gaussian Basis Set." Wiley, New York, 1972.

[S42] P. A. Snyder and L. B. Clark, Polarization assignments in the electronic spectra of mono-olefins, *J. Chem. Phys.* **52**, 998 (1970).

[S43] P. A. Snyder, P. M. Vipond, and W. C. Johnson, Jr., Circular dichroism of the alkyl amino acids in the vacuum ultraviolet, *Biopolymers* **12**, 975 (1973).

[S44] S. Solony, F. W. Birss, and J. B. Greenshields, Semiempirical SCF-LCAO-MO treatment of thiophene, furan, and pyrrole, *Can. J. Chem.* **43**, 1569 (1965).

[S45] S. P. Sood and K. Watanabe, Absorption and ionization coefficients of vinyl chloride, *J. Chem. Phys.* **45**, 2913 (1966).

[S46] B. L. Sowers, E. T. Arakawa, and R. D. Birkhoff, Optical properties of six-member carbon ring organic liquids in the vacuum ultraviolet, *J. Chem. Phys.* **54**, 2319 (1971).

[S47] B. L. Sowers, Optical Properties of Liquid CCl_4, C_6H_{14}, C_6H_{12}, C_6H_{10}, C_6H_8, and C_6H_6 in the Vacuum Ultraviolet, Thesis, Univ. of Tennessee, 1972.

[S48] F. H. Stillinger, private communication, 1973.

[S49] S. Stokes and L. W. Pickett, Absorption of bicycloheptane and bicycloheptene in the vacuum ultraviolet, *J. Chem. Phys.* **23**, 258 (1955).

[S50] G. Stühmer and E. Rieflin, Ultraviolet reflection spectra of NH_4Br in phases II and III, *Opt. Commun.* **6**, 247 (1972).

[S51] T. M. Sugden and A. D. Walsh, Ionization potentials of butadiene, *Trans. Faraday Soc.* **41**, 76 (1945).

[S52] L. H. Sutcliffe and A. D. Walsh, The vacuum ultraviolet spectrum of allene, *J. Chem. Phys.* **19**, 1210 (1951).

[S53] L. H. Sutcliffe and A. D. Walsh, The absorption spectrum of allene in the vacuum ultraviolet, *J. Chem. Soc.* 899 (1952).

[S54] J. R. Swenson and R. Hoffmann, The interaction of nonbonding orbitals in carbonyls, *Helv. Chim. Acta* **53**, 2331 (1970).

[T1] K. Tabei and S. Nagakura, Near and vacuum ultraviolet absorption spectra and electronic structures of nitrosobenzene and its derivatives, *Bull. Chem. Soc. Japan* **38**, 965 (1965).

[T2] J. Tanaka, Effect of substituents on the electronic spectra of organic compounds. II. Electronic structure and spectra of nitro group and nitro compounds, *Nippon Kagaku Zasshi* **78**, 1643 (1957).

[T3] J. Tanaka, The electronic spectra of aromatic molecular crystals. I. Substituted benzene molecules, *Bull. Chem. Soc. Japan* **36,** 833 (1963).

[T4] J. Tanaka, The electronic spectra of pyrene, chrysene, azulene, coronene, and tetracene crystals, *Bull. Chem. Soc. Japan* **38,** 86 (1965).

[T5] M. Tanaka, J. Tanaka, and S. Nagakura, The electronic structures and electronic spectra of some aliphatic nitroso compounds, *Bull. Chem. Soc. Japan* **39,** 766 (1966).

[T6] M. Tanaka and S. Nagakura, Electronic structures and spectra of adenine and thymine, *Theoret. Chim. Acta* **6,** 320 (1966).

[T7] K. N. Tanner and A. B. F. Duncan, Raman effect and ultraviolet absorption spectra of molybdenum and tungsten hexafluorides, *J. Amer. Chem. Soc.* **73,** 1164 (1951).

[T8] J. P. Teegan and A. D. Walsh, The absorption spectrum of 1,1-dichloroethylene in the vacuum ultraviolet, *Trans. Faraday Soc.* **47,** 1 (1951).

[T9] R. K. Thomas and H. Thompson, Photoelectron spectra of carbonyl halides and related compounds, *Proc. Roy. Soc. London* **A327,** 13 (1972).

[T10] H. W. Thompson, Absorption spectra of some polyatomic molecules containing methyl and ethyl radicals, *Proc. Roy. Soc. London* **150A,** 603 (1935).

[T11] H. W. Thompson and J. W. Linnett, Absorption spectra of polyatomic molecules containing methyl and ethyl radicals. III, *Proc. Roy. Soc. London* **156A,** 108 (1936).

[T12] E. W. Thulstrup, Assignment of the lowest electronic transitions in benzene, *Int. J. Quantum Chem.* **3S,** 641 (1970).

[T13] B. Tinland, CNDO-CI study of the electronic structure of nitromethane, *Spectrosc. Lett.* **1,** 407 (1968).

[T14] I. Tinoco, Jr., A. Halpern, and W. T. Simpson, The relationship between conformation and light absorption in polypeptides and proteins, *in* "Polyamino Acids, Polypeptides, and Proteins" (M. A. Stahmann, ed.), p. 157. Univ. of Wisconsin Press, Madison, Wisconsin, 1962.

[T15] S. Trajmar, J. K. Rice, P. S. P. Wei, and A. Kuppermann, Triplet states of acetylene by electron impact, *Chem. Phys. Lett.* **1,** 703 (1968).

[T16] A. Trombetti and C. Zauli, Molecular spectra and structure of selenophene, *J. Chem. Soc. A* 1106 (1967).

[T17] A. Trombetti, Infrared and ultraviolet absorption spectra of diimide (N_2H_2), *Can. J. Phys.* **46,** 1005 (1968).

[T18] H. Tsubomura and T. Sakata, A new electronic transition found for the aniline derivatives, *Chem. Phys. Lett.* **21,** 511 (1973).

[T19] D. W. Turner, Spectrophotometry in the far-ultraviolet region. II. Absorption spectra of steroids and triterpenoids, *J. Chem. Soc.* 30 (1959).

[T20] D. W. Turner, Far and vacuum ultraviolet spectroscopy, *in* "Determination of Organic Structures by Physical Methods" (F. C. Nachod and W. D. Phillips, eds.), Vol. II, p. 339. Academic Press, New York, 1962.

[T21] D. W. Turner, C. Baker, A. D. Baker, and C. R. Brundle, "Molecular Photoelectron Spectroscopy." Wiley, New York, 1970.

[U1] A. Udvarhazi and M. A. El-Sayed, Vacuum-ultraviolet spectra of cyclic ketones, *J. Chem. Phys.* **42,** 3335 (1965).

[U2] D. W. Urry, Optical rotation, *Ann. Rev. Phys. Chem.* **19,** 477 (1968).

[U3] "UV Atlas of Organic Compounds," Vol. III. Plenum Press, New York, 1967.

[U4] E. Uzan, H. Damany, and V. Chandrasekharan, Optical properties of calcite in the Schumann region, *Opt. Commun.* **1**, 221 (1969).

[U5] E. Uzan, H. Damany, and V. Chandrasekharan, Optical properties of dolomite in the Schumann region, *Opt. Commun.* **2**, 6 (1970).

[U6] E. Uzan, H. Damany, and V. Chandrasekharan, Optical properties of sodium nitrate in the Schumann region, *Opt. Commun.* **2**, 84 (1970).

[U7] E. Uzan, H. Damany, and V. Chandrasekharan, Determination of the absorption coefficients of calcite in the dichroic absorption region (1910–1690 Å), *Opt. Commun.* **2**, 273 (1970).

[U8] E. Uzan, H. Damany, and V. Chandrasekharan, Optical properties of magnesite in the Schumann region, *Opt. Commun.* **2**, 452 (1971).

[V1] I. P. Vinogradov and N. Ya. Dodonova, Absorption spectra of aliphatic amino acids and simple peptides in the vacuum ultraviolet, *Opt. Spectrosc.* **30**, 14 (1971).

[V2] I. P. Vinogradov, I. V. Aleshin, and N. Ya. Dodonova, Spectral luminescence studies of benzene and some of its derivatives in the 120–300 nm region at 77°K, *Opt. Spectrosc.* **29**, 256 (1970).

[W1] A. D. Walsh, Absorption spectra of the chloroethylenes in the vacuum ultraviolet, *Trans. Faraday Soc.* **41**, 35 (1945).

[W2] A. D. Walsh, Absorption spectra of acrolein, crotonaldehyde and mesityl oxide in the vacuum ultraviolet, *Trans. Faraday Soc.* **41**, 498 (1945).

[W3] A. D. Walsh, Absorption spectra of furfuraldehyde and benzaldehyde in the vacuum ultraviolet, *Trans. Faraday Soc.* **42**, 62 (1946).

[W4] A. D. Walsh, Absorption spectrum of glyoxal in the vacuum ultraviolet, *Trans. Faraday Soc.* **42**, 66 (1946).

[W5] A. D. Walsh, Absorption spectrum of acetaldehyde in the vacuum ultraviolet, *Proc. Roy. Soc. London* **185A**, 176 (1946).

[W6] A. D. Walsh, The absorption spectra of benzene derivatives in the vacuum ultraviolet. II, *Proc. Roy. Soc. London* **191A**, 32 (1947).

[W7] A. D. Walsh, Electronic orbitals, shapes, and spectra of polyatomic molecules. IX. Hexatomic molecules: Ethylene, *J. Chem. Soc.* 2325 (1953).

[W8] A. D. Walsh and P. A. Warsop, Ultraviolet absorption of *trans*-dichloroethylene, *Trans. Faraday Soc.* **63**, 524 (1967).

[W9] A. D. Walsh and P. A. Warsop, Ultraviolet absorption spectrum of *cis*-1,2-dichloroethylene, *Trans. Faraday Soc.* **64**, 1418 (1968).

[W10] A. D. Walsh and P. A. Warsop, Ultraviolet absorption spectrum of trichloroethylene, *Trans. Faraday Soc.* **64**, 1425 (1968).

[W11] A. D. Walsh, P. A. Warsop, and J. A. B. Whiteside, Ultraviolet absorption spectrum of 1,1-dichloroethylene, *Trans. Faraday Soc.* **64**, 1432 (1968).

[W12] J. C. Ward, Measurements on ultraviolet dichroism, *Proc. Roy. Soc. London* **228A**, 205 (1955).

[W13] A. Warshel and M. Karplus, Vibrational structure of electronic transitions in conjugated molecules, *Chem. Phys. Lett.* **17**, 7 (1972).

[W14] E. Wasserman, Nonconjugated Interactions in Pi-Electron Systems, Ph.D. Thesis, Harvard Univ., Cambridge, Massachusetts, 1958.

[W15] K. Watanabe and T. Namioka, Ionization potential of propyne, *J. Chem. Phys.* **24**, 915 (1956).

[W16] K. Watanabe and T. Nakayama, Absorption and photoionization coefficients of furan vapor, *J. Chem. Phys.* **29**, 48 (1958).

[W17] F. H. Watson, Jr., A. T. Armstrong, and S. P. McGlynn, Electronic transitions in mono-olefinic hydrocarbons. I. Computational results, *Theoret. Chim. Acta* **16**, 75 (1970).

[W18] F. H. Watson, Jr. and S. P. McGlynn, Electronic transitions in mono-olefinic hydrocarbons. II. Experimental results, *Theoret. Chim. Acta* **21**, 309 (1971).

[W19] M. J. Weiss, C. E. Kuyatt, and S. Mielczarek, Inelastic electron scattering from formaldehyde, *J. Chem. Phys.* **54**, 4147 (1971).

[W20] W. West and R. B. Killingsworth, The vibration spectra and electric moments of azomethane, N,N'-dimethylhydrazine and acetaldazine, *J. Chem. Phys.* **6**, 1 (1938).

[W21] R. F. Whitlock and A. B. F. Duncan, Electronic spectrum of cyclobutanone, *J. Chem. Phys.* **55**, 218 (1971).

[W22] J. L. Whitten and M. Hackmeyer, Configuration interaction studies of ground and excited states of polyatomic molecules. I. The CI formulation and studies of formaldehyde, *J. Chem. Phys.* **51**, 5584 (1969).

[W23] J. L. Whitten, Remarks on the description of excited electronic states by configuration interaction theory and a study of the $^1(\pi \rightarrow \pi^*)$ state of H_2CO, *J. Chem. Phys.* **56**, 5458 (1972).

[W24] C. F. Wilcox, Jr., S. Winstein, and W. G. McMillan, Neighboring carbon and hydrogen. XXXIV. Interaction of nonconjugated chromophores, *J. Amer. Chem. Soc.* **82**, 5450 (1960).

[W25] P. G. Wilkinson and R. S. Mulliken, Far ultraviolet absorption spectra of ethylene and ethylene-d_4, *J. Chem. Phys.* **23**, 1895 (1955).

[W26] P. G. Wilkinson, Absorption spectra of benzene and benzene-d_6 in the vacuum ultraviolet, *Can. J. Phys.* **34**, 596 (1956).

[W27] P. G. Wilkinson, Absorption spectra of ethylene and ethylene-d_4 in the vacuum ultraviolet. II, *Can. J. Phys.* **34**, 643 (1956).

[W28] P. G. Wilkinson, Absorption spectra of acetylene and acetylene-d_2 in the vacuum ultraviolet, *J. Mol. Spectrosc.* **2**, 387 (1958).

[W29] M. W. Williams, R. A. MacRae, R. N. Hamm, and E. T. Arakawa, Collective oscillations in pure liquid benzene, *Phys. Rev. Lett.* **22**, 1088 (1969).

[W30] C. Willis and R. A. Back, Stability of diimide, *Nature (London)* **241**, 43 (1973).

[W31] E. G. Wilson, J. Jortner, and S. A. Rice, A far-ultraviolet spectroscopic study of xenon difluoride, *J. Amer. Chem. Soc.* **85**, 813 (1963).

[W32] G. L. Workman and A. B. F. Duncan, Electronic spectrum of carbonyl fluoride, *J. Chem. Phys.* **52**, 3204 (1970).

[W33] G. L. Workman and C. W. Mathews, Correlation of bands attributed to F_2CO (1800–2100 Å) with the Cameron bands of CO, *J. Chem. Phys.* **53**, 857 (1970).

[W34] H. Wynberg, A. De Groot, and D. W. Davies, A nonconjugated 1,3-diene, *Tetrahedron Lett.* 1083 (1963).

[Y1] T. Yamada and H. Fukutome, Vacuum ultraviolet absorption spectra of sublimed films of nucleic acid bases, *Biopolymers* **6**, 43 (1968).

[Y2] M. Yamakawa, T. Kubota, and H. Akazawa, Electronic structures of aromatic amine N-oxides, *Theoret. Chim. Acta* **15**, 244 (1969).

[Y3] H. Yamashita and R. Kato, Vacuum ultraviolet absorption in alkali nitrites and alkali nitrates, *J. Phys. Soc. Japan* **29**, 1557 (1970).

[Y4] H. Yamashita and R. Kato, Optical properties of sodium nitrate in the vacuum ultraviolet region, *Int. Symp. Vacuum Ultraviolet Spectrosc.*, Tokyo (August 1971).

[Y5] H. Yamashita, Optical properties of sodium nitrate in the vacuum ultraviolet region, *J. Phys. Soc. Japan* **33**, 1407 (1972).

[Y6] M. Yaris, A. Moscowitz, and R. S. Berry, Low-lying excited states of mono-olefins, *J. Chem. Phys.* **49**, 3150 (1968).

[Y7] D. L. Yeager and V. McKoy, The equations of motion method: Excitation energies and intensities in formaldehyde, *J. Chem. Phys.* **60**, 2714 (1974).

[Y8] W. A. Yeranos, Semiempirical molecular órbital energy levels of XeF_4, *Mol. Phys.* **11**, 85 (1966).

[Y9] A. Yogev, D. Amar, and Y. Mazur, On the chirality of the isolated double-bond chromophore, *Chem. Commun.* 339 (1967).

[Y10] A. Yogev, J. Sagiv, and Y. Mazur, Linear and circular dichroic study of mono-olefins, *Chem. Commun.* 411 (1972).

[Y11] A. Yogev, J. Sagiv, and Y. Mazur, Studies in linear dichroism. VI. On the polarization of electronic transitions in isolated double bonds, *J. Amer. Chem. Soc.* **94**, 5122 (1972).

[Y12] M. Yoshino, J. Takeuchi, H. Suzuki, and K. Wakiya, Absorption and photoionization cross-sections of aromatic hydrocarbons in the 600–2200 Å wavelength region, *Int. Symp. Vacuum Ultraviolet Spectrosc., Tokyo* (August 1971).

[Z1] E. Zeeck, *Ab initio* calculations on the hindering of free rotation of the methyl group in propylene, *Theoret. Chim. Acta* **16**, 155 (1970).

[Z2] M. Zelikoff and K. Watanabe, Absorption coefficients of ethylene in the vacuum ultraviolet, *J. Opt. Soc. Amer.* **43**, 756 (1953).

[AD1] L. J. Aarons, M. F. Guest, M. B. Hall, and I. H. Hillier, Theoretical study of the geometry of PH_3, PF_3, and their ground ionic states, *J. Chem. Soc. Faraday Trans. II* **69**, 643 (1973).

[AD2] S. D. Allen and O. Schnepp, Circular dichroism of an optically active benzene chromophore—1-methyl indan, *J. Chem. Phys.* **59**, 4547 (1973).

[AD3] D. R. Armstrong and P. G. Perkins, Calculation of $\sigma \rightarrow \pi^*$ and $\pi \rightarrow \sigma^*$ transitions in vinylboranes, *Theoret. Chim. Acta* **9**, 412 (1968).

[AD4] H. E. Auer, Far-ultraviolet absorption and circular dichroism spectra of L-tryptophane and some derivatives, *J. Amer. Chem. Soc.* **95**, 3003 (1973).

[AD5] P. S. Bagus, M. Krauss, and R. E. LaVilla, The threshold region of the methane carbon K-absorption spectrum, *Chem. Phys. Lett.* **23**, 13 (1973).

[AD6] C. Baker and D. W. Turner, High resolution molecular photoelectron spectroscopy. III. Acetylenes and aza-acetylenes, *Proc. Roy. Soc. London* **308A**, 19 (1968).

[AD7] G. Baldini, M. Cottini, and E. Grilli, Ultraviolet reflection and absorption of KH_2PO_4 and $NH_4H_2PO_4$ crystals, *Solid State Commun.* **11**, 1257 (1972).

[AD8] G. Baldini and L. Rigaldi, Thin films in vacuum ultraviolet spectroscopy, *Thin Solid Films* **13**, 143 (1972).

[AD9] A. A. Ballman, D. M. Dodd, N. A. Kuebler, R. A. Laudise, D. L. Wood, and D. W. Rudd, Synthetic quartz with high ultraviolet transmission, *Appl. Opt.* **7**, 1387 (1968).

[AD10] V. I. Baranovskii and M. S. Nakhmanson, Semiquantitative interpretation of molecular X-ray absorption spectra, *Bull. Acad. Sci. SSSR Phys. Ser.* **36**, 260 (1973).

[AD11] V. I. Baranovskii and M. S. Nakhmanson, Features of the X-ray spectra of tetrahedral and octahedral molecules and complexes, *Bull. Acad. Sci. SSSR Phys. Ser.* **36**, 356 (1973).

[AD12] H. Basch, Orbital nature in the electronic states of cyclopropene, *Mol. Phys.* **23**, 683 (1972).

[AD13] H. Basch, Open-shell multi-configurational self-consistent-field results for the lowest energy $^1(\pi, \pi^*)$ state of planar ethylene, *Chem. Phys. Lett.* **19**, 323 (1973).

[AD14] B. Bates and D. J. Bradley, Interference filters for the far ultraviolet (1700 Å to 2400 Å), *Appl. Opt.* **5**, 971 (1966).

[AD15] E. R. Bernstein and J. P. Reilly, Absorption spectrum of the 2000 Å system of borazine in the gas phase, *J. Chem. Phys.* **57**, 3960 (1972).

[AD16] D. A. Bird and J. H. Callomon, Further observations on the polarized absorption spectrum of benzene crystals at 2000 Å, *Mol. Phys.* **26**, 1317 (1973).

[AD17] R. D. Birkhoff, R. N. Hamm, M. W. Williams, and E. T. Arakawa, Optical properties of liquids in the vacuum UV, *in* "Chemical Spectroscopy and Photochemistry in the Vacuum Ultraviolet" (C. Sandorfy, P. J. Ausloos and M. B. Robin, eds.), p. 129. Reidel, Dordrecht, Holland, 1974.

[AD18] J. B. Birks, E. Pantos, and T. D. S. Hamilton, The $^1B_{1u} \leftarrow {}^1A_{1g}$ (0–0) transition in crystal benzene, *Chem. Phys. Lett.* **21**, 426 (1973).

[AD19] M. Boring, J. H. Wood, J. W. Moskowitz, and J. W. D. Connolly, Rydberg states in H_2O, *J. Chem. Phys.* **58**, 5163 (1973).

[AD20] N. Bosco and R. D. Morse, Analysis of the Rydberg transitions in ethylene sulfide, *Chem. Phys. Lett.* **20**, 404 (1973).

[AD21] C. Breton and R. Papoular, A continuous source of intense radiation in the ultraviolet region, *C. R. Acad. Sci. Ser. B* **275**, 129 (1972).

[AD22] C. Breton and R. Papoular, Vacuum ultraviolet radiation of laser-produced plasmas, *J. Opt. Soc. Amer.* **63**, 1225 (1973).

[AD23] A. Brillante, C. Taliani, and C. Zauli, The 2000 Å absorption system of the benzene single crystal in polarized light, *Mol. Phys.* **25**, 1263 (1973).

[AD24] F. Brogli and E. Heilbronner, The competition between spin-orbit coupling and conjugation in alkyl halides and its repercussion on their photoelectron spectra, *Helv. Chim. Acta* **54**, 1423 (1971).

[AD25] H. H. Brongersma, private communication, 1973.

[AD26] W. M. Burton and B. A. Powell, Fluorescence of tetraphenyl butadiene in the vacuum ultraviolet, *Appl. Opt.* **12**, 87 (1973).

[AD27] L. R. Canfield, R. G. Johnston, and R. P. Madden, NBS detector standards for the far ultraviolet, *Appl. Opt.* **12**, 1611 (1973).

[AD28] E. P. Carr and H. Stücklen, An electronic transition of the Rydberg series type in the absorption spectra of hydrocarbons, *J. Chem. Phys.* **7**, 631 (1939).

[AD29] V. Chandrasekharan and H. Damany, Birefringence of sapphire, magnesium fluoride and quartz in the vacuum ultraviolet, and retardation plates, *Appl. Opt.* **5**, 939 (1968).

[AD30] "Chemical Spectroscopy and Photochemistry in the Vacuum Ultraviolet" (C. Sandorfy, P. J. Ausloos, and M. B. Robin, eds.). Reidel, Dordrecht, Holland, 1974.

[AD31] L. G. Christophorou, R. P. Blaustein, and D. Pittman, Mobilities of thermal electrons in gases and liquids, *Chem. Phys. Lett.* **18**, 509 (1973).

[AD32] G. A. Clough and H. A. Koehler, A simple camera attachment for a vacuum ultraviolet monochromator, *Rev. Sci. Instrum.* **43**, 1836 (1972).

[AD33] F. J. Comes, R. Haensel, U. Nielsen, and W. H. E. Schwarz, Spectra of the xenon fluorides XeF_2 and XeF_4 in the far UV region, *J. Chem. Phys.* **58**, 516 (1973).

[AD34] P. A. Cox, P. Evans, A. Hamnett, and A. F. Orchard, The helium-I photoelectron spectrum of vanadium tetrachloride, *Chem. Phys. Lett.* **7**, 414 (1970).

[AD35] S. Cradock and R. A. Whiteford, Photoelectron spectra of the methyl, silyl, and germyl derivatives of the group VI elements, *J. Chem. Soc. Faraday II* **68**, 281 (1972).

[AD36] S. Cradock, E. A. V. Ebsworth, W. J. Savage, and R. A. Whiteford, Photoelectron spectra of some methyl, silyl, and germyl amines, phosphines and arsines, *J. Chem. Soc. Faraday II* **68**, 934 (1972).

[AD37] S. Cradock and D. W. H. Rankin, Photoelectron spectra of PF_2H and some substituted difluorophosphines, *J. Chem. Soc. Faraday II* **68**, 940 (1972).

[AD38] D. F. Dance and I. C. Walker, Threshold electron energy loss spectra for some unsaturated molecules, *Proc. Roy. Soc. London* **334A**, 259 (1973).

[AD39] R. Davidson, J. Høg, P. A. Warsop, and J. A. B. Whiteside, Electronic spectrum of tetrahydrofuran in the 205–185 nm region, *J. Chem. Soc. Faraday II* **68**, 1652 (1972).

[AD40] P. W. Deutsch and A. B. Kunz, Core excitation spectra of CH_4 and SiH_4, *J. Chem. Phys.* **59**, 1155 (1973).

[AD41] C. F. Dickinson and E. Ellis, Vacuum ultraviolet spectrophotometer system, *J. Sci. Instrum.* **6**, 527 (1973).

[AD42] E. M. Dicoum, Fr. T. Minh, and J. Bensimon, Excitation spectrum of methane by electron impact, and observation of a vibrational excitation between 1.5 and 3 eV, *Proc. Phys. Soc., London (At. Mol. Phys.)* **6**, L27 (1973).

[AD43] E. Diemann and A. Müller, The He(I) photoelectron spectra of OsO_4 and RuO_4, *Chem. Phys. Lett.* **19**, 538 (1973).

[AD44] J. Doucet, P. Sauvageau, and C. Sandorfy, The vacuum ultraviolet spectrum of tetrahydrofuran, *Chem. Phys. Lett.* **17**, 316 (1972).

[AD45] J. Doucet, P. Sauvageau, and C. Sandorfy, Vacuum ultraviolet and photoelectron spectra of fluorochloro derivatives of methane, *J. Chem. Phys.* **58**, 3708 (1973).

[AD46] A. B. F. Duncan, V. R. Ells, and W. A. Noyes, Jr., The absorption spectrum of ethyl methyl ketone, *J. Amer. Chem. Soc.* **58**, 1454 (1936).

[AD47] T. H. Dunning, Jr., R. P. Hosteny, and I. Shavitt, Low-lying π-electron states of *trans*-butadiene, *J. Amer. Chem. Soc.* **95**, 5067 (1973).

[AD48] L. O. Edwards and W. T. Simpson, Polarizations of electronic transitions from azimuthal variation in fluorescence intensity with application to biphenyl, *J. Chem. Phys.* **53**, 4237 (1970).

[AD49] A. W. Ehler and G. L. Weissler, Vacuum ultraviolet radiation from plasmas formed by a laser on metal surfaces, *Appl. Phys. Lett.* **8**, 89 (1966).

[AD50] E. Ehrhardt and F. Linder, Coincidence experiments for investigation of the energy transfer by collision of low-energy electrons with molecules, *Z. Naturforsch.* **22A**, 444 (1967).

[AD51] L. R. Elias, R. Flach, and W. M. Yen, Variable bandwidth transmission filter for the vacuum ultraviolet: $La_{1-x}Ce_xF_3$, *Appl. Opt.* **12**, 138 (1973).

[AD52] E. T. Fairchild, Interference filters for the VUV (1200–1900 Å), *Appl. Opt.* **12**, 2240 (1973).

[AD53] U. Fano, Virtual electronic levels in molecules and crystals, *Comments At. Mol. Phys.* **3**, 75 (1972).

[AD54] P. Finn, R. K. Pearson, J. M. Hollander, and W. L. Jolly, Chemical shifts in core electron binding energies for some gaseous nitrogen compounds, *Inorg. Chem.* **10**, 378 (1971).

[AD55] C. F. Fischer, Average-energy-of-configuration Hartree–Fock results for the atoms helium to radon, *At. Data* **4**, 301 (1972).

[AD56] I. Fischer-Hjalmars and J. Kowalewski, Simplified non-empirical excited state calculations. II. Interpretation of the electronic transitions in the vacuum-UV spectrum of ethylene, *Theoret. Chim. Acta* **29**, 345 (1973).

[AD57] G. B. Fisher, W. E. Spicer, P. C. McKernan, V. F. Pereskok, and S. J. Wanner, A standard for ultraviolet radiation, *Appl. Opt.* **12**, 799 (1973).

[AD58] P. D. Foo and K. K. Innes, Spectrum of acetylene: 1650–1950 Å, *Chem. Phys. Lett.* **22**, 439 (1973).

[AD59] J. N. Fox and J. E. G. Wheaton, The BRV continuum source, *J. Sci. Instrum.* **6**, 655 (1973).

[AD60] M. F. Fox, Far ultraviolet solution spectrophotometry, *Appl. Spectrosc.* **27**, 155 (1973).

[AD61] D. C. Frost, F. G. Herring, A. Katrib, and C. A. McDowell, The photoelectron spectrum of ethylene sulfide, *Chem. Phys. Lett.* **20**, 401 (1973).

[AD62] A. Gedanken, Z. Karsch, B. Raz, and J. Jortner, Wannier states of a molecular impurity in a liquid rare gas, *Chem. Phys. Lett.* **20**, 163 (1973).

[AD63] A. Gedanken, B. Raz, and J. Jortner, Extravalence molecular excitations in inert matrices, *J. Chem. Phys.* **58**, 1178 (1973).

[AD64] N. G. Gerasimova, V. N. Kreyskop, I. V. Panova, and V. Shin, Transmission and reflection spectra of natural and synthetic quartz in the vacuum ultraviolet, *Sov. J. Opt. Technol.* **40**, 4 (1972).

[AD65] F. A. Gianturco, C. Guidotti, and U. Lamanna, Electronic properties of sulfur hexafluoride. II. Molecular orbital interpretation of its X-ray absorption spectra, *J. Chem. Phys.* **57**, 840 (1972).

[AD66] G. D. Gray, J. Høg, P. A. Warsop, and J. A. B. Whiteside, Analysis of the 193-nm system of tetrahydropyran, *Trans. Faraday Soc.* **66**, 2130 (1970).

[AD67] N. J. Greenfield and G. D. Fasman, The circular dichroism of 3-methyl-pyrrolidin-2-one, *J. Amer. Chem. Soc.* **92**, 177 (1970).

[AD68] R. E. Grojean, Double beam VUV spectrometer and logarithmic ratio-meter, *Appl. Opt.* **12**, 139 (1973).

[AD69] Y. Hamada, A. Y. Hirakawa, and M. Tsuboi, The structure of the triethylenediamine molecule in an excited electronic state, *J. Mol. Spectrosc.* **47**, 440 (1973).

[AD70] W. R. Harshbarger, N. A. Kuebler, and M. B. Robin, Electronic structure and spectra of small rings. V. Photoelectron and electron impact spectra of cyclopropenone, *J. Chem. Phys.* **60**, 345 (1974).

[AD71] E. Haselbach and A. Schmelzer, On the correlation between photoelectron and electronic spectra in *trans*-azomethane, *Helv. Chim. Acta* **54**, 1575 (1971).

[AD72] J. A. Hashmall and E. Heilbronner, n-ionization potentials of alkyl bromides, *Angew. Chem. Int. Ed.* **9**, 305 (1970).

[AD73] P. J. Hay and I. Shavitt, Large-scale configuration interaction calculations on the π-electron states of benzene, *Chem. Phys. Lett.* **22**, 33 (1973).

[AD74] J. M. Heller, Jr., R. D. Birkhoff, M. W. Williams, and L. R. Painter, Optical properties of liquid glycerol in the vacuum ultraviolet, *Radiat. Res.* **52**, 25 (1972).

[AD75] W. P. Helman, The quenching of alkane fluorescence by CCl_4 measured by nanosecond decay time techniques, *Chem. Phys. Lett.* **17**, 306 (1972).

[AD76] G. Herzberg and H. C. Longuet-Higgins, Intersection of potential energy surfaces in polyatomic molecules, *Discuss. Faraday Soc.* **35**, 77 (1963).

[AD77] K. Hiraoka and W. H. Hamill, Characteristic energy losses by slow electrons in organic molecular thin films at 77°K, *J. Chem. Phys.* **57**, 3870 (1972).

[AD78] F. Hirayama and S. Lipsky, The effect of crystalline phase on the fluorescence characteristics of solid cyclohexane and bicyclohexyl, *Chem. Phys. Lett.* **22**, 172 (1973).

[AD79] R. Hoffmann, P. D. Mollère, and E. Heilbronner, Orbital noninteraction in bridged cyclohexanes, *J. Amer. Chem. Soc.* **95**, 4860 (1973).

[AD80] F. Hopfgarten and R. Manne, Molecular orbital interpretation of X-ray emission and photoelectron spectra. IV. Chloromethanes, *J. Elect. Spectrosc. Related Phenomena* **2**, 13 (1973).

[AD81] R. P. Hosteny, A. R. Hinds, A. C. Wahl, and M. Krauss, MC SCF calculations on the lowest triplet state of H_2O, *Chem. Phys. Lett.* **23**, 9 (1973).

[AD82] R. H. Huebner, W. F. Frey, and R. N. Compton, Threshold electron excitation of azulene, *Chem. Phys. Lett.* **23**, 587 (1973).

[AD83] W. Hug and I. Tinoco, Jr., Electronic spectra of nucleic acid bases. I. Interpretation of the in-plane spectra with the aid of all valence electron MO–CI calculations, *J. Amer. Chem. Soc.* **95**, 2803 (1973).

[AD84] W. R. Hunter and S. A. Malo, The temperature dependence of the short wavelength transmittance limit of vacuum ultraviolet window materials. I. Experiment, *J. Phys. Chem. Solids* **30**, 2739 (1969).

[AD85] W. R. Hunter, Optics in the vacuum ultraviolet, *Electro-Optical Systems Design*, p. 16, Nov. (1973).

[AD86] W. R. Hunter, The vacuum ultraviolet: Detectors and sources, *Electro-Optical Systems Design*, p. 24, Nov. (1973).

[AD87] T. Inagaki, Optical absorption of pure liquid benzene in the vacuum ultraviolet, *J. Chem. Phys.* **59**, 5207 (1973).

[AD88] A. A. Iverson, B. R. Russell, and P. R. Jones, The vacuum ultraviolet spectra of methyl vinyl silane and propylene, *Chem. Phys. Lett.* **17**, 98 (1972).

[AD89] A. A. Iverson and B. R. Russell, Electronic spectra of $TiCl_4$, VCl_4 and $SnCl_4$ in the vacuum ultraviolet, *Spectrochim. Acta* **29A**, 715 (1973).

[AD90] S. Iwata, K. Fuke, M. Sasaki, S. Nagakura, T. Otsubo, and S. Misumi, Electronic spectra and electronic structures of [2.2] paracyclophane and related compounds, *J. Mol. Spectrosc.* **46**, 1 (1973).

[AD91] V. A. Job and G. W. King, The electronic spectrum of cyanoacetylene, Part I. Analysis of the 2600-Å system, *J. Mol. Spectrosc.* **19**, 155 (1966).

[AD92] V. A. Job and G. W. King, The electronic spectrum of cyanoacetylene, Part II. Analysis of the 2300-Å system, *J. Mol. Spectrosc.* **19**, 178 (1966).

[AD93] D. E. Johnson, The interactions of 25 KeV electrons with guanine and cytosine, *Radiat. Res.* **49**, 63 (1972).

[AD94] D. E. Johnson and M. Isaacson, Cytosine reflectance measurements using electron energy loss spectra and synchrotron radiation, *Opt. Commun.* **8**, 406 (1973).

[AD95] W. C. Johnson, Jr. and I. Tinoco, Jr., Circular dichroism of polypeptide solutions in the vacuum ultraviolet, *J. Amer. Chem. Soc.* **94**, 4389 (1972).

[AD96] C. K. Jorgensen and J. S. Brinen, Far ultraviolet absorption spectra of cerium (III) and europium (III) aqua ions, *Mol. Phys.* **6**, 629 (1963).

[AD97] M. Jungen, A new interpretation of the Rydberg series of acetylene, *Chem. Phys.* **2**, 367 (1973).

[AD98] D. H. Katayama, R. E. Huffman, and C. L. O'Bryan, Absorption and photoionization cross sections for H_2O and D_2O in the vacuum ultraviolet, *J. Chem. Phys.* **59**, 4309 (1973).

[AD99] J. J. Kaufman, E. Kerman, and W. S. Koski, Implications of photoelectron spectroscopic measurements for compounds which produce no parent ion, *Int. J. Quantum Chem.* **4**, 391 (1971).

[AD100] K. Kaya and S. Nagakura, Electronic absorption spectra of hydrogen bonded amides, *Theoret. Chim. Acta* **7**, 124 (1967).

[AD101] G. D. Kerr, R. N. Hamm, M. W. Williams, R. D. Birkhoff, and L. R. Painter, Optical and dielectric properties of water in the vacuum ultraviolet, *Phys. Rev.* **5A**, 2523 (1972).

[AD102] K. Kimura, S. Katsumata, Y. Achiba, H. Matsumoto, and S. Nagakura, Photoelectron spectra and orbital structures of higher alkyl chlorides, bromides, and iodides, *Bull. Chem. Soc. Japan* **46**, 373 (1973).

[AD103] E. Kloster-Jensen, Preparation of pure triacetylene, tetraacetylene, and pentaacetylene and investigation of their electronic spectra, *Angew. Chem. Int. Ed.* **11**, 438 (1972).

[AD104] F. W. E. Knoop and L. J. Oosterhoff, Low-energy electron impact excitation of 1,3,5-*trans*-hexatriene, *Chem. Phys. Lett.* **22**, 247 (1973).

[AD105] T. Kobayashi and S. Nagakura, Photoelectron spectra of tetrahydropyran, 1,3-dioxane and 1,4-dioxane, *Bull. Chem. Soc. Japan* **46**, 1558 (1973).

[AD106] E. E. Koch, A. Otto, and K. Radler, The vacuum ultraviolet spectrum of naphthalene vapour for photon energies from 5 to 30 eV, *Chem. Phys. Lett.* **16**, 131 (1972).

[AD107] E. E. Koch, A. Otto, and K. Radler, The absorption spectrum of the anthracene molecule in the vacuum ultraviolet, *Chem. Phys. Lett.* **21**, 501 (1973).

[AD108] E. E. Koch and A. Otto, Optical properties of anthracene single crystals in the excitonic region of the spectrum between 4 and 10.5 eV, *Chem. Phys.* (to be published).

[AD109] I. V. Kosinskaya and L. P. Polozova, A continuum source for the near vacuum region of the spectrum, *J. Appl. Spectrosc.* **11**, 1561 (1969).

[AD110] N. Kristianpoller and D. Dutton, Optical properties of "liumogen": A phosphor for wavelength conversion, *Appl. Opt.* **3**, 287 (1964).

[AD111] G. Kuehnlenz and H. H. Jaffé, Use of the CNDO method in spectroscopy. VIII. Molecules containing boron, fluorine and chlorine, *J. Chem. Phys.* **58**, 2238 (1973).

[AD112] C. Kunz, Soft X-ray absorption in solids: Atomic or band approach—Which is better? *Comments Solid State Phys.* **5**, 31 (1973).

[AD113] S. Lange and W. H. Turner, Rapid transmission loss in vacuum ultraviolet irradiated Suprasil W, *Appl. Opt.* **12**, 1733 (1973).

[AD114] H. Larzul, F. Gelebart, and A. Johannin-Gilles, Concerning the absorption spectra of water and heavy water in the ultraviolet, *C. R. Acad. Sci. Paris* **261**, 4701 (1965).

[AD115] E. N. Lassettre and M. A. Dillon, Singlet-triplet energy differences calculated from generalized oscillator strengths, *J. Chem. Phys.* **59**, 4778 (1973).

[AD116] R. E. LaVilla, Carbon and fluorine X-ray emission and fluorine K absorption spectra of the fluoromethane molecules, $CH_{4-n}F_n$ ($0 \leq n \leq 4$). II, *J. Chem. Phys.* **58**, 3841 (1973).

[AD117] E. M. Layton, Jr., Spectral characteristics of several series of more unusual aromatic hydrocarbons, *J. Mol. Spectrosc.* **5**, 181 (1960).

[AD118] H. Lefebvre-Brion, Nature of the resonant states of NO^-, *Chem. Phys. Lett.* **19**, 456 (1973).

[AD119] C. Leibovici, Valence shell calculations of the structure and spectrum of vinyl cyanide, *J. Mol. Struct.* **9**, 177 (1971).

[AD120] D. G. Lewis and W. C. Johnson, Jr., Circular dichroism of DNA in the vacuum ultraviolet (to be published).

[AD121] E. Lindholm, Dissociation of molecules and molecule ions in molecular orbital theory. II. Dissociation of NH_3 and NH_3^+, *Ark. Fys.* **37**, 49 (1968).

[AD122] E. Lindholm, Rydberg series in small molecules. I. Quantum defects in Rydberg series, *Ark. Fys.* **40**, 97 (1969).

[AD123] D. R. Lloyd and N. Lynaugh, Photoelectron studies of boron compounds, Part 3. Complexes of borane with Lewis bases, *J. Chem. Soc. Faraday* **II**, 947 (1972).

[AD124] R. Macaulay, L. A. Burnelle, and C. Sandorfy, Theoretical investigation of the carbon nitrogen double bond, *Theoret. Chim. Acta* **29**, 1 (1973).

[AD125] H. J. Maria, J. R. McDonald, and S. P. McGlynn, Electronic absorption spectrum of nitrate ion and boron trihalides, *J. Amer. Chem. Soc.* **95**, 1050 (1973).

[AD126] H. J. Maria, J. L. Meeks, P. Hochmann, J. F. Arnett, and S. P. McGlynn, Low-energy Rydberg transitions, *Chem. Phys. Lett.* **19**, 309 (1973).

[AD127] M. Mashima and F. Ikeda, UV spectra of hydrazides, *Chem. Lett.* 209 (1972).

[AD128] W. R. Mason, Electronic structure and spectra of linear dicyano complexes, *J. Amer. Chem. Soc.* **95**, 3573 (1973).
[AD129] A. Matsui and W. C. Walker, Polarization of three vacuum-ultraviolet monochromators measured with biotite polarizers, *J. Opt. Soc. Amer.* **60**, 64 (1970).
[AD130] T. McAllister, Electron impact excitation spectra in an ion cyclotron resonance mass spectrometer, *J. Chem. Phys.* **57**, 3353 (1972).
[AD131] H. Metcalf and J. C. Baird, Circular polarization of vacuum ultraviolet light by piezobirefringence, *Appl. Opt.* **5**, 1407 (1966).
[AD132] W. Meyer, PNO-CI studies of electron correlation effects. I. Configuration expansion by means of nonorthogonal orbitals, and application to the ground state and ionized states of methane, *J. Chem. Phys.* **58**, 1017 (1973).
[AD133] G. Milazzo, Absorption spectra of heterocyclic compounds. VII. Spectra of saturated derivatives: Tetrahydro-N-methyl pyrrole, tetrahydrothiophene and tetrahydroselenophene, *Advan. Mol. Spectrosc.* **I**, 471 (1962).
[AD134] G. Milazzo, Absorption spectra of heterocyclic compounds. VI. Tentative interpretations of the spectra of cyclic pentatomic molecules, *Gazz. Chim. Ital.* **83**, 787 (1973).
[AD135] O. A. Mosher, W. M. Flicker, and A. Kuppermann, Electronic spectroscopy of s-$trans$-1,3-butadiene by electron impact, *J. Chem. Phys.* **59**, 6502 (1973).
[AD136] O. A. Mosher, W. M. Flicker, and A. Kuppermann, Triplet states in 1,3-butadiene, *Chem. Phys. Lett.* **19**, 332 (1973).
[AD137] R. S. Mulliken, Mixed V states, *Chem. Phys. Lett.* **25**, 305 (1974).
[AD138] R. S. Mulliken, private communication, 1974.
[AD139] J. N. Murrell and W. Schmidt, Photoelectron spectroscopic correlation of the molecular orbitals of methane, ethane, propane, isobutane and neopentane, *Mol. Phys.* **25**, 1709 (1972).
[AD140] B. Narayan, Spectra and ionization potential of cyanoacetylene, *Proc. Indian Acad. Sci.* **75A**, 92 (1972).
[AD141] B. Narayan, Electronic spectra of paraffin hydrocarbons, *Mol. Phys.* **23**, 281 (1972).
[AD142] S. Nishikawa and T. Watanabe, Superexited states of CH_4, *Chem. Phys. Lett.* **22**, 590 (1973).
[AD143] H. M. O'Bryan, The absorption and dispersion of celluloid between 300 and 1000 Å, *J. Opt. Soc. Amer.* **22**, 739 (1932).
[AD144] H. Ogata, H. Onizuka, Y. Nihei, and H. Kamada, On the first bands of the photoelectron spectra of amines, alcohols and mercaptans, *Chem. Lett.* 895 (1972).
[AD145] H. Okabe, A. H. Laufer, and J. J. Ball, Photodissociation of $OCCl_2$ in the vacuum ultraviolet: Production and electronic energy of excited Cl_2, *J. Chem. Phys.* **55**, 373 (1971).
[AD146] H. Okabe and V. H. Dibeler, Photon impact studies of C_2HCN and CH_3CN in the vacuum ultraviolet; Heats of formation of C_2H and CH_3CN, *J. Chem. Phys.* **59**, 2430 (1973).
[AD147] K. Osafune, S. Katsumata, and K. Kimura, Photoelectron spectroscopic study of hydrazine, *Chem. Phys. Lett.* **19**, 369 (1973).
[AD148] J. F. Osantowski and A. R. Toft, Side-viewing detector for a vacuum ultraviolet reflectometer, *Appl. Opt.* **12**, 2976 (1972).

[AD149] L. R. Painter, R. D. Birkhoff, and E. T. Arakawa, Optical measurements of liquid water in the vacuum ultraviolet, *J. Chem. Phys.* **51**, 243 (1969).

[AD150] L. R. Painter, "Electronic Properties of Liquids," Ann. Progr. Rep. to At. Energy Comm., 1972.

[AD151] A. N. Patyshev and N. S. Trush, Source of continuous spectrum in the vacuum ultraviolet, *J. Appl. Spectrosc.* **10**, 590 (1969).

[AD152] F. Pauzat, J. Ridard, and Ph. Millié, Ab initio calculation of the first ionization potential in linear alkanes using exciton theory, *Mol. Phys.* **24**, 1039 (1972).

[AD153] C. W. Peterson and G. E. Palma, BeO as a window in the vacuum UV, *J. Opt. Soc. Amer.* **63**, 387 (1973).

[AD154] S. D. Peyerimhoff and R. J. Buenker, Theoretical comparison of formic acid and the formate ion, *J. Chem. Phys.* **50**, 1846 (1969).

[AD155] L. Piela, L. Pietronero, and R. Resta, Electron band structure of solid methane: Ab initio calculations, *Phys. Rev. B* **7**, 5321 (1973).

[AD156] A. A. Planckaert, P. Sauvageau, and C. Sandorfy, The vacuum ultraviolet absorption spectra of boron trihalides, *Chem. Phys. Lett.* **20**, 170 (1973).

[AD157] J. W. Rabalais, L. Karlsson, L. O. Werme, T. Bergmark, and K. Siegbahn, Analysis of vibrational structure and Jahn–Teller effects in the electron spectrum of ammonia, *J. Chem. Phys.* **58**, 3370 (1973).

[AD158] D. Reinke, R. Kraessig, and H. Baumgärtel, Photoreactions of small organic molecules. I. Mass-spectrometric study of vinyl chloride, vinyl fluoride and 1,1-difluoroethylene in the vacuum ultraviolet, *Z. Naturforsch.* **28A**, 1021 (1973).

[AD159] M. B. Robin, Electronic structures of the azo, azoxy, and hydrazo groups, in "The Chemistry of Hydrazo, Azo, and Azoxy Groups" (S. Patai, ed.), p. 1. Wiley, New York, 1974.

[AD160] L. L. Robinson, L. C. Emerson, J. G. Carter, and R. D. Birkhoff, Optical reflectivities of liquid water in the vacuum ultraviolet, *J. Chem. Phys.* **46**, 4548 (1967).

[AD161] W. Rothman, F. Hirayama, and S. Lipsky, Fluorescence of saturated hydrocarbons. III. Effect of molecular structure, *J. Chem. Phys.* **58**, 1300 (1973).

[AD162] B. R. Russell, private communication, 1973.

[AD163] L. Sanche and G. J. Schulz, Electron transmission spectroscopy: Resonances in triatomic molecules and hydrocarbons, *J. Chem. Phys.* **58**, 479 (1973).

[AD164] C. Sandorfy, Chemical spectroscopy in the vacuum ultraviolet, *J. Mol. Struct.* **19A**, 183 (1974).

[AD165] P. Sauvageau, R. Gilbert, P. P. Berlow, and C. Sandorfy, Vacuum ultraviolet absorption spectra of fluoromethanes, *J. Chem. Phys.* **59**, 762 (1973).

[AD166] W. Schäfer, A. Schweig, S. Gronowitz, A. Taticchi, and F. Fringuelli, Reversal in the sequence of two highest occupied molecular orbitals in the series thiophene, selenophene and tellurophene, *J. Chem. Soc. Chem. Commun.* 541 (1973).

[AD167] J. D. Scott, G. C. Causley, and B. R. Russell, The vacuum ultraviolet absorption spectra of dimethylsulfide, dimethylselenide, and dimethyltelluride, *J. Chem. Phys.* **59**, 6577 (1974).

[AD168] P. R. Scott, J. Raftery, and W. G. Richards, Energy levels in linear molecules. II, *Proc. Phys. Soc., London (At. Mol. Phys.)* **6**, 881 (1973).
[AD169] A. F. Simonenko, A helium standard light source for the vacuum ultraviolet, *J. Appl. Spectrosc.* **9**, 880 (1968).
[AD170] W. L. Smith, The absorption spectrum of diacetylene in the vacuum ultraviolet, *Proc. Roy. Soc. London* **300A**, 519 (1967).
[AD171] P. A. Snyder and W. C. Johnson, Jr., Circular dichroism of (+)-2-butanol in the vacuum ultraviolet. A comparison of theoretical and experimental values, *J. Chem. Phys.* **59**, 2618 (1973).
[AD172] J. L. Stanley, H. W. Bentley, and M. B. Denton, Radiation exposure considerations when employing microwave excited spectroscopic sources, *Appl. Spectrosc.* **27**, 265 (1973).
[AD173] C. Sugiura, Electron transfer band in the chlorine K-X-ray absorption spectra of some transition metal chlorides, *J. Chem. Phys.* **59**, 4907 (1973).
[AD174] A. M. Taleb, I. H. Munro, and J. B. Birks, The $^1E_{2g}{}^- \leftarrow {}^1A_{1g}{}^-$ transition in benzene, *Chem. Phys. Lett.* **21**, 454 (1973).
[AD175] W.-C. Tam and C. E. Brion, Rydberg states of HCN observed by electron impact spectroscopy, *J. Electron Spectrosc. Related Phenomena* **3**, 281 (1974).
[AD176] W.-C. Tam and C. E. Brion, Electron impact spectra of some alkyl derivatives of water and related compounds, *J. Electron Spectrosc. Related Phenomena* **3**, 263 (1974).
[AD177] K. Tanaka, On the electronic structure of the planar ethylene, *Int. J. Quantum Chem.* **6**, 1087 (1972).
[AD178] J. W. Taylor, Synchrotron radiation as a light source, *in* "Chemical Spectroscopy and Photochemistry in the Vacuum Ultraviolet" (C. Sandorfy, P. J. Ausloos, and M. B. Robin, eds.), p. 543. Reidel, Dordrecht, Holland, 1974.
[AD179] R. Tousey, The extreme ultraviolet—past and future, *Appl. Opt.* **6**, 679 (1962).
[AD180] S. Trajmar, W. Williams, and A. Kuppermann, Electron impact excitation of H_2O, *J. Chem. Phys.* **58**, 2521 (1973).
[AD181] D. G. Truhlar, Application of the configuration-interaction method and the random phase approximation to the *ab initio* calculation of electronic excitation energies of H_2O, *Int. J. Quantum Chem.* **7**, 807 (1973).
[AD182] D. W. Turner, Spectrophotometry in the far-ultraviolet region. Absorption spectra of some amides and cyclic imides, *J. Chem. Soc.* 4555 (1958).
[AD183] M. J. Van der Wiel and C. E. Brion, Partial oscillator strengths for ionization of the three valence orbitals of NH_3, *J. Elect. Spectrosc. Related Phenomena* **1**, 443 (1972).
[AD184] Z. S. Vasilina and N. A. Romanyuk, Vacuum ultraviolet reflection spectra for large angles of incidence, *Opt. Spectrosc.* **32**, 95 (1972).
[AD185] C. H. Warren and C. Ching, Theoretical study of the electronic spectrum of carbonyl cyanide, *Theoret. Chim. Acta* **30**, 1 (1973).
[AD186] I. Watanabe, Y. Yokoyama, and S. Ikeda, Vibrational structures in the photoelectron spectrum of formic acid, *Chem. Phys. Lett.* **19**, 406 (1973).
[AD187] N. Wiberg, Bis(trimethylsilyl) diimine: Preparation, structure and reactivity, *Angew. Chem. Int. Ed.* **10**, 374 (1971).

[AD188] S. D. Worley, G. D. Mateescu, C. W. McFarland, R. C. Fort, Jr., and C. F. Sheley, Photoelectron spectra and MINDO-SCF-MO calculations for adamantane and some of its derivatives, *J. Amer. Chem. Soc.* **95,** 7580 (1973).

[AD189] T. Yamazaki and K. Kimura, On the far ultraviolet absorption spectrum of ethyl fluoride, *Chem. Phys. Lett.* **22,** 616 (1973).

[AD190] D. Yeager, V. McKoy, and G. A. Segal, Assignments in the electronic spectrum of water (to be published).

[AD191] M. Yoshino, J. Takeuchi, and H. Suzuki, Absorption cross sections and photoionization efficiencies of benzene and styrene vapor in the vacuum ultraviolet, *J. Phys. Soc. Japan* **34,** 1039 (1973).

[AD192] M. A. Young and E. S. Pysh, Vacuum ultraviolet circular dichroism of poly (L-alanine) films, *Macromolecules* **6,** 790 (1973).

[AD193] T. M. Zimkina and A. S. Vinogradov, Features of inner atomic shell photoionization in molecules, *J. Phys.* **32,** C4-3 (1971).

[AD194] T. M. Zimkina and A. S. Vinogradov, Photoionization absorption in the ultrasoft X-ray region by atoms in polyatomic molecules, *Bull. Acad. Sci. SSSR, Phys. Ser.* **36,** 229 (1973).

Index

A

A bands, 1
  in allyl iodide, 68
  in carbon tetrachloride, 277
  in chlorobenzene, 230
  in chloroethylenes, 57, 58
  in o-dichlorobenzene, 230
  in difluorochlorophosphine, 316
  in difluoromethyl chloride, 307
  in fluoromethyl chloride, 307
  in haloethylenes, 68
  in hexachlorobenzene, 230
  in iodoacetylene, 116
  in iodoethylene, 67
  in methyl bromide, 307
  in methyl chloride, 307
  in methyl iodide, 116, 307
  in nitrosyl chloride, 162
  in phosgene, 99
  as Rydberg/valence shell conjugates, 68
  in titanium tetrachloride, 277
  in trichloroacetaldehyde, 102
  in 1,3,5-trichlorobenzene, 230
  in trifluoromethyl bromide, 307
  in trifluoromethyl chloride, 307
  in trifluoromethyl iodide, 307
  in vanadium tetrachloride, 345

Ab initio calculations
  on acetonitrile, 118
  on acetylene, 106
  on acrolein, 104
  on allene, 202
  on azulene, 267
  on benzene, 336
  on butadiene, 168, 335
  on carbonyl fluoride, 97, 98
  on cyclopropene, 45, 326
  on cyclopropenone, 328
  on diazomethane, 205
  on diimide, 73
  on ethylene, 9, 10, 18, 324
  on formaldazine, 327
  on formaldehyde, 85, 86, 123, 125, 327, 328
  on formamide, 123–125, 128, 131–133, 138, 140, 148
  on formate anion, 138, 156
  on formic acid, 138, 143, 144, 148, 152, 334
  on formyl fluoride, 138, 148
  on furan, 181
  on glyoxal, 104
  on isobutane, 300, 301
  on ketene, 203
  on naphthalene, 259

on neopentane, 300
on phosphine, 315
on phosphorus trifluoride, 315
on propane, 300
on propylene, 29
on pyrazine, 243
on tetrafluoroethylene, 53, 55, 61
on trimethyl borane, 309
on urea, 140
on water, 316, 317
on xenon difluoride, 270–272
on xenon hexafluoride, 272
Acetaldehyde
 core splitting in, 81
 crystal spectrum of, 77, 82
 high pressure experiment on, 81
 ionization potentials of, 80, 104
 oscillator strength in, 78, 83, 94, 96
 SF_6 scavenger spectrum of, 78
 spectrum of, 75–82, 88, 162
  in solution, 81
 term values in, 75, 76, 80, 203
 vibrations of, 76, 78
Acetaldehyde-d_1
 spectrum of, 77
 vibrations in, 78
Acetaldehyde-d_3
 spectrum of, 77
 vibrations in, 78
Acetaldehyde-d_4
 spectrum of, 77, 79
 vibrations in, 78
Acetamide
 crystal spectrum of, 139, 150
 ionization potentials in, 134
 spectrum of, 134, 136, 333
 term values in, 134
Acetaniline, spectrum of, 251
Acetate anion, spectrum of, 156
Acetic acid
 condensed phase effect in, 153
 intramolecular charge transfer theory of, 148
 ionization potential of, 134
 spectrum of, 134, 149, 150, 152, 153
 term values in, 134
Acetone
 condensed phase effect in, 88, 89
 core splitting in, 87
 electron-impact spectrum of, 87, 328

 high pressure effect on, 86, 88
 ionization potentials of, 80
 oscillator strengths in, 83, 86, 93
 spectrum of, 76, 80, 86–90
 term values in, 80, 103, 160, 328
 vibrations in, 33, 92
Acetone-d_6, spectrum of, 86, 90
Acetonitrile
 calculation on, 118
 ionization potential of, 117
 spectrum of, 117, 118, 332
Acetophenone, spectrum of, 251
Acetyl chloride
 D band in, 160
 spectrum of, 160, 162
 term values in, 160
Acetylene
 autoionization in, 112
 calculations on, 106, 112
 core splitting in, 108
 crystal spectrum of, 113, 115, 329
 electron-impact spectrum of, 109, 112, 113
 electronic structure of, 106
 intermediate exciton in, 329
 matrix shift in CF_4, 113, 293
  in rare gases, 113, 293
 oscillator strengths in, 107, 111, 112, 330
 Rydberg series in, 331
 Rydberg/valence shell conjugates in, 116
 spectrum of, 106–109, 115, 116, 329
  in rare gas matrices, 113, 114, 329
 term values in, 12, 107, 112
 triplet states in, 112, 113
 vibrations in, 106–108
Acetylene-d_2
 oscillator strengths in, 112
 spectrum of, 107
Acrolein
 calculation on, 104
 core splitting in, 103, 104
 ionization potential of, 80, 104
 spectrum of, 80, 102, 103, 105
 term values in, 80, 103
Acrylonitrile
 calculation on, 332
 ionization potential of, 119

oscillator strengths in, 118
spectrum of, 117–119, 332
Adamantane
  fluorescence in solid, 302, 303
  Jahn-Teller effect in, 302
  orbitals in, 301, 302
Adenine
  oscillator strengths in, 288
  spectrum of, 286–288
Alanine
  circular dichroism spectrum of, 158, 159
  crystal spectrum of, 157
DL-α-Alanyl-DL-α-alanine, spectrum of solid, 148
DL-α-Alanyl-glycyl glycine, spectrum of solid, 150
DL-α-Alanyl-DL-methionine, spectrum of solid, 148
DL-α-Alanyl-DL-proline, spectrum of solid, 148
DL-α-Alanyl serine, spectrum of solid, 148
Alcohols, electron-impact spectra of, 320, 321
Alkanes
  electron mobility in liquid, 304
  luminescences of, 301
Alkyl hydrazides, spectra of, 333
Alkyl red shift, 29
Allene
  calculation on, 202
  geometry of, 199, 201
  ionization potentials of, 199, 201, 202
  Jahn-Teller effect in, 199–201
  oscillator strengths in, 200
  spectrum of, 199–201, 204
  term values in, 200, 201, 203
  vibrations in, 199–201
Allene-d_4, vibrations in, 201
Allyl anion, resonance in, 122, 148
Allyl iodide
  A bands in, 68
  B,C bands in, 67, 68
  D band in, 68
  spectrum of, 66–68
  spin-orbit coupling in, 68
Amides
  hydrogen bonding in, 333
  orbitals in, 123
  spectra of crystals, 227

Amino acids, spectra of, 143, 157
3-Amino-pyrrolidin-2-one, spectrum of, 142
Ammonia
  electron-impact spectrum of, 213, 214, 312
  oscillator strengths in, 312, 313
  predissociation in, 312
  Rydberg/valence shell conjugates in, 312
  spectrum of, 312–314
  term values in, 312
Ammonium bromide, spectrum of, 278
Ammonium dihydrogen phosphate, spectrum of, 316
n-Amyl azide, spectrum of, 195, 196
Aniline, spectrum of, 253
Anthracene
  antiresonances in, 264
  condensed phase effect on, 264
  core splitting in, 263
  electron-impact spectrum of crystals, 265, 266
  ionization potential of, 263
  oscillator strengths in, 264, 265
  plasmons in, 265, 266
  reflection spectrum of, 264, 265, 340
  spectrum of, 258, 262–264, 339, 340
Antiresonances
  in anthracene, 264
  in benzene, 216, 217, 221
  in iodoethylene, 67
  in ketene, 204
  in naphthalene, 261
  in pyrazine, 243
  in pyrrole, 185
  in tetrafluoroethylene, 55
Argon
  electron mobility in liquid, 304
  solvent for ethylene, 20
Argon lamp, medium pressure, 294
Argon matrix
  spectrum of acetylene in, 293, 329
  benzene in, 212, 220, 293, 337
  of ethylene in, 293, 325
  of methyl iodide in, 293, 304
  of xenon in, 293
Aurous dicyanide ion
  spectrum of, 345
  spin-orbit coupling in, 345

Autoionization
  in acetylene, 112
  in bromobenzene, 236
  in chloroethylene, 327
  in cyanoacetylene, 331
  in ethylene, 18
  in fluorobenzene, 234
  in formaldehyde, 85
  in methane, 300
Azide ion, electronic states of, 119, 194, 195, 199, 204
Azo-t-butane, spectrum of, 327
trans-Azoethane, spectrum of, 70
trans-Azoisobutane, spectrum of, 70, 72
trans-Azoisopropane, spectrum of, 70
trans-Azomethane
  core splitting in, 72
  high pressure effect on, 69, 71
  ionization potentials of, 68, 69
  spectrum of, 69, 70, 72
Azo trimethylsilane, spectrum of, 327
Azo trimethyl silyl t-butane, spectrum of, 327
Azulene
  calculation on, 267
  condensed phase effect on, 267
  ionization potential of, 267
  SF_6 scavenger spectrum of, 340
  spectrum of, 266, 267
  term values in, 267

B

B bands
  in boron trichloride, 311
  in chloroform, 311
  in iodobenzene, 237
  in titanium tetrachloride, 277
B,C bands
  in allyl iodide, 67, 68
  in bromobenzene, 236
  in dichlorodifluoromethane, 308
  in difluoromethyl chloride, 308
  in ethyl fluoride, 308
  in fluorodichloromethane, 308, 309
  in fluoromethyl chloride, 308
  in fluorotrichloromethane, 308
  in iodoethylene, 67
  in methyl bromide, 308
  in methyl chloride, 307, 308
  in methylene chloride, 307
  in tin tetrachloride, 345
  in trifluoromethyl bromide, 308
  in trifluoromethyl chloride, 308
  in vanadium tetrachloride, 345
Barium fluoride
  cutoff wavelength in, 298
  fluorescence from, 297
Benzaldehyde
  ionization potential of, 254
  spectrum of, 251, 254
  term values in, 254
1,2-Benzanthracene
  electron-impact spectra of crystals of, 266
  plasmons in, 266
Benzene
  antiresonance in, 216, 217, 221
  calculations on, 211, 336
  condensed phase effect in, 215
  core splitting in, 218, 235
  crystal spectrum of, 212, 213, 220, 227, 337
  diffuse orbitals in, 336
  electron-impact spectrum of, 210, 213–216, 219, 221
  crystal, 223
  electronic structure of, 210
  ionization potentials of, 221–223
  in rare gas matrices, 337
  Jahn-Teller effect in, 215, 244, 337
  luminescence of, 215, 227
  matrix shift in nitrogen, 293
  in rare gases, 293
  oscillator strengths in, 209, 214, 215, 219, 220, 228, 235
  perfluoro-effect in, 234, 235
  plasmons in, 220, 223
  reflection spectrum of liquid, 219, 220
  SF_6 scavenger spectrum of, 221
  spectrum of, 211–224, 228, 229, 231, 244, 250, 256, 262, 331, 336–338
    in argon matrix, 212, 220
    in krypton matrix, 212, 220, 221
    of liquid, 174, 220, 336, 337
    in nitrogen matrix, 212
    in rare gas matrices, 337
    in xenon matrix, 220, 221
  term values in, 213, 223
    in rare gas matrices, 293

triplet states in, 215, 216
vibrations in, 211–213, 217–219, 244
Wannier excitons in, 220, 221
Benzene-d_3, spectrum in xenon matrix, 212
Benzene-d_6
  intensities in, 220
  spectrum of, 216
    in krypton matrix, 212
    in nitrogen matrix, 212
  vibrations in, 215, 217–219
Benzoic acid
  intramolecular charge transfer theory of, 250, 251, 253
  spectrum of, 250, 251, 253
Benzonitrile
  ionization potentials of, 254
  spectrum of, 254
  term values in, 254
Benzotrifluoride
  oscillator strengths in, 228
  spectrum of, 224, 228, 235
1,2-Benzpyrene
  electron-impact spectrum of crystals of, 266
  plasmons in, 266
Beryllium carbonate, crystal spectrum of, 275
Beryllium fluoride as window material, 296
Beryllium oxide as window material, 296
Biacetyl
  ionization potentials in, 80, 104
  spectrum of, 80, 104, 105
  term values in, 80, 105
Bicyclohexane, fluorescence in solid, 302
Bicyclohexylidene
  crystal spectrum of, 47, 48
  ionization potential of, 27
  spectrum of, 27, 47, 48
  term values in, 27
  vibrations in, 28
Bicyclooctene, spectrum of, 31
Biotite polarizer, 295
Biphenyl, spectrum of, 341
Bistrimethylsilyl acetylene, spectrum of, 111
Bis-1,4-trimethylsilyl butadiene, absorption intensities in, 173
Borazine, spectrum of, 311, 312

Boron tribromide
  ionization potentials of, 311
  spectrum of, 310, 311
  term values in, 311
Boron trichloride
  B bands in, 311
  D bands in, 311
  spectrum of, 310, 311
  term values in, 310, 311
  triplet state in, 311
Boron trifluoride
  spectrum of, 311
  X-ray spectrum of, 311
Bromide ion, spectrum of, 279
Bromobenzene
  autoionization in, 236
  B,C bands in, 236
  D band in, 236
  intramolecular charge transfer theory of, 237
  ionization potentials of, 236
  oscillator strengths in, 228
  spectrum of, 228, 230, 236, 237
  spin-orbit coupling in, 236
  term values in, 236
BRV continuum source, 294, 296
Butadiene
  calculations on, 168, 170, 335
  condensed phase effect on, 168, 169, 174, 335
  core splitting in, 173, 174
  crystal spectrum of, 168, 335
  diffuse orbitals in, 168, 335
  electron-impact spectrum of, 335
  electronic structure of, 166
  geometry of, 166, 174
  ionization potentials of, 175, 176
  oscillator strengths in, 168, 172, 173
  perfluoro-effect in, 175
  π-π splitting in, 179, 180
  spectrum of, 102, 105, 156, 166–170, 173–175, 202
  term values in, 168, 173, 174
  triplet states in, 335
  vibrations in, 168, 173, 176
Butane
  luminescence of, 301
  term values in, 174
sec-2-Butanol
  circular dichroism spectrum of, 318, 319

ionization potentials of, 319
spectrum of, 318, 319
Butene-1
  ionization potential of, 24
  negative ion states of, 326
  spectrum of, 24, 326
  term values in, 24
  triplets in, 36
  vibrations in, 28
cis-Butene-2
  ionization potential of, 24, 29
  negative-ion states of, 326
  spectrum of, 24, 29, 37
  term values in, 24, 29, 35
  triplets in, 36
  vibrations in, 28, 38
trans-Butene-2
  ionization potential of, 24, 29
  negative-ion states of, 326
  spectrum of, 24, 29, 37
  term values in, 24, 29, 35
  triplets in, 36
  vibrations in, 28, 38
trans-Butene-2-d_8
  C-type band in, 36
  vibrations in, 28
N-n-Butyl acetamide, spectrum of, 136
t-Butyl acetylene, spectrum of, 108, 110
n-Butyl benzene
  oscillator strengths in, 228
  spectrum of, 228
t-Butyl benzene
  oscillator strengths in, 228
  spectrum of, 228
sec-Butyl benzene
  oscillator strengths in, 228
  spectrum of, 228
t-Butyl cyanate
  oscillator strengths in, 199
  spectrum of, 199
t-Butyl ethylene, spectrum of, 25
t-Butyl nitrile, spectrum of, 118
Butyraldehyde, spectrum of, 95

C

Calcium fluoride
  cutoff wavelength in, 298
  fluorescence from, 297

Calf thymus nucleic acid
  electron-impact spectrum of, 288
  plasmons in, 288
Calcium carbonate, crystal spectrum of, 275
Carbon monoxide, Cameron bands of, 99
Carbon suboxide
  electronic states in, 207
  high pressure effect on, 205, 206, 208
  ionization potentials of, 205, 207
  oscillator strengths in, 205, 206
  spectrum of, 205–208
  term values in, 205–207
  vibrations in, 206–208
Carbon tetrachloride
  A bands in, 277
  D bands in, 277
  ionization potentials of, 276, 306
  as quencher, 301
  spectrum of, 102, 277, 305, 343
Carbonate ion, spectral resemblance to nitrate ion, 275
Carbonyl cyanide
  ionization potential of, 102
  spectrum of, 102, 328
Carbonyl fluoride
  calculation on, 97, 98
  ionization potential of, 80, 98
  oscillator strengths in, 83, 97, 98
  perfluoro-effect in, 98
  spectrum of, 80, 96–99
  term values in, 80, 98
Cellulose nitrate as window material, 297
Central-cell corrections, 292, 293
Cerous ion, spectrum of, 343
Cesium nitrate, crystal spectrum of, 273, 274
Chloride ion, spectrum of, 279
Chloroacetaldehyde
  D bands in, 101, 102
  ionization potential of, 80
  spectrum of, 80, 101
  term values in, 80
Chlorobenzene
  A band in, 230
  D band in, 63, 235, 236
  intramolecular charge transfer theory of, 237
  ionization potential of, 236
  oscillator strengths in, 228

spectrum of, 228, 231, 236, 237, 338
term values in, 236
Chloroethylene
  autoionization in, 327
  core splitting in, 52, 60
  D band in, 63
  ionization potential of, 52, 57, 327
  photoionization spectrum of, 326, 327
  spectrum of, 52, 56–58, 62
  term values of, 52
  triplets in, 64
  X-ray spectrum of, 66
Chloroform
  B bands in, 311
  D bands in, 311
  orbital ordering in, 304, 305
  spectrum of, 102, 305
  X-ray spectrum of, 304
1-Chloronaphthalene
  D band in, 263
  SF_6 scavenger spectrum of, 263
1-Chloro-1-nitrosocyclohexane
  D band in, 161
  spectrum of, 161
Chloroprene
  D band in, 63, 175
  spectrum of, 174, 175
Chromate ion, spectrum of, 280, 282
Chromium hexacarbonyl, spectrum of, 282–284
Chrysene
  electron-impact spectrum of crystals of, 266
  plasmons in, 266
trans-Cinnamic acid
  intramolecular charge transfer theory of, 253
  spectrum of, 253
Circular dichroism spectrum
  of alanine, 158, 159
  of sec-2-butanol, 318, 319
  of trans-cyclooctene, 22, 40–42
  of DNA, 346
  of D-galactose, 289
  of D-glucose, 289
  of isoleucine, 158, 159
  of ketones, 96
  of leucine, 158, 159
  of 3-methyl cyclopentanone, 96, 97
  of 3-methyl diketopiperazine, 141, 142
  of 1-methyl indan, 338
  of 3-methyl isopropylene cyclopentane, 43
  of 3-methyl pyrrolidin-2-one, 333
  of methylene steroids, 43
  of 2-phenyl-3,3-dimethyl butane, 229, 230
  of α-pinene, 40, 42–44
  of β-pinene, 40, 44
  of poly-L-alanine, 333
  of poly-N-methyl glutamate, 333
  of proline, 158, 159
  of steroids, 289
  of triterpenoids, 289
  of L-tryptophan, 346
  of valine, 158, 159
  of D-xylose, 289
Cobalt salicylaldimine, X-ray spectrum of, 285
Cobaltous ion, spectrum of, 343
Condensed phase effect
  on acetaldehyde, 77, 81, 82
  on acetic acid, 153
  on acetone, 88, 89
  on anthracene, 264
  on azulene, 267
  on benzene, 215
  on butadiene, 168, 169, 174, 335
  on 1,3-cyclohexadiene, 176
  on 1,3-cyclooctadiene, 176
  on cyclopentadiene, 173
  on N,N-diethyl aniline, 255, 256
  on diethyl ketone, 88
  on N,N-dimethyl acetamide, 129
  on N,N-dimethyl aniline, 255, 256
  on 2,3-dimethyl butadiene, 168, 169, 174
  on ethylene, 12
  on formamide, 129, 130
  on furan, 184, 185
  on hexamethyl benzene, 226
  on methyl isobutyl ketone, 89
  on methyl isopropyl ketone, 89
  on N-methyl maleimide, 192
  on naphthacene, 266
  on naphthalene, 259, 262
  on norbornadiene, 177, 178, 180
  on osmium tetroxide, 282
  on pyrrole, 184, 185
  on tetrakisdimethylaminoethylene, 50
  on tetramethyl ethylene, 33, 34

on tricyclo[3.3.0.0.^{2,6}]oct-3-ene, 45, 46
on trifluoroacetic acid, 153
on X-ray spectra, 293, 294
Copper phthalocyanine
  crystal spectrum of, 284, 285
  chlorinated, 284, 285
Copper salicylaldimine, X-ray spectrum of, 285
Core splitting
  in acetaldehyde, 81
  in acetone, 87
  in acetylene, 108
  in acrolein, 103, 104
  in anthracene, 263
  in *trans*-azomethane, 72
  in benzene, 218, 235
  in butadiene, 173, 174
  in chloroethylene, 52, 60
  in diazomethane, 204, 205
  in 1,1-dichloroethylene, 52, 60, 62
  in *cis*-dichloroethylene, 52, 60, 63
  in ethylene, 17
  in formaldehyde, 84, 328
  in formic acid, 154
  in ketene, 204
  in 3-methyl cyclopentanone, 96
  in pyrazine, 243
  in pyridine, 239, 240, 243
  in trichloroethylene, 52, 60
  in xenon difluoride, 342
  in xenon tetrafluoride, 342
Coronene
  electron-impact spectrum of, 341
  of crystals of, 266
Crotonaldehyde
  ionization potentials of, 80, 104
  spectrum of, 80, 102, 103
  term values in, 80, 103
Crotonic acid, spectrum of, 156, 166
Cubane
  ionization potential of, 302
  spectrum of, 302
  term values in, 302
Cupric chloride, X-ray spectrum of, 346
Cupric ion, spectrum of, 343
Cuprous dicyanide ion, spectrum of, 345
Cyanamide, electron-impact spectrum of, 120
Cyanoacetylene
  autoionization in, 331
  ionization potentials of, 331
  Rydberg/valence shell conjugates in, 332
  spectrum of, 331, 332
  term values in, 331, 332
Cyanoazide, spectrum of, 208
Cyanogen
  ionization potential of, 120
  spectrum of, 117, 120
Cyclobutanone
  ionization potential of, 80, 92
  oscillator strengths in, 93
  spectrum of, 80, 92, 93
  term values in, 80
  vibrations in, 92, 93
Cyclobutene
  calculations on, 48
  ionization potential of, 26
  spectrum of, 26
  term values in, 26
  vibrations in, 28
Cycloheptanone, vibrations in, 92
1,3,5-Cycloheptatriene, electron-impact spectrum of, 189
1,3-Cyclohexadiene
  absorption intensities in, 172
  condensed phase effect in, 176
  crystal spectrum of, 176
  high-pressure effect on, 176
  spectrum of, 170, 171, 173, 176, 181
    liquid, 174
  vibrations in, 176
1,4-Cyclohexadiene
  π-π splitting in, 179, 180
  spectrum of, 171, 173
Cyclohexane
  fluorescence in solid, 302
  orbitals in, 302
  spectrum of liquid, 174
Cyclohexanone
  ionization potential of, 80
  oscillator strengths in, 93
  spectrum of, 80, 93
  term values in, 80
  vibrations in, 92
Cyclohexene
  high-pressure effect on, 33
  ionization potential of, 26
  spectrum of, 26, 31, 38, 48
    liquid, 33, 174

term values in, 26
vibrations in, 28
1,3-Cyclooctadiene
  condensed phase effect in, 176
  crystal spectrum of, 176
  high-pressure effect on, 176
  spectrum of, 176
1,4-Cyclooctadiene, spectrum of, 171
Cyclooctatetraene, electron-impact spectrum of, 189
trans-Cyclooctene
  circular dichroism spectrum of, 22, 40–42
  ionization potential of, 26
  spectrum of, 26, 40, 41
  term values in, 26, 42–44
Cyclopentadiene
  condensed phase effect on, 173
  ionization potential of, 175
  oscillator strengths in, 170, 172
  spectrum of, 170, 171, 173, 175, 176, 181, 183
  term values of, 175, 176
  vibrations in 173, 176
Cyclopentanone
  ionization potential of, 80
  oscillator strengths in, 93
  spectrum of, 80, 93
  term values of, 80
  vibrations in, 92
Cyclopentene
  ionization potential of, 26
  spectrum of, 26, 31, 38, 39, 48, 179
  vibrations in, 38, 39
  term values in, 26
Cyclopropane, spectrum of, 73, 74
Cyclopropene
  absorption intensity in, 137
  calculation on, 45, 326
  diffuse orbitals in, 326
  ionization potential of, 25, 45
  photoelectron spectrum of, 45
  spectrum of, 25, 32, 45, 56, 326
  term values in, 25
  vibrations in, 28
Cyclopropenone
  calculation on, 328
  electron-impact spectrum of, 328
  term values in, 328

Cyclopropyl methyl ketone, spectrum of, 94
Cytosine
  electron-impact spectrum of, 346
  optical constants of, 346
  oscillator strengths in, 288
  spectrum of, 286–288

D

D bands
  in acetyl chloride, 160
  in allyl iodide, 68
  in boron trichloride, 311
  in bromobenzene, 236
  in carbon tetrachloride, 277
  in chloroacetaldehyde, 101, 102
  in chlorobenzene, 63, 235, 236
  in chloroethylene, 63
  in chloroform, 311
  in 1-chloronaphthalene, 263
  in 1-chloro-1-nitrosocyclohexane, 161
  in chloroprene, 63, 175
  in *o*-dichlorobenzene, 235, 236
  in 1,1-dichloroethylene, 57, 63, 64
  in *trans*-dichloroethylene, 57, 63, 64
  in difluorochlorophosphine, 316
  in difluorodichloromethane, 308
  in difluoromethyl chloride, 308
  in ethyl fluoride, 308
  in fluorodichloromethane, 308
  in fluoromethyl chloride, 308
  in fluorotrichloromethane, 308
  in iodoacetylene, 117
  in iodobenzene, 237
  in iodoethylene, 67
  in methyl chloride, 175
  in phosgene, 100, 101
  in tetrachloroethylene, 57, 63, 64
  in tin tetrachloride, 345
  in titanium tetrachloride, 277, 343, 345
  in trifluoromethyl bromide, 308
  in trifluoromethyl chloride, 308
  in vanadium tetrachloride, 345
Decalin, luminescence of, 301
Deuterium, matrix spectrum of xenon in, 293
Diacetylene
  high-pressure effect on, 331

ionization potentials of, 330
spectrum of, 330, 331
Diacetylene-d_2, spectrum of, 330
1,4-Diazabicyclooctane (DABCO), spectrum of, 313, 314
Diazirine, spectrum of, 45
Diazomethane
  calculation on, 205
  core splitting in, 204, 205
  ionization potential of, 205
  spectrum of, 204
  vibrations in, 205
Diazomethane-d_2, spectrum of, 204
Diborane
  term values in, 309
  X-ray spectrum of, 309
Di-t-butyl acetylene
  oscillator strengths in, 111
  spectrum of, 110
2,3-Di-t-butyl butadiene, absorption intensities in, 173
cis-Di-t-butyl ethylene
  ionization potential of, 25
  spectrum of, 25
  term values in, 25
$trans$-Di-t-butyl ethylene
  ionization potential of, 25
  spectrum of, 25
  term values in, 25
Di-t-butyl ketone, oscillator strengths in, 83, 93
Dichloroacetaldehyde, spectrum of, 101, 102
m-Dichlorobenzene
  A band in, 230
  oscillator strengths in, 228, 232
  spectrum of, 228, 231
o-Dichlorobenzene
  A band in, 230
  D bands in, 235, 236
  oscillator strengths in, 228, 232
  spectrum of, 228, 231
p-Dichlorobenzene
  A band in, 230
  oscillator strengths in, 228, 232
  spectrum of, 228, 231
Dichloroethylenes, X-ray spectra of, 66
1,1-Dichloroethylene
  core splitting in, 52, 60, 62
  D bands in, 57, 63, 64

ionization potential of, 52, 57
spectrum of, 52, 56, 57, 62
term values in, 52, 64
cis-Dichloroethylene
  core splitting in, 52, 60, 63
  ionization potential of, 52, 57
  spectrum of, 52, 56, 57, 62
  term values in, 52
$trans$-Dichloroethylene
  D bands in, 57, 63, 64
  ionization potential of, 52, 57
  spectrum of, 52, 56–59, 62
  term values in, 52
Dielectric filters, 296
N,N-Diethyl acetamide, spectrum of, 136
N,N-Diethyl aniline
  condensed phase effect on, 255, 256
  spectrum of, 255
Diethyl ether, term values in, 181
N,N-Diethyl formamide, spectrum of, 137, 138
Diethyl ketone
  condensed phase effect on, 88
  ionization potentials in, 80
  oscillator strengths in, 83, 93
  spectrum of, 76, 80, 88, 91
  term values in, 80
Diffuse orbitals
  in benzene, 336
  in butadiene, 168, 335
  in cyclopropene, 326
  in ethylene, 9–11, 86, 324, 335
  in formaldehyde, 86
  in oxygen, 10
1,1-Difluorochloroethylene
  ionization potential of, 52
  spectrum of, 52, 65, 66
  term values in, 52
Difluorochlorophosphine
  A bands in, 316
  D bands in, 316
  ionization potential of, 315
$trans$-Difluorodiazine
  ionization potential of, 73
  spectrum of, 73
Difluorodiazirine, spectrum of, 73, 74, 327
1,1-Difluorodichloroethylene
  ionization potential of, 52
  spectrum of, 52, 64
  term values of, 52, 64

Difluorodichloromethane
  B,C bands of, 308
  D bands of, 308
  ionization potential of, 308
  term values in, 308
1,1-Difluoroethylene
  ionization potentials of, 52
  photoelectron spectrum of, 50
  spectrum of, 51–54, 56, 64
  term values in, 52–55, 64
  triplets in, 56
cis-Difluoroethylene
  ionization potential of, 52
  spectrum of, 51–53
  term values in, 35, 52–55
trans-Difluoroethylene
  ionization potential of, 52
  spectrum of, 51–55
  term values in, 35, 52–55
Difluoromethyl chloride
  A bands in, 307
  B,C bands in, 308
  D bands in, 308
  ionization potential of, 308
  term values in, 308
Diglycyl, spectrum of, 142
Diimide
  calculation on, 73
  ionization potential of, 72
  spectrum of, 69, 72
  triplets in, 73
Diimide-d_2, spectrum of, 72
Diisobutylene, spectrum of, 48
Diisopropyl ketone
  oscillator strengths in, 83, 93
  spectrum of, 91
Diketopiperazine
  crystal spectrum of, 141, 143, 145
  spectrum of, 142
N,N-Dimethyl acetamide
  condensed phase effect on, 129
  crystal spectrum of, 129
  ionization potential of, 134
  spectrum of, 129, 134–136, 145
  term values in, 134, 135, 193
Dimethyl acetylene
  oscillator strengths in, 111
  spectrum of, 110
1,1-Dimethyl allene, spectrum of, 202

N,N-Dimethyl aniline
  condensed phase effect on, 255, 256
  intramolecular charge transfer theory of, 250, 253
  spectrum of, 250, 253, 255
1,3-Dimethyl butadiene, spectrum of, 174
2,3-Dimethyl butadiene
  crystal spectrum of, 169, 174
  spectrum of, 169
2,3-Dimethyl butene-1, spectrum of, 25
Dimethyl carbonate
  as solvent, 334
  spectrum of, 334
1,1-Dimethyl-3-chlorobutadiene, absorption intensities in, 172, 173
Dimethyl cyanamide
  intramolecular charge transfer theory of, 120
  spectrum of, 117, 119, 120
2,2-Dimethyl-3-cyclopentene-1,3-dione, spectrum of, 190, 191, 193, 194
2,2-Dimethyl cyclopropene, spectrum of, 26
Dimethyl ether, term value in, 203
N,N-Dimethyl formamide
  ionization potentials of, 131–134
  oscillator strengths in, 163
  spectrum of, 128, 133–135
  term values in, 128, 134, 135
  vibrations in, 131, 133
N,N-Dimethyl nitrosamine
  intramolecular charge transfer theory of, 163, 164
  spectrum of, 163
Dimethyl oxalate
  ionization potential of, 160
  spectrum of, 159, 160
Dimethyl-*s*-triazine, spectrum of, 247
n-Dioctyl ketone, crystal spectrum of, 87, 88
Dioxadiene, spectrum of, 188, 189
1,3-Dioxane, orbitals in, 320
1,4-Dioxane, orbitals in, 320
Dioxene, spectrum of, 188
N,N-Di-*n*-propyl acetamide, spectrum of, 136
Di-*n*-propyl ketone
  oscillator strengths in, 83, 93
  spectrum of, 91
Dissymmetry parameter, 41

Divinyl acetylene
  ionization potentials of, 117
  spectrum of, 117
Divinyl ether, spectrum of, 188
DNA, circular dichroism spectrum of, 346
Dodecane, luminescence of, 301

E

Electron-impact spectrum
  of acetone, 87, 328
  of acetylene, 109, 112, 113
  of alcohols, 320, 321
  of alkanes, 300, 301
  of ammonia, 213, 214, 312
  of anthracene crystals, 265, 266
  of 1,2-benzanthracene crystals, 266
  of benzene, 210, 213–216, 219, 221
  of benzene crystal, 223
  of 1,2-benzpyrene crystals, 266
  of butadiene, 335
  of calf thymus nucleic acid, 288
  of chrysene crystals, 266
  of coronene, 341
    crystals, 266
  of cyanamide, 120
  of 1,3,5-cycloheptatriene, 189
  of cyclooctatetraene, 189
  of cyclopropenone, 328
  of cytosine, 346
  of ethers, 320, 321
  of ethyl acetylene, 109
  of ethylene, 4, 17–21
  of ferrocene, 284
  of formaldehyde, 82, 85
  of guanine, 346
  of n-heptane, 301
  of hexabenzocoronene, 341
    crystals, 266
  of 1,3,5-hexatriene, 189, 336
  of hydrogen cyanide, 332
  of hydrogen sulfide, 318
  of mesitylene crystal, 229
  of methane, 299
  of methyl acetylene, 109
  of naphthalene, 260, 262
  of nitromethane, 334, 335
  of nucleic acid bases, 287
  of perylene, 341
  of phenanthrene, 265
  of picene crystals, 266
  of polystyrene, 256, 257
  of pyrazine, 240, 243, 245, 247
  of pyrene, 341
    crystals, 266
  of pyridazine, 247
  of pyridine, 238, 240
  of pyrimidine, 247
  of tetracene crystals, 266
  of tetrafluoroethylene, 53
  of s-tetrazine, 249
  of toluene crystal, 229
  of s-triazine, 247, 248
  of trifluoronitrosomethane, 161
  of triphenylene, 341
  of water, 316–318
Electron mobilities
  in liquids, 304
  in xenon, 342
  in xenon difluoride, 342
Ethane
  luminescence of, 301
  orbital ordering in, 300, 305
  spectrum of, 256, 300, 314
  term values in, 12, 107, 315
Ethane-d_6, spectrum of, 300
Ethers, electron-impact spectra of, 320, 321
N-Ethyl acetamide, spectrum of, 136
Ethyl acetate
  ionization potential of, 134
  spectrum of, 134, 149
Ethyl acetylene
  electron-impact spectrum of, 109
  spectrum of, 108
Ethyl alcohol, term value in, 203
Ethyl allene, spectrum of, 202
Ethyl benzene
  oscillator strengths in, 228
  spectrum of, 224, 228, 229
Ethyl N-n-butyl aldimine, spectrum of, 74
Ethyl fluoride
  B,C bands in, 308
  D bands in, 308
  spectrum of, 305, 306, 308
  term values in, 306, 308
Ethyl iodide, spectrum of, 66, 67
Ethyl isocyanate, spectrum of, 197, 198

N-Ethyl moleimide, spectrum of, 193
Ethyl nitrate
  intramolecular charge transfer theory of, 165
  oscillator strength in, 164, 274
  spectrum of, 164, 165, 273
Ethyl nitrite, ionization potential of, 163
Ethyl pentafluoropropionate
  ionization potential of, 134
  spectrum of, 134, 155
  term values in, 134, 155
Ethyl propionate, term values in, 155
Ethylene
  autoionization in, 18
  C-type band in, 14, 36
  calculations on, 8–11, 18, 19, 324
  condensed phase effect, 12
  core splitting in, 17
  crystal spectrum, 7, 20
  deuterated, spectra of, 27
  diffuse orbitals in, 9–11, 324, 326, 335
  electron-impact spectrum of, 4, 17–21
  electronic structure of, 3
  geometries in excited states, 3–5, 7, 9, 14, 168
  high-pressure effect on, 19, 20
  ionization potential of, 24, 29, 52
  in matrices, 325
  Jahn–Teller effect in, 14
  matrix shift in rare gases, 293
  negative ion spectrum of, 325
  oscillator strengths in, 5, 6, 18, 19, 22, 24, 137, 325
  photoelectron spectrum of, 14–16, 45
  plasmon in, 19
  Rydberg/valence shell conjugates in, 9
  SF_6 scavenger spectrum, 17
  spectrum of, 3, 4, 6, 8, 24, 29, 48, 52, 69, 202, 204
    in krypton matrix, 20, 21
    liquid, 6–8
    in liquid argon, 20
    in liquid krypton, 20
    in rare gas matrices, 325
  term values in, 24, 29, 33–35, 52, 107, 290
    in rare gas matrices, 293
  triplets in, 16, 17, 21, 36, 324, 325
  vibrational structure in, 5, 7, 8, 11–14, 17, 21, 33, 61
  Wannier excitons in, 20
Ethylene-d_1, vibrational structure in, 13
Ethylene-d_4
  photoelectron spectrum of, 14–16
  spectrum of, 6–8, 12
    in krypton matrix, 20
  vibronic structure in, 6, 8, 12–14, 17
Ethylene oxide, term values in, 321
Ethylene sulfide
  spectrum of, 321
  term values in, 321
Ethylidene cyclohexane
  ionization potential of, 27
  spectrum of, 27
  term values in, 27
  vibrations in, 28
Ethylidene cyclopentane
  ionization potential of, 26
  spectrum of, 26
  term values in, 26
Europium ion, spectrum of, 343
Excited precursors, 291

F

Ferric chloride, X-ray spectrum of, 346
Ferrocene, electron-impact spectrum of, 284
Ferrous ion, spectrum of, 343
Fluorenone, spectrum of, 268
Fluoroacetamide, spectrum of, 125, 126
Fluorobenzene
  autoionization in, 234
  intramolecular charge transfer theory of, 237, 250
  ionization potentials of, 232–235
  oscillator strengths in, 228
  spectrum of, 53, 224, 228, 232–235, 249
  term values in, 233
1-Fluoro-1-chloroethylene, spectrum of, 52, 65
Fluorodichloromethane
  B,C bands in, 308, 309
  D bands in, 308
  ionization potential of, 308
  term values in, 308, 309
Fluoroethylene
  ionization potential of, 52
  photoelectron spectrum of, 50

SF$_6$ scavenger spectrum of, 56
  spectrum of, 51-53
  term values in, 35, 52-55
  triplets in, 56
Fluoroform, X-ray spectrum of, 306, 307
Fluoromethanes
  spectra of, 306
  X-ray spectra of, 306, 307
Fluoromethyl chloride
  A band in, 307
  B,C bands in, 308
  D band in, 308
  ionization potential of, 308
  term values in, 308
m-Fluorotoluene
  oscillator strengths in, 228
  spectrum of, 224, 228
o-Fluorotoluene
  oscillator strengths in, 228
  spectrum of, 224, 228
p-Fluorotoluene
  oscillator strengths in, 228
  spectrum of, 224, 228, 235
Fluorotrichloromethane
  B,C bands in, 308
  D bands in, 308
  ionization potentials of, 308
  term values in, 308
Formaldazine, calculation on, 327
Formaldehyde
  autoionization in, 85
  calculations on, 85, 86, 123, 327, 328
  core splitting in, 84, 328
  electron-impact spectrum of, 82, 85
  ionization potential of, 80
  orbitals in, 81, 97
    diffuse, 86
  oscillator strengths in, 82-85, 96
  photoelectron spectrum of, 84
  spectrum of, 76, 80, 82-85, 88
  term values in, 80, 92, 98, 99
  vibrations in, 84
Formamide
  calculation on, 123, 124, 128, 131-133, 138, 140, 148
  condensed phase effect on, 129, 130
  crystal spectrum of, 129, 130
  high-pressure effect in, 130
  intramolecular charge transfer theory of, 128, 135, 136

  ionization potentials of, 124, 130, 131, 133, 134
  oscillator strength in, 137, 138
  spectrum of, 125, 126, 128-130, 133, 134, 139, 143, 333
  term values in, 128, 134, 135, 197
  triplet states in, 138, 140
  vibrations in, 131, 133
Formate anion
  calculation on, 138, 156
  oscillator strength in, 138
  spectrum of, 138, 156
  triplet states in, 138
Formic acid
  calculation on, 134, 143, 144, 148, 152, 334
  core splitting in, 154
  intramolecular charge transfer theory of, 150
  ionization potentials of, 131, 134, 153
  oscillator strengths in, 138, 154, 155
  spectrum of, 134, 143, 149-154, 159, 334
  term values in, 134, 135, 154
  triplet states in, 138, 153
  vibrations in, 131, 151, 153, 154
Formyl fluoride
  calculation on, 138, 148
  oscillator strengths in, 138
  spectrum of, 138
  triplet states in, 138
Furan
  calculation on, 181
  condensed phase effect on, 184, 185
  ionization potential of, 181, 183
  oscillator strengths in, 183
  photoionization spectrum of, 182
  spectrum of, 181-184, 186
  term values in, 181, 183
  vibrations in, 181, 183
Furfuraldehyde, spectrum of, 328, 329

G

D-Galactose, circular dichroism spectrum of, 289
D-Glucose, circular dichroism spectrum of, 289
Glutamic acid, crystal spectrum of, 157
L-Glutamyl-L-α-alanine, spectrum of solid, 148

Glycerol
  plasmon in, 319
  reflection spectrum of, 319
Glycine
  crystal spectrum of, 156, 157
Glycyl-β-alanine, spectrum of solid, 149
Glycyl-DL-leucine, spectrum of solid, 149
Glycyl-DL-methionine, spectrum of solid, 149
Glycyl-DL-norvaline, spectrum of solid, 149
Glycyl-DL-valine, spectrum of solid, 149
Glyoxal
  calculation on, 104
  ionization potentials of, 80, 104
  spectrum of, 80, 104, 105
  term values in, 80, 105
Graphite, plasmons in, 220, 265
Guanine
  electron-impact spectrum of, 346
  spectrum of, 286, 287

H

Halide ions, spectra of, 278, 279
He II effect in thiophene, 187
Helium lamp
  for Hopfield continuum, 294
  using flowing gas, 294
Heptadecane, reflection spectrum of, 301
2-Heptadecenoic acid, spectrum of, 166
n-Heptane
  electron-impact spectrum of, 301
  as solvent for acrylonitrile, 332
  superexcited states in, 301
Heptene-1, vibrations in, 28
Heptene-3, spectrum of, 25
Hexabenzocoronene
  electron-impact spectrum of, 341
  crystals, 266
Hexachlorobenzene
  A band in, 230
  oscillator strengths in, 228, 232
  spectrum of, 228, 230, 231
Hexachlorobutadiene, spectrum of, 172
$trans$-2,4-Hexadiene, spectrum of, 170
Hexaethyl benzene
  oscillator strengths in, 228
  spectrum of, 228

Hexafluoroacetone
  ionization potential of, 80
  spectrum of, 80
  term values in, 80, 98, 101
Hexafluorobenzene
  absorption intensities in, 235
  spectrum of, 233–235
  term values in, 234
Hexafluorobutadiene
  absorption intensities in, 172
  independent systems theory of, 172
  spectrum of, 167, 172, 173
Hexahydro-1,3,5-trinitro-s-triazine, spectrum of, 165
Hexamethyl benzene
  condensed phase effect on, 226
  crystal spectrum of, 226, 227
  ionization potential of, 226
  oscillator strengths in, 228, 229
  spectrum of, 226–228, 256, 337
  term value in, 227
1,3,5-Hexatriene
  absorption intensities in, 220
  calculation on, 189, 190
  electron-impact spectrum of, 189, 336
  ionization potential of, 189
  spectrum of, 117, 189
  triplet states in, 336
Hexene-1
  ionization potential of, 25
  spectrum of, 25
cis-Hexene-3, spectrum of, 25
$trans$-Hexene-3
  ionization potential of, 25
  spectrum of, 23, 25
  term values in, 25
High-pressure effect
  on acetaldehyde, 81
  on acetone, 86, 88
  on $trans$-azomethane, 69, 71
  on carbon suboxide, 205, 206, 208
  on 1,3-cyclohexadiene, 176
  on cyclohexene, 33
  on 1,3-cyclooctadiene, 176
  on diacetylene, 331
  on ethylene, 19, 20
  on formamide, 130
  on N-methyl maleimide, 190–192
  on norbornadiene, 177, 178
  on tetramethyl ethylene, 33

L-Histidyl-L-leucine, spectrum of solid, 148
Hopfield continuum, 294
Hydrazine
 spectrum of, 314, 315
 term values in, 314, 315
Hydrazoic acid
 ionization potentials in, 196
 oscillator strengths in, 194–196
 spectrum of, 194–197
 term values in, 196
 vibrations in, 195
Hydrogen cyanide, electron-impact spectrum of, 332
Hydrogen sulfide
 electron-impact spectrum of, 318
 negative-ion states in, 318

I

Imides, spectra of, 333
Independent systems theory
 of biphenyl, 341
 of hexafluorobutadiene, 172
 of norbornadiene, 180
Inner-well states, 291, 292
Intermediate exciton
 in acetylene, 329
 in CF_4 matrix, 293
 in ethylene/rare gas matrices, 325
 in neon matrix, 293
 in xenon matrix, 293
Intramolecular charge transfer theory
 of acetic acid, 148
 applied to acids, 334
  to acyl fluorides, 334
  to amides, 334
 of benzoic acid, 250
 of bromobenzene, 237
 of chlorobenzene, 237
 of *trans*-cinnamic acid, 253
 of N,N-dimethyl aniline, 250
 of dimethyl cyanamide, 120
 of N,N-dimethyl nitrosamine, 163, 164
 of ethyl nitrate, 165
 of fluorobenzene, 237
 of formamide, 128, 135, 136
 of formic acid, 150
 of iodobenzene, 237
 of methyl nitrite, 163

 of nitramide, 165
 of 4-nitroaniline, 250
 of nitrobenzene, 250
 of nitrogen oxides, 162
 of nitrosobenzene, 250
 of nitrosyl chloride, 162
 of styrene, 250
 of vinyl ketones, 103, 104
Iodide ion, spectrum of, 279
Iodoacetylene
 A band in, 116
 B,C bands in, 116
 D band in, 117
 ionization potential of, 117
 spectrum of, 66, 67, 116
 spin-orbit coupling in, 116, 117
Iodobenzene
 B band in, 237
 D band in, 237
 intramolecular charge transfer theory of, 237
 oscillator strengths in, 228, 232
 spectrum of, 228, 230, 232, 236, 237
Iodoethylene
 A band in, 67
 antiresonance in, 67
 B,C bands in, 67
 D bands in, 67
 ionization potential of, 52
 spectrum of, 52, 66–68, 116
 spin-orbit splitting in, 67
Ionization potentials in matrices, 303, 304, 325, 337
Isobutane, calculation on, 300, 301
Isobutene
 deuterated, spectra of, 27
 ionization potential of, 24, 29
 spectrum of, 24, 29, 30, 37
 term values in, 24, 29, 30
Isobutyl methyl ketone, oscillator strengths in, 83, 93
Isobutylaldehyde
 ionization potential of, 80
 spectrum of, 80, 95
 term values of, 80
Isocyanic acid
 ionization potentials in, 197
 spectrum of, 196–198
 term values in, 197

Isoleucine, circular dichroism spectrum of, 158, 159
Isoprene, spectrum of, 173, 174
Isopropyl benzene, spectrum of, 224
Isopropyl N-ethyl aldimine, spectrum of, 74
Isopropyl ethylene
  ionization potential of, 25
  spectrum of, 25
  term values in, 25
Isopropyl methyl ketone
  oscillator strengths in, 83, 93
  spectrum of, 88, 89
Isoquinoline, SF_6 scavenger spectrum of, 263
Isothiocyanic acid, spectrum of, 197

J

Jahn-Teller effect
  in adamantane, 302
  in allene, 199–201
  in benzene, 215, 244, 337
  in ethylene, 14
  in methane, 299
  in sodium nitrate, 273
  in tetramethyl allene, 202
  in titanium tetrachloride, 277
  in s-triazine, 248
  in trimethylamine borane, 311
  in trimethyl borane, 309, 310

K

Ketene
  antiresonance in, 204
  calculation on, 203
  core splitting in, 204
  ionization potentials in, 202
  oscillator strengths in, 203, 204
  spectrum of, 202–204
  term values in, 203, 204
  vibrations in, 203
Ketones
  circular dichroism spectra of, 96
  orbitals in, 123
  oscillator strengths in, 93–96
Krypton, solvent for ethylene, 20, 21
Krypton matrix spectrum
  of acetylene, 293, 329
  of benzene, 212, 220, 221, 293, 337
  of benzene-d_6, 212
  of ethylene, 20, 21, 293, 325
  of methyl iodide, 293, 303, 304
  of naphthalene, 261
  of toluene, 225, 226
  of toluene-d_8, 226
  of xenon, 293
  of m-xylene, 226
  of p-xylene, 225, 226

L

Lanthanum fluoride
  with Ce^{3+} as filter, 296
  cutoff wavelength in, 296, 298
  fluorescence from, 297
  spectrum of, 296
Laser spark, as light source, 294, 295
Leucine
  circular dichroism spectrum of, 158, 159
  crystal spectrum of, 157
L-Leucine-glycyl glycine, spectrum of solid, 150
Linoleic acid, spectrum of, 155, 156
Liumogen as wavelength shifter, 295
Lithium fluoride
  cutoff wavelength in, 298
  retardation plate, 295
Lithium nitrate, crystal spectrum of, 273, 274
Luminescence of alkanes, 301–303

M

Magnesium calcium carbonate, crystal spectrum of, 275
Magnesium carbonate, crystal spectrum of, 275
Magnesium fluoride
  cutoff wavelength in, 298
  retardation plate, 295
Maleic anhydride
  ionization potential of, 134, 190, 193
  spectrum of, 134, 190, 191, 193
  term values in, 134, 193
Manganous ion, spectrum of, 343
Matrices, ionization potentials in, 303, 304, 325, 337
Matrix polarization, 293

Matrix spectra and term values, 292
Mesityl oxide
  ionization potential of, 80
  spectrum of, 80, 102, 103
  term values in, 80, 103
Mesitylene
  electron-impact spectrum of crystal, 229
  oscillator strengths in, 228, 229
  spectrum of, 228
Metal alkyls, spectra of, 284
Methane
  antiresonances in, 299
  autoionization in, 300
  band structure of solid, 299
  electron-impact spectrum of, 299
  electron mobility in liquid, 304
  Jahn-Teller effect in, 299
  matrix spectrum of methyl iodide in, 304
  superexcited states in, 299, 301
  term values in, 315
  trapped electron spectrum of, 299
  triplet states in, 297, 298
  X-ray spectrum of, 299, 300, 306
Methane-d_4, X-ray spectrum of, 299, 300
Methanol
  spectrum of, 319
  term values in, 92
Methionine, crystal spectrum of, 157
N-Methyl acetamide
  spectrum of, 136
    liquid, 139, 143
Methyl acetate
  ionization potential of, 134
  spectrum of, 134, 155
  term values in, 134
  vibrations in, 154
Methyl acetylene
  electron-impact spectrum of, 109
  spectrum of, 108
  term value in, 200
Methyl amine, orbitals in, 319
Methyl bromide
  A band in, 307
  B,C term value in, 308
  spin-orbit coupling in, 236
2-Methyl butene-1
  ionization potential of, 25

  spectrum of, 25
  term values of, 25
2-Methyl butene-2, spectrum of, 48
Methyl n-butyl ketone
  oscillator strength in, 83
  spectrum of, 90
Methyl sec-butyl ketone
  oscillator strengths in, 83
  spectrum of, 91
  vibrations in, 92
Methyl t-butyl ketone
  ionization potential of, 80
  oscillator strengths in, 83
  spectrum of, 80, 91
  term values in, 80
Methyl chloride
  A band in, 307
  B,C bands in, 307
  D band in, 175
  spectrum of, 62, 305
  term values in, 308
1-Methyl cyclobutene
  spectrum of, 26, 48
  vibrations in, 28
Methyl N-cyclohexyl aldimine-N-oxide, spectrum of, 75
3-Methyl cyclopentanone
  circular dichroism spectrum of, 96, 97
  core splitting in, 96
  spectrum of, 96, 97
3-Methyl diketopiperazine
  circular dichroism spectrum of, 141, 142
  spectrum of, 142
Methyl ethyl ketone
  ionization potential of, 80
  oscillator strengths in, 83, 93
  spectrum of, 80, 90, 328
  term values in, 80
  vibrations in, 92
Methyl fluoride, X-ray spectrum of, 306
N-Methyl formamide
  ionization potentials of, 131–134
  spectrum of, 133, 134, 137
  term values in, 134
  vibrations in, 131, 133
Methyl formate, spectrum of, 152, 162
9-Methyl guanine
  oscillator strengths in, 288
  spectrum of, 288
9-Methyl hypoxanthine

oscillator strengths in, 288
spectrum of, 288
1-Methyl indan
  circular dichroism spectrum of, 338
  spectrum of, 338
Methyl iodide
  A band in, 307
  B,C bands in, 116, 117, 237
  ionization potentials in matrices, 303, 304
  matrix shift in CF_4, 293, 304
    in rare gases, 293, 304
  spectrum of, 86
    in krypton matrix, 303, 304
    in methane matrix, 304
  term values in rare gas matrices, 293
  Wannier excitons in, 303
Methyl isobutyl ketone
  condensed phase effect in, 89
  spectrum of, 89
Methyl isocyanate
  ionization potential of, 197
  spectrum of, 197
Methyl isopropyl ketone
  condensed phase effect on, 89
  vibrations in, 92
3-Methyl isopropylene cyclopentane, circular dichroism spectrum of, 43
N-Methyl maleimide
  condensed phase effect in, 192
  high-pressure effect on, 190–192
  ionization potential of, 134, 190
  oscillator strength in, 193
  spectrum of, 134, 140, 190–193
  term values in, 134, 192, 193
Methyl mercaptan, orbitals in, 319
2-Methyl naphthalene, spectrum of, 258
Methyl nitrite
  intramolecular charge transfer theory of, 163
  oscillator strength in, 163, 164
  spectrum of, 162, 163
2-Methyl pentene-2, spectrum of, 48
3-Methyl pentene-2, vibrations in, 28
Methyl n-pentyl ketone
  oscillator strengths in, 83
  spectrum of, 90
Methyl phosphine
  ionization potential of, 315

Rydberg/valence shell conjugates in, 315
  spectrum of, 315
Methyl N-n-propyl aldimine, spectrum of, 74
Methyl n-propyl ketone
  oscillator strengths in, 83, 93
  spectrum of, 90
  vibrations in, 92
N-Methyl pyrrole
  fluorescence from, 185, 186
  spectrum of, 185, 186
1-Methyl-2-pyrrolidone, spectrum of, 125, 126, 135, 313
3-Methyl pyrrolidine-2-one, circular dichroism spectrum of, 333
Methyl selenide
  ionization potential of, 321
  Rydberg/valence shell conjugates in, 323
  spectrum of, 321, 322
  term values in, 321, 323
Methyl telluride
  ionization potential of, 321
  Rydberg/valence shell conjugates in, 323
  spectrum of, 321, 322
  term values in, 321, 323
Methyl telluride
  ionization potential of, 321
  Rydberg/valence shell conjugates in, 323
  spectrum of, 321, 322
  term values in, 321, 323
Methyl-s-triazine, spectrum of, 247
Methyl vinyl silane, spectrum of, 326
Methylene chloride
  B,C bands in, 307
  orbital ordering in, 305
  spectrum of, 305
Methylene cyclobutane
  ionization potential of, 26
  spectrum of, 26, 30, 48
  term values in, 26
Methylene cyclohexane
  ionization potential of, 27
  spectrum of, 27, 30
  term values in, 27
  vibrations in, 28

Methylene cyclopentane
  ionization potential of, 26
  spectrum of, 26, 30
  term values in, 26
  vibrations in, 28
Methylene fluoride, X-ray spectrum of, 307
Methylene steroids, circular dichroism spectra of, 43
Molybdenum hexacarbonyl, spectrum of, 282–284
Molybdenum hexafluoride, spectrum of, 278, 280
Myristamide, crystal spectrum of, 127, 128, 137, 143, 144

N

Naphthacene
  condensed phase effect on, 266
  ionization potential of, 266
  spectrum of, 266
  term values in, 266
Naphthalene
  antiresonances in, 261
  calculation on, 259
  condensed phase effect on, 259, 262
  electron-impact spectrum of, 260, 262
  ionization potentials of, 258, 259, 262
  orbital structure of, 257
  oscillator strengths in, 259, 262
  plasmons in, 262
  SF_6 scavenger spectrum of, 263
  spectrum of, 258–260, 339, 341
    in krypton matrix, 261
  term values in, 259, 260
  triplet states in, 262, 263
  Wannier excitons in, 262
Neon
  matrix spectrum of acetylene, 293, 329, 330
    of benzene, 293, 337
    of ethylene, 293, 325
    of methyl iodide, 293, 304
    of xenon in, 293
Neopentane
  calculation on, 300
  electron mobility in liquid, 304
Nickel dimethyl glyoxime, crystal spectrum of, 285

Nickel ion, spectrum of, 343
Nitramide
  intramolecular charge transfer theory of, 165
  oscillator strengths in, 164
  spectrum of, 164, 165
Nitrate ion
  calculation on, 273
  oscillator strength in, 274
  spectral resemblance to carbonate ion, 275
    to nitrite ion, 275
  spectrum of, 164–166, 273
Nitric oxide, negative-ion states of, 325
Nitrite ion
  spectral resemblance to nitrate ion, 275
  spectrum of, 164, 166
4-Nitroaniline
  intramolecular charge transfer theory of, 250, 254
  spectrum of, 250, 254
Nitrobenzene
  calculations on, 253
  intramolecular charge transfer theory of, 250–252
  oscillator strengths in, 252
  spectrum of, 250–253
Nitroethylene
  oscillator strengths in, 165
  spectrum of, 165
Nitrogen
  matrix spectrum of benzene in, 212, 293
    of benzene-d_6, 212
Nitrogen trifluoride
  term values in, 315
  X-ray spectrum of, 315
Nitromesitylene, spectrum of, 252, 253
Nitromethane
  calculations on, 164, 165
  electron-impact spectrum of, 334, 335
  ionization potentials of, 163, 164
  oscillator strengths in, 164
  spectrum of, 164, 251
Nitrones, spectra of, 199
3-Nitro-propene-1, spectrum of, 165
Nitrosamine, calculation on, 164
Nitrosobenzene
  intramolecular charge transfer theory of, 250, 253
  spectrum of, 250, 253

Nitrosyl chloride
  A bands in, 162
  intramolecular charge transfer theory of, 162
  spectrum of, 162
Nonorthogonality of Rydberg and core orbitals, 292
Norbornadiene
  condensed phase effect on, 177, 178, 180
  crystal spectrum, 177, 178
  high-pressure effect on, 177, 178
  independent systems theory of, 180
  π–π splitting in, 179, 180
  spectrum of, 177–180
  term values in, 177
  vibrations in, 177
Norbornene
  ionization potential of, 27
  spectrum of, 27, 29, 31, 179, 180
  term values in, 27, 44
  vibrations in, 28, 33
Norvaline, crystal spectrum of, 157
Nylon, spectra of, 143
Nylon 3, spectrum of, 146, 147
Nylon 610, spectrum of, 146, 147
Nucleic acid bases
  electron-impact spectro of, 287
  spectra of, 286–288, 346
  spectral resemblance to benzene, 286–288

O

Octane, orbital ordering in, 300
1,3,5,7-Octatetraene, calculation on, 189, 190
Octene-1
  ionization potential of, 25
  spectrum of, 23, 25
cis-Octene-2, spectrum of, 25
trans-Octene-2, spectrum of, 25
Octyne-1, spectrum of, 116, 330
Octyne-2, spectrum of, 116, 330
Olefinic steroids, spectra of, 48
Outer-well states, 291, 292
Osmium tetroxide
  calculations on, 282
  oscillator strengths in, 281, 282
  spectrum of, 280–282
  term values in, 281

Oxamide, spectrum of, 127
Oxygen
  diffuse orbitals in, 10
  Schumann–Runge bands, 6

P

Paracyclophanes, spectra of, 338
Pentacene
  ionization potential of, 266
  orbital structure of, 257
  spectrum of, 266
Pentaerythritol tetranitrate, spectrum of, 165
Pentafluorobenzene, spectrum of, 233, 234
Pentene-1
  ionization potential of, 25
  spectrum of, 25
  term values in, 25
  vibrations in, 28
cis-Pentene-2, spectrum of, 25
trans-Pentene-2, spectrum of, 23, 25
Perchlorate ion, spectrum of, 280
cis-Perfluorobutene-2
  ionization potential of, 52
  spectrum of, 52, 56
  term values in, 35, 52, 54
Perfluoro-n-hexane, as solvent for benzene, 337
Perfluorotoluene
  oscillator strengths in, 228
  spectrum of, 228
Permanganate ion, spectrum of, 280, 282
Perrhenate ion, spectrum of, 280
Pertechnetate ion, spectrum of, 280
Perylene, electron-impact spectrum of, 341
Phenanthrene
  electron-impact spectrum of, 265
  oscillator strengths in, 209
  spectrum of, 265
Phenol, spectrum of, 224, 249, 250
Phenothiazine, spectrum of, 265
Phenyl acetylene, spectrum of, 251
2-Phenyl-3,3-dimethyl butane
  circular dichroism spectrum of, 229, 230
  spectrum of, 229
Phenyl isocyanate, spectrum of, 251

Phosgene
  A bands in, 99
  D bands in, 100, 101
  ionization potentials of, 80, 99, 100
  spectrum of, 80, 99, 100, 101, 160, 328
  term values in, 80, 160
Phosphine
  calculation on, 315
  inversion barrier in, 315
  term values in, 315
Phosphorus trifluoride
  calculation on, 315
  inversion barrier in, 315
Phthalocyanine, crystal spectrum of, 284, 285
Picene
  electron-impact spectrum of crystals of, 266
  plasmons in, 266
α-Pinene
  circular dichroism spectrum of, 40, 42–44
  ionization potential of, 27, 44
  spectrum of, 27, 40, 43, 44
  term values in, 27, 44
β-Pinene
  circular dichroism spectrum of, 40, 44
  ionization potential of, 44
  spectrum of, 27, 40
  term values in, 44
Plasmons
  in anthracene, 265, 266
  in 1,2-benzanthracene, 266
  in benzene, 220, 223
  in 1,2-benzpyrene, 266
  in calf thymus nucleic acid, 288
  in chrysene, 266
  in cytosine, 346
  in ethylene, 19
  in glycerol, 319
  in graphite, 220
  in guanine, 346
  in naphthalene, 262
  in picene, 266
  in polystyrene, 257, 338
  in pyrene, 266
  in tetracene, 266
Polyacrylonitrile, spectrum of, 118
Poly-L-alanine
  circular dichorism spectrum of, 333

  spectrum of, 139, 142–146
Poly-γ-ethyl-L-glutamate, spectrum of, 145
Poly-L-leucine, spectrum of, 143
Poly-L-methionine, spectrum of, 146
Poly-N-methyl glutamate, circular dichroism spectrum of, 333
Poly-γ-methyl-L-glutamate, spectrum of, 145
Polymethyl methracrylate, spectrum of, 155
Poly-L-proline I, spectrum of, 142
Poly-L-proline II, spectrum of solid, 142
Polyribonucleotides, spectra of, 288
Poly-L-serine, spectrum of, 142, 146
Polystyrene
  electron-impact spectrum of, 256, 257
  phenyl-chlorinated, spectra of, 338
  plasmons in, 257, 338
  reflection spectrum of, 256, 257
  spectrum of, 256, 257
    atactic, 256
    isotactic, 256
Poly(U + A), spectrum of, 288
Poly-L-valine, spectrum of, 142
Polyvinyl carbazole, spectrum of, 341
Potassium dihydrogen phosphate, spectrum of, 316
Potassium nitrate, crystal spectrum of, 274
Potassium nitrite, crystal spectrum of, 274
Praseodymium ion, spectrum of, 343
Proline, circular dichroism spectrum of, 158, 159
Propane
  calculation on, 300
  luminescence of, 301
  orbital ordering in, 300
  spectrum of, 300
  term values in, 200, 306
n-Propanol, term values in, 103
Propionaldehyde
  crystal spectrum of, 95
  ionization potential of, 80
  spectrum of, 76, 80, 95
  term values in, 80, 103
Propionitrile
  ionization potential of, 118
  spectrum of, 117, 118

Propyl-N-cyclohexyl aldimine, spectrum of, 74
Propylene
  calculation on, 29
  ionization potential of, 24, 29, 104
  negative-ion states of, 326
  spectrum of, 24, 29, 37, 68, 103, 180, 326
  term values in, 24, 29, 35, 200
  vibrations in, 28
Propylene-d_6, oscillator strengths in, 112
Purine
  oscillator strengths in, 288
  spectrum of, 288
Pyrazine
  antiresonance in, 243
  calculation on, 243
  core splitting in, 243
  electron-impact spectrum of, 240, 243, 245, 247
  ionization potentials of, 243, 245
  oscillator strengths in, 240, 243, 244
  spectrum of, 240–244, 249
  term values in, 240, 244
  vibrations in, 237, 238, 243, 244
Pyrazine-d_4
  rotational contour in, 244
  vibrations in, 243
Pyrene
  electron-impact spectrum of, 341
    crystals of, 266
  plasmons in, 266
Pyridazine
  electron-impact spectrum of, 247
  ionization potentials of, 247
  spectrum of, 241, 242, 247, 249
  term values of, 247
  triplet states of, 247
  vibrations in, 247
Pyridine
  calculations on, 239
  core splitting in, 239, 240, 243
  electron-impact spectrum of, 238, 240
  ionization potential of, 239, 240
  oscillator strengths in, 239
  spectrum of, 238–240
  term values in, 240
  triplet states in, 240
  vibrations in, 237–239, 244

Pyrimidine
  electron-impact spectrum of, 247
  ionization potentials in, 245, 246
  oscillator strengths in, 245
  perfluoro-effect in, 245, 246
  spectrum of, 239, 241, 242, 245, 246, 249
  term values in, 245, 246
  triplet states of, 246, 247
  vibrations in, 245, 246
Pyrrole
  absorption intensities in, 184
  antiresonance in, 185
  condensed phase effect on, 184, 185
  ionization potential of, 184, 185
  spectrum of, 184–186
  vibrations in, 185
N-d_1-Pyrrole, vibrations in, 185
Pyrrolidine, spectrum of, 313, 314

Q

Quartz
  cutoff wavelength in, 298
  retardation plate, 295
  spectrum of, 297
  as window material, 296, 297
Quinoline, SF_6 scavenger spectrum of, 263

R

Recapitulation, 291
Reflection spectra
  of ammonium dihydrogen phosphate, 316
  of anthracene, 340
  of glycerol, 319
  of heptadecane, 301
  of liquid benzene, 219, 220
  of liquid water, 316
  of liquids, 296
  of polystyrene, 256, 257
  of potassium dihydrogen phosphate, 316
  of tetradecane, 301
  of triglycine sulfate, 333
  of undecane, 301
Retardation plates, 295
Rhenium hexafluoride
  ionization potential of, 279

spectrum of, 278, 279
D-Ribose-5-phosphate, spectrum of, 288
Rotatory strength, 39
Rubidium atom, radial wave functions of, 291
Rubidium nitrate, crystal spectrum of, 274
Ruthenium tetroxide
  calculations on, 282
  oscillator strengths in, 281, 282
  spectrum of, 280–282
  term values in, 281
Rydberg state, fluorescence from, 185, 186
Rydberg term value, constancy of, 92, 112, 290, 292, 342
Rydberg/valence shell conjugates
  in acetylene, 116
  in ammonia, 312
  in diacetylene, 332
  in ethylene, 9
  in haloethylenes, 68
  in methyl phosphine, 315
  in methyl selenide, 323
  in methyl sulfide, 323
  in methyl telluride, 323
  in water, 207

S

Sapphire
  cutoff wavelength in, 298
  retardation plate, 295
Selenophene
  orbitals in, 335, 336
  spectrum of, 188
Serine, crystal spectrum of, 157
SF_6 scavenger spectra
  of acetaldehyde, 78
  of azulene, 340
  of benzene, 221
  of 1-chloronaphthalene, 263
  of ethylene, 17
  of fluoroethylene, 56
  of isoquinoline, 263
  of naphthalene, 263
  of quinoline, 263
  of urea, 140
  of xenon hexafluoride, 273
  of xenon tetrafluoride, 273

Silane
  calculation on, 323
  X-ray spectrum of, 323
Silver dicyanide ion, spectrum of, 345
Sodium carbonate, solution spectrum of, 275
Sodium formate, cyrstal spectrum of, 156
Sodium nitrate
  crystal spectrum of, 273, 274
  Jahn–Teller interaction in, 273
Sodium nitrite, crystal spectrum of, 274
Solution spectrophotometry, 342
Solvents
  dimethyl carbonate, 334
  perfluoro-n-hexane, 337
  trifluoroethanol, 346
  trimethyl phosphate, 346
Spin-orbit coupling
  in allyl iodide, 68
  in aurous dicyanide ion, 345
  in bromobenzene, 236
  in halide ions, 278, 279
  in iodoacetylene, 116, 117
  in iodoethylene, 67
  in methyl bromide, 236
  in xenon difluoride, 271, 341, 342
  in xenon tetrafluoride, 272, 342
Stelline, spectrum of solid, 150
Steroids, circular dichroism spectra of, 289
Styrene
  intramolecular charge transfer theory of, 250, 251
  oscillator strengths in, 251
  spectrum of, 250, 251, 254, 256, 338
Succinimide, spectrum of, 333
Sulfate ion, spectrum of, 280
Sulfur hexafluoride
  term values in, 280
  X-ray spectrum of, 323
Superexcited states
  in n-heptane, 301
  in methane, 299
Synchrotron radiation, 221, 265, 294, 339, 346

T

Tellurophene, orbitals in, 336
Terbium ion, spectrum of, 343

Term values
  independence of, MO's in acetylene, 112
  in ketones, 92
  in XeF$_2$, 342
  in XeF$_4$, 342
  trends, 1
m-Terphenyl, crystal spectrum of, 267
o-Terphenyl, crystal spectrum of, 267
p-Terphenyl, crystal spectrum of, 267
Tetraacetylene, spectrum of, 330
Tetra-n-butyl ammonium tetrabromomanganate, spectrum of, 279
Tetra-n-butyl ammonium tetrabromozincate, spectrum of, 279
Tetra-n-butyl ammonium tetrachlorozincate, spectrum of, 279
Tetra-n-butyl ammonium tetrahalomanganate, spectra of, 278, 279
Tetra-n-butyl ammonium tetrahalozincate
  charge transfer to solvent, 278
  spectra of, 278, 279
Tetra-n-butyl ammonium tetraiodozincate, spectrum of, 279
Tetracene
  electron-impact spectrum of crystals of, 266
  plasmons in, 266
Tetrachloroethylene
  D bands in, 57, 63, 64
  ionization potential of, 52, 57
  spectrum of, 52, 56–59, 61, 62
  term values in, 52
Tetracyclopropyl ethylene, spectrum of, 25
Tetradecane, reflection spectrum of, 301
1,1,4,4-Tetrafluorobutadiene
  ionization potential in, 175
  spectrum of, 167, 172, 175
Tetrafluoroethylene
  antiresonances in, 55
  calculation on, 53, 55, 61
  electron-impact spectrum of, 53
  ionization potentials of, 52
  photoelectron spectrum of, 50
  spectrum of, 51–55, 66, 67
  term values in, 35, 52–55
Tetrafluoromethane
  ionization potential of, 306
  matrix spectrum of acetylene in, 293, 329
  of methyl iodide in, 273, 304
  X-ray spectrum of, 307
Tetrahydrofuran
  spectrum of, 314, 320, 335
  term values in, 181
Tetrahydropyran
  C-type band in, 320
  spectrum of, 319, 320
Tetrahydroselenophene, spectrum of, 323
Tetrahydrothiophene, spectrum of, 323
Tetrakisdimethylamino ethylene, spectrum of, 48–50
Tetramethyl allene
  ionization potentials of, 202
  Jahn–Teller effect in, 202
  oscillator strengths in, 202
  term values in, 202
1,1′,3,3′-Tetramethyl-$\Delta^{2,2'}$-bisimidazolidine, spectrum of, 48–50
Tetramethyl ethylene
  crystal spectrum of, 33, 34
  geometry of, 48
  high-pressure effect on, 33
  ionization potential of, 24, 29
  spectrum of, 24, 29, 32–34, 38, 48
  term values in, 24, 29, 34, 35
  vibrations in, 28
N,N,N',N'-Tetramethyl urea, spectrum of, 140
1,1,4,4-Tetraphenyl butadiene as wavelength shifter, 295
s-Tetrazine
  electron-impact spectrum of, 249
  ionization potentials of, 249
  spectrum of, 248, 249
Thiophene
  He-II effect in, 187
  ionization potentials of, 186, 187
  orbitals in, 335
  spectrum of, 181, 186–188
  term values in, 187, 188
  vibrations in, 187, 188
Thiophene-d_4, spectrum of, 186, 188
Thymine, spectrum of, 287
Thymus DNA, spectrum of, 288
Tin tetrachloride
  B,C bands in, 345

D bands in, 345
spectrum of, 343-345
Titanium tetrabromide, spectrum of, 277
Titanium tetrachloride
  A bands in, 277
  B bands in, 277
  calculations on, 276
  D bands in, 277, 343
  ionization potentials of, 276
  Jahn-Teller effect in, 277
  spectrum of, 276-278, 282, 343-345
  term values in, 277
Toluene
  electron-impact spectrum of crystal, 229
  matrix spectrum in krypton, 225, 226
  oscillator strengths in, 228, 229
  spectrum of, 224, 228, 229, 236, 249, 250
  triplet states in, 229
Toluene-d_8, matrix spectrum in krypton, 226
p-Tosyl-DL-valine, spectrum of solid, 148
Transition-metal hexahalides, oscillator strengths in, 280
Triacetylene, spectrum of, 330
s-Triazine
  electron-impact spectrum of, 247, 248
  ionization potentials of, 247, 248
  Jahn-Teller effect in, 248
  perfluoro-effect in, 248
  spectrum of, 241, 247-249
  term values in, 248
  vibrations in, 247, 248
s-Triazine-d_3, spectrum of, 247
Trichloroacetaldehyde
  A bands in, 102
  crystal spectrum of, 101, 102
  spectrum of, 101, 102
1,3,5-Trichlorobenzene
  A bands in, 230
  oscillator strengths in, 228
  spectrum of, 228, 231
Trichloroethylene
  core splitting in, 52, 60
  spectrum of, 52, 56, 57, 59, 62
  term values in, 52
Trichlorophosphine oxide
  spectrum of, 316
  term values in, 316
α,α,α-Trichlorotoluene, spectrum of, 236

Tricyclo[3.3.0.0^{2,6}]oct-3-ene
  crystal spectrum of, 45, 46
  spectrum of, 27, 45, 46
Trifluoroacetaldehyde, term values in, 101
Trifluoroacetamide
  absorption intensity in, 154
  calculation on, 332, 333
  ionization potential of, 134
  spectrum of, 134, 146, 151, 334
  term values in, 134, 153
Trifluoroacetic acid
  absorption intensity in, 154
  condensed phase effect in, 153
  crystal spectrum of, 153
  ionization potential of, 134
  spectrum of, 134, 146, 151, 153, 334
  term values in, 134, 153
Trifluoroacetyl fluoride
  absorption intensity in, 154
  spectrum of, 146, 151, 334
  term values in, 153
1,3,5-Trifluorobenzene
  ionization potential of, 234
  spectrum of, 234
Trifluorochloroethylene
  ionization potentials of, 52
  spectrum of, 52, 65, 66
  term values in, 52
Trifluoroethanol as solvent, 346
Trifluoroethylene
  ionization potential of, 52
  spectrum of, 51-53, 55, 59, 66
  term values in, 35, 52-55
Trifluoromethyl bromide
  B,C bands in, 308
  D bands in, 308
  ionization potential of, 308
  term values in, 308
Trifluoromethyl chloride
  A bands in, 307
  B,C bands in, 308
  D bands in, 308
  ionization potential of, 308
  term values in, 308
Trifluoromethyl iodide, A band in, 307
Trifluoronitrosomethane
  electron-impact spectrum of, 161
  term values in, 161

Trifluorophosphine oxide
  spectrum of, 316
  term values in, 316
2,4,6-Trifluoropyrimidine, photoelectron spectrum of, 246
Triglycine sulfate, reflection spectrum of, 333
Triglycyl, spectrum of, 142
Triglycyl glycine
  crystal spectrum of, 149
  spectrum of, 333
Trimethylamine borane
  Jahn–Teller effect in, 311
  spectrum of, 311
Trimethylamine-N-oxide, spectrum of, 316
Trimethyl borane
  calculation on, 309
  Jahn–Teller effect in, 309, 310
  spectrum of, 309, 310
Trimethyl borane-d_3, spectrum of, 311
1,1,3-Trimethyl butadiene, absorption intensities in, 172, 173
Trimethyl ethylene
  ionization potential of, 24, 29
  spectrum of, 24, 29, 38
  term values in, 24, 29, 35, 44
  vibrations in, 28
Trimethyl phosphate as solvent, 346
Trimethyl-s-triazine
  ionization potential of, 248
  spectrum of, 247, 248
Trimethylene oxide, term values in, 103
Trimethylene sulfide, spectrum of, 320
Trimethylsilyl acetylene, spectrum of, 108, 109, 111
Trimethylsilyl t-butyl acetylene, spectrum of, 111
Trimethylsilyl ethylenes, spectra of, 46
Trimethylsilyl tetramethylsilanyl acetylene, spectrum of, 111
Triphenylene
  electron-impact spectrum of, 341
  orbital structure of, 257
Triterpenoids, circular dichorism spectra of, 289
Tropolone, spectrum of, 105
L-Tryptophan, circular dichroism spectrum of, 346

Tungsten hexacarbonyl, spectrum of, 282–284
Tungsten hexafluoride, spectrum of, 278, 280

U

Undecane, reflection spectrum of, 301
Uracil
  oscillator strengths in, 288
  spectrum of, 245, 286–288
Urea
  calculation on, 140
  orbitals in, 123
  SF_6 scavenger spectrum of, 140
  spectrum of, 125, 140

V

Valine
  circular dichroism spectrum of, 158, 159
  crystal spectrum of, 157
Vanadium tetrachloride
  A bands in, 345
  B,C bands in, 345
  D bands in, 345
  ionization potentials in, 345
  spectrum of, 343–345
Virtual precursors, 291

W

Wannier excitons
  in acetylene, 113
  in amides, 139
  in benzene, 220, 221, 337
  in cupric chloride, 346
  in ethylene, 20
  in ferric chloride, 346
  in ketones, 90
  in methyl iodide, 303
  in naphthalene, 262
Water
  calculations on, 316, 317
  dissociation in, 130, 312
  electron-impact spectrum of, 316–318
  negative-ion states of, 317, 318
  reflection spectrum of liquid, 316

spectrum of, 207, 316
term values in, 312
triplet states in, 316, 317
Water-d_2, spectrum of, 316
Wavelength shifters
  liumogen, 295
  1,1,4,4-tetraphenyl butadiene, 295
Windowless cell, construction, 294

X

X-ray spectra
  of boron trifluoride, 311
  of chloroethylene, 66
  of chloroform, 304
  of cobalt salicylaldimine, 285
  condensed phase effect and, 293
  of copper salicylaldimine, 285
  of cupric chloride, 346
  of diborane, 309
  of dichloroethylenes, 66
  of ferric chloride, 346
  of fluoroform, 306, 307
  of inorganic ions, 345, 346
  of metal chlorides, 346
  of methane, 299, 300, 306
  of methane-d_4, 299, 300
  of methyl fluoride, 306
  of methylene fluoride, 307
  of nitrogen trifluoride, 315
  of silane, 323
  of sulfur hexafluoride, 323
  of tetrafluoromethane, 307
Xenon
  matrix spectrum of benzene in, 220, 221, 293, 337
    of benzene-d_3 in, 212
    of ethylene in, 293, 325
  term values in, 271, 293
Xenon difluoride
  calculations on, 270-272
  core splitting in, 342
  crystal spectrum of, 342
  electron mobility in, 342
  oscillator strength in, 272
  spectrum of, 269-272, 341, 342
  spin-orbit coupling in, 271, 341, 342
  term values in, 271
Xenon hexafluoride
  ionization potential of, 273
  SF_6 scavenger spectrum of, 273
  spectrum of, 269, 272
Xenon lamp, medium pressure, 294
Xenon tetrafluoride
  core splitting in, 342
  crystal spectrum of, 342
  ionization potentials of, 273
  SF_6 scavenger spectrum of, 273
  spectrum of, 269, 270, 272, 273, 341, 342
  spin-orbit coupling in, 272, 342
m-Xylene
  crystal spectrum of, 227
  matrix spectrum in krypton, 226
  oscillator strengths in, 228
  spectrum of, 224, 228
o-Xylene
  oscillator strengths in, 228
  spectrum of, 224, 228, 229
p-Xylene
  oscillator strengths in, 228
  spectrum of, 224, 228, 243
    in krypton matrix, 225, 226
  vibrations in, 226
D-Xylose, circular dichroism spectrum of, 289

Y

Ytterbium ion, spectrum of, 343

Z

Zinc phthalocyanine, crystal spectrum of, 284, 285